Functionally Graded Materials

Functionally Graded Materials: Analysis and Applications to FGM, FG-CNTRC and FG Porous Structures aims to disseminates knowledge between users, manufacturers, designers, and researchers involved in structures or structural components manufactured using functionally graded materials.

Despite a number of existing texts on the theory and analysis of plates and/or shells, there is not a single book that is devoted entirely to the analysis of functionally graded-based materials including FGMs, FG-CNTRCs and FG porous plates and shells. *Functionally Graded Materials* provides the reader with current research conducted in the field of composite materials, numerical simulations, computer methods and mechanical responses of structural components under different scenarios.

This book is a valuable reference source for postgraduate students, engineers, scientists, professors, researchers, and applied mathematicians in this field.

Functionally Graded Materials

Analysis and Applications to FGM, FG-CNTRC and FG Porous Structures

Souhir Zghal and Fakhreddine Dammak

CRC Press
Taylor & Francis Group
Boca Raton London New York

CRC Press is an imprint of the
Taylor & Francis Group, an **informa** business

Designed cover image: www.shutterstock.com

First edition published 2025
by CRC Press
2385 NW Executive Center Drive, Suite 320, Boca Raton FL 33431

and by CRC Press
4 Park Square, Milton Park, Abingdon, Oxon, OX14 4RN

CRC Press is an imprint of Taylor & Francis Group, LLC

ISBN: 978-1-032-76783-3 (hbk)
ISBN: 978-1-032-77564-7 (pbk)
ISBN: 978-1-003-48378-6 (ebk)

DOI: 10.1201/9781003483786

Typeset in Nimbus Roman
by KnowledgeWorks Global Ltd.

Dedication

This book is dedicated to my parents, my brother ***Ahmed*** *and my sister* ***Chourouk*** *for their constant encouragement and moral support along with patience and understanding.*

Contents

Preface

This book aims to disseminate knowledge between users, manufacturers, designers and researchers involved in structures or structural components manufactured using functionally graded materials. It is devoted to the analysis of functionally graded-based structures with several numerical tests and applications of FGM, FG-CNTRC and FG porous structures using the finite element method. Thus, the book should be considered not only as a graduate textbook but also as a reference book to those working or interested in areas of finite element, structural analysis under mechanical and/or thermal loads and materials science.

In recent years, the advances in materials science and computer methods for engineering structures have pushed the great progress in the modeling of structural components like beams, plates and shells made of new candidates of materials such as functionally graded materials (FGMs), functionally graded carbon nanotubes reinforced composites (FG-CNTRC) and FG porous structures. The main feature of these materials is the continuous transition of the composition through the thickness. This gradient property improves rigidity, strength and durability, resistance to fatigue, corrosion resistance and thermal conductivity of structures and thus avoid delamination, cracks at the interfaces and any kind of damage observed in conventional materials. The mechanical analysis of structural elements made from such materials have therefore attracted increasing attention from both research and engineering communities due to their great importance in both theoretical and practical aspects.

The book deals with our research made in the field of composite materials, numerical simulations, computer methods and mechanical responses of structural components under different solicitations. The research content of this book includes several aspects: The numerical modeling using different theories and computer methods, the implementation of new material behavior law's for the first time and even new investigations covering static bending, free vibration, dynamic, buckling and post-buckling under mechanical and thermal loads of FGM, FG-CNTRC and FG porous beams, plates and shells.

In the analysis of the behavior of shell structures, three mainly kinematic models have been developed and implemented in a house computer software namely (MPEF) by our creative master head Professor Fakhreddine Dammak on the basis of Abaqus commands, FORTRAN and visual basic interface. The kinematic models, known as Kirchhoff-Love, first-order shear deformation (FSDT) and Double Directors high-order models have been implemented and ameliorated to bring up a high level of mechanical analysis covering static bending, dynamic, free vibration, buckling and post-buckling with thermal and mechanical aspects. The new axis of FG-CNTRC and FG porous materials has been conceptualized and proposed in our team group by Dr.Ing-Souhir Zghal with a vast range of structural applications and numerical implementation and tests. The book makes valuable contributions in the subject area of functionally gradient-based structures: modeling and analysis by providing the first-ever-known results, which are also helpful in better understanding the mechanical behavior of such functionally graded-based structures. To present this thematic, manuscript is composed of eight chapters:

The first chapter contains a detailed kinematic description of the different developed models: the governing equations and the finite element formulation. An update of the Kirchhoff-Love kinematic model is presented with the consideration of large displacement and finite rotations. Equally, the first-order shear deformation theory (FSDT) is ameliorated via an improvement of the shear part of strains. The improvement consists of the introduction of a quadratic function that is able to describe accurately the transverse shear strains and to satisfy the zero condition of the transverse shear stresses at the top and bottom surfaces. Then, the numerical framework of a Double Directors finite element shell model is developed and presented. The applicability of these numerical developments is the subject of the subsequent chapters.

The second chapter deals with thermal analysis of FG-based plates and shells including free vibrations, buckling and post-buckling responses of these structural components. The governing equations are elaborated within the improved (I-FSDT) finite shell model where its detailed kinematics and developments are presented in Chapter 1. Various numerical applications of FGM plates and shells under different thermal loads are presented and discussed. Uniform and non-uniform (or non-linear) thermal loads are introduced in the model and their effects are deeply examined and discussed. Convergence, comparison and parametric studies are conducted in each case of structural analysis in order to show the effectiveness and robustness of the developed formulation in one hand and the influence of several geometrical and mechanical characteristics on thermal responses of FGM structures on the other hand.

The third chapter is devoted to buckling and post-buckling analysis of FGM plates and shells subjected to either thermal or mechanical loadings. For thermal buckling and post-buckling analysis, the results are performed using the extended Kirchhoff love-model in order to outline the capability of this developed model to predict accurately the buckling and post-buckling behaviors of thin FGM plates and shells. For buckling and post-buckling of FGM structures subjected to mechanical loadings, the results are performed using the double directors finite shell model (DDST). Various numerical tests are presented with different case studies and parametric variations to trace the effect of FG power-index, boundary conditions, aspect ratios, type of applied loads and others on buckling and post-buckling responses of FGM plates and shells.

In Chapter 4, functionally graded carbon nanotubes reinforced composites (FG-CNTRC) structures are explored. These new trends of materials are estimated using the modified rule of mixture with the inclusion of efficiency parameters to take into account the scale dependency of the carbon nanotubes (CNTs). Various mechanical responses such as linear static and stress responses, free vibration, dynamic, buckling and post-buckling behaviors are investigated and presented. Different forms of CNTs, known as UD, FG-V, FG-X and FG-O distributions are included in the models with a focus on the effects of carbon nanotubes volume fractions and the types of distributions on mechanical responses of FG-CNTRC plates and shells. The analysis covers static bending, deflections and free vibration and dynamic responses of FG-CNTRC structures with various numerical tests and illustrations. All the analyses are performed using the double directors finite shell (DDST) model and all governing equations and solutions strategies are detailed in Chapter 1.

Non-linear analysis of functionally graded carbon nanotubes reinforced composite (FG-CNTRC) plates and shells using various kinematics models is the subject of the fifth chapter. An extended version of the Love-Kirchhoff model is developed for FG-CNTRC structures with finite rotations and large deflections. An improved or modified FSDT-theory (I-FSDT) is also used for large deflection responses of these types of functional structures. Finally, the non-linear responses of this material category are explored using the double directors finite shell element. Several numerical tests are conducted on FG-CNTRC structures including rectangular, square and ring plates, cylindrical panels, hemispherical, open ended cylindrical and hyperboloid shells. Different analyses on non-linear behaviors of FG-CNTRC plates and shells are treated with the inclusion of convergence, comparison and parametric studies. The effects of several parameters and notably CNTs volume fractions and distributions are examined and interpreted and the chapter concludes with the discussion of CNTs as new candidate with strong mechanical properties for composite reinforcements.

Chapter 6 brings the buckling and post-buckling of FG-CNTRC plates and shells, where stability and safety of engines and structures is a crucial subject for mechanical engineers and designers. The governing equations such as the kinematic relations, the constitutive lawn the equations of motion are derived in Chapter 1. Herein, the analysis is performed based on the double directors finite shell (DDST) model. The solution strategies are also presented in this chapter. The effects of CNTs volume fractions, distribution patterns and geometrical parameters are investigated and discussed in detail. Proper examples of FG-CNTRC plates and shells are given in this chapter and it is concluded that CNTs reinforcements enhance strongly the stiffness of FG-CNTRC plates and shells.

In Chapter 7, bending and free vibration analysis of functionally graded porous beams is first investigated using a refined mixed finite element beam model. The material properties of functionally graded porous beams are estimated using a modified power law distribution with two different types of porosity namely even and uneven distributions. The potential of the proposed model is highlighted via a comparison study. Then, a parametric study is carried out to show the effects of power law index, porosity coefficient, boundary conditions and types of porosity distributions on deflections, stresses and natural frequencies of the studied FG porous beams.

Chapter 8 involves the mechanical behavior of functionally graded porous structures in which the porosity can spread uniformly or randomly according to two patterns, namely even and uneven porosity distributions, is investigated. The governing equations are derived using the I-FSDT model where its development steps are as highlighted in Chapter 1. The effective material properties are determined using a modified power law distribution taking into account the porosity volume fraction. Various mechanical responses are examined and presented leading hence to trace the effects of porosity coefficient and patterns of distributions on vibrational, dynamic and buckling responses of FG porous plates and shells.

Despite a number of existing texts on the theory and analysis of plates and/or shells, there is not a single book that is devoted entirely to the analysis of functionally graded-based materials including FGMs, FG-CNTRCs and FG porous plates and shells. It is hoped that this book will fill the gap to some extent and be used as a valuable reference source for postgraduate students, engineers, scientists, professors, researchers and applied mathematicians in this field.

Author

Dr. Souhir Zghal, is an Associate professor at high Institute of Applied Science and Technology of University of Gabes, Gabes, Tunisia. She was born in Sfax where she received the high school degree at Majida Boulila lyceum, in 2004. In 2009, she has earned her mechanical engineering degree from National Engineering School of Tunis (ENIT). She received her P.h.d in mechanical engineering in 2015 from National Engineering School of Tunis (ENIT) and FEMTO-ST Institute, University of Franche-Comté, Besancon-France. Her research interests include finite element analysis, composite structures, computer methods and simulations, advanced materials like functionally graded materials (FGM), carbon nanotubes, porosity, and graphene platelets, structural dynamics, vibrations, viscoelasticity, sandwich structures, non-linear dynamics, computational structural mechanics, stress analysis and algorithms. Her group is particularly interested in research problems related to structural advanced materials by developing finite element codes and computer methods that are able to investigate various geometries with different shapes and with a vast area of analysis like bending, vibrations, dynamics, mechanical and thermal buckling and post-buckling responses of FGM, FG-CNTRC and porous-FGM based beams, plates and shells. An ambitious group with a creative master head has published more than 200 scientific articles and communications in the most recognized mechanical engineering journals and conferences proceedings. As a young researcher in mechanical engineering field, she has received over 1500 citations and an h-index of 17 (Data from *Scopus* January 2024). She is also a recognized reviewer in the most impactful international journals through Elsevier, Springer, Taylor & Francis and Sage Publishing Company.

Prof. Fakhreddine Dammak, Full professor at National School of Engineers of Sfax (ENIS), Sfax, Tunisia. He was born on November 16, 1962 in Sfax and he obtained his Ph.D. degree in mechanical engineering from Laval university, Quebec, Canada in 1997. During the period of 1998 to 2004 he was an assistant at National Institute of Applied Science and Technologies of Tunis, Tunisia. Then, he joined the National School of Engineers of Sfax (ENIS), in 2005 as an associate professor in mechanical engineering. He has taught over 20 different graduate and undergraduate courses in mechanical engineering and has conducted 4 courses covering the finite element and plate/shell modeling codes for engineering students. He has over 250 technical/scientific papers published in international journals and conferences with an h-index of 36 (Data from Scopus January 2024). He is expert in finite element modeling, structural analysis and materials science covering new materials like functionally graded materials, carbon-nanotubes reinforced composite (FG-CNTRC), FG porous structures, smart materials, nanocomposite, sheet forming and more on. He is also a member of the scientific communities of Design and Modeling of Mechanical Systems international conference with its all editions and a reviewer of over 40 different international journals. He is a distinguished professor of finite element codes of the Ministry of Higher Education and Research in Tunisia.

1 Kinematic models of functionally gradient-based shells

This chapter presents the kinematics of a continuum based two-dimensional shell element, which are applied for various mechanical analysis including functionally gradient-power based structures (FGM), carbon nanotubes reinforced composites shell structures (FG-CNTRC) and functionally porous shells (FG-porous). The elaborated finite element formulation is based on a general kinematic model, where the classical shell theory (CST), the first-order shear deformation theory (FSDT) and the high-order shear theories (HSDT) can be derived. The basis of the present finite element formulation is the standard four-node shell element with bi-linear shape functions. The implementation is applicable to the analysis of isotropic, functionally graded materials (FGM) and functionally graded carbon nanotubes-reinforced composites (FG-CNTRC) under different boundary conditions and external loads. Furthermore, to avoid shear locking problems at the thin shell limit, the assumed natural strain (ANS) method is used to improve the finite element formulation. Material properties of the shells are assumed to be graded in the thickness direction according to a power-law distribution for FGM structures or an extended rule of mixture with some efficiency parameters for (FG-CNTRC) shells and a modified power-law for FG-porous structures.

1.1 INTRODUCTION

In this part, numerical frameworks and solution strategies for the modeling of the mechanical behavior of structural components like beams, plates and shells have been developed and implemented during my researches made at National Engineering School of Sfax (ENIS) in collaboration with Professor Fakhreddine Dammak. These elaborations consist on the amelioration of the numerical developments devoted to shell-type of structures and covering the plates ones. In particular, the kinematic models of continuum-based two-dimensional shell elements are presented and described with a focus on the improvement and enhancement made on these theories.

The engineering structures are often subjected to thermal and mechanical loads during their service lives. Therefore, the mechanical analysis of functionally gradient-based structures as FGM, FG-CNTRC and FG porous structures under thermal and mechanical loads is of utmost importance for their various applications in many engineering areas such as in aerospace, space craft, nuclear automobile and civil engineering domains. Three main models are often used to approximate the kinematics of such structures which are the Kirchhoff model, the Reissner-Mindlin model and the high-order models. Several studies have been reported in the literature regarding these theories. The Kirchhoff model constitutes one of the most known and useful models which has improved its applicability especially for thin shells [100, 174, 297]. Even this model neglects the effect of shear deformations, it represents a convenient model for thin structures favored by its simple implementation in the most finite element codes and low computational cost. The first-order shear deformation theory (FDST) was introduced by Reissner and Mindlin [210, 169] to take into account the effect of shear deformations and handle thus the limitations of Kirchhoff model. However, this model becomes insufficient when thick shell structures are investigated. In addition, this model requires the introduction of shear coefficients which can be prohibitive for such applications. In fact, shear coefficients are easy to obtain for linear isotropic material (5/6) while it appears complex for composites mainly for non-linear behavior law. Hence, to avoid these shear coefficients, several

DOI: 10.1201/9781003483786-1

high-order models [207, 208, 75, 43, 50, 279, 278] have been developed in the last years. For these types of shell theories, a discretization of the reference surface is necessary. The nodal parameters are the displacement vector and the shell director vector of the reference surface. Multiplicative formulations based on the interpolation of the inextensible director vector have been reported in the literature [240, 241, 243, 244, 78, 21]. Linear and non-linear computational aspects are presented with discrete approximations of the displacement and strain fields. Further aspects of linear applications are discussed in [9, 130]. Geometrical non-linearity can be included in two-dimensional (2D) problems and three-dimensional (3D) finite element analysis as reported in [238, 239]. However, 3D elements can combine generally the enhanced assumed strain (EAS) and the assumed natural strain (ANS) methods in order to avoid locking problems, which can generate numerical costs for complex geometries. So, 2D models appears more convenient in most structural applications due to their low computational effort compared to 3D models. It should be pointed that locking phenomena occurs generally when the formulation considers transverse shear deformations or incompressibility in 3D constitutive material law. In this context, several interpolation schemes have been proposed to ameliorate the well-known shear locking effect which typically arises as the thickness of the plate goes to zero. Hence, to avoid the shear locking phenomenon that dominates the performance of displacement finite element formulations in the thin shell limit, several methods have been developed where the assumed natural transverse shear strain (ANS) and the EAS method (EAS) are the most popular in the finite element method [149, 22, 242, 119].

In this chapter, the parametrization of the shell reference surface, as well as the kinematic relations related to each model are presented. Then, the governing equations are established within the variational framework. Next, the discretization of the displacement and strain fields is performed using the finite element method leading to the construction of the geometrical and material parts of the tangent operator. For the Kirchhoff-love model, an extension of the model with large displacements and finite rotations is achieved by the development of three and four-nodes finite shell element based on the work of [54]. Continuously, the (FSDT) shell model is improved using a quadratic function that is able to describe accurately the transverse shear strains and satisfy the zero condition of the transverse shear stresses at the upper and bottom surfaces. This contribution constitutes an amelioration of the standard (FSDT) theory without need of shear correction factors. For thick structures, high-order models have been also developed within a double directors shell model which include geometrical nonlinearities in the sense of large displacements.

1.2 KINEMATIC RELATIONS

To distinguish the initial configuration C_0 from the deformed C_t, capital letters (resp. lowercase letters) are used for quantities relative to the configuration C_0 (resp. C_t).

The 3D position of any point (q) of the structure is obtained by a power series expansion in the vicinity of the reference mean surface:

$$x_q(\xi^1,\xi^2,z) = x_p(\xi^1,\xi^2) + \sum_{k=1}^{n} z^k d_k(\xi^1,\xi^2), \tag{1.1}$$

where d_k $(k = 1,2,...,n)$ are the n unit vectors field at each point of the surface, referred to as the directors field. For $n = 0$, corresponding to membrane theory, the shell is represented by the mean surface. The case $n = 1$, corresponding to single director shell model or first-order theory, includes the classical shell theory (CST) based on Kirchhoff-Love assumptions and the first-order shear deformation theory (FSDT) based on Reissner-Mindlin assumptions. The case $n = 2$ leads to two director vectors, which allow to have a better representation of the transverse shear strain and remove the shear correction factors associated to (FSDT) theory.

1.2.1 PARAMETRIZATION OF THE REFERENCE SURFACE

The reference surface in an arbitrary configuration, initial or deformed, is described by the same parametric coordinates ξ^1 and ξ^2. The position vector of an arbitrary point (p) of the mean reference surface, in the initial configuration C_0 is defined by its components in the Cartesian global basis (E_i):

$$X_p = X_i(\xi^1,\xi^2)E_i \ , \ (\xi^1,\xi^2) \in A_\xi \subset R^2, \tag{1.2}$$

where A_ξ is the parametric surface. A differential element at point p is given by:

$$dX_p = A_\alpha d\xi^\alpha \ , \ A_\alpha = X_{p,\alpha} \ , \ \alpha = 1,2. \tag{1.3}$$

The vectors of the covariant basis A_1 and A_2 are tangents at coordinate lines ξ^1 and ξ^2. The square of the incremental element length is given by:

$$dS_p^2 = dX_p \cdot dX_p = A_{\alpha\beta} d\xi^\alpha d\xi^\beta \ , \ A_{\alpha\beta} = A_\alpha \cdot A_\beta. \tag{1.4}$$

dS_p^2 is known as the first fundamental form (quadratic form) and the coefficients $A_{\alpha\beta}$ are the components of the covariant metric tensor A. The unit normal vector at the mean reference surface, in the initial configuration C_0 is given by:

$$N = \frac{A_1 \wedge A_2}{\|A_1 \wedge A_2\|} \ , \ A_\alpha \cdot N = 0 \ , \ \|N\| = 1. \tag{1.5}$$

The parametric basis (A_1, A_2, N) is covariant at the point (p) of the reference surface in the initial configuration. Two unit vectors (T_1, T_2), located in the tangent plane can be defined at any point on the middle surface. These vectors, with the normal vector N, make an orthonormal basis (T_1, T_2, N). This basis can be obtained from the unit vectors N and E_3:

$$N = Q \cdot E_3, \tag{1.6}$$

where $Q = T_\alpha \otimes E_\alpha + N \otimes E_3$.

The position vector at any point (p), of the reference mean surface in the deformed configuration C_t, is defined by its components in the Cartesian basis (E_i):

$$x_p = x_p(\xi^1,\xi^2) = x_i(\xi^1,\xi^2)E_i. \tag{1.7}$$

A differential element at point (p) is given by:

$$dx_p = a_\alpha d\xi^\alpha \ , \ a_\alpha = x_{p,\alpha} \ , \ \alpha = 1,2, \tag{1.8}$$

where a_1 and a_2 are the covariant vectors. The square of the incremental element length is given by:

$$ds_p^2 = dx_p \cdot dx_p = a_{\alpha\beta} d\xi^\alpha d\xi^\beta \ , \ a_{\alpha\beta} = a_\alpha \cdot a_\beta. \tag{1.9}$$

The coefficients $a_{\alpha\beta}$ are the covariant components of metric tensor a.

1.2.2 INITIAL CONFIGURATION

In the initial configuration, the shell geometry description is done with a single vector. The position of any point (q) of shell domain is given by (show Fig. 1.1):

$$X_q(\xi^1,\xi^2,z) = X_p(\xi^1,\xi^2) + zD(\xi^1,\xi^2). \tag{1.10}$$

The covariant basis at point (q) is obtained from the position vector by $(G_1, G_2, G_3) = (\partial X_q/\partial\xi^1, \partial X_q/\partial\xi^2, \partial X_q/\partial z)$, which yields in base vectors relative to the initial state:

$$\begin{cases} G_\alpha = A_\alpha + zD_{,\alpha}, \ \ \alpha = 1,2 \\ G_3 = D \end{cases} \tag{1.11}$$

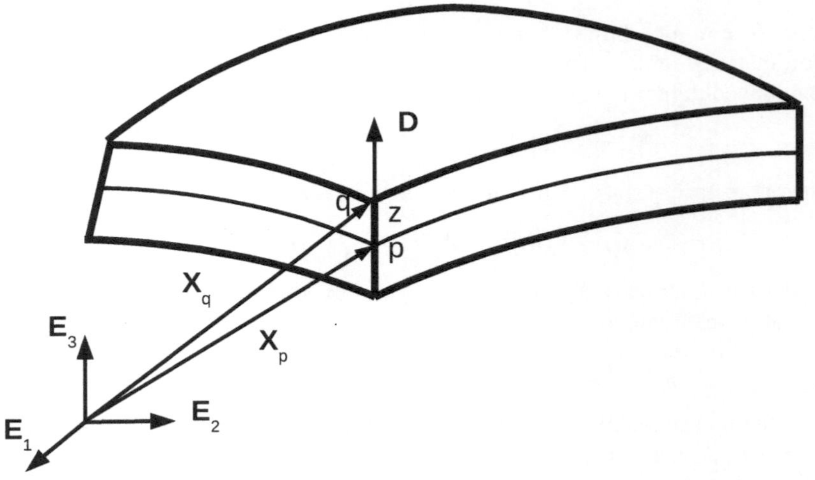

Figure 1.1 Kinematic description of the initial configuration.

1.3 GOVERNING EQUATIONS OF NON-LINEAR DISCRETE KIRCHHOFF SHELL MODEL

In this section, the geometry and the kinematic of non-linear discrete Kirchhoff shell model are briefly described.

1.3.1 DISPLACEMENT AND STRAIN FIELDS

The displacement field of any point q of the shell in the current C_t configuration can be expressed in terms of convective curvilinear coordinates $\xi = (\xi^1, \xi^2, \xi^3 = z)$ as follows:

$$x_q = x_p + zd, \tag{1.12}$$

where z and d denote the thickness variable and the director vector of the shell, respectively.

On the other hand, the Lagrangian strain field E can be expressed as follows [54]:

$$\begin{cases} E_{\beta\gamma} = e_{\beta\gamma} + z\chi_{\beta\gamma} \\ 2E_{\beta 3} = \gamma_\beta \end{cases}, \tag{1.13}$$

where $e_{\beta\gamma}$, $\chi_{\beta\gamma}$ and γ_β denote the membrane, the bending and the shear strains. Their corresponding variation are given as indicated in [54]. In matrix form, these components are given by:

$$e = \begin{bmatrix} e_{11} \\ e_{22} \\ 2e_{12} \end{bmatrix}, \quad \chi = \begin{bmatrix} \chi_{11} \\ \chi_{22} \\ 2\chi_{12} \end{bmatrix}, \quad \gamma = \begin{Bmatrix} \gamma_1 \\ \gamma_2 \end{Bmatrix}. \tag{1.14}$$

It should be noted that according to the Kirchhoff hypothesis, the shear contribution of strain vanishes in discrete form over the element boundaries side (I,J) and becomes hence equal to zero. As well as, we propose here two types of finite elements: four-node and three-node elements which constitutes an amelioration of Kirchhoff model compared to the model presented in [54].

1.3.2 WEAK FORMULATION

The weak form of the equilibrium equation with the Kirchhoff-Love assumption can be written as:

$$G = \int_A (N \cdot \delta e + M \cdot \delta \chi) \, dA - G_{ext} = 0, \tag{1.15}$$

where δe and $\delta \chi$ are the variations of the shell strains. N and M are the membrane and bending stresses resultants whereas the shear stress is annulled in discrete form of the variational formulation. The expression of N and M are given by:

$$N = \begin{bmatrix} N^{11} \\ N^{22} \\ N^{12} \end{bmatrix}, \quad M = \begin{bmatrix} M^{11} \\ M^{22} \\ M^{12} \end{bmatrix} \tag{1.16}$$

with:

$$(N^{\alpha\beta}, M^{\alpha\beta}) = \int_{-h/2}^{h/2} (1, z) S^{\alpha\beta} \, dz, \quad (\alpha, \beta) = (1, 2). \tag{1.17}$$

$S^{\alpha\beta}$ represents the components of the second Piola-Kirchhoff stress tensor. The expressions of generalized resultant of stress R and strain vectors Σ are given as:

$$R = \begin{bmatrix} N \\ M \end{bmatrix}, \quad \Sigma = \begin{bmatrix} e \\ \chi \end{bmatrix}. \tag{1.18}$$

As a result, the weak form of equilibrium becomes:

$$G(\Phi, \delta\Phi) = \int_A \delta\Sigma^T \cdot R dA - G_{ext}(\Phi, \delta\Phi) = 0. \tag{1.19}$$

where $\Phi = (u, d)$ is the displacement and shell vector. Equation (7.19) describes the non-linear shell model which can be solved by the Newton iterative algorithm. The consistent tangent operator for Newton's solution procedure can be constructed by the directional derivative of the weak form in the direction of the increment $\Delta\Phi = (\Delta u, \Delta d)$. For large displacement analysis, it is a conventional practice to split the tangent operator into geometric and material parts, denoted by $D_G G \cdot \Delta\Phi$ and $D_M G \cdot \Delta\Phi$, respectively, as follows:

$$D_G \cdot \Delta\Phi = D_G G \cdot \Delta\Phi + D_M G \cdot \Delta\Phi \tag{1.20}$$

1.3.3 MATERIAL PART

The material part of the tangent operator results from the variations of the generalized resultant of stress R while maintaining the strain constant and thus takes the form:

$$D_M G \cdot \Delta\Phi = \int_A (\delta\Sigma^T \cdot \Delta R) dA, \tag{1.21}$$

Thereby, the material part of the tangent modulus can be expressed as:

$$\Delta R = H_T \Delta\Sigma, \quad H_T = \begin{bmatrix} H_{11} & H_{12} \\ H_{12} & H_{22} \end{bmatrix}. \tag{1.22}$$

The components (H_{11}, H_{12}, H_{22}) read as:

$$(H_{11}, H_{12}, H_{22}) = \int_{-h/2}^{h/2} (1, z, z^2) \, H dz. \tag{1.23}$$

Herein, H represents the in-plane elastic matrix, which is defined according to the used material constitutive equations.

1.3.4 GEOMETRICAL PART

The geometric part results from the variation of the virtual strains while holding stress resultants constant. Its form can be given as:

$$D_G G \cdot \Delta\Phi = \int_A (\Delta\delta\Sigma^T \cdot R) dA, \tag{1.24}$$

This form can be divided into membrane and bending parts as follows:

$$D_G G \cdot \Delta\Phi = D_G G_m \cdot \Delta\Phi + D_G G_b \cdot \Delta\Phi, \tag{1.25}$$

The discretization of the weak form of equilibrium as well as the material and geometrical parts are developed in the following section.

1.3.5 FINITE ELEMENT FORMULATION

1.3.5.1 Approximation of the displacement field

In this section, the numerical implementation of the present shell theoretical formulation based on three or four-node non-linear Kirchhoff shell element is presented. In this approach, the shear contribution is neglected where the values of shear deformations vanish on the boundaries in discrete from as indicated in [282, 283, 306]. Hence, based on the isoparametric concept, the interpolation of the displacement vector $(u = x - X)$ and shell vector d can be written as:

$$\begin{aligned} \delta u &= \sum_{I=1}^{n} N^I \delta u_I \ , \quad \delta d = \sum_{I=1}^{n} N^I \delta d_I + \sum_{K=n+1}^{2n} P_K \delta\alpha_K t_K \\ \Delta u &= \sum_{I=1}^{n} N^I \Delta u_I \ , \quad \Delta d = \sum_{I=1}^{n} N^I \Delta d_I + \sum_{K=n+1}^{2n} P_K \Delta\alpha_K t_K, \end{aligned} \tag{1.26}$$

where (I) denotes a node of the element. K and $\delta\alpha_k$ are the mid-point of the element boundaries and the variables associated to δd on the boundaries. The vector t_K is unit and its direction in defined by the position of the nodes couple (I, J) as depicted in Fig. 1.2. Its expression is given by:

$$t_K = (x_J - x_I)/L_K, \quad L_K = \|x_J - x_I\|, \tag{1.27}$$

where L_K is the $I - J$ side length. N^I are the standard isoparametric shape functions [57, 23]. The expressions of shape functions N^I and P_K are given respectively in Tab. 1.1 with ($\xi = \xi^1$, $\eta = \xi^2$, $\lambda = 1 - \xi - \eta$). The three and four-nodes shell elements in parametric coordinate systems are illustrated in Fig. 1.3.

1.3.5.2 Local cartesian coordiantes

The shape functions derivatives in local Cartesian system are given by:

$$\left\{ \begin{array}{c} \overline{N}^I_{,1} \\ \overline{N}^I_{,2} \end{array} \right\} = [J]^{-1} \left\{ \begin{array}{c} N^I_{,\xi} \\ N^I_{,\eta} \end{array} \right\}, \tag{1.28}$$

where $N^I_{,\xi}$ and $N^I_{,\eta}$ are the parametric shape function derivatives. The expression of the Jacobian transformation J in local Cartesian system with base vectors $\{n_1^0, n_2^0, n^0\}$ is given as:

$$J = \begin{bmatrix} n_1^0 \cdot A_1 & n_2^0 \cdot A_1 \\ n_1^0 \cdot A_2 & n_2^0 \cdot A_2 \end{bmatrix}, \tag{1.29}$$

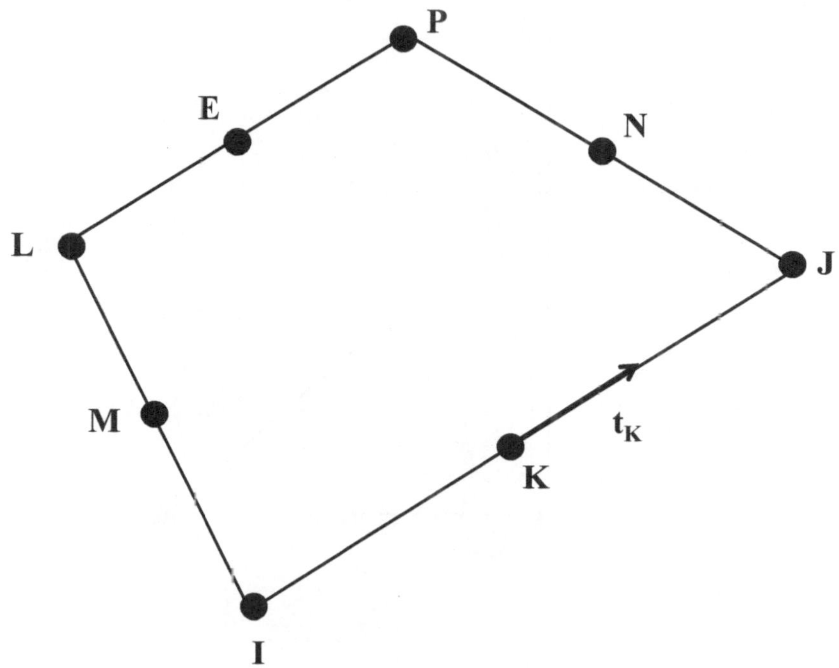

Figure 1.2 Position of the nodes couple (I, J).

Table 1.1
shape functions N^I and P_K for triangular and quadrilateral element.

Element	Triangular element ($n = 3$)	Quadrilateral element ($n = 4$)
N^I	$N_1 = 1 - \xi - \eta$ $N_2 = \xi$ $N_3 = \eta$	$N_1 = 1/4(1-\xi)(1-\eta)$ $N_2 = 1/4(1+\xi)(1-\eta)$ $N_3 = 1/4(1+\xi)(1+\eta)$ $N_4 = 1/4(1-\xi)(1+\eta)$
P_K	$P_4 = 4\xi\lambda$ $P_5 = 4\xi\eta$ $P_6 = 4\eta\lambda$	$P_5 = 0.5(1-\xi^2)(1-\eta)$ $P_6 = 0.5(1+\xi)(1-\eta^2)$ $P_7 = 0.5(1-\xi^2)(1+\eta)$ $P_8 = 0.5(1-\xi)(1-\eta^2)$

where A_1 and A_2 are the covariant base vectors of the mid-surface. The expression of the normal field to the mid-surface n^0 in the initial configuration C_0 is given by:

$$n^0 = A_1 \wedge A_2 / \|A_1 \wedge A_2\|. \tag{1.30}$$

1.3.5.3 Approximation of the strain field

The discretization of the membrane and bending parts of the strain field can be given by:

$$\delta e = B_m \delta\Phi_n \ , \ \ \delta\chi = B_b \cdot \delta\Phi_n, \tag{1.31}$$

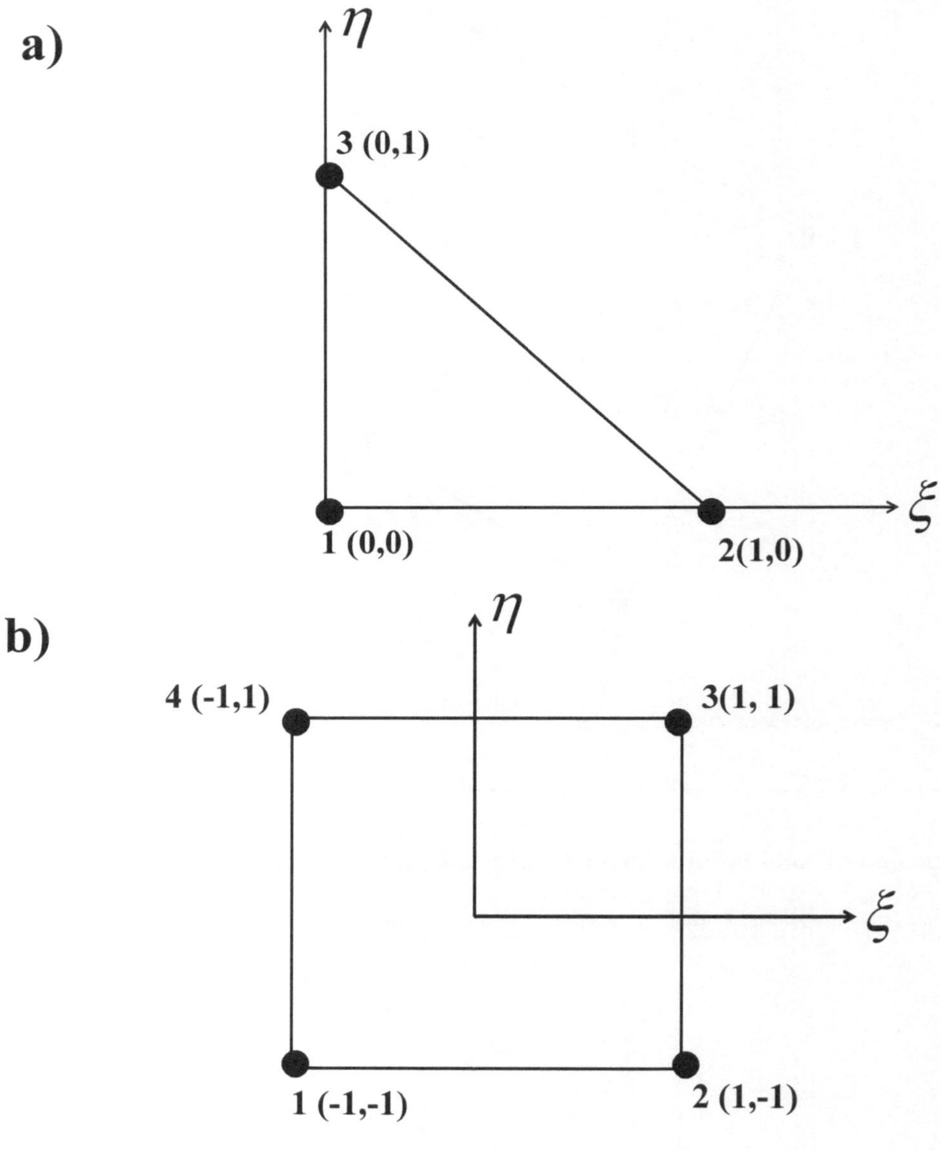

Figure 1.3 Finite shell element with (a) three-nodes and (b) four-nodes in parametric coordinates systems.

where B_m and B_b are the discrete strain-displacement matrices. The membrane and bending matrices, at nodal level I, can be expressed as follows:

$$B_m^I = \begin{bmatrix} n_1^T \overline{N}_{,1}^I & 0 \\ n_2^T \overline{N}_{,2}^I & 0 \\ n_1^T \overline{N}_{,2}^I + n_2^T \overline{N}_{,1}^I & 0 \end{bmatrix}$$

$$B_b^I = \begin{bmatrix} d_{,1}^T \overline{N}_{,1}^I + n_{,1}^T \cdot M_{d,1}^I & n_1^T \cdot M_{r,1}^I \\ d_{,2}^T \overline{N}_{,2}^I + n_{,2}^T \cdot M_{d,2}^I & n_2^T \cdot M_{r,2}^I \\ d_{,1}^T \overline{N}_{,2}^I + d_{,2}^T \overline{N}_{,1}^I + n_{,1}^T \cdot M_{d,2}^I + n_{,2}^T \cdot M_{d,1}^I & n_1^T \cdot M_{r,2}^I + n_2^T \cdot M_{r,1}^I \end{bmatrix}, \quad (1.32)$$

The shape functions N^I are as indicated in Tab. 1.1 while the unit vectors (n_1, n_2) are given by:

$$n_1 = \Sigma \overline{N}^I_{,1} (X+u)_I, \quad n_2 = \Sigma \overline{N}^I_{,2} (X+u)_I. \tag{1.33}$$

The matrices M^I_d and M^I_r are expressed by:

$$M^I_d = P_K td^I_K + P_M td^I_M, \quad td^I_K = \frac{3}{2L_K} t_K \otimes d_K$$
$$M^I_r = N^I I + P_K tt^I_K + P_M tt^I_M, \quad tt^I_K = \frac{3}{4} t_K \otimes t_K. \tag{1.34}$$

After the discretization of the strain field using the non-linear discrete Kirchhoff shell model, the weak form of equilibrium leads to the following element residual which can be written as:

$$r = \int_A B^T R dA. \tag{1.35}$$

where the matrix B takes the following form:

$$B = \begin{bmatrix} B_m \\ B_b \end{bmatrix} \tag{1.36}$$

As a result, the material tangent operator can be deduced as:

$$K_M = \int_A B^T H_T B dA. \tag{1.37}$$

For the geometric tangent operator, its expression can be written in matrix form as:

$$D_G G \cdot \Delta\Phi = \delta\Phi^T_n (K_G) \Delta\Phi_n. \tag{1.38}$$

where K_G represent the global geometric tangent operator which regroups the membrane and the bending contributions. In fact, for couple of nodes (I, J), the expression of the matrix K_G is given by:

$$K_{G_{IJ}} = \begin{bmatrix} UUM_{IJ} I + UUF_{IJ} & UBF_{IJ} \\ BUF_{IJ} & BBF_{IJ} \end{bmatrix}. \tag{1.39}$$

where UUM and (UUF, UBF, BBF) correspond to membrane and bending parts, respectively. Their expressions are given in Appendix A.

1.3.5.4 Nodal transformation

The relation between the variation of the director vector δd_k and its derivative $\delta d_{k,\alpha}$ is given by:

$$\delta d = \delta\theta \wedge d = \Lambda \delta\theta, \quad \overline{\Lambda} = -\widetilde{d}, \tag{1.40}$$

where $\widetilde{d}$ is the skew-symmetric tensor such that $\widetilde{d}d = 0$, or in material description:

$$\delta d = \Lambda \delta \overline{\Theta} E_3 = \overline{\Lambda} \delta\Theta, \quad \overline{\Lambda} = -\Lambda \widetilde{E}_3, \tag{1.41}$$

where we assumed that $d = \Lambda E_3$. It is shown in [239] that with $E_3 = [0 \ 0 \ 1]^t$ a spatial description leads to a shell problem with 6DOF/node and the material description leads to a shell problem with 5DOF/node, where the transformation $\overline{\Lambda}$ take the following form:

$$\overline{\Lambda} = [-d_2 \ \ d_1]_{3\times 2}. \tag{1.42}$$

The expressions of the directors and their corresponding matrices for both spatial and material description are given in Tab. 1.2.

As well as, the global transformation matrix Π_n which allows the passage from elementary to global coordinates system can be given as:

$$\Pi_n = \begin{bmatrix} \Pi_1 & 0 & 0 & 0 \\ 0 & \Pi_2 & 0 & 0 \\ 0 & 0 & \Pi_3 & 0 \\ 0 & 0 & 0 & \Pi_4 \end{bmatrix}, \quad \Pi_I = \begin{bmatrix} I & 0 \\ 0 & \overline{\Lambda} \end{bmatrix}, \quad I = 1,..,4. \tag{1.43}$$

At this point, we recall that the five or six degree-of-freedom (DOF) formulation is obtained by introducing the transformation Π_I and setting hence:

$$\delta\Phi_I = \Pi_I \cdot \delta U_I, \quad I = 1,..,4, \tag{1.44}$$

where the vectors $\delta\Phi_I$ and δU_I are defined as follows:

$$\delta\Phi_I = \begin{bmatrix} \delta u & \delta d \end{bmatrix}_I^T, \quad \delta U_I = \begin{bmatrix} \delta u & \delta\Theta \end{bmatrix}_I^T \tag{1.45}$$

As a result, the global tangent stiffness matrix can be rewritten as follows:

$$K_T = \Pi_n^T (K_M + K_G) \Pi_n \tag{1.46}$$

It is worth mentioning that the developed non-linear finite element Kirchhoff model does not present any locking problem which ensures its effectiveness and usefulness.

1.4 GOVERNING EQUATIONS OF THE FIRST-ORDER SHEAR DEFORMATION SHELL THEORY

In this section, the governing equations derived from the first-order shear deformation shell theory based on Mindlin/Reissner kinematic assumptions are presented. As well as, the geometry and the kinematics of this model are described.

1.4.1 GEOMETRY AND KINEMATICS

The position vector of mid-surface in the deformed configuration C_t is given by:

$$x_p = X_p + u \ , \tag{1.47}$$

where u is the displacement vector and x_p is the position vector of a point p located at the mid-surface of the deformed configuration C_t as shown in Fig. 1.4. The covariant vectors are obtained

Table 1.2
Nodal descriptions

Spatial description with 6DOF/node	Material description with 5DOF/node
$\Delta d = \Delta\theta \wedge d^k$ $d^{k+1} = \cos(\Delta d) d^k + \frac{\sin(\Delta d)}{\Delta d} \Delta d, \ \Delta d = \|\Delta d\|$ $\overline{\Lambda}^{k+1} = -\tilde{d}^{k+1}$	$\Delta d = \overline{\Lambda}^k \delta\Theta$ $d^{k+1} = \cos(\Delta d) d^k + \frac{\sin(\Delta d)}{\Delta d} \Delta d, \ \Delta d = \|\Delta d\|$ $\Delta\theta = d^k \wedge \Delta d, \ \overline{\Lambda}^{k+1} = \exp(\overline{\Delta\theta}) \overline{\Lambda}^k$

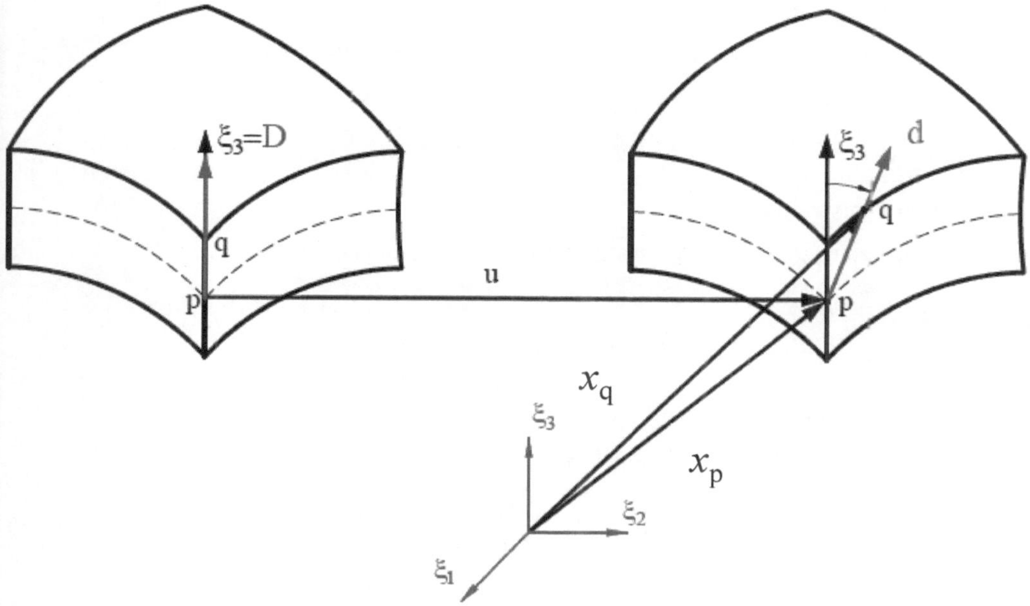

Figure 1.4 Description of the shell body in initial C_0 and deformed C_t configurations.

from the position vector in both initial C_0 configuration by $G_\alpha = \partial X_q/\partial \xi^\alpha$ and in deformed C_t configuration by $g_\alpha = \partial x_q/\partial \xi^\alpha$. Their expressions are given as follows:

$$\begin{cases} G_\alpha = A_\alpha + zD_{,\alpha} \;, G_3 = D \\ g_\alpha = a_\alpha + zd_{,\alpha} \;, g_3 = d \end{cases} \quad \alpha = 1,2\ . \tag{1.48}$$

A_α and a_α represent the covariant base vectors in the initial and the deformed configuration, respectively. The volume elements dV and the geometrical variable $Det(G)$ are given by:

$$dV = Det(G)d\xi^1 d\xi^2 dz \;, Det(G) = \sqrt{(G)} = \sqrt{(|G_{ij}|)},\ i,j = 1,2,3. \tag{1.49}$$

$G_{\alpha\beta}$ refers to the metric tensor in initial configuration C_0. The components of the metric tensor are separated to an in-plane and out of plane part. Their expressions are given by:

$$\begin{cases} G_{\alpha\beta} = A_{\alpha\beta} + zB_{\alpha\beta} \\ G_{3\alpha} = \omega_\alpha^0 \;,\; G_{33} = 1 \end{cases}, \tag{1.50}$$

where the kinematic variables $A_{\alpha\beta}$, $B_{\alpha\beta}$ and ω_α^0 can be given as:

$$A_{\alpha\beta} = A_\alpha \cdot A_3 \;,\; B_{\alpha\beta} = A_\alpha \cdot D_{,\beta} + A_\beta \cdot D_{,\alpha} \;,\; \omega_\alpha^0 = A_\alpha \cdot D\ . \tag{1.51}$$

Similar expressions of the metric tensors can be obtained for the deformed configuration C_t.

1.4.2 MEASURE OF DEFORMATIONS

The strain field is described by means of Green's strain tensor which can be partitioned into in-plane and shear strains as follows:

$$E_{ij} = \tfrac{1}{2}(g_{ij} - G_{ij}),\ \begin{cases} E_{\alpha\beta} = e_{\alpha\beta} + z\chi_{\alpha\beta} \\ \gamma_{\alpha 3} = 2E_{\alpha 3} = g_{\alpha 3} - G_{\alpha 3} \end{cases},\ \alpha,\beta = 1,2, \tag{1.52}$$

where $e_{\alpha\beta}$, $\chi_{\alpha\beta}$ and $\gamma_{\alpha 3}$ denote membrane, bending and shear strains, respectively.

1.4.2.1 Membrane strains

The membrane strains $e_{\alpha\beta}$ and its derivative variation $\delta e_{\alpha\beta}$ can be computed as:

$$\left\{ \begin{array}{l} e_{\alpha\beta} = 1/2(a_{\alpha\beta} - A_{\alpha\beta}) \\ \delta e_{\alpha\beta} = 1/2(a_{\alpha} \cdot \delta u_{,\beta} + a_{\beta} \cdot \delta u_{,\alpha}) \end{array} \right. , \quad \alpha,\beta = 1,2 \tag{1.53}$$

1.4.2.2 Bending strains

The membrane strains $\chi_{\alpha\beta}$ and its derivative variation $\delta\chi_{\alpha\beta}$ can be computed as:

$$\left\{ \begin{array}{l} \chi_{\alpha\beta} = \frac{1}{2}\left(b_{\alpha\beta} - B_{\alpha\beta}\right) \\ \delta\chi_{\alpha\beta} = 1/2(a_{\alpha} \cdot \delta d_{,\beta} + a_{\beta} \cdot \delta d_{,\alpha} + \delta u_{,\alpha} \cdot d_{,\beta} + \delta u_{,\beta} \cdot d_{,\alpha}) \end{array} \right. , \quad \alpha,\beta = 1,2 \tag{1.54}$$

1.4.2.3 Shear strains

Similarly, the shear strains can be obtained as:

$$\left\{ \begin{array}{l} \gamma_{\alpha 3} = a_{\alpha} \cdot d - A_{\alpha} \cdot D \\ \delta\gamma_{\alpha 3} = a_{\alpha} \cdot \delta d + \delta u_{,\alpha} \cdot d \end{array} \right. , \quad \alpha,\beta = 1,2 \tag{1.55}$$

The components of the strain vectors can be condensed, in matrix form, as:

$$e = \left[\begin{array}{c} e_{11} \\ e_{22} \\ 2e_{12} \end{array} \right], \quad \chi = \left[\begin{array}{c} \chi_{11} \\ \chi_{22} \\ 2\chi_{12} \end{array} \right], \quad \gamma = f(z) \left[\begin{array}{c} \gamma_{13} \\ \gamma_{23} \end{array} \right]. \tag{1.56}$$

where $f(z)$ is the function which defines the distribution of the transverse shear strains through the thickness of the shell and takes the following form:

$$f(z) = \frac{5}{4}\left(1 - 4\left(\frac{z}{h}\right)^2\right) \tag{1.57}$$

At this stage, we recall that choosing of this quadratic function is inspired from the work of [237] where a parabolic distribution of the transverse shear stresses along the thickness is assumed, in contrast to the standard (FSDT) theory which introduces shear correction factors to comply with the linear distribution of the shear transverse strain distribution assumption. As well as, further works have proposed a correction on this distribution [257] where a shear function is implemented in the FSDT formulation leading to a parabolic distribution of transverse shear stresses across the thickness. A brief summary on these functions is given in Tab. 1.3. It is also noteworthy, that the shear function in Eq. (1.57), is able to yield the exact value of the transverse shear correction factor evaluated from an energetic equivalence with 3D elasticity theory.

Table 1.3
Transverse shear deformation functions.

Model	$f(z)$
FSDT [257]	$1-4(z/h)^2$
TSDT [237]	$5(1-4(z/h)^2)/4$
Present ($f(z)$)	$5(1-4(z/h)^2)/4$

1.4.3 WEAK FORM AND VARIATIONAL FORMULATION

For a total Lagrangian formulation, the virtual work is expressed by:

$$G(\Phi,\delta\Phi)=\int_V \delta E_{ij}:S^{ij}\ dV-G_{ext}(\Phi,\delta\Phi)=0 \tag{1.58}$$

where δE_{ij} are the covariant components of the virtual Green-Lagrange strain tensor. S^{ij} are the contravariant components of the stress tensor which is related to the second Piola-Kirchhoff tensor. dV is the shell volume element in the initial configuration and G_{ext} is the external virtual work. By utilizing Eqs. (1.52),(1.53),(1.54) and (1.55), the weak form becomes:

$$G=\int_A (N\cdot\delta e+M\cdot\delta\chi+T\cdot\delta\gamma)dA-G_{ext}=0, \tag{1.59}$$

(N, M, T) represent the membrane, bending and shear stresses resultants, expressed as:

$$N=\begin{bmatrix} N^{11}\\ N^{22}\\ N^{12}\end{bmatrix},\quad M=\begin{bmatrix} M^{11}\\ M^{22}\\ M^{12}\end{bmatrix},\quad T=\left\{\begin{matrix} T^{13}\\ T^{23}\end{matrix}\right\}, \tag{1.60}$$

where

$$N^{\alpha\beta}=\int_{-h/2}^{h/2} S^{\alpha\beta}\ dz,\ \ M^{\alpha\beta}=\int_{-h/2}^{h/2} zS^{\alpha\beta}\ dz,\ \ T^{\alpha 3}=\int_{-h/2}^{h/2} f(z)S^{\alpha 3}\ dz \tag{1.61}$$

with S^{α_β} is the second Piola-Kirchhoff stress tensor. The generalized stress resultant R and strain vectors Σ reads as:

$$R=\begin{bmatrix} N\\ M\\ T\end{bmatrix},\quad \Sigma=\begin{bmatrix} e\\ \chi\\ \gamma\end{bmatrix}. \tag{1.62}$$

By replacing Eq. (1.62) into Eq. (1.59), the weak form of equilibrium becomes:

$$G(\Phi,\delta\Phi)=\int_A \delta\Sigma^T\cdot R dA-G_{ext}(\Phi,\delta\Phi)=0. \tag{1.63}$$

In Eq. (1.63), $\Phi=(u,d)$ represents the displacement and shell vectors. Its resolution can be achieved by the Newton iterative algorithm for which the consistent tangent operator can be constructed by the directional derivative of the weak form in the direction of the increment $\Delta\Phi=(\Delta u,\Delta d)$. On the other hand, the tangent operator can be partitioned into geometrical and material stiffness matrices as follows:

$$D_G\cdot\Delta\Phi=D_G G\cdot\Delta\Phi+D_M G\cdot\Delta\Phi \tag{1.64}$$

1.4.4 THERMO-ELASTIC CONSTITUTIVE EQUATIONS

In order to describe the thermo-elastic behavior of (FGM) structures, working in a thermal environment, the thermal stresses are introduced in the constitutive equations as:

$$R=H_T\Sigma-R^{th}, \tag{1.65}$$

where H_T is the elastic modulus expressed as:

$$H_T=\begin{bmatrix} H_m & H_{mb} & 0\\ H_{mb} & H_b & 0\\ 0 & 0 & H_s\end{bmatrix} \tag{1.66}$$

with:

$$(H_m, H_{mb}, H_b) = \int_{-h/2}^{h/2} (1, z, z^2)\, H dz, \;\; H_s = \int_{-h/2}^{h/2} (f(z))^2 H_\tau dz. \tag{1.67}$$

In Eq. (1.64), R^{th} is the thermal stress resultant induced by the temperature change ΔT and it is given by:

$$R^{th} = \begin{bmatrix} N^{th} & M^{th} & 0 \end{bmatrix}^T, \tag{1.68}$$

These components can be expressed as:

$$(N^{th}, M^{th}) = \int_{-h/2}^{h/2} (1, z)\, H \alpha(T, z) \Delta T \begin{bmatrix} 1 \\ 1 \\ 0 \end{bmatrix} dz. \tag{1.69}$$

where H, H_τ, $\alpha(T,z)$ and ΔT represent the in-plane and out-of-plane linear elastic matrices, the thermal expansion coefficient and the temperature change, respectively. Their expressions depend on the material behavior laws and the constitutive equations of the studied structures which will be developed in the next sections.

1.4.5 FINITE ELEMENT PROCEDURE

For numerical analysis purpose, the development steps of the finite element procedure related to the proposed improved first-order shear deformation theory (I-FSDT) are presented here. By using the isoparametric concept, the interpolations of the displacement and strain fields are given below.

1.4.5.1 Finite displacements

The interpolation of the displacement field is achieved using a four-node shell element as follows:

$$u = \sum_{I=1}^{4} N^I u_I \;, \;\; d = \sum_{I=1}^{4} N^I d_I \;, \tag{1.70}$$

N^I are the standard isoparametric shape functions at nodal level I. The orthogonal transformation from the basis $\{n_1^0, n_2^0\}$ to $\{A_1, A_2\}$ base vectors is done via the Jacobian matrix J which is given by:

$$J = \begin{bmatrix} n_1^0 \cdot A_1 & n_2^0 \cdot A_1 \\ n_1^0 \cdot A_2 & n_2^0 \cdot A_2 \end{bmatrix}, \tag{1.71}$$

Herein, n^0 is the normal vector in the initial configuration C_0. It can be defined as:

$$n^0 = A_1 \wedge A_2 / \|A_1 \wedge A_2\|, \tag{1.72}$$

The single layer shell with covariant base vectors is illustrated in Fig. 1.5. The relation between the derivation of N^I in the local Cartesian and the local elementary systems can be written as:

$$\begin{Bmatrix} \overline{N}^I_{,1} \\ \overline{N}^I_{,2} \end{Bmatrix} = [J]^{-1} \begin{Bmatrix} N^I_{,\xi} \\ N^I_{,\eta} \end{Bmatrix}, \tag{1.73}$$

1.4.5.2 Finite strain-displacement relations

The finite strain-displacement relations are obtained via the discretization of the membrane, the bending and the transverse shear strains which are given as follows:

$$\begin{cases} \delta e = B_m \cdot \delta U_n \\ \delta \chi = B_b \cdot \delta U_n \\ \delta \gamma = B_s \cdot \delta U_n. \end{cases} \tag{1.74}$$

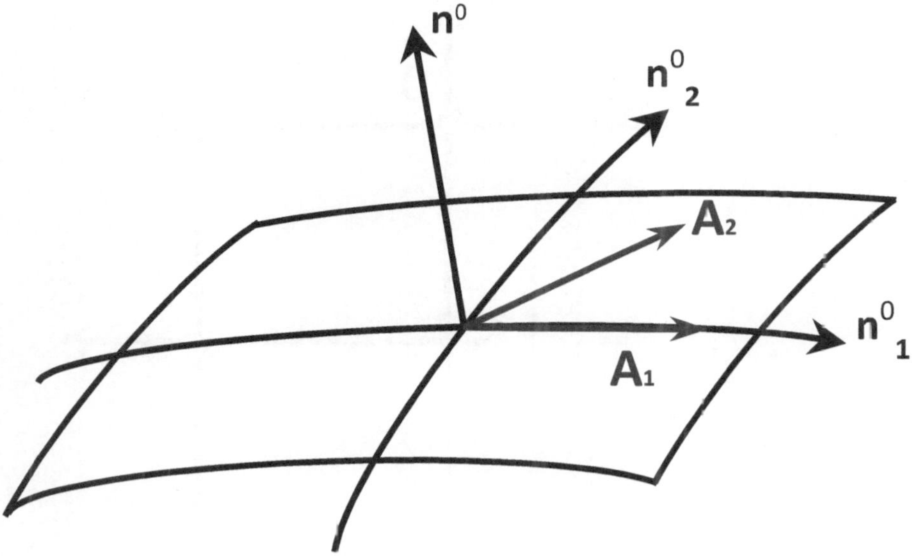

Figure 1.5 Single layer shell with covariant base vectors.

where B_m, B_b and B_s are the discrete strain-displacement matrices, given as:

$$\left\{\begin{array}{c} B_m^I = \left[\begin{array}{cc} B_{mm}^I & 0 \end{array}\right], \quad B_{mm}^I = \left[\begin{array}{c} n_1^T \overline{N}_{,1}^I \\ n_2^T \overline{N}_{,2}^I \\ n_1^T \overline{N}_{,2}^I + n_2^T \overline{N}_{,1}^I \end{array}\right] \\ B_b^I = \left[\begin{array}{cc} B_{bm}^I & B_{bb}^I \Lambda^I \end{array}\right], \quad B_{bm}^I = \left[\begin{array}{c} d_{,1}^T \overline{N}_{,1}^I \\ d_{,2}^T \overline{N}_{,2}^I \\ d_{,1}^T \overline{N}_{,2}^I + d_{,2}^T \overline{N}_{,1}^I \end{array}\right], \quad B_{bb}^I = B_{mm}^I \\ \mathbf{B}_s = J^{-1} \mathbf{B}_{s\xi} \\ \mathbf{B}_{s\xi} = \left[\begin{array}{cccccccc} N_{,1}^1 \mathbf{d}_{1B}^T & N_{,1}^2 \mathbf{a}_{1B}^T & N_{,1}^2 \mathbf{d}_{1B}^T & N_{,1}^2 \mathbf{a}_{1B}^T & N_{,1}^3 \mathbf{d}_{1D}^T & N_{,1}^3 \mathbf{a}_{1D}^T & N_{,1}^4 \mathbf{d}_{1D}^T & N_{,1}^3 \mathbf{a}_{1D}^T \\ N_{,2}^1 \mathbf{d}_{1A}^T & N_{,2}^4 \mathbf{a}_{2A}^T & N_{,2}^2 \mathbf{d}_{1C}^T & N_{,2}^3 \mathbf{a}_{2C}^T & N_{,2}^3 \mathbf{d}_{1C}^T & N_{,2}^3 \mathbf{a}_{2C}^T & N_{,2}^4 \mathbf{d}_{1A}^T & N_{,2}^4 \mathbf{a}_{2A}^T \end{array}\right] \end{array}\right. \tag{1.75}$$

It should be mentioned that the membrane and bending matrices are given at nodal level I while the shear strain is provided for all nodes via the assumed natural transverse shear strain method of [20]. A, B, C and D are the mid-points of the element boundaries set as shown in Fig. 1.6. ($a_\alpha, \alpha = 1, 2$) are the covariant base vectors in the deformed configuration C_t, while n_1 and n_2 are the vectors of the actual basis, given as:

$$\begin{array}{l} n_1 = \Sigma \overline{N}_{,1}^I (X + u)_I, \\ n_2 = \Sigma \overline{N}_{,2}^I (X + u)_I. \end{array} \tag{1.76}$$

The variation of the generalized strain $\delta\Sigma$ in a discrete form can be hence given as:

$$\delta\Sigma = B \cdot \delta U_n, \quad B = \left[\begin{array}{c} B_m \\ B_b \\ B_s \end{array}\right]. \tag{1.77}$$

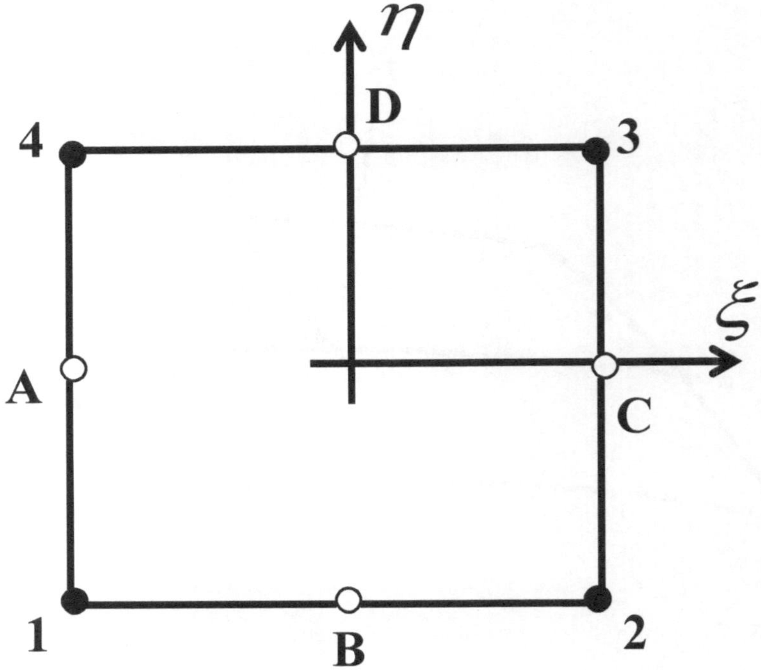

Figure 1.6 Four-node shell element with assumed strain construction.

1.4.5.3 Linearization of the weak form

The linearization of the weak form of equilibrium is achieved as mentioned previously in section 1.3.2 (see Eq. (7.19)) with added shear part and the material and geometrical operators are obtained as:

Material tangent operator

The material part of the tangent operator results from the variation in the stress resultants and thus takes the form:

$$D_M G \cdot \Delta\Phi = \int_A (\delta\Sigma^T \cdot \Delta R) dA, \tag{1.78}$$

In discrete form, the material tangent operator can be rewritten as follows:

$$D_M G \cdot \Delta\Phi = \delta\Phi^T K_M \Delta\Phi, \;\; K_M = \int_A B^T H_T B dA. \tag{1.79}$$

where K_M is the material stiffness matrix and B regroups the membrane, bending and shear strain-displacement matrices as mentioned in Eq. (1.77).

Geometrical tangent operator

The geometrical part of the tangent operator results from the variation of the virtual strain while maintaining constant the stress resultants:

$$D_G G \cdot \Delta\Phi = \int_A (\Delta\delta\Sigma^T \cdot R) dA. \tag{1.80}$$

Equally, the geometrical tangent operator can be rewritten in discrete form as:

$$D_G G \cdot \Delta\Phi = \delta\Phi^T K_G \Delta\Phi = \delta U_n^T \cdot \Pi_n^T \cdot K_G \cdot \Pi_n \cdot \Delta U_n. \tag{1.81}$$

where K_G represents the global geometrical stiffness matrix which regroups the membrane, the bending and the shear contributions. Π_n and U_n represent the global transformation matrix and the discrete displacement vector, respectively. It is worth mentioning that the nodal transformation matrix of the (FSDT) shell model is identical to the Kirchhoff-Love model as indicated in section 2.2.2. For a couple of nodes (I,J), the expression of the matrix K_G is given by:

$$K_{G_{IJ}} = \begin{bmatrix} UUM_{IJ}I & (UBF_{IJ} + UBC_{IJ})I \\ (BUF_{IJ} + BUC_{IJ})I & (BBF_{IJ} + BBC_{IJ})I \end{bmatrix}, \tag{1.82}$$

where UUM, (UBF, BBF) and (UBC, BBC) are the membrane, bending and shear sub-matrices. Their expressions are given in Appendix. B.

1.5 GOVERNING EQUATIONS OF NON-LINEAR DOUBLE DIRECTORS SHELL MODEL

In this section, the geometry and the kinematic of non-linear double directors shell model are briefly described.

1.5.1 GEOMETRY AND KINEMATICS

The double directors shell model is based on three generic assumptions: the first one, known as the plane curve fiber, reflects a high-order distribution of the position vector in the deformed configuration. The second one, known as cubic fiber, assumes that the position vector is cubic along the thickness direction. The third assumption consists of imposing a zero shear stress at the top and bottom surfaces.

1.5.1.1 Initial configuration

In the initial configuration C_0, the covariant components of the metric tensor can be divided into in-plane and out-of-plane components as follows:

$$\begin{cases} G_{\alpha\beta} = A_{\alpha\beta} + zB_{\alpha\beta} + z^2 C_{\alpha\beta} \\ G_{3\alpha} = \omega_\alpha^0 + zB_\alpha \ , \ G_{33} = \lambda_0^2 \end{cases}, \tag{1.83}$$

where the kinematic variables $A_{\alpha\beta}$, $B_{\alpha\beta}$, $C_{\alpha\beta}$, ω_α^0, B_α and λ_0^2 are given by:

$$\begin{cases} A_{\alpha\beta} = A_\alpha \cdot A_\beta \ , \ B_{\alpha\beta} = A_\alpha \cdot D_{,\beta} + A_\beta \cdot D_{,\alpha} \ , \ C_{\alpha\beta} = D_{,\alpha} \cdot D_{,\beta} \\ \omega_\alpha^0 = A_\alpha \cdot D \ , \ B_\alpha = D \cdot D_{,\alpha} = \lambda_0 \lambda_{0,\alpha} \ , \ \lambda_0^2 = D \cdot D \end{cases}, \tag{1.84}$$

$A_{\alpha\beta}$, $B_{\alpha\beta}$ and $C_{\alpha\beta}$ are the curvature tensors. ω_α^0 is initial shear term. B_α and λ_0^2 are the variation of the thickness in the initial configuration. With the assumption of a constant thickness along the element, the metric tensor becomes:

$$\begin{cases} G_{\alpha\beta} = A_{\alpha\beta} + zB_{\alpha\beta} + z^2 C_{\alpha\beta} \\ G_{3\alpha} = \omega_\alpha^0 \ , \ G_{33} = 1 \end{cases}. \tag{1.85}$$

1.5.1.2 Deformed configuration

The double directors shell model is based on the following kinematic assumptions: The fiber, which is assumed initially straight, becomes curve after deformation. This plane curve is defined in a plane formed by two director vectors d_1 and d_2. These vectors are initially identical and equal to D. Vectors d_1, d_2 and D are unit vectors if the thickness is assumed constant. The angle between the both director vectors d_1 and d_2, noted by $\gamma = (d_1, d_2)$, is assumed small. This assumption makes us to neglect the thickness variation viewed through both fibers (d_1, d_2). The curve fiber is tangent to one from both director vectors (d_1), in the vicinity of the reference surface. With these geometric assumptions, the real position (q) after deformation is given by (see, Fig. 1.7). As a result, the position vector for any point q of the shell can be written in terms of convective curvilinear coordinates $\xi = (\xi^1, \xi^2, \xi^3 = z)$ as follows:

$$x_q(\xi^1, \xi^2, z) = x_p(\xi^1, \xi^2) + f_1(z)d_1(\xi^1, \xi^2) + f_2(z)d_2(\xi^1, \xi^2), \tag{1.86}$$

where h is the thickness, p is the normal projection point on mid-surface of the material point q (see Fig. 1.7) and D is the initial shell director vector. d_1 and d_2 denote the first and the second deformed shell directors vectors, respectively. The functions $f_1(z)$ and $f_2(z)$ which reflect a high-order distribution of the displacement field in the function of the thickness variable can be expressed as follows:

$$f_1(z) = z - f_2(z), \quad f_2(z) = \frac{4z^3}{3h^2}. \tag{1.87}$$

It is worth mentioning that the double directors model is an expansion of the Reissner-Mindlin

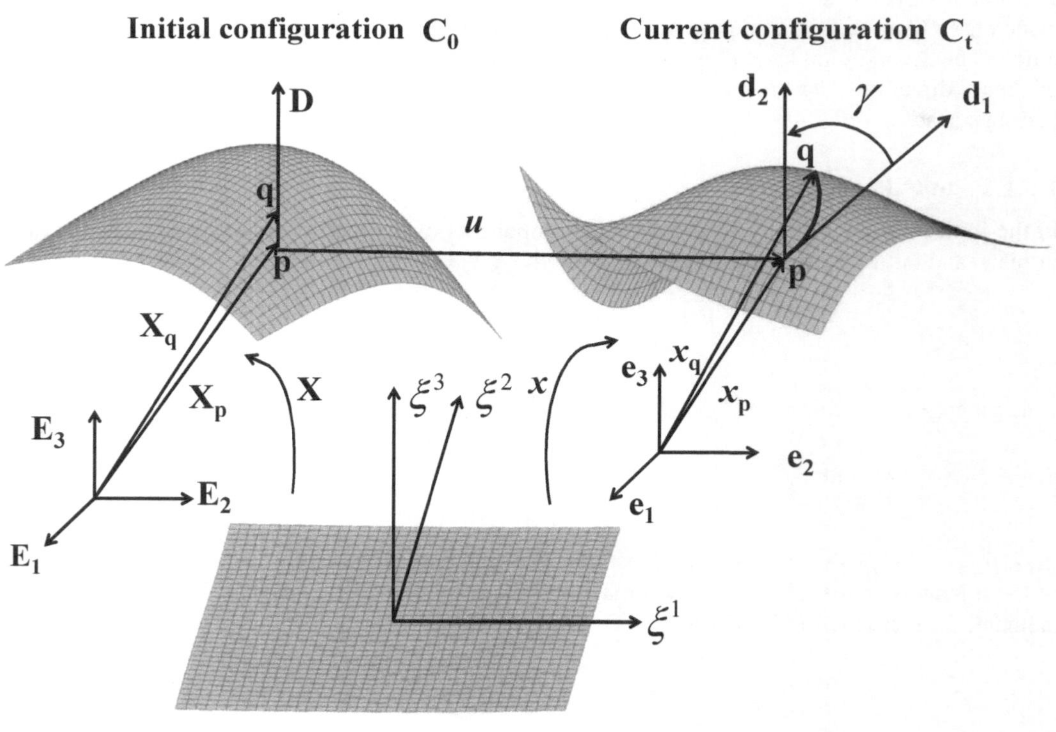

Figure 1.7 Geometry of the double directors shell model.

Table 1.4
Shell theories functions f_1 and f_2.

Theory	Membrane	Kirchhoff-Love	Reissner-Mindlin	Double Director model
$f_1(z)$	0	0	z	$z-4z^3/(3h^2)$
$f_2(z)$	0	z	0	$4z^3/(3h^2)$

Table 1.5
Different shell theories functions.

Theory	$f_1(z)$	$f_2(z)$
[140], [30], [204]	$z-f_2(z)$	$4z^3/3h^2$
[173], [237]	$z-f_2(z)$	$-z/4+5z^3/3h^2$

model, which includes the Kirchhoff-Love model. Furthermore, the Reissner-Mindlin, Kirchhoff-Love and double directors shell models have the same relations between the functions $f_1(z)$ and $f_2(z)$, where:

$$f_1(z)+f_2(z)=z \ , \ f_1'(z)+f_2'(z)=1. \tag{1.88}$$

A summary of these functions can be found in Tab. 1.4.

The general model with functions f_1 and f_2, defined in Tab. 1.4, can be reduced in linear case to obtain plate model by projection in the tangent plane. The following displacement field is obtained:

$$\begin{cases} u=u_0+\left(z-4\frac{z^3}{3h^2}\right)\beta_1+\frac{4z^3}{3h^2}\alpha_1 \\ v=v_0+\left(z-4\frac{z^3}{3h^2}\right)\beta_2+\frac{4z^3}{3h^2}\alpha_2 \\ w=w_0 \end{cases} , \quad \alpha_k=-w_{,k} \ , \quad k=1,2 \ , \tag{1.89}$$

where u_0, v_0 and w_0 are the mean surface displacements. β_1 and β_2 are the rotations and α_1 and α_2 are the derivative of transverse displacement w. This displacement field is initially proposed in linear case [140, 30, 204]. Using an energy approach, other functions $f_1(z)$ and $f_2(z)$ were investigated in [173, 237] (Tab. 1.5).

1.5.2 STRAIN MEASURE

Using the kinematic assumptions, the Lagrangian strain E can be written as follows [282, 283]:

$$E=\tfrac{1}{2}(g-G) \ , \quad E_{ij}=\tfrac{1}{2}(g_{ij}-G_{ij}) \ , \quad \begin{cases} E_{\alpha\beta}=e_{\alpha\beta}+f_1(z)\chi^1_{\alpha\beta}+f_2(z)\chi^2_{\alpha\beta} \\ 2E_{\alpha 3}=f_1'(z)\gamma^1_\alpha+f_2'(z)\gamma^2_\alpha \\ E_{33}=\frac{1}{2}\left[\left(f_1'+f_2'\right)d-1\right] \end{cases} , \tag{1.90}$$

where $e_{\alpha\beta}$, $\chi^k_{\alpha\beta}$ and γ^k_α denote the membrane, the bending and the shear strains, which can be computed as:

$$\begin{cases} \delta e_{\alpha\beta}=\frac{1}{2}\left(a_{\alpha\beta}-A_{\alpha\beta}\right), \ \gamma^k_\alpha=c^k_\alpha-C^k_\alpha \\ \chi^k_{\alpha\beta}=\frac{1}{2}\left(b^k_{\alpha\beta}-B^k_{\alpha\beta}\right) \end{cases} , \quad k=1,2, \tag{1.91}$$

$a_{\alpha\beta}$, $b^k_{\alpha\beta}$, $B^k_{\alpha\beta}$, c^k_α, C^k_α and d are defined as follows:

$$\begin{cases} a_{\alpha\beta} = a_\alpha \cdot a_\beta \;,\; b^k_{\alpha\beta} = a_\alpha \cdot d^k_{,\beta} + a_\beta \cdot d^k_{,\alpha} \;,\; B^k_{\alpha\beta} = A_\alpha \cdot D^k_{,\beta} + A_\beta \cdot D^k_{,\alpha} \\ c^k_\alpha = a_\alpha \cdot d_k \;,\; C^k_\alpha = A_\alpha \cdot D_k \;,\; d = d_1 \cdot d_1 \;,\; k = 1,2. \end{cases} \tag{1.92}$$

In matrix notation, the membrane, bending and shear strains vectors are given by:

$$e = \begin{bmatrix} e_{11} \\ e_{22} \\ 2e_{12} \end{bmatrix}, \quad \chi^k = \begin{bmatrix} \chi^k_{11} \\ \chi^k_{22} \\ 2\chi^k_{12} \end{bmatrix}, \quad \gamma^k = \begin{bmatrix} \gamma^k_1 \\ \gamma^k_2 \end{bmatrix}, \quad k = 1,2. \tag{1.93}$$

Moreover, the transverse shear stress vanishes on the top and bottom shell surfaces ($\sigma_{13}(\pm h/2) = \sigma_{23}(\pm h/2) = 0$) which verify the following condition:

$$\gamma^2_\alpha = 0, \quad 2E_{\alpha 3} = f_1'(z)\gamma^1_\alpha. \tag{1.94}$$

This kinematic constraint will be imposed in a discrete form in the finite element approximation. The generalized strain and virtual strain vectors Σ and $\delta\Sigma$ are defined by:

$$\Sigma = \begin{bmatrix} e \\ \chi^1 \\ \chi^2 \\ \gamma^1 \end{bmatrix}_{11\times 1} \qquad \delta\Sigma = \begin{bmatrix} \delta e \\ \delta\chi^1 \\ \delta\chi^2 \\ \delta\gamma^1 \end{bmatrix}_{11\times 1} \qquad \gamma^2 = \begin{bmatrix} 0 \\ 0 \end{bmatrix}, \tag{1.95}$$

where the virtual strains can be written as:

$$\begin{cases} \delta e_{\alpha\beta} = 1/2(a_\alpha \cdot \delta x_{,\beta} + a_\beta \cdot \delta x_{,\alpha}), \;\; \delta\gamma^k_\alpha = a_\alpha \cdot \delta d_k + \delta x_{,\alpha} \cdot d_k \\ \delta\chi^k_{\alpha\beta} = 1/2(a_\alpha \cdot \delta d_{k,\beta} + a_\beta \cdot \delta d_{k,\alpha} + \delta x_{,\alpha} \cdot d_{k,\beta} + \delta x_{,\beta} \cdot d_{k,\alpha}) \end{cases}, \quad k = 1,2. \tag{1.96}$$

1.5.3 VARIATIONAL PRINCIPLE AND LINEARIZATION

Finite element solutions are obtained from variational formulation which is based on the weak form of equilibrium which takes the following form:

$$G(\Phi, \delta\Phi) = \int_A \left(N \cdot \delta e + T_1 \cdot \delta\gamma^1 + \sum_{k=1}^{2} \left(M_k \cdot \delta\chi^k \right) \right) dA - G_{ext}(\Phi, \delta\Phi) = 0, \tag{1.97}$$

where $\delta\Phi = (\delta u, \delta d_1, \delta d_2)$ is an arbitrary variation. G_{ext} is the external virtual work. δe, $\delta\chi^k$ and $\delta\gamma^1$ are the variations of shell strains. N, M_k and T_1 are the membrane, bending and shear stresses resultants respectively which can be expressed as [79, 282]:

$$N = \begin{Bmatrix} N^{11} \\ N^{22} \\ N^{12} \end{Bmatrix}, \quad M_k = \begin{Bmatrix} M^{11}_k \\ M^{22}_k \\ M^{12}_k \end{Bmatrix}, \quad T_1 = \begin{Bmatrix} T^1_1 \\ T^2_1 \end{Bmatrix}, \quad k = 1,2, \tag{1.98}$$

where their components are defined as follows:

$$(N^{\alpha\beta}, M^{\alpha\beta}_1, M^\alpha_2) = \int_{-h/2}^{h/2} (1, f_1, f_2) S^{\alpha\beta}\, dz, \quad T^\alpha_1 = \int_{-h/2}^{h/2} f_1'(z) S^{\alpha 3}\, dz. \tag{1.99}$$

$S^{\alpha\beta}$ are the components of the second Piola-Kirchhoff stress tensor. Furthermore, by defining the generalized resultant vectors of stress R and strain Σ with:

$$R = \left\{ \begin{array}{c} N \\ M_1 \\ M_2 \\ T_1 \end{array} \right\}_{11\times 1}, \quad \Sigma = \left\{ \begin{array}{c} e \\ \chi^1 \\ \chi^2 \\ \gamma^1 \end{array} \right\}_{11\times 1} \tag{1.100}$$

the weak form of equilibrium equation can be written as follows:

$$G(\Phi, \delta\Phi) = \int_A \delta\Sigma^T \cdot R dA - G_{ext}(\Phi, \delta\Phi) = 0. \tag{1.101}$$

where $\Phi = (u, d_1, d_2)$ is the displacement and the directors vectors. Equation (1.101) indicates the non-linear shell model which can be solved by the Newton iterative algorithm. The consistent tangent operator for Newton solution procedure can be constructed by the directional derivative of the weak form in the direction of the increment $\Delta\Phi = (\Delta u, \Delta d_1, \Delta d_2)$. For large displacement analysis, it is a conventional practice to split the tangent operator into geometric and material contributions, denoted by $D_G G \cdot \Delta\Phi$ and $D_M G \cdot \Delta\Phi$, respectively, as follows:

$$DG \cdot \Delta\Phi = D_G G \cdot \Delta\Phi + D_M G \cdot \Delta\Phi \tag{1.102}$$

1.5.3.1 Material contribution

As indicated previously in section 1.4.5.3 ,the material contribution of the tangent operator can be obtained as:

$$D_M G \cdot \Delta\Phi = \int_A (\delta\Sigma^T \cdot \Delta R) dA, \tag{1.103}$$

The material tangent modulus can be expressed as:

$$R = H_T \Sigma, \quad H_T = \begin{bmatrix} H_{11} & H_{12} & H_{13} & 0 \\ H_{12} & H_{22} & H_{23} & 0 \\ H_{13} & H_{23} & H_{33} & 0 \\ 0 & 0 & 0 & H_{44} \end{bmatrix}. \tag{1.104}$$

where:

$$(H_{11}, H_{12}, H_{13}, H_{22}, H_{23}, H_{33}) = \int_{-h/2}^{h/2} (1, f_1, f_2, f_1^2, f_1 f_2, f_2^2)\, H dz. \tag{1.105}$$

and

$$H_{44} = \int_{-h/2}^{h/2} \left(f_1'\right)^2 H_\tau dz. \tag{1.106}$$

Herein, H and H_τ represent the in plane and out-of-plane linear elastic sub-matrices for a single layer shell structure.

1.5.3.2 Geometrical contribution

The geometrical contribution results from the variation of the virtual strain while maintaining constant the stress resultants:

$$D_G G \cdot \Delta\Phi = \int_A (\Delta\delta\Sigma^T \cdot R) dA, \tag{1.107}$$

This form can be divided into membrane, bending and shear parts as follows:

$$D_G G \cdot \Delta\Phi = D_G G_m \cdot \Delta\Phi + D_G G_{b_1} \cdot \Delta\Phi + D_G G_{b_2} \cdot \Delta\Phi + D_G G_s \cdot \Delta\Phi, \tag{1.108}$$

1.5.4 FINITE ELEMENT APPROXIMATION

In this section, the interpolation of the displacement and the strain fields related to the proposed non-linear double directors shell theoretical formulation are presented.

1.5.4.1 Approximation of the displacement field

In this section, the numerical implementation of the presented shell theoretical formulation based upon a four-node shell element is established. Using the isoparametric concept, the displacement field $(u = x - X)$ and the first director d_1 are approximated by:

$$u = \sum_{I=1}^{4} N^I u_I, \quad \Delta u = \sum_{I=1}^{4} N^I \Delta u_I, \quad \delta d_1 = \sum_{I=1}^{4} N^I \delta d_{1I}, \quad \Delta d_1 = \sum_{I=1}^{4} N^I \Delta d_{1I}, \tag{1.109}$$

where N^I are the standard bilinear isoparametric shape functions.

For the second director vector, its interpolation is given by:

$$\delta d_2 = \sum_{I=1}^{4} N^I \delta d_I + \sum_{K=5}^{8} P_K \delta \alpha_K t_K, \quad \Delta d_2 = \sum_{I=1}^{4} N^I \Delta d_I + \sum_{K=5}^{8} P_K \Delta \alpha_K t_K \tag{1.110}$$

where (I) denotes a node of the element. K and $\delta\alpha_k$ are the mid-point of the element boundaries and the variables associated to δd_2 on the boundaries. The vector t_K is unit and its direction in defined by the position of the nodes couple (I,J). The variable $\delta\alpha_k$ will be replaced by introducing the constraint condition over the element boundaries which is expressed under integral form as follows:

$$\int \gamma_\alpha^2 = 0 \tag{1.111}$$

The shape functions P_K are quadratic and are given as indicated in Tab. 1.1.

1.5.4.2 Approximation of the strain field

The approximation of the strain field is carried out using the proposed four-node shell element formulation. Thereby, the approximation of the membrane, first and second bending parts relative to the first director vector d_1 and the second director vector d_2 respectively and the shear part can be given as follows:

$$\delta e = B_m \cdot \delta U_n, \quad \delta\chi^1 = B_1 \cdot \delta U_n, \quad \delta\chi^2 = B_2 \cdot \delta U_n, \quad \delta\gamma^1 = B_s \cdot \delta U_n. \tag{1.112}$$

where B_m, B_1, B_2 and B_s are the discrete strain-displacement matrices corresponding to membrane, first bending, second bending and shear parts of the strain field and their expressions are as mentioned in Tab. 1.6. As well as, in Tab. 1.6 the shear matrix is given for all nodes while the membrane and bending matrices are given at node I. A typical isoparametric finite element is considered as shown in Fig. 1.8 where A, B, C and D denote the mid-points of the element boundaries set, respectively. The variation of the generalized strain $\delta\Sigma$ can be given then as follows:

$$\delta\Sigma = \begin{Bmatrix} \delta e \\ \delta\chi^1 \\ \delta\chi^2 \\ \delta\gamma^1 \end{Bmatrix} = B \cdot \delta U_n, \quad B = \begin{bmatrix} B_m \\ B_1 \\ B_2 \\ B_s \end{bmatrix}. \tag{1.113}$$

The material tangent operator can be determined from Eqs. (1.100) and (1.101) as follows:

$$k_m = \int_A B^T H_T B dA. \tag{1.114}$$

Table 1.6
Strain-displacement matrices.

- Membrane

$$B_m^I = \begin{bmatrix} B_{mm}^I & 0 & 0 \end{bmatrix}, \quad B_{mm}^I = \begin{bmatrix} n_1^T \overline{N}_{,1}^I \\ n_2^T \overline{N}_{,2}^I \\ n_1^T \overline{N}_{,2}^I + n_2^T \overline{N}_{,1}^I \end{bmatrix}$$

- First bending

$$B_1^I = \begin{bmatrix} B_{1m}^I & B_{1b}^I \Lambda_1^I & 0 \end{bmatrix}, \quad B_{1m}^I = \begin{bmatrix} d_{1,1}^T \overline{N}_{,1}^I \\ d_{1,2}^T \overline{N}_{,2}^I \\ d_{1,1}^T \overline{N}_{,2}^I + d_{1,2}^T \overline{N}_{,1}^I \end{bmatrix}, \quad B_{1b}^I = \begin{bmatrix} n_1^T \overline{N}_{,1}^I \\ n_2^T \overline{N}_{,2}^I \\ n_1^T \overline{N}_{,2}^I + n_2^T \overline{N}_{,1}^I \end{bmatrix}$$

- Second bending [307, 310]

$$B_2^I = \begin{bmatrix} B_{2m}^I & 0 & B_{2b}^I \Lambda_2^I \end{bmatrix}$$

$$B_{2m}^I = \begin{bmatrix} d_{2,1}^T \overline{N}_{,1}^I + n_1^T M_{d,1}^I \\ d_{2,2}^T \overline{N}_{,2}^I + n_2^T M_{d,2}^I \\ d_{2,1}^T \overline{N}_{,2}^I + n_1^T M_{d,2}^I + d_{2,2}^T \overline{N}_{,1}^I + n_2^T M_{d,1}^I \end{bmatrix}, \quad B_{2b}^I = \begin{bmatrix} n_1^T \cdot M_{r,1}^I \\ n_2^T \cdot M_{r,2}^I \\ n_1^T \cdot M_{r,2}^I + n_2^T \cdot M_{r,1}^I \end{bmatrix}$$

$$M_d^I = P_K td_K^I + P_M td_M^I, \quad td_K^I = \frac{3}{2L_K} t_K \otimes d_K$$

$$M_r^I = N^I I + P_K tt_K^I + P_M tt_M^I, \quad tt_K^I = \frac{3}{4} t_K \otimes t_K.$$

- Shear

$$\mathbf{B}_s = J^{-1} \mathbf{B}_{s\xi}$$

$$\mathbf{B}_{s\xi} = \begin{bmatrix} N_{,1}^1 \mathbf{d}_{1B}^T & N_{,1}^2 \mathbf{a}_{1B}^T & \mathbf{0} & N_{,1}^2 \mathbf{d}_{1B}^T & N_{,1}^2 \mathbf{a}_{1B}^T & \mathbf{0} & N_{,1}^3 \mathbf{d}_{1D}^T & N_{,1}^3 \mathbf{a}_{1D}^T & \mathbf{0} & N_{,1}^4 \mathbf{d}_{1D}^T & N_{,1}^3 \mathbf{a}_{1D}^T & \mathbf{0} \\ N_{,2}^1 \mathbf{d}_{1A}^T & N_{,2}^4 \mathbf{a}_{2A}^T & \mathbf{0} & N_{,2}^2 \mathbf{d}_{1C}^T & N_{,2}^3 \mathbf{a}_{2C}^T & \mathbf{0} & N_{,2}^3 \mathbf{d}_{1C}^T & N_{,2}^3 \mathbf{a}_{2C}^T & \mathbf{0} & N_{,2}^4 \mathbf{d}_{1A}^T & N_{,2}^4 \mathbf{a}_{2A}^T & \mathbf{0} \end{bmatrix}$$

In addition, the element residual r can be deduced from the weak form of equilibrium Eq. (1.100) as:

$$r = \int_A B^T R dA. \tag{1.115}$$

For the geometric tangent part, it can be written in matrix form as:

$$D_G G \cdot \Delta\Phi = \delta\Phi^T k_g \Delta\Phi = \delta U_n^T \cdot \Pi_n^T \cdot k_g \cdot \Pi_n \cdot \Delta U_n. \tag{1.116}$$

where k_g is the geometric stiffness matrix.Its expression is given by:

$$k_{g_{IJ}} = \begin{bmatrix} UUM_{IJ} I + UUF_{2IJ} & (UBF_{1IJ} + UBC_{1IJ}) I & UBF_{2IJ} \\ (BUF_{1IJ} + BUC_{1IJ}) I & (BBF_{1IJ} + BBC_{1IJ}) I & 0 \\ BUF_{2IJ} & 0 & BBF_{2IJ} \end{bmatrix}, \tag{1.117}$$

where UUM, $(UBF_1, BBF_1, UUF_2, UBF_2, BBF_2)$ and (UBC_1, BBC_1) represent the membrane, the bending and shear parts of the strain field, respectively and their corresponding expressions are given as indicated in [79].

1.5.5 NODAL TRANSFORMATION IN LOCAL CARTESIAN COORDINATES

The transformation from the curvilinear system coordinates to the Cartesian ones can be achieved through the Jacobian matrix J by setting:

$$\begin{Bmatrix} \overline{N}_{,1}^I \\ \overline{N}_{,2}^I \end{Bmatrix} = [J]^{-1} \begin{Bmatrix} N_{,1}^I \\ N_{,2}^I \end{Bmatrix}, \quad J = \begin{bmatrix} n_1^0 \cdot A_1 & n_2^0 \cdot A_1 \\ n_1^0 \cdot A_2 & n_2^0 \cdot A_2 \end{bmatrix}. \tag{1.118}$$

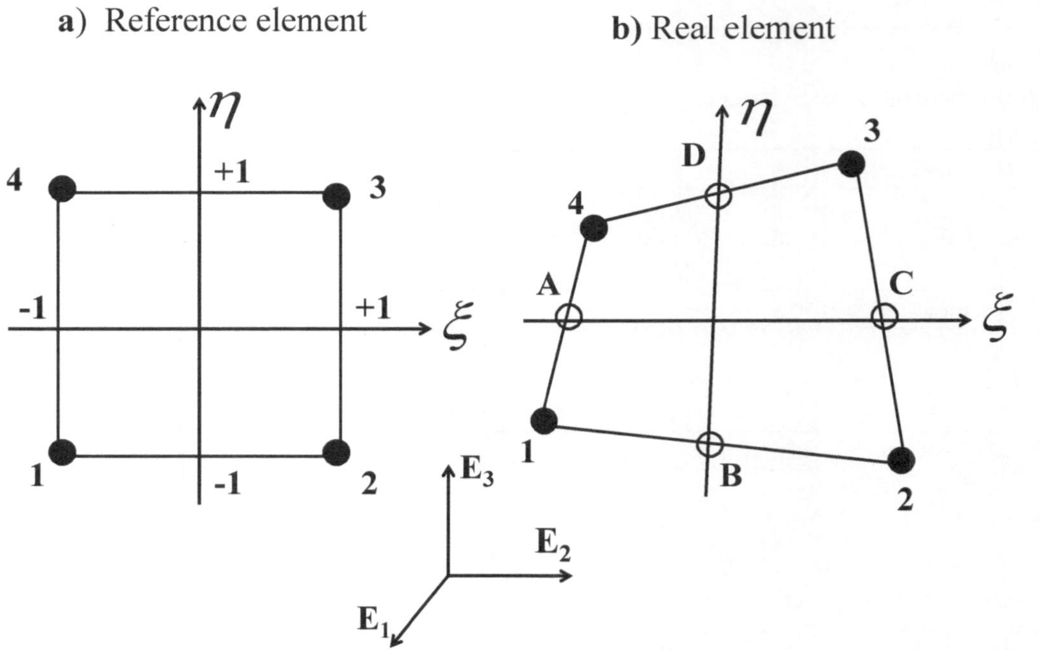

Figure 1.8 Isoparametric four-nodes shell element.

where $(\overline{N}^I_{,\alpha}, N^I_{,\alpha})$ with $(\alpha = 1,2)$ denote the shape function derivatives in the local Cartesian and elementary system coordinates, respectively. The Jacobian matrix J reflects the passage from the local Cartesian base $\{n_1^0, n_2^0, n^0\}$ to the covariant base $\{A_1, A_2, A_3\}$ with $n^0 = A_1 \wedge A_2 / \|A_1 \wedge A_2\|$ is the initial normal vector to the mid-surface.

Consequently, the nodal transformation can be defined as follows:

$$\delta\Phi_I = \left\{ \begin{array}{c} \delta u \\ \delta d_1 \\ \delta d_2 \end{array} \right\} = \Pi_I \cdot \left\{ \begin{array}{c} \delta u \\ \delta\Theta_1 \\ \delta\Theta_2 \end{array} \right\}, \quad \Pi_I = \left[\begin{array}{ccc} I_{3\times3} & 0 & 0 \\ 0 & \Lambda_1^I & 0 \\ 0 & 0 & \Lambda_2^I \end{array} \right]_{9\times7}, \quad I = 1,..,4, \tag{1.119}$$

The transformation Λ which allows either a spatial description with 9DOF/node or a material description with 7DOF/node is expressed as:

$$\Lambda_k = \left[\begin{array}{cc} -t_{2k} & t_{1k} \end{array} \right]_{3\times2}, \quad k = 1,2. \tag{1.120}$$

Thereby, the global interpolation of the displacement field can be written as:

$$\delta\Phi_n = \Pi_n \cdot \delta U_n, \quad \Pi_n = diag(\Pi_1, \Pi_2, \Pi_3, \Pi_4), \quad n = 1,...,4 \tag{1.121}$$

For the assessment of numerical results, the material description with 7DOF/node is retained. The four-nodes shell element with 7DOF/node and (ANS) collocation points is illustrated in Fig. 1.9.

1.6 CONSTITUTIVE RELATIONS OF FGM AND FG-CNTRC STRUCTURES

In this section, the material behavior laws of studied materials are presented. In my works, three types of materials, FGM perfect, FGM imperfect (porous) and FG-CNTRC shell structures, are

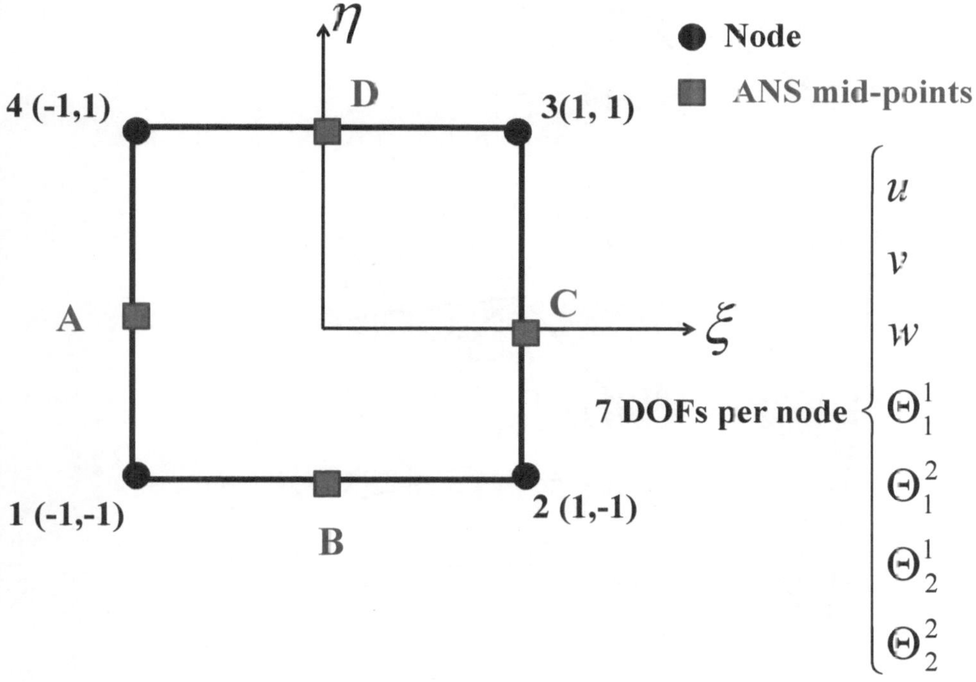

Figure 1.9 Four-nodes shell element with 7DOF/node and (ANS) collocation points.

considered. The two first types obey to an isotropic law while the second type follows an orthotropic law. To distinguish these two material laws, H^{FGM} and H_τ^{FGM} are used to identify the constitutive material matrices of FGM shell while H^{CNT} and H_τ^{CNT} refer to constitutive material matrices of FG-CNTRC shell.

1.6.1 CONSTITUTIVE RELATIONS OF FGM STRUCTURES

The expressions of the in plane and out-of-plane linear elastic sub-matrices of an FGM material, in a Cartesian system, are given by:

$$H^{FGM} = \frac{E(z)}{1-\nu^2}\begin{bmatrix} 1 & \nu & 0 \\ \nu & 1 & 0 \\ 0 & 0 & (1-\nu)/2 \end{bmatrix}, \quad H_\tau^{FGM} = \frac{E(z)}{2(1+\nu(z))}\begin{bmatrix} 1 & 0 \\ 0 & 1 \end{bmatrix}, \tag{1.122}$$

where $E(z)$ and ν are the Young's modulus and the Poisson's ratio of the isotropic phases of the FGM material, respectively. It should be noted that, for perfect or imperfect porous FGM materials, the Young's modulus can depend, further of z coordinate, on the temperature or the porosity parameters α and it can be hence noted as, $E(z,T)$, for thermo-elastic analysis and $E(z,\alpha)$ for porosity-dependent porosity analysis.

1.6.2 CONSTITUTIVE RELATIONS OF FG-CNTRC STRUCTURES

An orthotropic single layer FG-CNTRC shell is considered here as illustrated in Fig. 1.10. As a result, the in plane and out-of-plane linear elastic sub-matrices H^{CNT} and H_τ^{CNT} can be expressed

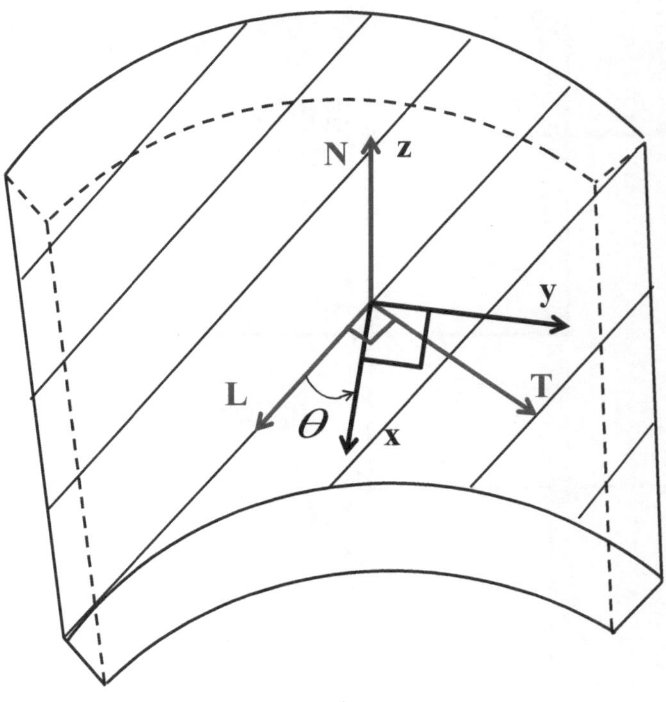

Figure 1.10 Orthotropic single layer FG-CNTRC shell.

as follows:

$$H^{CNT} = T_1{}^T H_L T_1, \quad H_\tau^{CNT} = T_2^T H_{\tau_L} T_2, \tag{1.123}$$

H_L and H_{τ_L} are the constitutive material matrices of the FG-CNTRC shell along the longitudinal and the transverse directions, respectively. T_1 and T_2 are the transformation matrices from the orthotropic system to Cartesian system for the FG-CNTRC shell. The expressions of these matrices, are given respectively as follows:

$$H_L = \begin{bmatrix} H_{11} & H_{12} & 0 \\ H_{12} & H_{22} & 0 \\ 0 & 0 & G_{12} \end{bmatrix}, \quad H_{\tau_L} = \begin{bmatrix} G_{13} & 0 \\ 0 & G_{23} \end{bmatrix}, \tag{1.124}$$

$$T_1 = \begin{bmatrix} C^2 & S^2 & CS \\ S^2 & C^2 & -CS \\ -2CS & 2CS & C^2 - S^2 \end{bmatrix}, \quad T_2 = \begin{bmatrix} C & S \\ -S & C \end{bmatrix}, \quad C = \cos\theta, \; S = \sin\theta. \tag{1.125}$$

The components of the matrices H_L and H_{τ_L} can be expressed as follows:

$$\begin{cases} H_{11} = \frac{E_{11}}{\Delta}; \quad H_{22} = \frac{E_{22}}{\Delta}; \quad H_{12} = \frac{\nu_{21} E_{11}}{\Delta} = \frac{\nu_{12} E_{22}}{\Delta} \\ \Delta = 1 - \nu_{21}\nu_{12} = 1 - \nu_{12}^2 \frac{E_{22}}{E_{11}} \end{cases}, \tag{1.126}$$

where E_{11}, E_{22}, ν_{12}, G_{12}, G_{13} and G_{23} are the Young's moduli, the Poisson's ratio and the shear's moduli of the CNTRC shell, respectively with $G_{12} = G_{13} = G_{23}$. Their corresponding expressions will be provided in Chapter 4.

1.7 CONCLUSION

In this chapter, the theoretical frameworks developed to derive the linear and non-linear governing equations of functionally gradient-based structures are presented. Three models are implemented in the finite element code (MPEF), which is developed essentially by our creative master head, the professor Fakhreddine Dammak based on the computer program packages of Abaqus. These developments make valuable contributions to the knowledge base in the shell subject area by providing numerical models and tools that are helpful in better modeling and analysis of the mechanical behavior of such advanced structures covering FGM, FG-CNTRC and FG porous materials. The proposed models constitute an alternative for modeling different types of geometries and various aspects of analysis covering bending static, free vibration, dynamic, thermal vibration, mechanical buckling, thermal buckling and post-buckling and also non-linear analysis of wide range of advanced materials. In conclusion, our contributions consists of the improvement and enhancement of the standard forms of the shell theories. In fact, we have developed an expansion of the Kirchhoff-Love model by the consideration of the geometrical non-linearity and large deflections. Since this model is limited to thin shells analysis, we have promoted to the first-order shear deformation theory (FSDT) with an improvement of the shear strains. For this model, our contributions consists of the enhancement of the transverse shear strains by the introduction of a quadratic function that able to describe the realistic behavior of transverse shear strains and stresses. The amelioration of the (FSDT) finite element shell model has been outlined via several numerical assessments covering mechanical and thermal analysis of FGM and recently FG porous plates and shells. Continuously, a double directors finite shell element have been developed and implemented in the range of high-order models to investigate various aspects of mechanical analysis for FGM, FG-CNTRC and FG porous structures. The applicability and the validity of these models will be highlighted in the next chapters.

2 Thermal analysis of FGM structures using (I-FSDT)

In this chapter thermal free vibration, buckling and post-buckling analysis of functionally gradient-based structures based on the numerical models developed in Chapter 1 are performed. In these analyses, the modified or improved (I-FSDT) model is used to investigate the mechanical behaviors of FGM plates and shells subjected to different thermal loadings, namely uniform and non-linear (or nonuniform) temperature rises. The convergence and the validity and applicability of the numerical finite element model are first checked by comparison studies. Parametric studies are also carried out to outline the influence of geometrical and material parameters on thermal and mechanical responses of FGM structures. These case studies can be useful for structural engineers, researchers, industrial members and scientists.

2.1 INTRODUCTION

Functionally graded materials (FGMs) are a new class of advanced materials, in the group of composites, which were first introduced in the late 1984s, by a group of scientists in Sendai, Japan [128, 129] during the discussion in the possibility of materials that can resist to severe thermo-mechanical loadings. The goal was to prepare a thermal barrier materials for aerospace structures and nuclear reactors. These materials are characterized by their continuous and smooth variation of the volume fraction of its constituents from one surface to another along a specific direction. Typically, these materials are made from metal and ceramic phases. For the surface that contacts high temperature, ceramics are used to provide heat resistance. For the opposite surface, metallic materials are used to supply the necessary thermal conductivity and mechanical strength. Day by day, these materials have been replaced the conventional materials which suffers from delamination, stress concentration and crack problems. Hence, due to the exotic properties of FGM materials, many researchers have been attracted to the study of their mechanical behaviors covering static, vibration and thermal buckling and post-buckling responses. Actually, these materials have been integrated into many structural components like beams, plates and shells and they are gaining potential applications in different areas due to their flexibility in making a particular composite material according to the requirement. For instance, we can find these FGMs materials in:

1. Combustion chambers, rocket engine components and space vehicles, fuselage tanks, stabilizers, rotor blades, aircraft wings, cryogenic propellant tanks, gas turbine engines, nozzles and compressor components of fighter planes and helicopters [139].
2. Medical applications including the replacement of living tissues in human body with biopolymer FGMs. Orthopedic and dental implants are usually composed of collagen hydroxyapatite (HAP) and titanium alloys [200]. High density polyethylene with a graded biopolymer coating is used in orthopedic implants like total hip, shoulders and knee joint replacements [197, 285].
3. Photo-electronic devices, anti-reflective layers, optical fibers, optical lenses, photo-detectors, solar cells, optical and piezoelectric sensors [141].
4. Automotive parts which require high strength with resistance to crack, fracture and thermal shocks, machine and forming tools and cutting tools [49].

Therefore, FGMs are found to be the most advantageous new materials which have different applications in various sectors. In addition, the mathematical modeling of the graded property of these

DOI: 10.1201/9781003483786-2

materials constitutes a key factor in the analysis of the FGM structural components. In recent years, significant research works have been published in the analysis of FGM beams, plates and shells using different theories and models. A knowledge of the thermal and mechanical behavior of these FGM structures is prominent for both academic and industrial researchers in order to provide high performance in the next generation of aerospace, naval, civil, nuclear and vehicle structures, where the development of reliable analysis tools is considerably required. Most of the advanced applications of FGMs are found in thermal analysis, where vibration, buckling and post-buckling behaviors of structural components exposed to different types of thermal loadings are investigated by many researchers using various theories and models. For thermal vibration analysis, various investigations are focused on evaluating the free and forced vibration frequencies of FGM plates and shells by solving the eigenvalue vibration problem [299, 126, 157, 155, 156]. Continuously, various analyses are made in the last years on buckling and post-buckling responses of FGM structures working in thermal environment [108, 170, 124, 105, 6, 253, 272, 41, 225, 228, 229]. Through these intensive investigations on thermal buckling and post-buckling of FGM plates and shells, the standard (FSDT) theory, the Reddy's high-order shear deformation theory (HSDT) or (3D) elasticity theory are often used to modeling the kinematics of the structures and to predict these behaviors. For that, we have developed a modification and amelioration on these models to overcome all the limitations of the need of shear correction factors in the (FSDT) or the several kinematic variables in the (HSDT) or the complex resolution for (3D) solutions. So, to ensure the compromise accurate results/fast computational time, we provide the first new results on thermal vibration, buckling and post-buckling analysis of FGM structures within the numerical frameworks presented in Chapter 1.

2.2 THERMO-ELASTIC PROPERTIES OF FUNCTIONALLY GRADED MATERIALS (FGMS)

2.2.1 EFFECTIVE MATERIAL PROPERTIES

A functionally graded (FG) shell structure subjected to thermal load is considered. The shell is made from a mixture of ceramic and metal phases for which the material composition varies continuously through the thickness direction of the shell such that the top surface ($z = +h/2$) is ceramic-rich whereas the bottom surface ($z = -h/2$) is metal-rich as shown in Fig. 2.1. The thermo-elastic material properties P such as the modulus of elasticity E, the mass density ρ, coefficient of thermal expansion α and the thermal conductivity K can be expressed as:

$$P(T,z) = (P_c(T) - P_m(T))V_c(z) + P_m(T), \quad V_c(z) = \left(\tfrac{z}{h} + \tfrac{1}{2}\right)^p \tag{2.1}$$

where the indices m and c refer to metal and ceramic phases while the exponent p defines the power law index which varies between zero and infinity. For ($p = 0$), the shell is fully ceramic whereas for ($p \approx \infty$), the shell is fully metallic. It should be noted that three main graded functions can be employed to describe the material properties of such structures which are: the power-law function (P-FGM), the sigmoid function (S-FGM) and the exponential function (E-FGM). Each function reflects the gradual variation of the constituents' volume fraction through the thickness direction. In fact, the E-FGM is applicable for the crack propagation analysis in materials, the S-FGM is used for layered FGM structures investigations while the P-FGM is the most useful for the researchers due its simplicity and convenience [255]. So, for the sake of simplicity, the P-FGM law is used in the current investigations. The dependency on temperature can be written in terms of non-linear function of temperature as follows:

$$P(T) = P_0\left(P_{-1}T^{-1} + 1 + P_1T + P_2T^2 + P_3T^3\right)\,, \tag{2.2}$$

where P_0, P_{-1}, P_1, P_2 and P_3 are the temperature coefficients and T is the temperature parameter in K. It should be also noted that the Young's modulus E and the thermal expansion coefficient α are

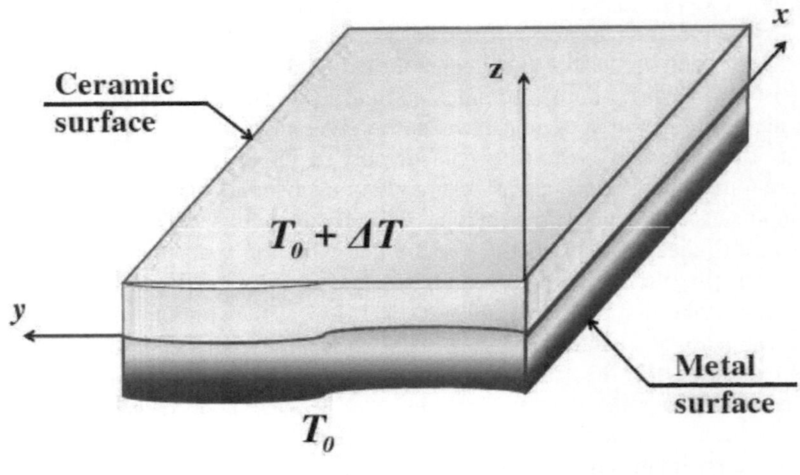

Figure 2.1 Geometry of FG shell subjected to thermal load.

considered as temperature dependent while the Poisson's ratio ν is assumed to be constant.

2.2.2 THERMAL LOADINGS

The FG shell can be subjected to uniformly distributed or non-linear (also known as, non-uniform) temperature rises through the thickness direction. For Uniform temperature rise, labeled (UTR), the temperature change is given by:

$$\Delta T = T_f - T_0 \, , \tag{2.3}$$

where T_0 is the initial temperature which can be uniformly raised to a final temperature T_f as shown in Fig. 2.1. For non-linear temperature distribution, labeled ($NLTD$), the temperature change can be obtained by solving the following steady-state heat transfer equation:

$$\begin{cases} \frac{d}{dz}\left[K(z)\frac{dT}{dz}\right] = 0, \quad -h/2 \prec z \prec h/2 \\ T = T_c, \quad z = h/2 \\ T = T_m, \quad z = -h/2 \end{cases} , \tag{2.4}$$

where $K(z)$, T_c and T_m are the thermal conductivity coefficient and the temperature changes at ceramic and metal sides, respectively. Similar to the coefficients of Young's modulus and thermal expansion, the coefficient of thermal conductivity can be expressed as a power-law distribution by:

$$K(z) = (K_c - K_m)\left(\frac{z}{h} + \frac{1}{2}\right)^p + K_m. \tag{2.5}$$

The analytical solution of Eq. (3.5) yields to:

$$\begin{cases} T(z) = T_m + \Delta T \cdot \eta(z) \\ \Delta T = T_c - T_m \\ \eta(z) = \frac{\int_{-h/2}^{z} \frac{dz}{K(z)}}{\int_{-h/2}^{h/2} \frac{dz}{K(z)}} \end{cases} , \tag{2.6}$$

where ΔT is the variation of temperature between the top and the bottom surfaces of the shell. The thickness function $\eta(z)$ is evaluates, using the Gauss numerical integration procedure as mentioned in [271].

2.3 THERMAL FREE VIBRATION OF FGM STRUCTURES

In this section, free vibration analysis of FGM structures subjected to different thermal loadings is investigated. First, the solution strategy adopted for the determination of the natural frequencies is presented. Then, different numerical simulations are carried out in order to show the validity of the used formulation and the effects of thermal loadings, gradient indexes, boundary conditions and other geometrical parameters on vibrational behavior of FGM plates and panels. It is worth to mention that the used formulation for thermal free vibration analysis is the improved FSDT theory (I-FSDT) where the governing equations are as presented in section.1.4, Chapter1.

2.3.1 SOLUTION STRATEGY FOR THERMAL FREE VIBRATION PROBLEM

After the derivation of the governing equations, the vibrational problem is formulated via the construction of the element stiffness and mass matrices, respectively. As a result, the vibrational eigenvalue problem can be written as:

$$(\mathbf{K}-\omega^2\mathbf{M})\cdot\{\Lambda\}=0. \tag{2.7}$$

where ω is the eigen-frequency of the FGM shell and Λ is the associated eigen-mode. (M, K) are the global mass and stiffness matrices of the FGM shell structure. Their expressions are given by:

$$M^I=\begin{bmatrix} m_{1I} & m_{1I} & m_{1I} & m_{2I} & m_{2I} \end{bmatrix}_{diag} \tag{2.8}$$

$$K=\int_A B^T H_T B dA. \tag{2.9}$$

where,

$$m_{kI}=\frac{\int_A \overline{\rho}_k dA \int_A \overline{\rho}_k N^I N^I dA}{\int_A (\Sigma \overline{\rho}_k N^I N^I) dA}, \quad k=1,2. \tag{2.10}$$

and

$$(\overline{\rho}_1,\overline{\rho}_2)=\int_A \rho(z)(1,f^2)dA. \tag{2.11}$$

The expressions of H_T and B are as mentioned in Eqs. (1.66) and (1.77) from (section.1.4, Chapter 1), respectively.

2.3.2 NUMERICAL APPLICATIONS

In this section, the applicability of the developed finite element model (I-FSDT) for thermal free vibration of FGM plates and cylindrical panels is outlined via several numerical tests. In the following, the material properties of the studied examples, which are temperature-dependent, are presented in Tabs. 2.1 and 2.2.

2.3.2.1 Validation of the formulation

To verify the improved (I-FSDT) shell model for thermal free vibration analysis, an isotropic clamped (CCCC) square ($a/b=1$) plate exposed to uniform temperature rise is first analyzed. The dimensions of the plate are taken as mentioned in [111] where the length side and total thickness are equal to: $a=1m$ and $h=0.01m$. The isotropic square plate is made of mild steel and its material properties, which are temperature independent, are considered as follows: Young's modulus $E=210GPa$, Poisson's ratio $\nu=0.3$, coefficient of thermal expansion $\alpha=12.6\times10^{-6}/^\circ C$ and density $\rho=7.850Kg/m^3$. The results are presented in terms of non-dimensional frequency parameter which is defined as: $\omega^*=\omega a^2\sqrt{\rho h/D}$ with ω is the circular natural frequency and $D=Eh^3/12(1-\nu^2)$ is the dynamic stiffness of the plate. Table 2.3 shows the present first four non-dimensional natural frequencies of the plate in comparison with those given by Jeyaraj [111] using ANSYS finite element Software and Rao and Ganesan [202] using a four-noded plate element. A good agreement with those reported in the literature [111, 202] is shown which draws the high accuracy of the proposed finite element shell model (I-FSDT).

2.3.2.2 Convergence property

The second step of validation consists of examining the convergence property of a fully clamped (CCCC) FGM square plate by varying the mesh sizes (4×4, 8×8, 16×16, 32×32 and 100×100). The geometrical properties of the FG plate are as indicated in [126, 142, 299] with a length side $a=0.2m$, length-to-thickness ratio $a/h=10$ and a power-law index $p=2$. The plate is composed of stainless steel metallic ($US304$) and silicon nitride ceramic (i_3N_4) phases, respectively. Its mechanical properties are listed in Tabs. 2.1 and 2.2. In this example, the normalized natural frequency parameter is defined as $\overline{\omega}=\frac{\omega a^2}{\pi^2}\sqrt{\frac{12\rho_m}{(1-\nu_m^2)}}$ with E_m, ρ_m and ν_m are the material properties of metal phase chosen at the reference temperature $T_0=300K$. The plate is subjected to different uniform temperature rises ($\Delta T=0,300,500K$) and the obtained results are mentioned in Tabs. 2.4 and 2.5. It can be remarked that as the number of elements per side rises, as the normalized frequency parameter converges well to the references solutions [126, 142, 299] for the first eight modes of vibration. Furthermore, the present results for a mesh size (32×32) are closer to the ones given by the 3D-elasticity theory [142] which draw the capacity of the present improved (I-FSDT) formulation in the prediction of the vibrational behavior of FG plates under thermal loads. It is also noteworthy that the relative error between the results does not exceed 3% which outlines the efficiency of the present formulation in terms of solution accuracy and rate of convergence with mesh refinement.

The consistency of the present finite element shell model has been also checked through the convergence analysis presented in Fig. 2.2. Indeed, the normalized frequency parameter relative to the reference solution versus the number of elements per size is depicted for the first eight modes of vibration. This graph shows that the present model provides a good convergence rate with mesh refinement for all possible cases of temperature rises ($\Delta T=0,300,500K$) and a (32×32) of elements are enough to obtain accurate vibration results. Thereby, this mesh size is considered to analyze vibrational behavior of the FG plates in the rest of numerical examples.

Table 2.1

Temperature-dependent thermo-elastic coefficients of studied ceramic materials [126, 142, 154, 206].

Material	Property	P_{-1}	P_0	P_1	P_2	P_3
Si_3N_4	$E_c(Pa)$	0	348.43×10^9	-3.070×10^{-4}	2.160×10^{-7}	-8.946×10^{-11}
	ν_c	0	0.24	0	0	0
	$\alpha_c(K^{-1})$	0	5.8723×10^{-6}	9.095×10^{-4}	0	0
	$K(W/mK)$	0	9.19	0	0	0
	$\rho_c(Kg/m^3)$	0	2370	0	0	0
ZrO_2	$E_c(Pa)$	0	132.20×10^9	-3.805×10^{-4}	-6.127×10^{-8}	0
	ν_c	0	0.3330	0	0	0
	$\alpha_c(K^{-1})$	0	13.300×10^{-6}	-1.42110×10^{-3}	9.549×10^{-7}	0
	$K(W/mK)$	0	1.78	0	0	0
	$\rho_c(Kg/m^3)$	0	3657	0	0	0

Table 2.2

Temperature-dependent thermo-elastic coefficients of considered metallic materials [126, 142, 154, 206].

Material	Property	P_{-1}	P_0	P_1	P_2	P_3
$SUS304$	$E_m(Pa)$	0	201.04×10^9	3.079×10^{-4}	-6.534×10^{-7}	0
	ν_m	0	0.3262	-2.002×10^{-4}	3.797×10^{-7}	0
	$\alpha_m(K^{-1})$	0	12.33×10^{-6}	8.086×10^{-4}	0	0
	$K(W/mK)$	0	12.04	0	0	0
	$\rho_m(Kg/m^3)$	0	8166	0	0	0
$Ti-6Al-4V$	$E_m(Pa)$	0	122.70×10^9	-4.605×10^{-4}	0	0
	ν_m	0	0.2888	1.108×10^{-4}	0	0
	$\alpha_m(K^{-1})$	0	7.4300×10^{-6}	7.483×10^{-4}	-3.621×10^{-7}	0
	$K(W/mK)$	0	6.10	0	0	0
	$\rho_m(Kg/m^3)$	0	3657	0	0	0

Table 2.3

Non-dimensional frequency parameter $\omega^* = \omega a^2\sqrt{\rho h/D}$ of clamped (CCCC) square isotropic plate subjected to uniform temperature rise.

Mode	Temperature rise (°C)								
	0			10			20		
	Present	Ref.[202]	Ref.[111]	Present	Ref.[202]	Ref.[111]	Present	Ref.[202]	Ref.[111]
1	36	37	38	28.6	28	30	18.3	15	18
2 and 3	73.4	78	78	65.3	69	69	56	57	58
4	108	111	112	99.5	101	102	90.3	89	91

2.3.2.3 Free vibration of FGM plates under thermal loadings

In this section, free vibration of functionally graded plates with thermal effects will be studied. The metal composition of the FG plate is titanium alloy ($Ti-6Al-4V$), whose physical properties are mentioned in Tab. 2.2, whereas, the ceramic composition is zirconium oxide (ZrO_2), whose physical properties are indicated in Tab. 2.1. The length-to-thickness ratio of the FG plate is $a/h = 10$ with $a = 0.2m$ ([126]). Furthermore, the plate is subjected first to uniform temperature rise (UTR) with ($\Delta T = 500K$) and secondly to non-linear temperature distribution ($NLTD$) with ($T_m = 300K$, $T_c = 800K$). The corresponding first eight natural frequencies of the FG plates are shown in Tabs. 2.6, 2.7 and 2.8, respectively. For uniform and non-linear thermal loadings, a good agreement is obtained. However, the little difference between the results, which can reach 7% in some cases of natural frequencies, for high modes of vibration can be explained by the different shell theories and solution techniques used in the two studies. In fact, in our paper, an improved (I-FSDT) shell model with the Gauss numerical integration is used while in [126] an analytical approach based on the (TSDT) theory with a double Fourier series expansion is adopted. In terms of material composition, it is observed that when $p = 0$ or tends to infinity, the FG plate is respectively reduced to pure ceramic and pure metal material. Between these extreme cases, the value of natural frequency decreases from ceramic to metal composition and this is due to high rigidity of ceramic phase compared to metal phase. The first eight mode shapes of vibration of clamped (CCCC) ($ZrO_2/Ti-6Al-4V$) FGM square plate ($a = 0.2m$, $a/h = 10$) and subjected to uniform temperature rise ($\Delta T = 300K$) are plotted in Fig. 2.3, respectively.

Table 2.4

Convergence and comparison of the normalized first four natural frequency parameter $\overline{\omega} = \frac{\omega a^2}{\pi^2}\sqrt{12\rho_m(1-\nu_m^2)/E_m h^2}$ of ($Si_3N_4/SUS304$) clamped (CCCC) FGM square plate subjected to uniform temperature rise with ($a = 0.2m$, $a/h = 10$ and $p = 2$).

$\Delta T(K)$	Source	Mode			
		1	2	3	4
0	4×4	4.3000	8.9200	8.9200	10.6200
	8×8	4.1700	8.0200	8.0200	10.7800
	16×16	4.1500	7.9200	7.9200	10.9800
	32×32	4.1400	7.9000	7.9000	11.0400
	100×100	4.1400	7.8900	7.9000	11.0500
	HSDT-semi-analytical.[299]	4.1062	7.8902	7.8902	11.1834
	%difference	0.8164	0.0025	0.0025	1.1928
	TSDT-analytical.[126]	4.1165	7.9696	7.9696	11.2198
	%difference	0.5676	0.9987	0.9987	1.5133
	3D-elasticity.[142]	4.1658	7.9389	7.9389	11.1212
	%difference	0.6231	0.6197	0.6197	0.6443
300	4×4	3.7300	8.3900	8.3900	10.0200
	8×8	3.7300	7.2800	7.2800	10.2000
	16×16	3.7000	7.2800	7.2800	10.2000
	32×32	3.6900	7.2600	7.2600	10.2500
	100×100	3.6900	7.2500	7.2500	10.2600
	HSDT-semi-analytical.[299]	3.6636	7.2544	7.2544	10.3924
	%difference	0.7154	0.0606	0.0606	1.2904
	TSDT-analytical.[126]	3.6593	7.3098	7.3098	10.4021
	%difference	0.8319	0.8248	0.8248	1.3849
	3D-elasticity.[142]	3.7202	7.3010	7.3010	10.3348
	%difference	0.8184	0.7034	0.7034	0.7290
500	4×4	3.5500	7.8300	7.8300	9.3800
	8×8	3.3100	6.8000	6.8000	9.3000
	16×16	3.2700	6.6500	6.6500	9.4100
	32×32	3.2600	6.6200	6.6200	9.4500
	100×100	3.2500	6.6100	6.6100	9.4600
	HSDT-semi-analytical.[299]	3.2357	6.6281	6.6281	9.5900
	%difference	0.4400	0.2738	0.2738	1.3742
	TSDT-analytical.[126]	3.2147	6.6561	6.6561	9.5761
	%difference	1.0861	0.6974	0.6974	1.2272
	3D-elasticity.[142]	3.2741	6.6509	6.6509	9.5192
	%difference	0.7415	0.6187	0.6187	0.6257

In second step, the influence of the power-law index (p) on the fundamental frequency f of a fully clamped (CCCC) ($ZrO_2/Ti-6Al-4V$) square plates ($a/h = 5$, $a = 0.2m$) subjected to uniform temperature rise (UTR) and non-linear temperature distribution ($NLTD$) is properly examined. The FG plate is exposed to different temperature rises with uniform ($\Delta T = 0, 200, 500, 800K$) and non-linear ($T_c = 0, 200, 500, 800K$) distributions. Results are determined in terms of the frequency ratio f/f_m where f_m and f are the fundamental frequency of metal part and other material composition in each thermal load as shown in Figs. 2.4(a) and 2.4(b), respectively. The notations UTR and $NLTD$ are used to denote the uniform and non-linear temperature distributions, respectively. Moving from ceramic to metal composition, the frequency ratio f/f_m decreases drastically for all temperature distributions. Moreover, in the case of non-linear temperature distribution, and when p is almost close to the ceramic phase ($p \in [0-0.15]$), the frequency ratio is maximum with ($NLTD = 800K$) and minimum with ($NLTD = 0K$). This effect is totally inversed, for a value higher than 0.15 of

Table 2.5

Convergence and comparison of the high four normalized natural frequency parameter $\overline{\omega} = \frac{\omega a^2}{\pi^2}\sqrt{12\rho_m(1-\nu_m^2)/E_m h^2}$ of ($Si_3N_4/SUS304$) clamped (CCCC) FGM square plate subjected to uniform temperature rise with ($a = 0.2m$, $a/h = 10$ and $p = 2$).

$\Delta T(K)$	Reference	Mode			
		5	6	7	8
0	4 × 4	14.5900	14.5900	14.7100	14.7100
	8 × 8	13.4000	13.5700	15.1600	15.1600
	16 × 16	13.1000	13.2300	15.5600	15.5600
	32 × 32	13.0400	13.1600	15.6200	15.6200
	100 × 100	13.0200	13.1400	15.6400	15.6400
	HSDT-semi-analytical.[299]	12.5881	13.1867	15.4530	16.0017
	%difference	3.3172	0.3554	1.1956	3.4635
	TSDT-analytical.[126]	13.1060	13.2089	15.9471	15.9471
	%difference	0.6605	0.5243	1.9635	1.9635
	3D-elasticity.[142]	13.0973	13.2234	15.3627	15.3627
	%difference	0.5937	0.6347	1.7730	1.7730
300	4 × 4	13.9500	13.9500	14.0300	14.0300
	8 × 8	12.2300	12.3700	14.6000	14.6000
	16 × 16	12.2300	12.3700	14.6000	14.6000
	32 × 32	12.1600	12.2900	14.7000	14.7000
	100 × 100	12.1400	12.2700	14.7400	14.7400
	HSDT-semi-analytical.[299]	11.7054	12.3175	14.4520	15.0019
	%difference	3.5799	0.3871	1.9538	1.7816
	TSDT-analytical.[126]	12.1982	12.3052	14.9090	14.9090
	%difference	0.4794	0.2868	1.4217	1.4217
	3D-elasticity.[142]	12.2256	12.3563	14.8112	14.8112
	%difference	0.7051	0.7033	0.4830	0.4830
500	4 × 4	13.0900	13.0900	13.1900	13.3800
	8 × 8	11.7300	11.8800	13.3000	13.3000
	16 × 16	11.3400	11.4800	13.5800	13.5800
	32 × 32	11.2600	11.3900	13.6700	13.6700
	100 × 100	11.2400	11.3700	13.6900	13.6900
	HSDT-semi-analytical.[299]	10.8285	11.4350	13.4412	13.9756
	%difference	3.6610	0.5716	1.8173	2.0861
	TSDT-analytical.[126]	11.2708	11.3812	13.8346	13.8346
	%difference	0.2740	0.0985	1.0562	1.0562
	3D-elasticity.[142]	11.3126	11.4468	13.7907	13.7907
	%difference	0.6459	0.6754	0.7355	0.7355

the gradient index ($p \geq 0.15$). The explanation of this phenomenon may be related to the elastic modulus of metal which linearly decreases with the temperature rise and that of ceramic which does quadratically. Therefore, the FG plate becomes more sensitive to the temperature changes.

The non-dimensional frequency parameter $\overline{\omega} = \frac{\omega a^2}{\pi^2}\sqrt{12\rho_m(1-\nu_m^2)/E_m h^2}$ versus the temperature change (in K) of ($ZrO_2/Ti-6Al-4V$) FGM clamped (CCCC) square plate subjected to different types of thermal loadings and for two sets of length-to-thickness ratios ($a/h = 5$ and $a/h = 10$) are plotted in Figs. 2.5(a) and 2.5(b), respectively. For both temperature distributions, the non-dimensional frequency parameter decreases as the temperature increases which reflects the significant effect of temperature change on vibrational behavior of FG plates. In fact, the stiffness of the structure becomes lower with the temperature rise which manifest by a decrease in the frequency parameter. As well as, it can be remarked that the non-dimensional frequency parameter is higher

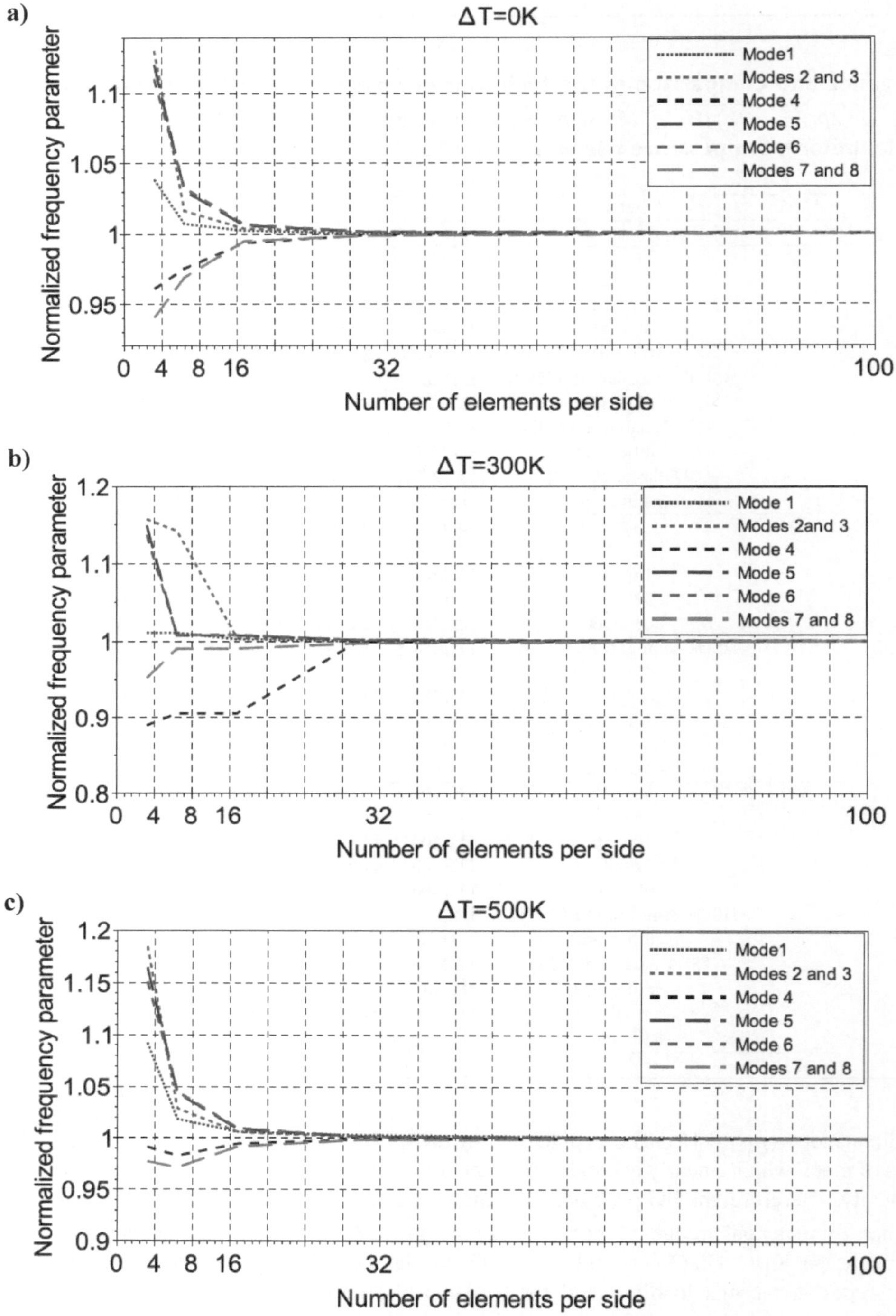

Figure 2.2 Convergence rate of frequency parameter of ($Si_3N_4/SUS304$) clamped (CCCC) square FGM plates ($a/h = 10$ and $p = 2$) under different uniform temperature rises.

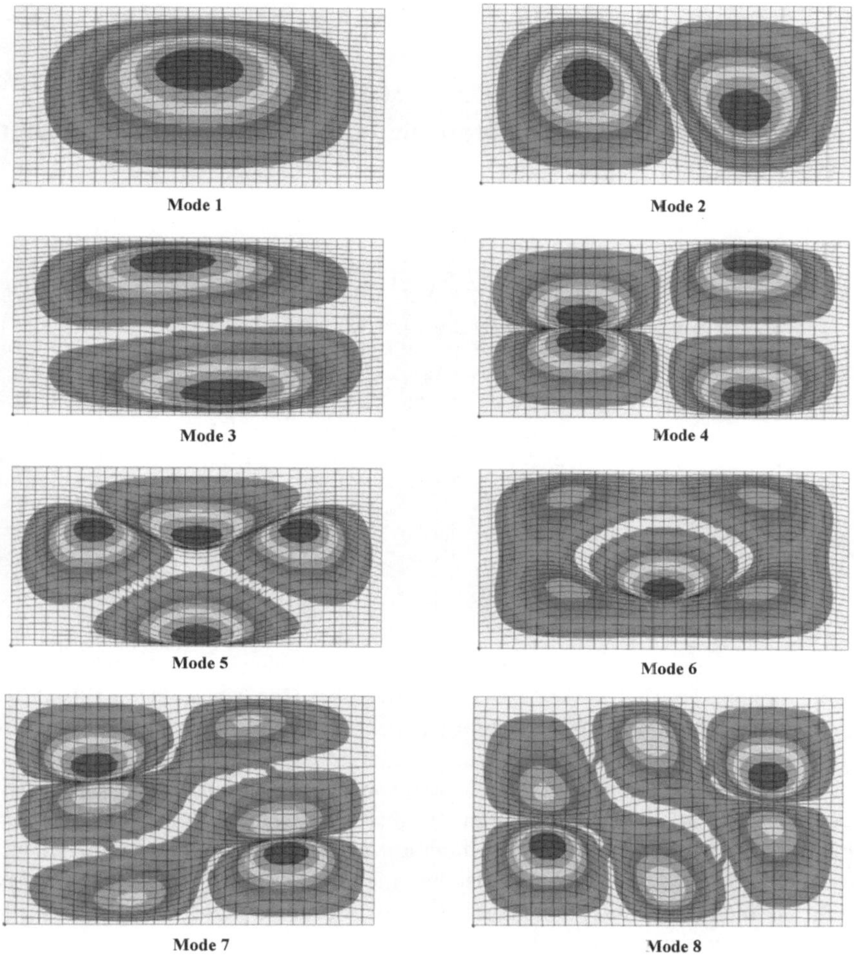

Figure 2.3 First eight mode shapes of $(ZrO_2/Ti-6Al-4V)$ FGM square clamped (CCCC) plate subjected to uniform temperature rise $(\Delta T = 300K)$.

for a FG plate with a length-to-thickness ratio $(a/h = 10)$ than that with a length-to-thickness ratio $(a/h = 5)$ and this is for all power-law index (p). Thus, it can be related to the stiffness reduction by temperature rise which is more prominent for $a/h = 5$ than that for $a/h = 10$. On the other hand, the effect of temperature becomes high as the plate becomes metal-rich which manifest by lower values of the non-dimensional frequency parameter and this is due to the diminution of the flexural rigidity of the FG plate with heated metal portions.

2.3.2.4 Free vibration of FGM cylindrical panels under thermal loadings

Now, free vibration analysis of functionally graded (FG) cylindrical panels subjected to different types of thermal loadings will be studied. A fully clamped (CCCC) FG cylindrical panel with length-to-radius ratio $L/R = 0.1$, length-to-thickness ratio $L/h = 10$ and equal straight and curved lengths $L = b$ is considered as shown in Fig. 2.6. The metal and ceramic components are respectively stainless steel $(SUS304)$ and Silicon nitride (Si_3N_4) materials. Other physical properties are consistent with those presented in Tabs. 2.1 and 2.2. The panel is discretized using 32×20 finite shell ele-

Table 2.6
Three first normalized natural frequency parameter $\overline{\omega} = \frac{\omega a^2}{\pi^2}\sqrt{12\rho_m(1-\nu_m^2)/E_m h^2}$ of clamped ($ZrO_2/Ti-6Al-4V$) FGM square plate ($a/h = 10$) subjected to different types of thermal loadings.

Material composition	Mode					
	1		2		3	
(a) UTR ($\Delta T = 500K$)	Present	Ref.[126]	Present	Ref.[126]	Present	Ref.[126]
ZrO_2	2.9100	3.0273	5.6600	5.9890	5.6600	5.9890
$p = 0.2$	2.8100	2.9224	5.4900	5.7980	5.4900	5.7980
$p = 1$	2.6400	2.7433	5.1800	5.4700	5.1800	5.4700
$p = 5$	2.5000	2.5931	4.9100	5.1934	4.9100	5.1934
$Ti-6Al-4V$	2.4000	2.4928	4.7500	5.0132	4.7500	5.0132
(b) NLTD ($T_m = 300K$, $T_c = 800K$)	Present	Ref.[126]	Present	Ref.[126]	Present	Ref.[126]
ZrO_2	3.4700	3.3704	6.7000	6.5920	6.7000	6.5920
$p = 0.2$	3.4100	3.3040	6.5800	6.4613	6.5800	6.4613
$p = 1$	3.2700	3.1584	6.3200	6.1792	6.3200	6.1792
$p = 5$	3.1300	2.9970	6.0300	5.8742	6.0300	5.8742
$Ti-6Al-4V$	2.9900	2.8498	5.7900	5.6098	5.7900	5.6098

ments. Two kinds of uniform temperature rises are considered, namely, $\Delta T = 0K$ and $\Delta T = 300K$. The results are compared with those of Yang and Shen [298]. One can see from Tabs. 2.9 and 2.10 that the results obtained by the (I-FSDT) finite element shell model are in good agreement with the results of literature. For some modes of vibration, the relative error between the results is about 7% due to the different resolution strategies in each work. In reference [298], a semi-analytical approach is used while in this paper a finite element method based on an improved (I-FSDT) shell model is employed. Therefore, the present formulation is highly consistent and convenient for thermal vibration analysis of FG structures. We also find that the non-dimensional frequency parameter decreases with the increase of the gradient index. This effect can be attributed to the reduction in metal composition for which the stiffness goes down.

The corresponding first eight mode shapes of ($Si_3N_4/SUS304$) cylindrical panels (CCCC) under uniform temperature rise ($\Delta T = 300K$) with geometrical characteristics ($L/R = 0.1$, $L/h = 30$) are depicted in Fig. 2.7. The effect of length-to-thickness ratio (L/h) on normalized frequency parameter $\omega^* = \omega L^2\sqrt{12\rho_m(1-\nu_m^2)/E_m h^2}$ of ($Si_3N_4/SUS304$) FG clamped (CCCC) cylindrical panel under uniform ($\Delta T = 800K$) and non-linear temperature ($T_m = 300K$, $T_c = 1100K$) rises is shown in Figs. 2.8(a) and 2.8(b).

As can be seen, the normalized frequency parameter is greatly influenced by the variation of the length-to-thickness ratio in that ω^* decreases drastically as the length-to-thickness ratio (L/h) increases. This decrease is more pronounced in the range ($L/h \in [5-50]$) and becomes stagnant for a value beyond 50 ($p \geq 50$). The reason for this observation is that the FG cylindrical panel behaves a thicker structure in the range ($L/h \in [5-50]$) and its flexural rigidity is significant which manifest by high values of vibration frequencies. For a fixed value of length-to-thickness ratio and moving from ceramic to metal material compositions, the values of natural frequencies go down. This phenomenon can be attributed to the high stiffness of ceramic portion compared to metal portion. Continuously, Figs. 2.9(a) and 2.9(b) show the effect of temperature on vibrational behavior of ($Si_3N_4/SUS304$) FG clamped (CCCC) cylindrical panel ($L/h = 10$, $L/R = 0.1$) subjected to uniform and non-linear temperature distributions, respectively and varying the gradient index (p). It is revealed that the increase in the power-law index (varying from ceramic to metal phases) induces a significant decrease in the frequency parameter ω^* and this decrease is more prominent with

Table 2.7
Three second normalized natural frequency parameter $\overline{\omega} = \frac{\omega c^2}{\pi^2}\sqrt{12\rho_m(1-\nu_m^2)/E_m h^2}$ of clamped ($ZrO_2/Ti-6Al-4V$) FGM square plate ($a/h=10$) subjected to various types of thermal loadings.

Material composition	Mode					
	4		5		6	
(a) UTR ($\Delta T = 500K$)	Present	Ref.[126]	Present	Ref.[126]	Present	Ref.[126]
ZrO_2	7.9700	8.4936	9.4400	9.9479	9.5400	10.0318
$p=0.2$	7.7300	8.2318	9.1600	9.6467	9.2600	9.7286
$p=1$	7.3000	7.7800	8.6700	9.1236	8.7600	9.2022
$p=5$	8.4700	8.3078	6.9600	7.3965	8.2400	8.6769
$Ti-6Al-4V$	6.7200	7.1522	7.9800	8.3985	8.0700	8.4723
(b) NLTD ($T_m = 0K$, $T_c = 500K$)	Present	Ref.[126]	Present	Ref.[126]	Present	Ref.[126]
ZrO_2	9.4000	9.3125	11.1300	10.8924	11.2400	10.9807
$p=0.2$	9.2300	9.1303	10.9300	10.6833	11.0400	10.7692
$p=1$	8.8600	8.7349	10.4900	10.2239	10.6000	10.3058
$p=5$	10.0200	9.7238	10.1200	9.8023	12.0900	9.8023
$Ti-6Al-4V$	8.1400	7.9475	9.6400	9.3104	9.7400	9.3865

raised temperatures which outlines that the FG panel becomes more sensitive to the temperature change especially for heated metal portions. Finally, the effect of boundary conditions on frequency parameter ω^* of ($Si_3N_4/SUS304$) FG cylindrical panel under uniform and non-linear temperature distributions is examined through Figs. 2.10(a) and 2.10(b). Five combinations of boundary conditions are considered where the letters "C", "S" and "F" denotes clamped, simply supported and free edges, respectively. The first and the third letters refer to constraint applied to curved sides of the panel while the second and the fourth letters refer to straight sides. It can be seen that a fully clamped

Table 2.8
Seventh and Eighth normalized natural frequency parameter $\overline{\omega} = \frac{\omega a^2}{\pi^2}\sqrt{12\rho_m(1-\nu_m^2)/E_m h^2}$ of clamped ($ZrO_2/Ti-6Al-4V$) FGM square plate ($a/h=10$) subjected to various types of thermal loadings.

Material composition	Mode			
	7		8	
(a) UTR ($\Delta T = 500K$)	Present	Ref.[126]	Present	Ref.[126]
ZrO_2	11.4000	12.1416	11.4000	12.1416
$p=0.2$	11.0600	11.7797	11.0700	11.7797
$p=1$	10.4800	11.1492	10.4800	11.1492
$p=5$	9.9600	10.6091	9.9600	10.6091
$Ti-6Al-4V$	9.6600	10.2762	9.6600	10.2762
(b) NLTD ($T_m = 0K$, $T_c = 500K$)	Present	Ref.[126]	Present	Ref.[126]
ZrO_2	13.4200	13.2730	13.4200	13.2730
$p=0.2$	13.1800	13.0201	13.1800	13.0201
$p=1$	12.6600	12.4627	12.6600	12.4627
$p=5$	12.0900	11.8553	12.0900	11.8553
$Ti-6Al-4V$	11.6300	11.3598	11.6300	11.3598

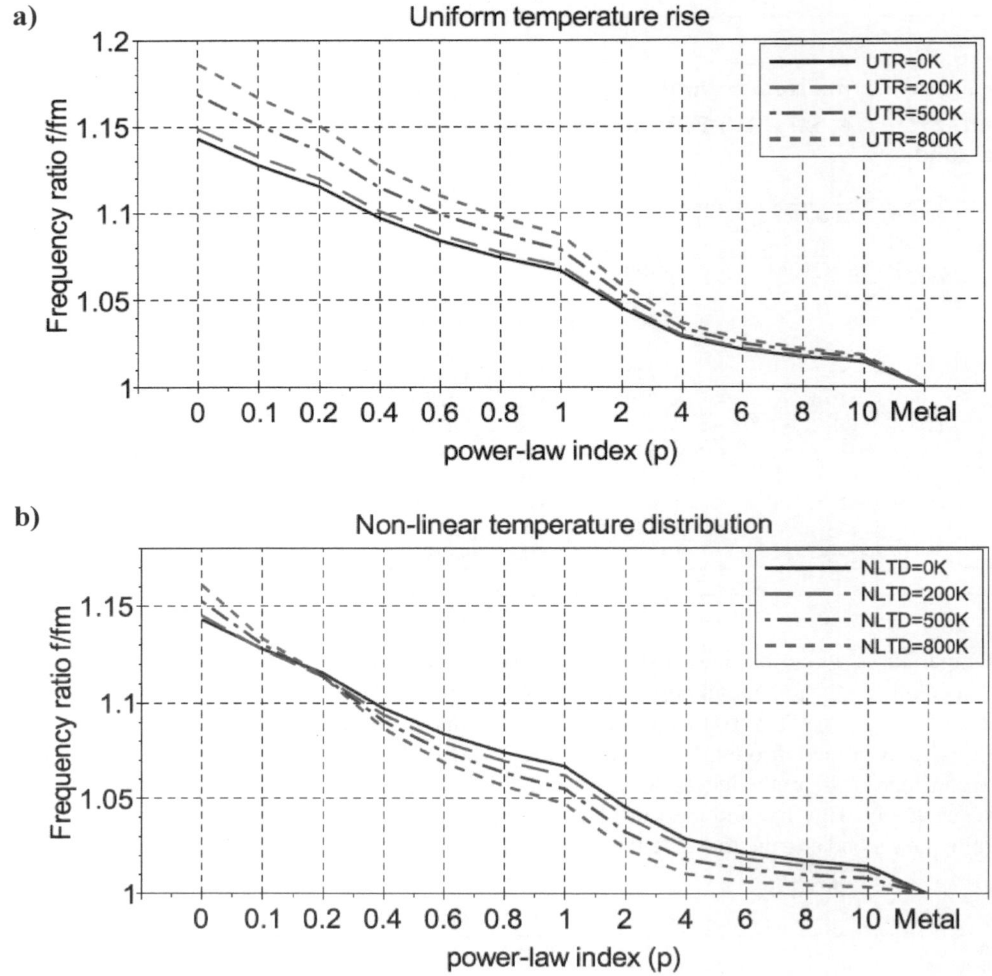

Figure 2.4 Frequency ratio f/f_m versus the gradient index (p) of fully clamped (CCCC) $(ZrO_2/Ti-6Al-4V)$ square plates $(a/h=5)$ subjected to (a) uniform temperature rise (UTR) and (b) non-linear temperature distribution $(NLTD)$.

boundary condition (CCCC) induces the highest frequency parameter whereas a simply supported on curved edges and free on straight edges (SFSF) boundary condition presents the lowest one. As well as, more the FG panel contains clamped constraints (CSCS, CFCF) more its stiffness is higher which manifest by high values of natural frequencies of the FG cylindrical panel. Moreover, with the temperature increases, a decrease in the natural frequency of the FG panel is observed which reflects that the vibrational behavior of FG cylindrical panel is influenced by the temperature changes and this parameter must be considered in the conception of FG structures.

2.4 THERMAL BUCKLING ANALYSIS OF FGM STRUCTURES

Another application of the (I-FSDT) shell model consists of the investigation of the thermal buckling and post-buckling behaviors of FGM structures, which is highlighted, in this section. The

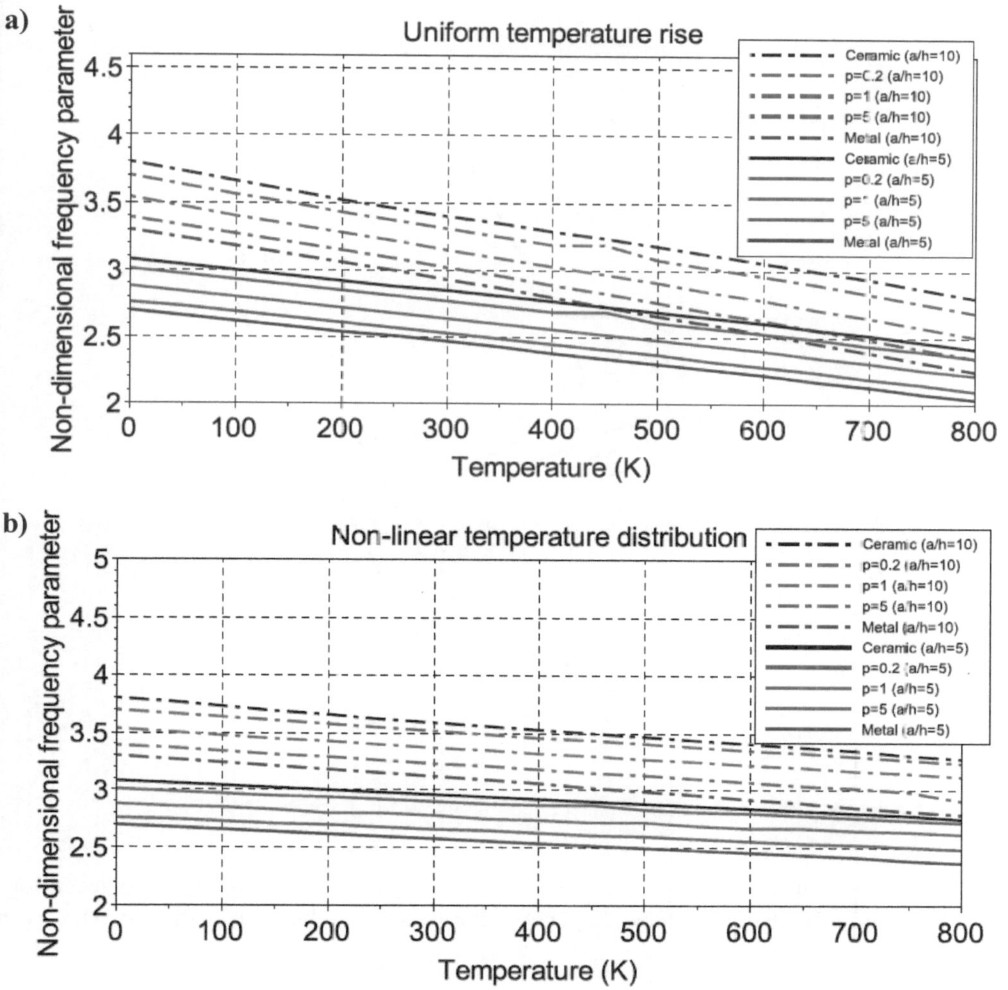

Figure 2.5 Non-dimensional frequency parameter $\overline{\omega} = \frac{\omega a^2}{\pi^2}\sqrt{12\rho_m(1-\nu_m^2)/E_m h^2}$ versus temperature of $(ZrO_2/Ti-6Al-4V)$ FGM square clamped (CCCC) plate $(a/b=1,\, a=0.2m)$ subjected to (a) uniform temperature rise (UTR) and (b) non-linear temperature distribution $(NLTD)$ and for different length-to-thickness ratios.

determination of the critical thermal buckling load is carried out by resolution of the buckling eigenvalue problem, given as:

$$(K_M + \lambda_{cr} K_G) \cdot U_n = 0 \ , \tag{2.12}$$

where K_M and K_G are the material and geometrical stiffness matrices, respectively. Their expressions are as mentioned in Eqs. (1.79) and (1.82) from (section.1.4, Chapter 1), respectively. λ_{cr} is the critical temperature eigenvalue.

2.4.1 THERMAL BUCKLING OF FGM PLATES

Numerical investigations are presented in this section to check the performance of the present finite element procedure based on improved FSDT (I-FSDT) for various shell structures including FGM

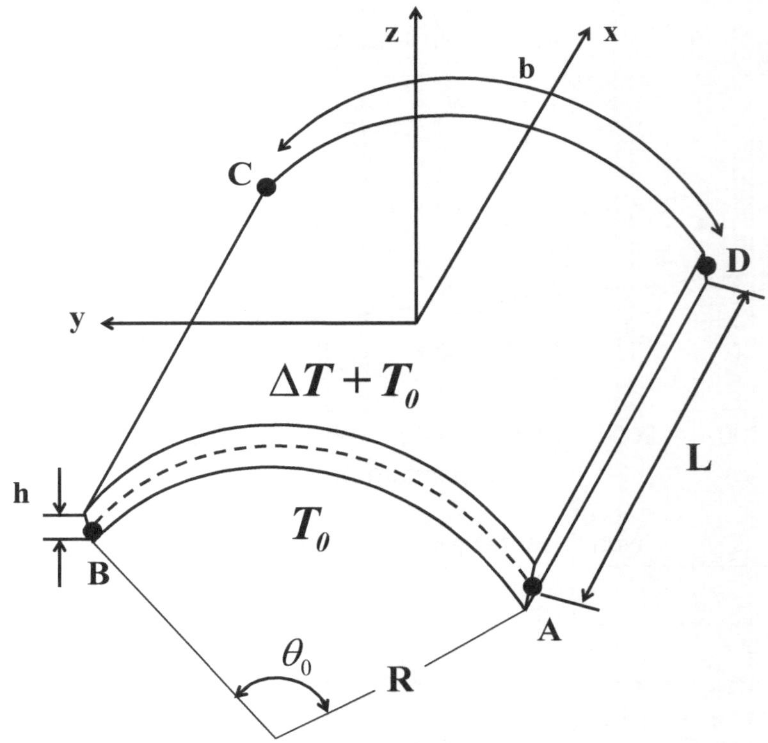

Figure 2.6 Geometry of the FG cylindrical panel under thermal load.

circular plate, square plate and cylindrical shells. Two types of temperature distributions are considered, which are uniform and non-linear temperature through the thickness as defined previously in section 2.2.2.

Table 2.9

Four first normalized natural frequency parameter $\omega^* = \omega L^2\sqrt{12\rho_m(1-\nu_m^2)/E_m h^2}$ of clamped (CCCC) ($Si_3N_4/SUS304$) cylindrical panel ($L/R = 0.1$, $L/h = 10$) subjected to uniform temperature change ($\Delta T = 0K$).

Material composition	$\Delta T = 0K$							
	1		2		3		4	
	Present	Ref.[298]	Present	Ref.[298]	Present	Ref.[298]	Present	Ref.[298]
Si_3N_4	77.100	74.518	146.000	144.663	146.000	145.740	203.000	206.992
$p = 0.2$	62.500	57.479	118.000	111.717	119.000	112.531	165.000	159.855
$p = 2$	40.600	40.750	76.300	78.817	76.600	79.407	106.000	112.457
$p = 10$	34.600	35.852	64.700	69.075	64.900	69.609	89.500	98.386
$SUS304$	32.000	32.761	60.200	63.314	60.400	63.806	83.700	90.370

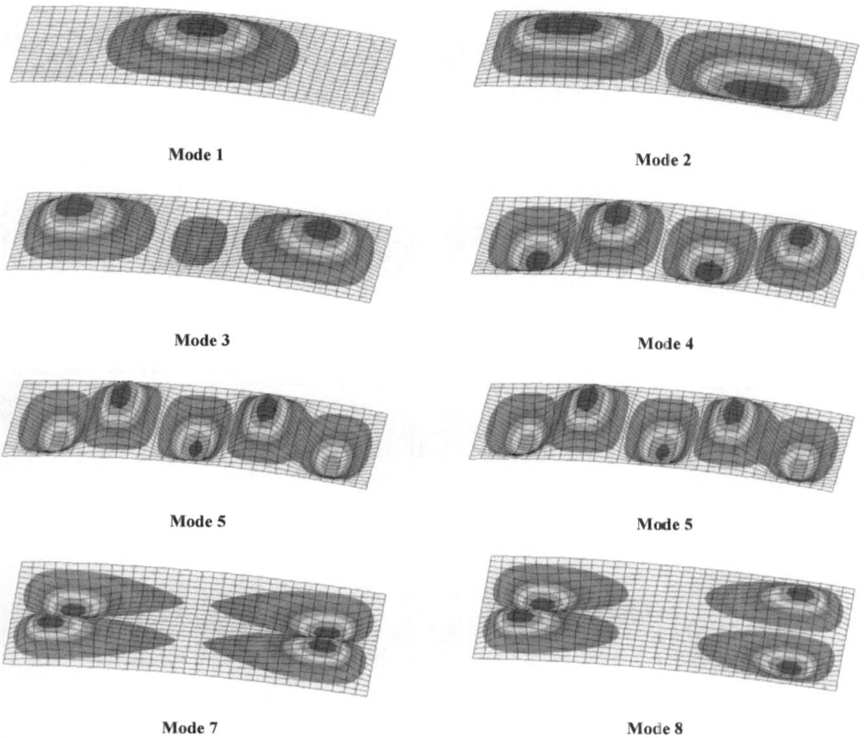

Figure 2.7 First eight mode shapes of ($Si_3N_4/SUS304$) clamped (CCCC) cylindrical panels ($L/R = 0.1$, $L/h = 30$) subjected ti uniform temperature rise ($\Delta T = 300K$).

2.4.1.1 Thermal buckling of FGM circular plate

1. Numerical Integration and convergence
   Herein, the integration of the temperature distribution function $\eta(z)$ can be solved using two different methods, as follows:

Table 2.10

Four first normalized natural frequency parameter $\omega^* = \omega L^2\sqrt{12\rho_m(1-\nu_m^2)/E_m h^2}$ of clamped (CCCC) ($Si_3N_4/SUS304$) cylindrical panel ($L/R = 0.1$, $L/h = 10$) subjected to uniform temperature change ($\Delta T = 300K$).

Material composition	$\Delta T = 300K$							
	1		2		3		4	
	Present	Ref.[298]	Present	Ref.[298]	Present	Ref.[298]	Present	Ref.[298]
Si_3N_4	70.100	72.446	136.000	140.642	136.000	141.687	191.000	201.238
$p = 0.2$	56.600	55.739	110.000	108.338	110.000	109.126	155.000	155.010
$p = 2$	36.800	39.369	71.700	76.106	72.000	76.675	101.000	108.541
$p = 10$	31.300	34.571	61.100	66.539	61.300	67.055	86.100	94.709
$SUS304$	28.700	31.511	56.600	60.846	56.800	61.321	80.000	86.805

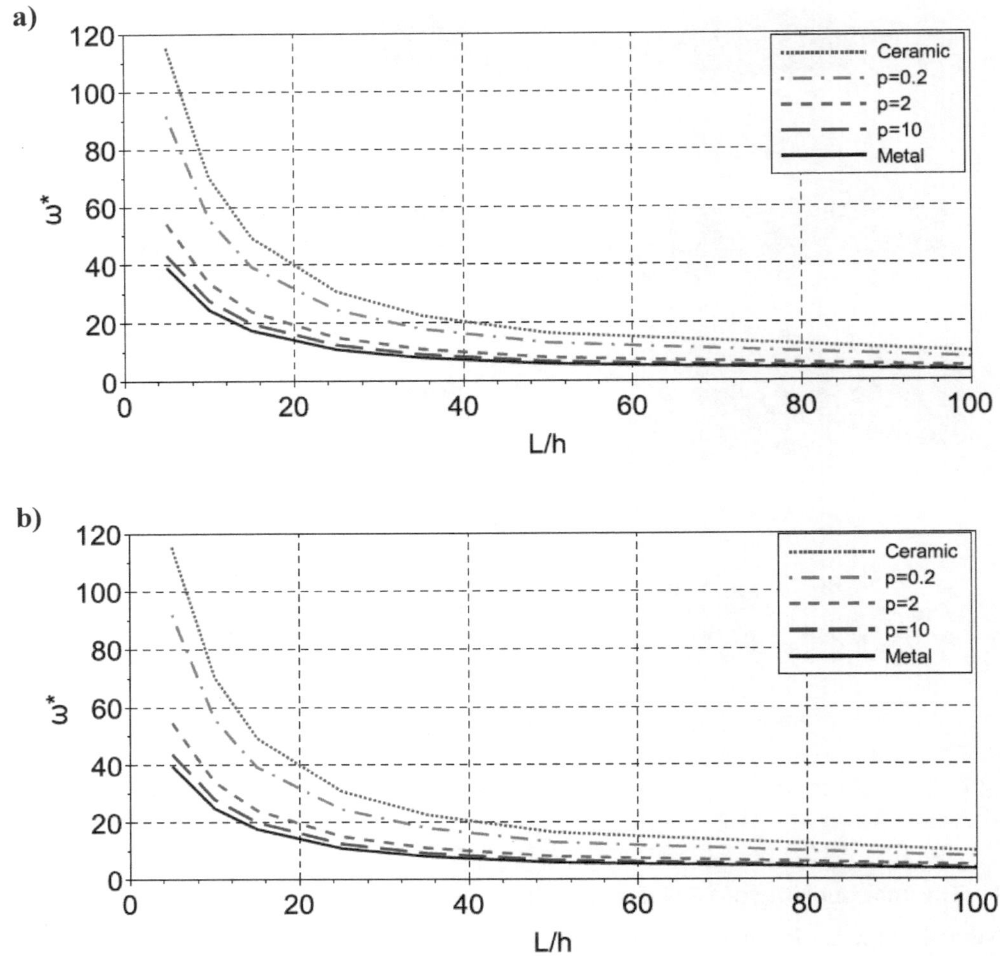

Figure 2.8 Normalized frequency parameter $\omega^* = \omega L^2\sqrt{12\rho_m(1-\nu_m^2)/E_m h^2}$ versus length-to-thickness ratio (L/h) of $(Si_3N_4/SUS304)$ FG clamped (CCCC) cylindrical panel subjected to (a) uniform temperature rise (UTR) and (b) non-linear temperature distribution $(NLTD)$ and for different gradient indexes.

Case 1: A polynomial series with n terms

$$\eta(z) = r \cdot \frac{\sum_{i=0}^{i=n} \frac{(-r^k \frac{K_{cm}}{K_m})^i}{ik+1}}{\sum_{n=0}^{i=n} \frac{(-\frac{K_{cm}}{K_m})^i}{ik+1}}, \quad r = \frac{z}{h} + \frac{1}{2} \tag{2.13}$$

For this case, the convergence of the series is assured by involving a specified number of terms (n). In fact, most current researches is the correct one [123, 103, 215, 177, 109] consider the first seven terms of series. According to the studies of [18, 273] the number of series terms is assumed to be 20 terms. As well as, [105] takes the first 91 terms of the series. Moreover, Malekzadeh [190] considered the first 10^3 terms.

Case 2: Numerical integration

By considering the Gauss integration method, $\eta(z)$ is defined by

$$\eta(z) = \frac{\int_{-h/2}^{z} \frac{dz}{K(z)}}{\int_{-h/2}^{h/2} \frac{dz}{K(z)}}. \tag{2.14}$$

Zhang et al. [311, 313] investigated the thermal buckling and the post-buckling using the numerical integration. In fact, papers dealing with numerical integration are few compared to those used the polynomial series method.

For convergence purpose, a clamped FGM circular plate made of Aluminum (Al) and Alumina (Al_2O_3) with $h/R = 0.01$ is considered. From the geometrical symmetry of the plate, the analysis is reduced to one-quarter. The quarter is meshed by 16×16 four-nodes element. The polynomial series method limited to seven or twenty terms of series is adopted

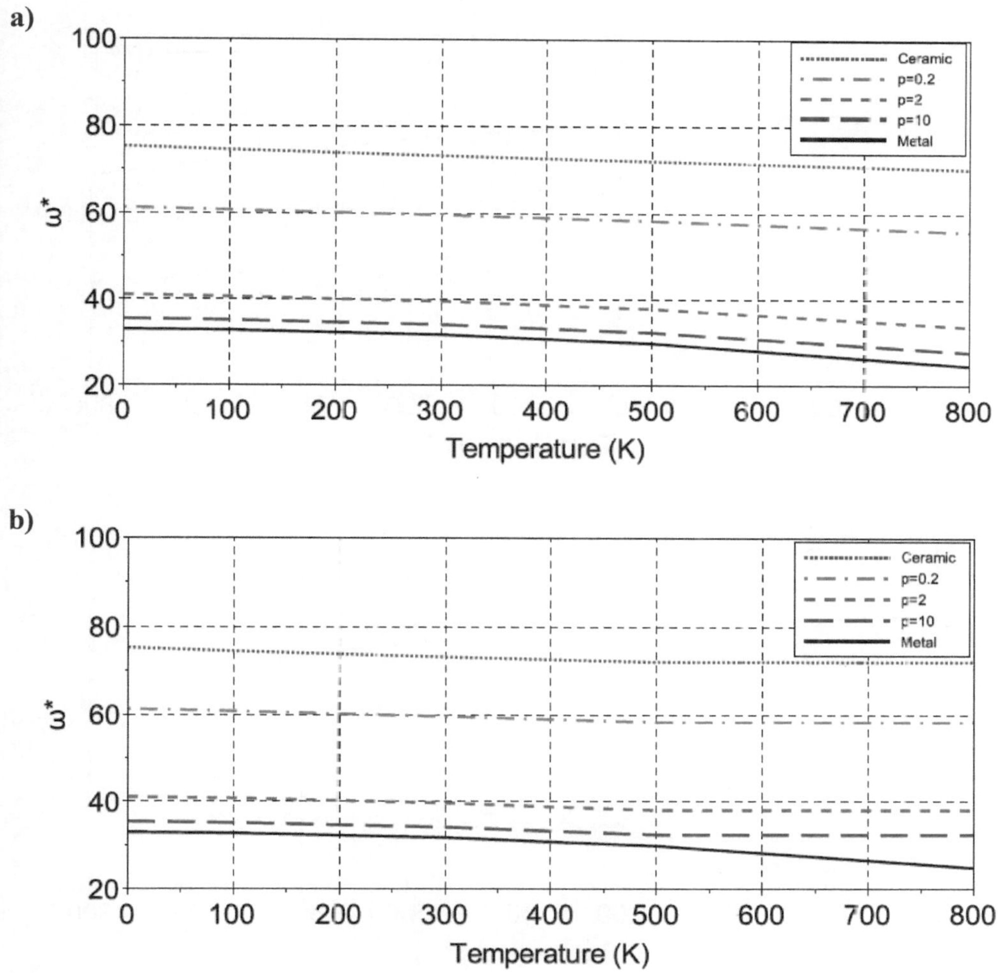

Figure 2.9 Normalized frequency parameter $\omega^* = \omega L^2\sqrt{12\rho_m(1-\nu_m^2)/E_m h^2}$ of ($Si_3N_4/SUS304$) versus temperature of FG clamped (CCCC) cylindrical panel ($L/h = 10, L/R = 0.1$) subjected to (a) uniform temperature rise (UTR) and (b) non-linear temperature distribution ($NLTD$) with different gradient indexes.

in literature [273, 18, 109] as a solution of thermal buckling of structures under non-linear temperature rise through the thickness. However, numerical Gauss integration method can also be used to compute the critical buckling temperature. Figure 2.11 shows the convergence of critical buckling of FGM circular plate versus the terms of series with a power-law index $p = 1$. For twenty terms of series ($n = 20$) the convergence is achieved with a discrepancy around 5.4%. Also, it is observed that as the number of terms increases, the obtained results converge to the exact solution. Furthermore, the polynomial series method converges with a discrepancy around 0.6% for ($n \geq 50$) which can explain the use of high number of series terms ($n = 91, 10^3$) by [105, 190]. However, the polynomial method requires a prohibitive computing time which is inappropriate for such studies.
Table 2.11 depicts the convergence of critical buckling of FGM circular plate versus the integration points with a power-law index $p = 1$. Here, the obtained results of critical

a)

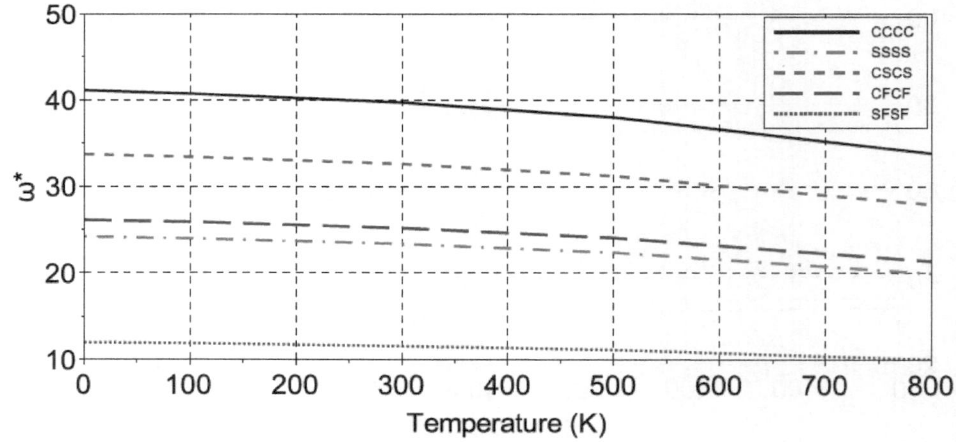

a)

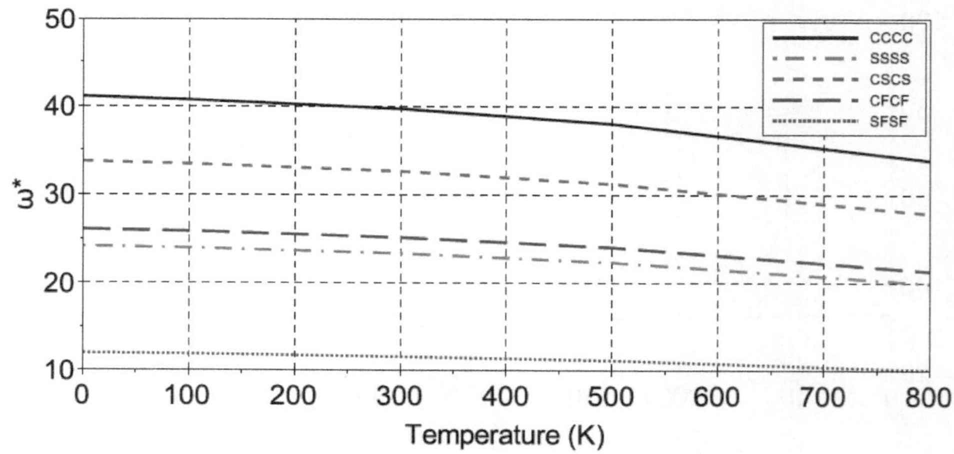

Figure 2.10 Effect of boundary conditions on normalized frequency parameter $\omega^* = \omega L^2 \sqrt{12\rho_m(1-\nu_m^2)/E_m h^2}$ of ($Si_3N_4/SUS304$) FG cylindrical panel ($L/h = 10$, $L/R = 0.1$, $p = 2$) subjected to (a) uniform temperature rise (UTR) and (b) non-linear temperature distribution ($NLTD$).

temperature, using only nine integration points ($n = 9$) induces a solution very close to the exact solution with a relative error around 0.1%. Thus, the Gauss integration method converges more rapidly than the polynomial series. Therefore, in the following studies, the Gauss integration procedure with 9 points is used to compute the thermal buckling.

2. Validation and comparison

The critical buckling temperature of clamped isotropic circular plate under uniformly and non-linear distributed temperature rises with different thickness-to-radius ratios (h/R) is first computed as mentioned in Tab. 2.12. Herein, only one-quarter of the geometry is modeled and meshed using 32×32 four-nodes element as shown in Fig. 2.12. The material properties are the Young's modulus $E = 1GPa$, the Poisson coefficient $\nu = 0.3$ and the thermal expansion coefficient $\alpha = 10^{-4}/°C$. These results are compared with the analytical method of Najafizadeh et al. [177]. A good agreement is obtained between the present results and those given from the literature [177] for both temperature rises which traces the convenience of the present (I-FSDT) formulation for the prediction of thermal buckling of isotropic circular plates efficiently.

3. Parametric analysis

Now, some parametric analyses are performed to show the effect of temperature dependency of material properties, FG power-law index, the material composition and the geometrical parameters on thermal buckling response of FGM circular plates. To this end, a clamped FGM circular plate

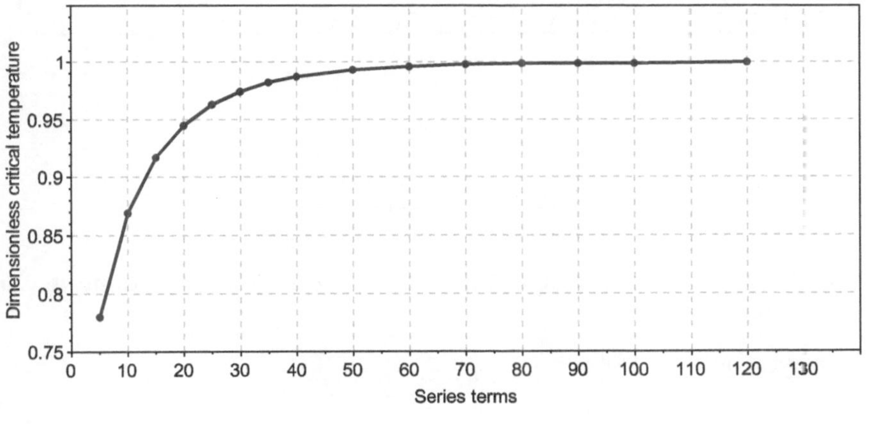

Figure 2.11 Convergence of the critical temperature versus series terms.

Table 2.11
Critical temperature of FGM circular plate versus integration points.

Integration points	5	6	7	8	9	10
Critical temperatures	19.578	19.707	19.761	19.783	19.791	19.795
Dimensionless critical temperatures	0.989	0.995	0.998	0.999	0.999	1

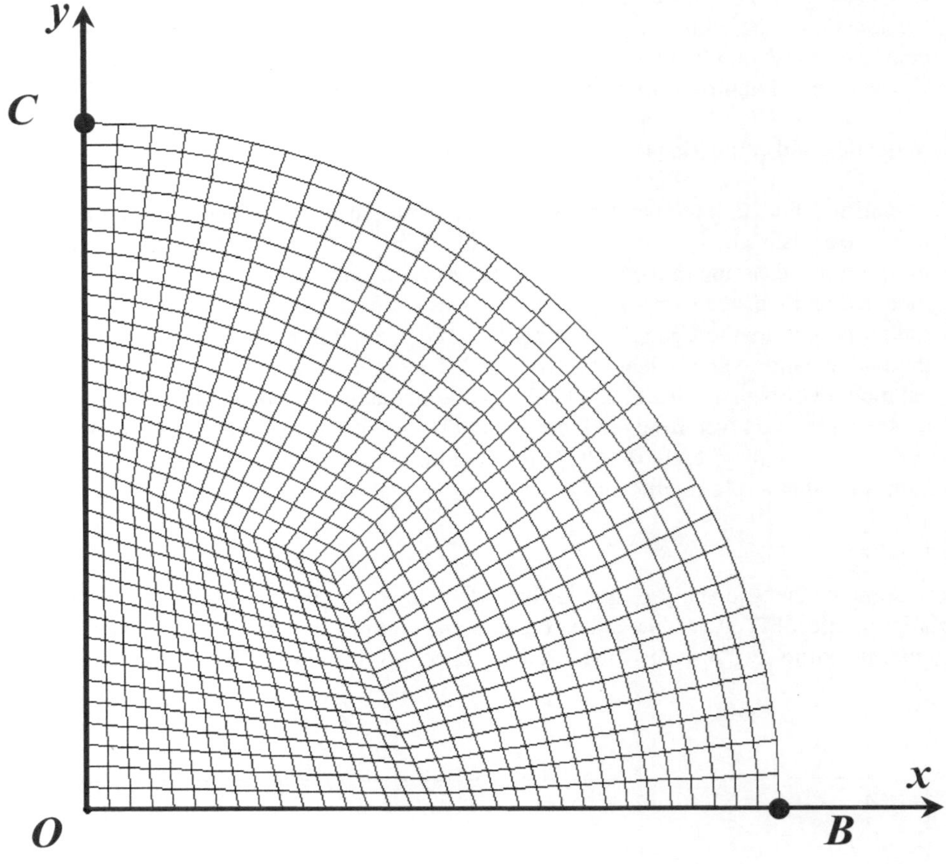

Figure 2.12 Mesh for circular plate.

Table 2.12

Critical buckling temperature of isotropic CCCC circular plate under (a) uniform and (b) non-linear temperature rises.

temperature rise	Method	h/R				
		0.01	0.02	0.03	0.04	0.05
(a) UTR	Present	0.941	3.762	8.464	14.974	23.309
	Analytical [177]	0.941	3.76	8.44	15	23.3
(b) NLTR	Present	1.883	7.524	16.929	29.94	46.61
	Analytical [177]	1.88	7.52	16.9	29.9	46.6

made of Al and Al_2O_3 is considered. The material properties are listed in Tab. 2.13 and the Poisson's coefficient across the whole plate is assumed to be constant and equal to 0.3.

Tables 2.14 and 2.15 lists the critical buckling temperature of the circular plate subjected to uniform and non-linear temperature rises for various thickness-to-radius ratios h/R and power index p, respectively. The present results are compared to those reported by Tran et al. [273] based on HSDT and analytical solution [177]. The obtained results match very well with the results from the references [177, 273]. The buckling temperature increases with an increased ratio h/R. Also, according to Tab. 2.15, for non-linear temperature distribution case, it can be observed that some discrepancies arise for power indexes ($p = 0.5$ and $p = 1$) and the relative error between the present results and those of [177] (based on FSDT) can reach 32% for $p = 0.5$ and $h/R = 0.05$. This could be explained by the difference in the computational method of non-linear temperature distribution across the thickness.

Figure 2.13 depicts the variation of the critical buckling temperature using the shear correction function f_3 with respect to power index p and thickness-to-radius ratio h/R, respectively. It is shown that the present results are similar to those proposed in [273]. Pure metal circular plate has the lowest buckling temperature and Ceramic (Al_2O_3) circular plate has the highest buckling temperature. Moreover, the critical buckling temperature decreases with the power index p. This can be explained by the bending stiffness of the plate which becomes higher with the ceramic reinforcements. Figure 2.14 shows that the critical buckling temperature of a FGM circular plate under non-uniform temperature distribution is higher than that of uniform distribution. The first mode shape of the FGM circular plate with ($R/h = 100$) is shown in Fig. 2.15.

2.4.1.2 Thermal buckling of FGM square plate

Firstly, the effectiveness and usefulness of the proposed modified FSDT are verified for thermal buckling analysis of laminate composite plates. For that, a simply supported rectangular composite laminated plate with eight-layered is considered. The plate is subjected to uniform temperature

Table 2.13
Material properties of Al and Al_2O_3 [273].

Material	Properties			
	E (GPa)	$Density(Kg/m^3)$	$Conductivity(W/mK)$	Thermal expansion (/°C)
Al	70	2707	204	23×10^{-6}
Al_2O_3	380	3800	10.4	7.4×10^{-6}

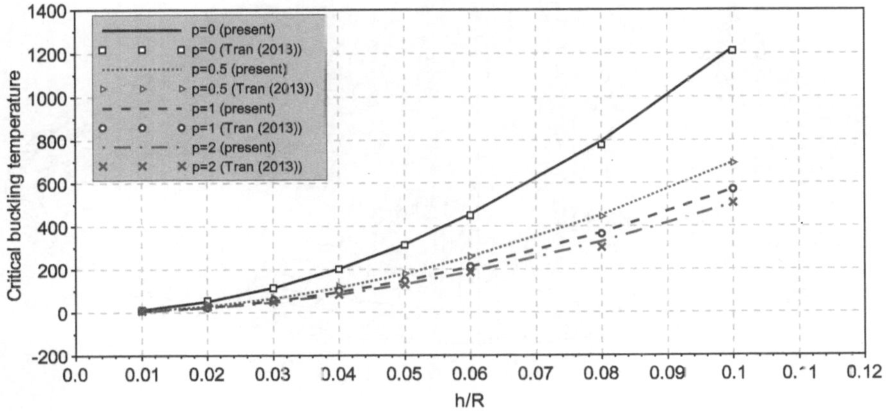

Figure 2.13 Critical buckling temperature ΔT_{cr} (in K) versus thickness-to-radius ratio (h/R) of clamped circular FGM plates under uniform temperature distribution.

rise and its geometrical and material properties are taken as indicated in [294]. Table 2.16 depicts the present critical buckling temperatures with different mesh sizes (4×4, 8×8, 12×12, 16×16,

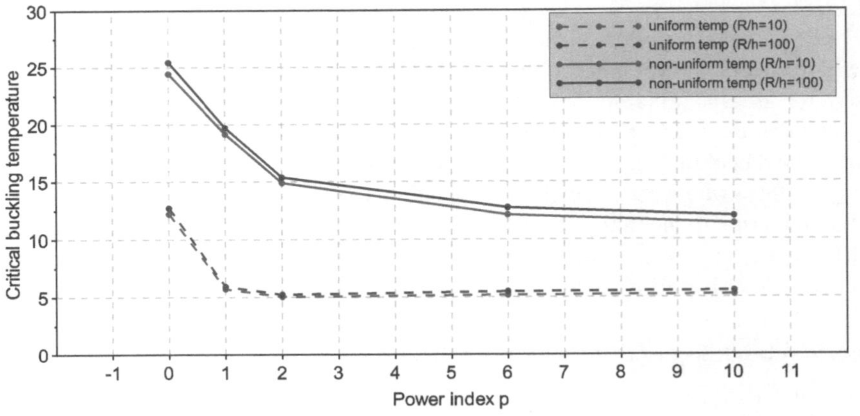

Figure 2.14 Critical buckling temperature ΔT_{cr} (in K) versus power index p of FGM circular plates, plotted results in form $\Delta T_{cr}/100$ for $R/h = 10$.

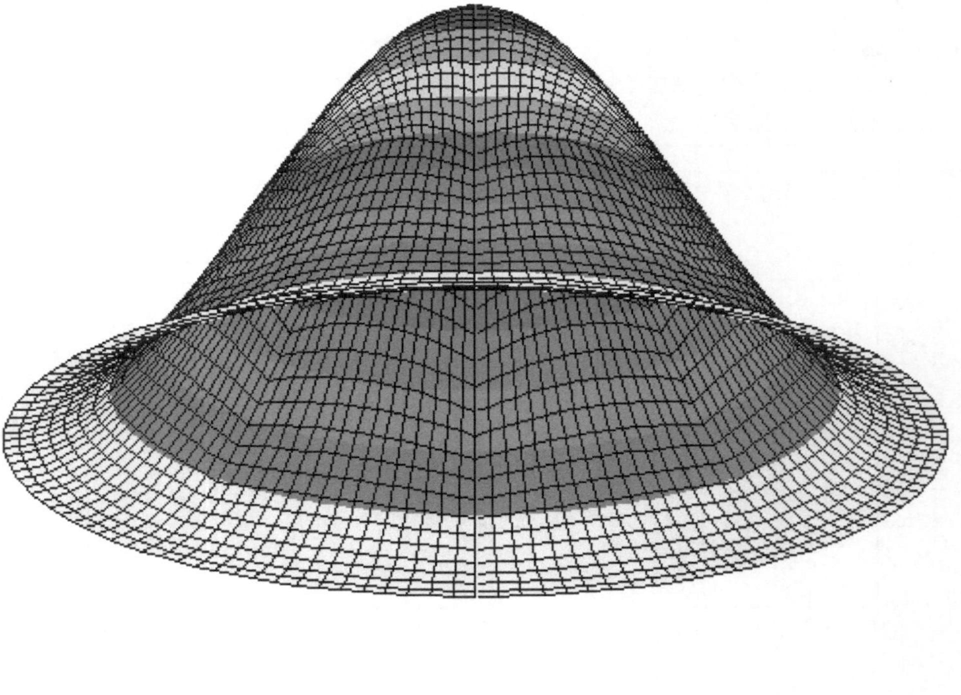

Figure 2.15 Thermal buckling shape of the FGM circular plates with $R/h = 100$.

Table 2.14

Critical buckling temperature of FGM circular plate subjected to uniform temperature distribution.

p	Method	h/R				
		0.01	0.02	0.03	0.04	0.05
0	Present	12.726	50.843	114.386	202.356	314.998
	HSDT [273]	13.3846	51.4591	114.556	202.201	313.7709
	FSDT [177]	12.712	50.795	114.09	202.333	315.16
	CPT [177]	12.716	50.866	114.449	203.465	317.914
0.5	Present	7.213	28.823	64.861	114.782	178.753
	HSDT [273]	7.6414	29.2259	65.0216	114.797	178.2466
	FSDT [177]	7.202	28.783	64.662	114.706	178.731
	CPT [177]	7.204	28.819	64.843	115.273	180.121
1	Present	5.912	23.625	53.168	94.096	146.551
	HSDT [273]	6.2761	23.9771	53.335	94.164	146.2188
	FSDT [177]	5.9064	23.603	53.029	94.081	146.815
	CPT [177]	5.907	23.63	53.169	94.52	147.694

Table 2.15

Critical buckling temperature of circular FGM plate under non-linear temperature rise.

p	Method	h/R				
		0.01	0.02	0.03	0.04	0.05
0	Present	25.45	101.687	228.772	404.711	629.995
	HSDT [273]	26.7693	102.9183	229.112	404.402	627.5418
	FSDT [177]	25.924	101.455	228.180	404.666	630.320
	CPT [177]	25.433	101.590	228.898	405.821	635.828
0.5	Present	25.156	100.527	226.273	400.560	624.069
	HSDT [273]	20.1554	77.0880	171.5049	302.7956	470.1542
	FSDT [177]	18.996	75.915	170.594	302.532	471.393
	CPT [177]	19.002	76.009	171.021	304.039	475.061
1	Present	19.73	78.85	177.504	314.234	489.767
	HSDT [273]	16.3386	62.42	138.8481	245.1389	380.6544
	FSDT [177]	15.373	61.440	138.037	244.897	381.646
	CPT [177]	15.378	61.512	138.433	246.050	384.453

32×32) and they are compared to the numerical solutions of ABAQUS software and those available in the literature [294, 133, 91]. An excellent agreement between the present results and those given by ABAQUS software is revealed which affirms the ability of the present model to restitute the thermal buckling behavior of FG plates with a good accuracy. Moreover, the relative error between the present results using 8×8 mesh and those given by [133] and [91] does not exceed 1%. However, this error can reach 2.28% for 12×12 mesh. The difference between results can be explained by the Kirchhoff assumptions used in [294, 133, 91] and on the other hand by the used computational technique.

The second example consists of a FGM square plate made of Aluminum (Al) and Alumina (Al_2O_3) with $a/b = 1$ and meshed using 32×32 four-nodes element as shown in Fig. 2.16. This plate is subjected to uniform temperature rise with different power indexes p and length-to-thickness ratios (a/h= 50, 100). By considering two boundary conditions: simply supported (SSSS) and clamped

Table 2.16

Critical buckling temperature ΔT_{cr} (in F) of simply supported composite laminated plates under uniform temperature rise.

Stacking sequence	ΔT_{cr} (in F)				
	Present	ABAQUS (S4)	Shi et al. [294]	Shiau et al. [133]	Ounis et al. [91]
$[0/90/90/0]_s$	$13.14(4\times4)$	$13.14(4\times4)$	–	$12.2616(4\times4)$	$12.2716(4\times4)$
	$12.15(8\times8)$	$12.11(8\times8)$	–	$12.2610(8\times8)$	$12.2617(8\times8)$
	$11.99(12\times12)$	$11.98(12\times12)$	$12.26(12\times12)$	$12.2610(12\times12)$	$12.2612(12\times12)$
	$11.92(16\times16)$	$11.92(16\times16)$	–	–	–
	$11.873(32\times32)$	$11.87(32\times32)$	–	–	–

Table 2.17

Comparison of the critical temperature rise ΔT_{cr} (in K) for the clamped Ni/Al_2O_3 square plates subjected to uniform temperature rise ($p = 1$, $a/b = 1$ and $a/h = 0.02$).

p	ΔT_{cr}		Percentage error %
	Present	3D-FEM [175]	
0.3	112.82	112.02	0.7
1	122.191	121.48	0.5
5	136.689	135.88	0.5

(CCCC) in all edges, the critical buckling temperatures are compared to those given by Zhao et al. [291] and Kandasamy et al. [114] in Tab. 2.18. The obtained results are in good agreement with those of the literature [291, 114]. In fact, for ($p \leq 1$), the values of the critical buckling temperature derived from the present formulation stick well with the results in [291, 114]. However, for ($p \succ 1$), the relative error is about 20% and this is due to the used transverse shear correction coefficient in [291, 114] which is insufficient to predict the thermal buckling of FGM plates. Therefore, the introduction of parabolic function f_3 has a significant effect in the improvement of thermal buckling behavior of such structures.

As well as, the critical buckling temperatures of a clamped FGM square plate are higher than those of a simply supported one. This is due to the improvement of the bending stiffness for a clamped plate. In addition, the critical buckling temperature decreases with an increased power index p. Hence, this must be related to the fact that the volume fraction of ceramic improves the bending stiffness of the FGM plate.

For the purpose of the validation of the present modified FSDT with 3-D solutions, the comparison for (Ni/Al_2O_3) square clamped plate subjected to uniform thermal loading is performed in Tab. 2.17. The geometrical ratios are given as indicated by Na and Kim [175]: ($a/b = 1$ and $a/h = 0.02$) and the material properties are listed in Tab. 2.13. The Poisson's coefficient is assumed to be constant ($\nu = 0.3$). An excellent agreement between the results is observed which proves again the effectiveness of the proposed modified FSDT model. In fact, the present formulation is general and convenient for the prediction of the thermal buckling behavior of FG structures. In addition, the present model has proved its ability of providing the thermal post-buckling responses of FG plates and cylindrical shell with good accuracy [212].

The effect of the length-to-thickness ratio (a/h) for a simply supported and a clamped FGM square plates is examined as shown in Figs. 2.17(a) and 2.17(b), respectively. It can be seen that, as

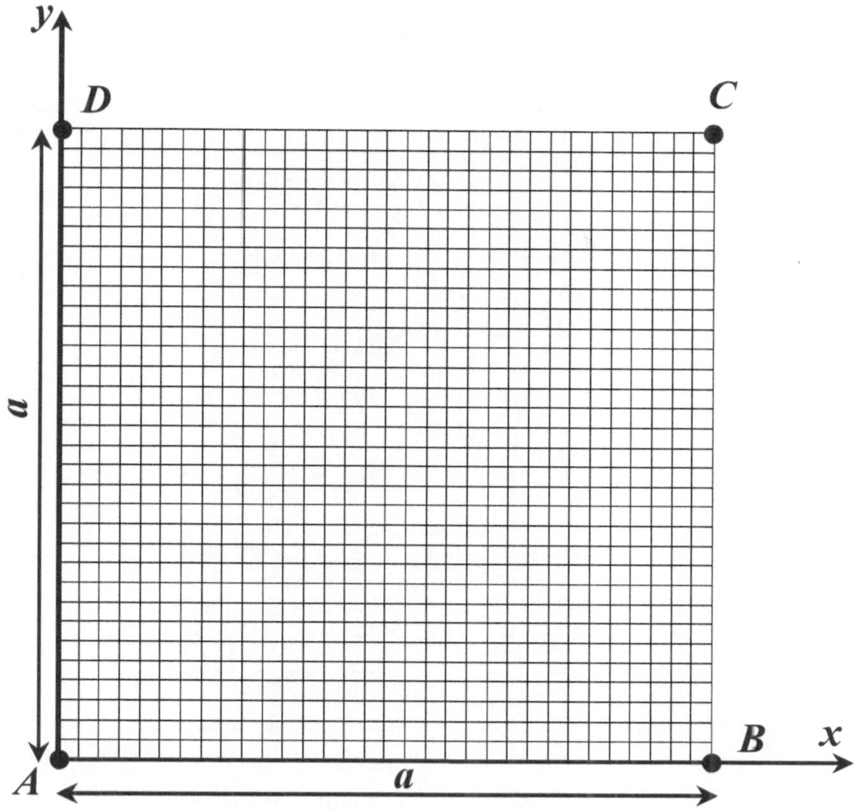

Figure 2.16 Mesh of the square plate.

the length-to-thickness ratio increases, as the critical buckling temperature decreases and this is for both applied boundary conditions. In fact, for a high value of the length-to-thickness ratio, the FGM plate behaves as a thin structure and becomes hence less sensitive to the variation of the thickness parameter due to the low flexural rigidity of the plate. Figure 2.18, provides the effects of applied boundary conditions on critical buckling temperature ΔT_{cr} of the FGM plates ($a/h = 50$). It can be seen that, clamped FGM plate has higher critical buckling temperature than simply supported one which outlines that when more constraints are applied, more the plate can resist to high values of critical buckling temperatures. The corresponding thermal buckling mode shapes of the FGM square plate, with an aspect ratio $a/h = 100$, are illustrated in Fig.2.19.

2.4.2 THERMAL BUCKLING OF FGM CYLINDRICAL SHELL

In the present subsection, the critical thermal buckling of FGM cylindrical shell using the (I-FSDT) model is investigated. The analyzed cylinder is composed of buckling temperature ratio with $h = 1mm, R/h = 400$ and $L^2/Rh = 300$. Their material properties are given in Tab. 2.19. The temperature gradient is uniformly distributed along the thickness of the shell. The analysis is based on 20×20 four-nodes elements as illustrated in Fig. 2.20. The curved edges of the cylinder are simply-supported (SS). The effect of power index p on critical thermal buckling of temperature-independent (TID) and temperature-dependent (TD) material properties are shown in Tab. 2.20.

The obtained results are in good agreement with the results provided by Kandasamy et al. [114] and Shen et al. [225]. In fact, the relative error between the present critical buckling temperatures with TD material properties using the present (I-FSDT) model and those given by the standard

Table 2.18
Critical buckling temperature of square FGM plate under non-linear temperature rise.

p	Methods	ΔT_{cr}			
		SSSS		CCCC	
		a/h		a/h	
		50	100	50	100
0	Present	68.33	17.11	181.02	45.49
	FSDT [291]	67.91	17.36	175.81	44.17
	FSDT [114]	70.1	17.5	186.3	46.8
0.5	Present	40.43	10.12	102.68	25.78
	FSDT [291]	38.27	9.80	99.16	24.89
	FSDT [114]	39.7	9.9	105.7	26.5
1	Present	36.45	9.15	84.18	21.13
	FSDT [291]	31.18	7.94	82.35	20.77
	FSDT [114]	32.3	8.0	86.0	21.5
2	Present	36.10	9.11	74.59	18.73
	FSDT [291]	29.65	6.92	71.01	18.48
	FSDT [114]	28.3	7.1	75.4	18.9
5	Present	36.62	9.25	76.76	19.32
	FSDT [291]	29.60	7.42	74.591	19.15
	FSDT [114]	28.7	7.2	76.4	19.2

Table 2.19
Material properties of Silicon Nitride ($Si3N4$) and stainless steel ($SUS304$) [218].

Material	Properties	P_0	P_{-1}	P_1	P_2	P_3
Si_3N_4	E (GPa)	348.43e9	0	-3.70e-4	2.160e-7	-8.946e-11
	ν	0.24	0	0	0	0
	$\alpha(K^{-1})$	5.8723e-6	0	9.095e-4	0	0
	$K(W/mK)$	13.723	0	0	0	0
	$\rho(Kg/m^3)$	2370	0	0	0	0
$SUS304$	$E(GPa)$	201.04e9	0	3.079e-4	-6.534e-7	0
	ν	0.3262	0	-2.002e-4	3.797e-7	0
	$\alpha(K^{-1})$	12.330e-6	0	8.086e-4	0	0
	$K(W/mK)$	15.379	0	0	0	0
	$\rho(Kg/m^3)$	8166	0	0	0	0

FSDT [114] is about 22% for $p = 5$. This error does not exceed 3% in comparison with results given by HSDT [225]. This can be explained by the ability of the introduced shear correction function to improve the critical buckling temperature. In addition, it is shown that the critical temperature of the shell increases progressively with an increased power index p. Moreover, the critical temperature of cylindrical shell with TD material are lower than those of TID results. Indeed, the buckling temperature is overestimated for TD FGM plates over TID ones.

Figures 2.21(a) and 2.21(b) show the critical buckling temperature of FGM cylindrical shell for various power index p and with (TD, TID) material properties for uniform and non-uniform temperature distributions, respectively. The critical buckling temperature increases by increasing the power index p and a fully metallic cylindrical shell ($p = 0$) has the lowest critical buckling temperature. In addition, it can be seen that the type of temperature material properties has a sig-

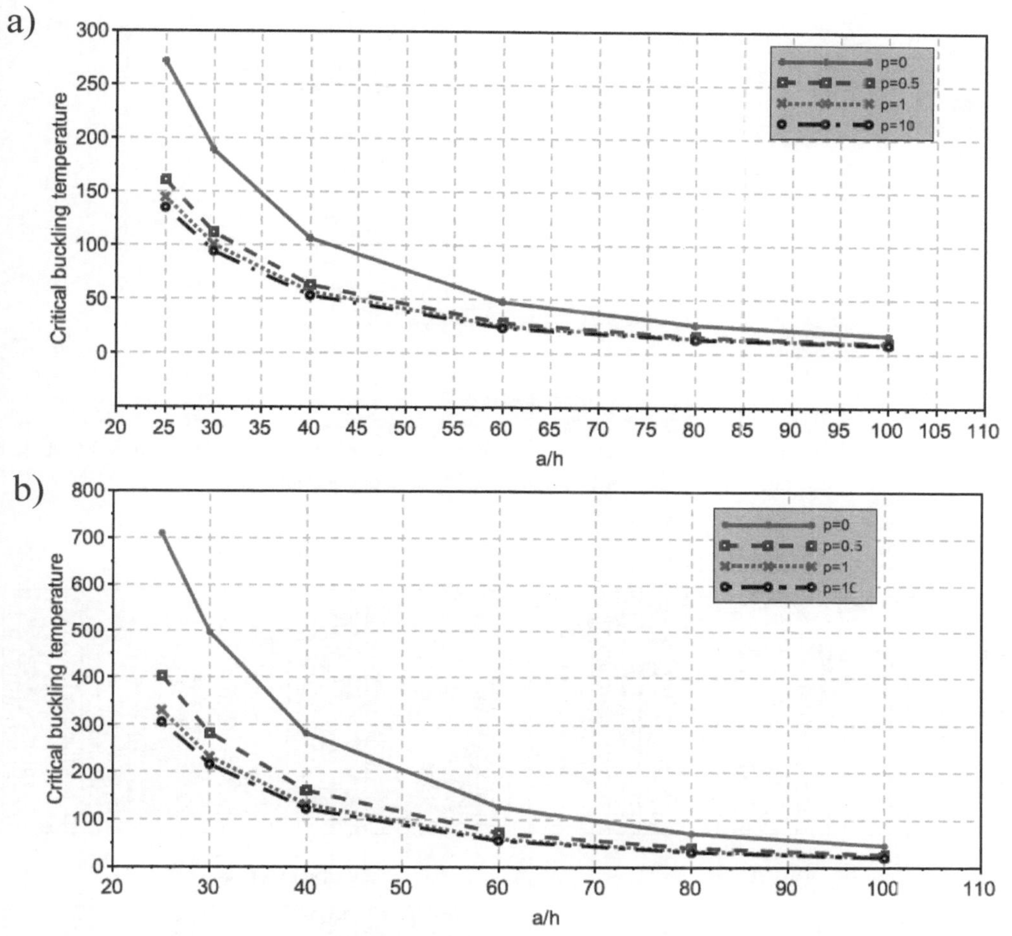

Figure 2.17 Critical buckling temperature ΔT_{cr} (in K) versus the length-to-thickness ratio (a/h) of a a) simply supported (SSSS) and b) clamped (CCCC) square plate.

Table 2.20
Critical buckling temperature ΔT_{cr} (in K) of FGM cylindrical shell under uniform temperature distribution with (TD) and (TID) material properties.

p	TID			TD		
	Present	FSDT [114]	HSDT [225]	Present	FSDT [114]	HSDT [225]
0.0	80.15	80.2	86.66	81.94	78.9	82.27
0.5	99.1	94.2	106.54	99.67	87.1	99.39
1	110.84	99.2	118.46	110.79	92.5	110.25
2	124.76	108.2	132.45	124.31	101.1	122.21
5	142.45	125.4	150.28	141.98	116.0	137.15

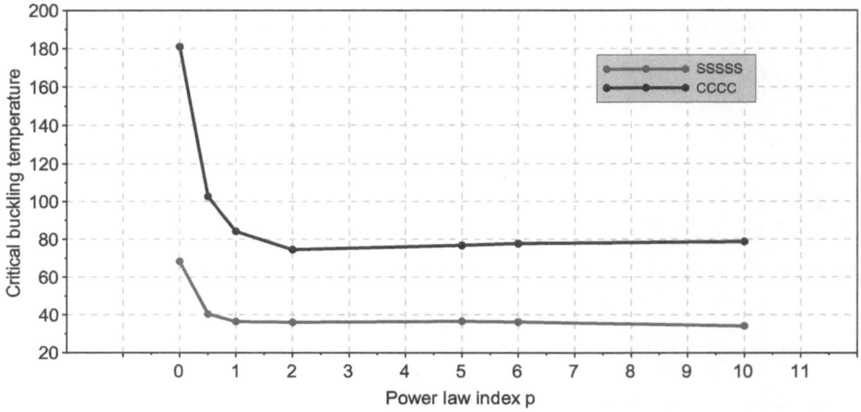

Figure 2.18 Critical buckling temperature ΔT_{cr} (in K) versus the power law index p of the square FGM plate with a length-to-thickness ratio ($a/h = 50$) for SSSS and CCCC boundary conditions.

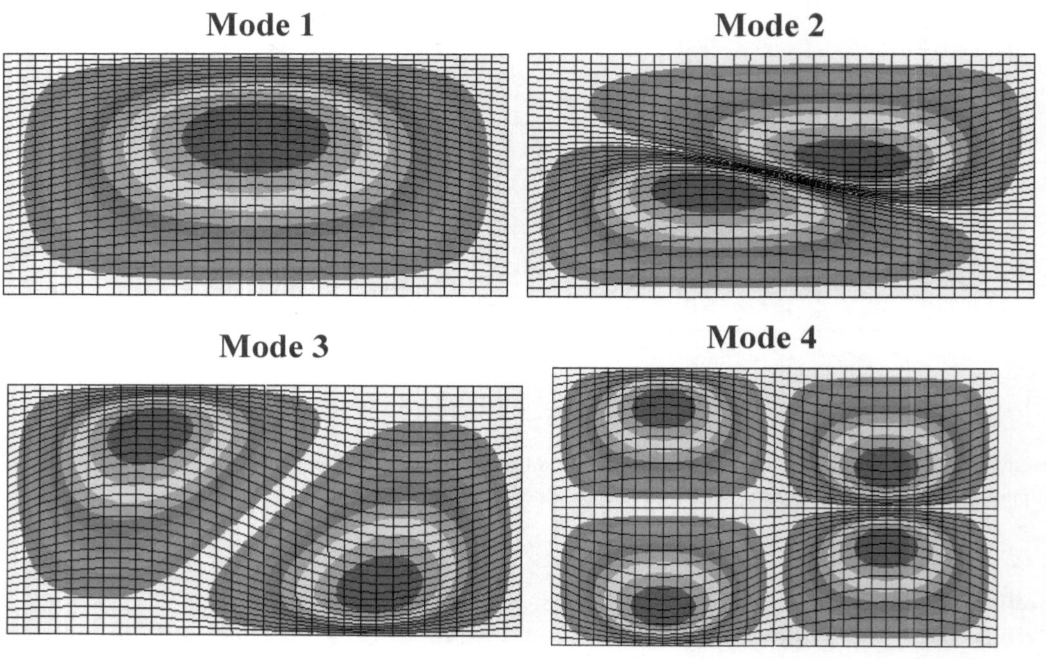

Figure 2.19 Thermal buckling mode shapes of the FGM square plates ($a/h = 100$).

nificant effect on critical buckling temperature for non-uniform temperature distribution compared to uniform distribution case. This can be explained by the flexural rigidity of the cylindrical shell which becomes sensitive to the variation of the temperature field along the thickness direction with non-uniform distribution. Figures 2.22(a) and 2.22(b) illustrate the variation of the buckling temperature of FGM cylindrical shell versus the thickness-to-radius ratio (h/R) with $p = 1$ for uniform and non-uniform temperature distributions through the thickness, respectively. It can be observed in each figure that, as the thickness-to-radius ratio increases, the buckling temperature increases. In

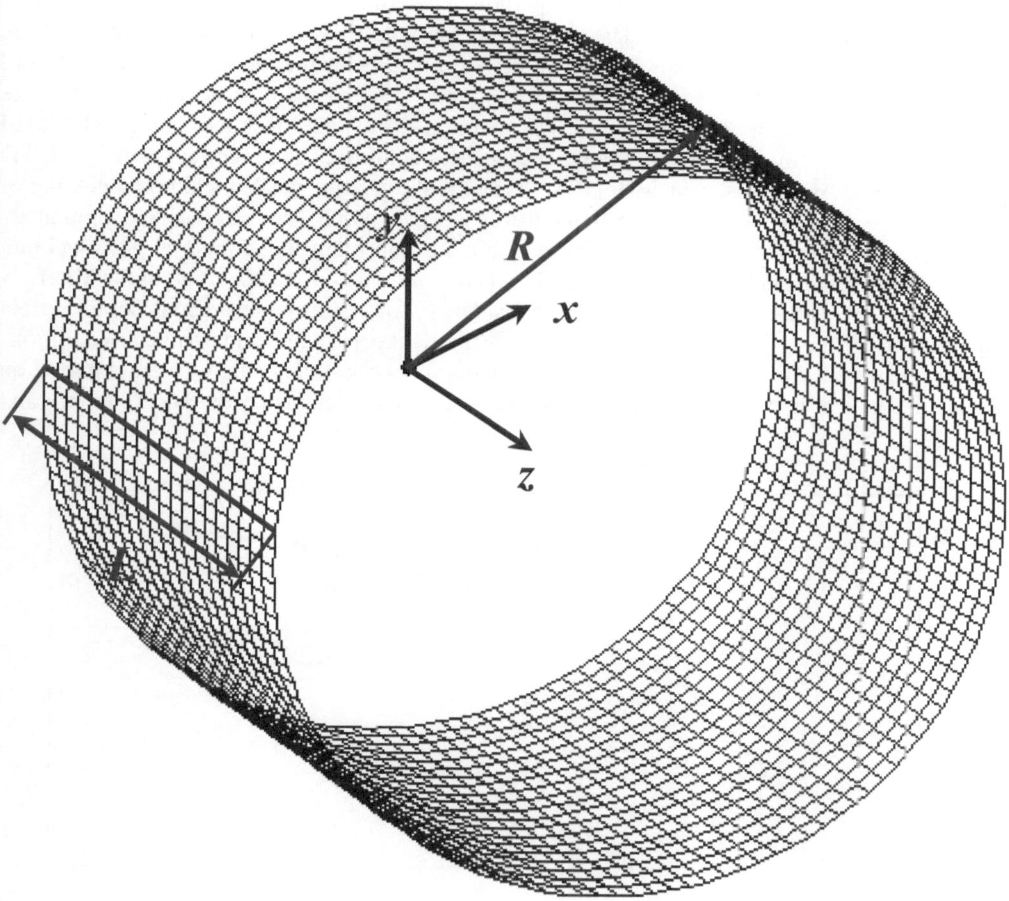

Figure 2.20 Geometry and coordinates system of the FGM cylindrical shell.

addition, the temperature dependence of material properties does not have a significant effect on thermal buckling in the case of uniform temperature distribution across the thickness, while this effect is more pronounced for the FGM cylindrical shell under non-uniform temperature change through the thickness.

2.5 THERMAL POST-BUCKLING ANALYSIS OF FGM STRUCTURES

2.5.1 STATE OF THE ART ON POST-BUCKLING OF FGM STRUCTURES

Thermal post-buckling analysis of FGM structures has been investigated in our works. At this stage, we recall that many studies have been reported on the buckling and postbuckling analysis of FGM plates and shells subjected to mechanical or thermal loadings. A brief literature review on the works made on this type of analysis is given. Based on the CST with the Von Karman assumptions, Ma and Wang [148] inspected the thermal post-buckling of FG circular plates subjected to mechanical, thermal and thermal-mechanical loadings. Parametric studies were conducted to show the effect of the boundary conditions and the power law coefficient on the large deflection and the thermal post-buckling responses. Woo et al. [290] studied the post-buckling of moderately thick FG plates and shallow shells exposed to compression and thermal loads using the HSDT theory and Von Karman assumptions. In this study, they provided a comparative study between results given by the HSDT, the Reissner-Mindlin's theory and the classical plate theory and they outlined the efficiency of the transverse shear deformation on the prediction of the non-linear behavior and deflection of plate and shell structures. Moreover, Park and Kim [194] investigated the thermal post-buckling of FG plate using the FSDT in conjunction with the Von Karman non-linear relationships. Zhang [312] used the HSDT and the Ritz method to study the thermal post-buckling of FG elliptical plates with TD material properties. As well as, Lee et al. [134] inspected the thermo-mechanical post-buckling of FG plates using the FSDT theory and the element-free Kp-Ritz procedure with the Von Karman assumptions. On the other hand, an analytical investigation of mechanical and thermal post-buckling of FG plates was elaborated by Duc et al. [180]. In this research, the FSDT theory with the Von Karman strains and the TD material properties were considered. The results revealed considerable effect of temperature dependency of material properties, geometrical parameters and boundary conditions on the non-linear behavior of FG plates. Then following his work, many studies were reported on non-linear mechanical and thermal responses of FG plates and shells using the HSDT theory and

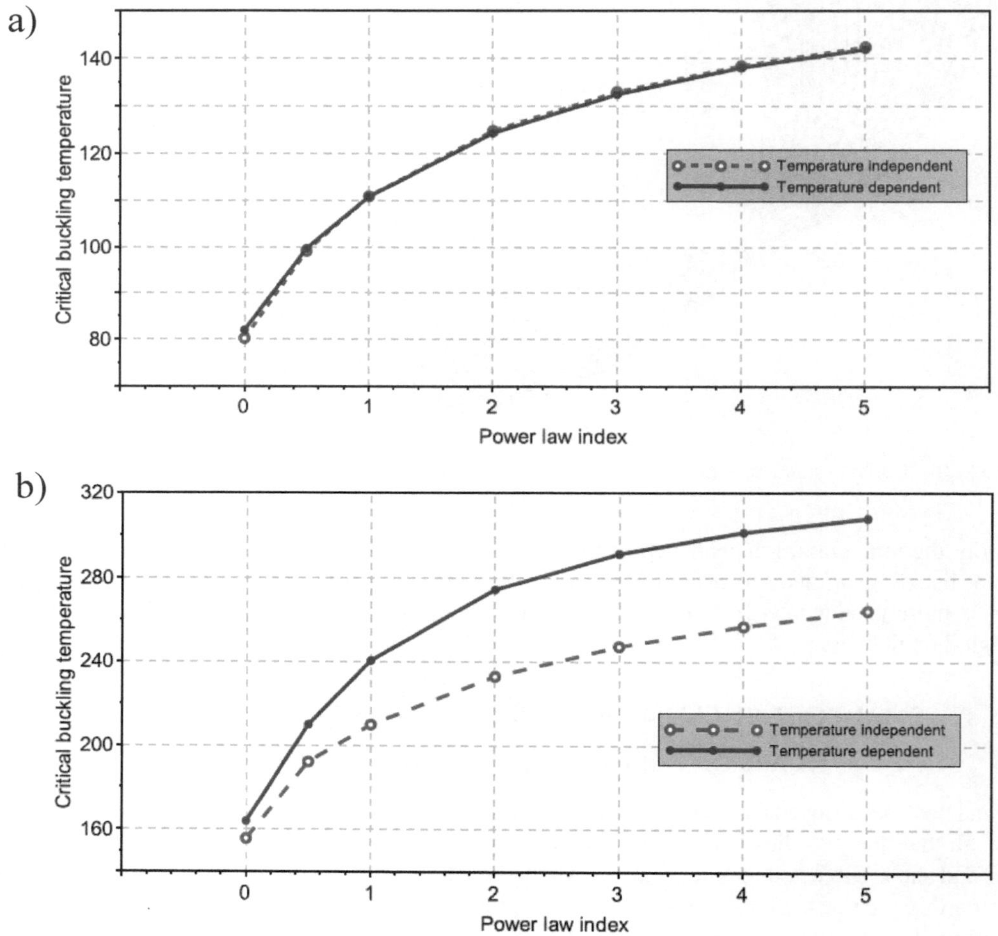

Figure 2.21 Critical buckling temperature ΔT_{cr} (in K) of FGM cylindrical shell versus p for a) uniform and b) non-uniform temperature rise with TD and TID material properties.

Von Karman assumptions [66, 63, 62, 64, 65, 186]. Moreover, the post-buckling problems of FG plates under mechanical, thermal or combined thermo-mechanical loads have been inspected by Dung et al. [68] using an analytical solution. The plate was reinforced by FGM stiffeners on elastic foundations and the material properties were assumed to be TD. The formulation is done with the third-order shear deformation (TSDT) theory and Von Karman strains. Many other investigations on the thermal post-buckling behavior of FG plates can be found in [264, 199, 189, 276, 313, 217, 229].

The thermal post-buckling of FG cylindrical shells with TD material properties was first examined by Shen [222]. This study, is based on the CST and the geometrical non-linearity is described by Von-Karman-Donnel strains. The thermal post-buckling response is obtained by considering an initial geometric imperfections of the structure. The thermal post-buckling response is limited to simply supported boundary condition. Then, an analytical solution for the thermal post-buckling of FG cylindrical shells and plates using a HSDT was given by shen [102, 101]. The non-linear kinematic was provided by Von-Karman-Donnel model. Comparative studies were given in order to show the effects of the heat conduction and uniform temperature field on the non-linear

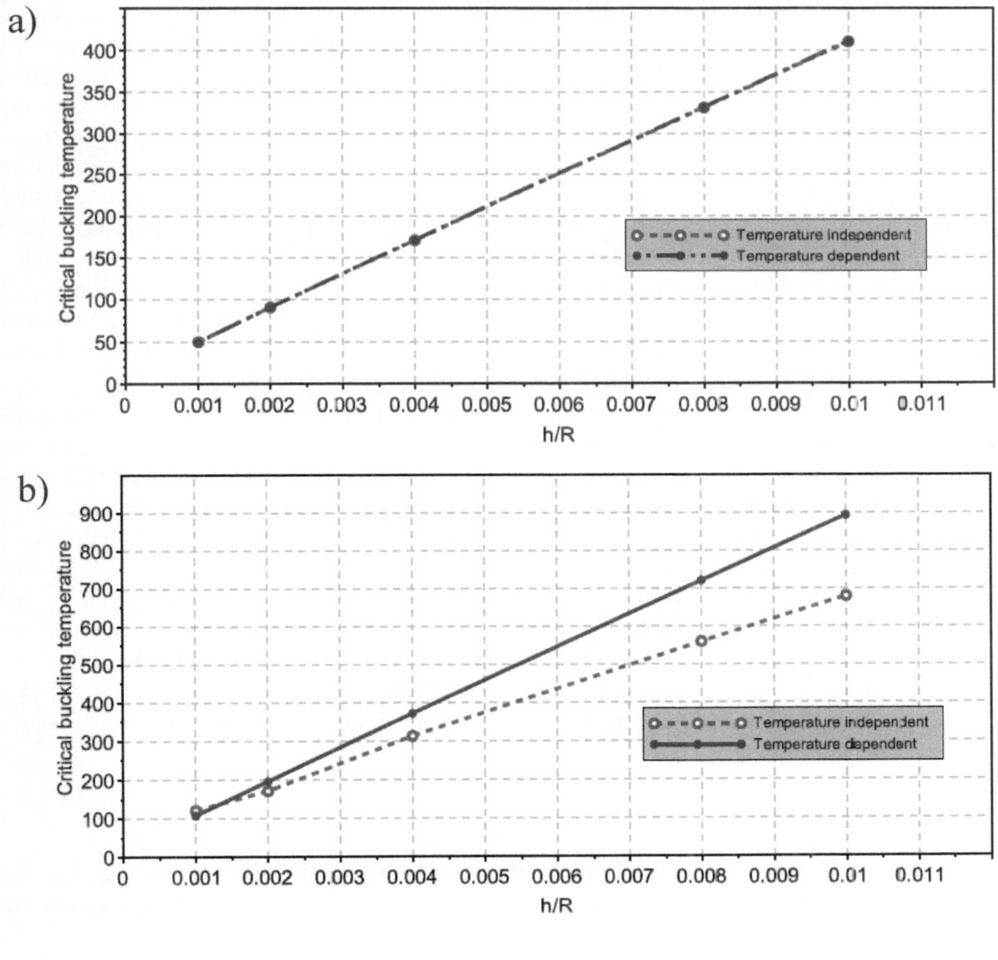

Figure 2.22 Critical buckling temperature ΔT_{cr} (in K) versus the thickness-to-radius (h/R) of FGM cylindrical shell ($p = 1$) for a) uniform and b) non-uniform temperature rise with TD and TID material properties.

behavior of such structures. The results show that the TD material properties and the volume fraction index have a significant influence on the post-buckling responses of FGM structures. Moreover, Shen and his co-authors presented various works on the post-buckling behavior of a FG cylindrical shell in thermal environments [220, 221, 230, 231, 232]. The theoretical formulation of the dynamic post-buckling of FG cylindrical shells under thermal load and applied actuator voltage were set up by Mirzavand and his co-workers, applying the CST [15] and the TSDT theory [14]. The Sanders non-linear kinematic relations were employed and TD material properties were considered. As a result, the application of the actuator voltage has a significant effect on the thermal post-buckling curves. On the other hand, Alijani et al. [1] presented a semi-analytical finite element procedure to investigate the non-linear thermal buckling behavior of isotropic, composite and FGM cylindrical shells. This formulation was based on the FSDT theory and the Fourier series were adopted to express the displacement of the middle surface and the radius of the cylinder. Eventually, the effects of geometrical parameters, imperfection ratio and power law index on the post-buckling response were studied. Moreover, several studies have been presented by Duc et al. in the context of the thermal and mechanical post-buckling of stiffened FGM circular cylindrical shells [182], cylindrical panels [184, 181] and double curved shallow shells [185]. In these studies, they used an analytical method based on the CST theory with the incorporation of Von Karman-Donnel geometrical non-linearity. As well as, the thermal buckling behavior of the FGM sandwich truncated conical shells was reported in [183] using the FSDT theory. Liew et al. [147] examined the post-buckling of FG cylindrical panels subjected to compression and thermal loads. In this study, Liew et al. used the FSDT theory and the element-free kp-Ritz method for numerical computations. Dealing with the geometrically non-linear analysis, Panda et al. [281, 280] were given a finite element solution of the buckling and post-buckling behavior of FG shell panels in thermal environment. In these works, the governing equations are based on the HSDT and the Green-Lagrange non-linear strains. These studies outlined the effects of the boundary conditions, the TD of material properties, the temperature distribution across the thickness and the power-law index on the buckling and post-buckling responses of FGM panels. In addition to these interesting researches, Panda and his co-authors conducted numerous studies on non-linear analysis of laminated composite shell panels concerning either thermal post-buckling and free vibration behaviors [249, 250, 251]. As well as, Mahapatra et al. conducted various studies on non-linear free vibration of laminated composite plates and shells under hydrothermal environment [269, 266, 268, 267, 270] using the HSDT theory in conjunction with the Green-Lagrange Kinematics. Based on the same strategy, Hirwani et al. studied the dynamic behavior of composite plates subjected to blast loads [52], the deflection responses of the damaged composite shell structures [53] and the linear/non-linear frequency analysis of composite shell structures [51].

As can be remarked from this report of the open literature, the most of investigations on the thermal post-buckling of FGM structures are based on the Von Karman assumptions, which are limited to moderately small deformations. Therefore, the evaluation of the excessive deflection and post-buckling should be achieved by a more general type of geometrical non-linearity. In our works, the improved or modified (I-FSDT) theory is used to investigate the thermal post-buckling analysis of FGM plates and shells.

2.5.2 SOLUTION STRATEGY

For the post-buckling analysis, the equilibrium paths or responses of FGM structures are obtained using iterative and incremental Newton–Raphson method where its detailed finite element procedure is depicted in the flowchart of Fig. 2.23.

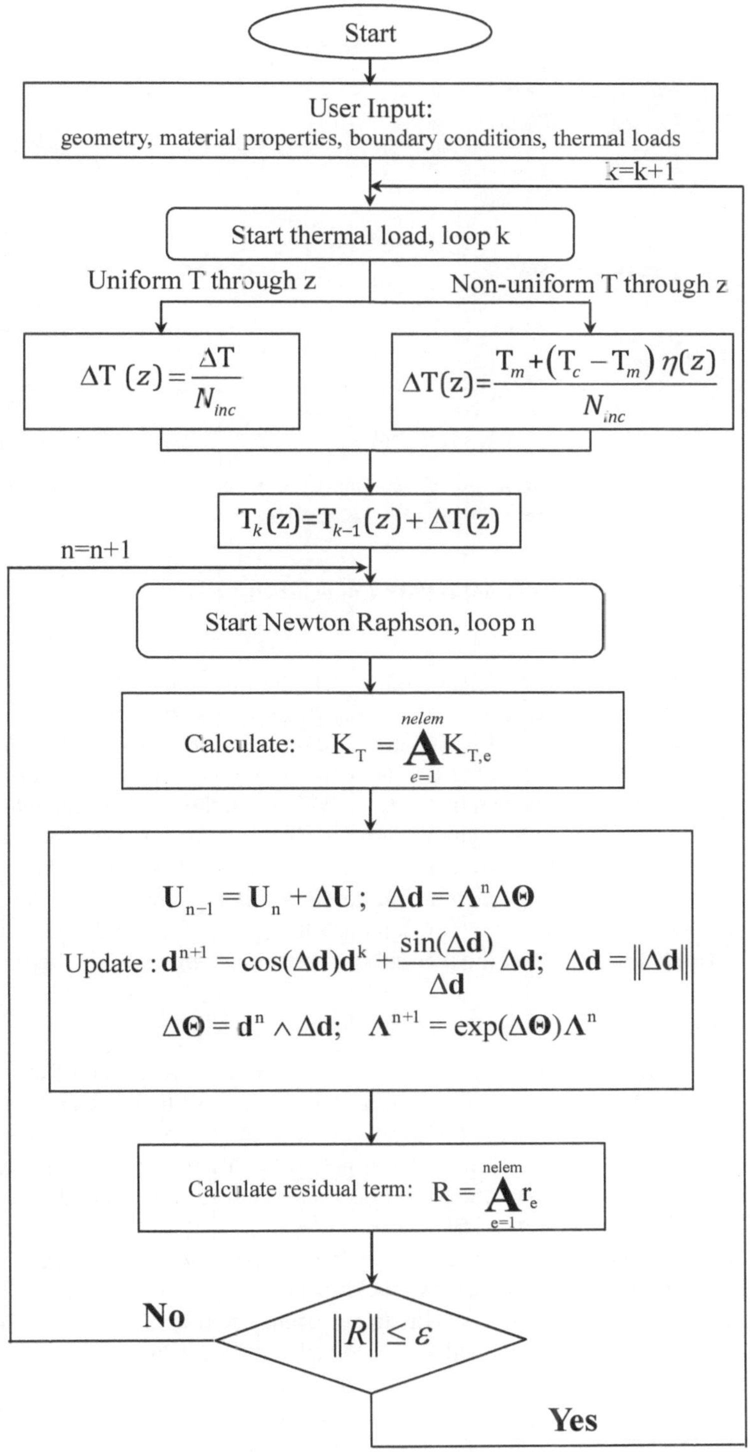

Figure 2.23 The flowchart of the finite element procedure for thermal post-buckling analysis of FGM structures.

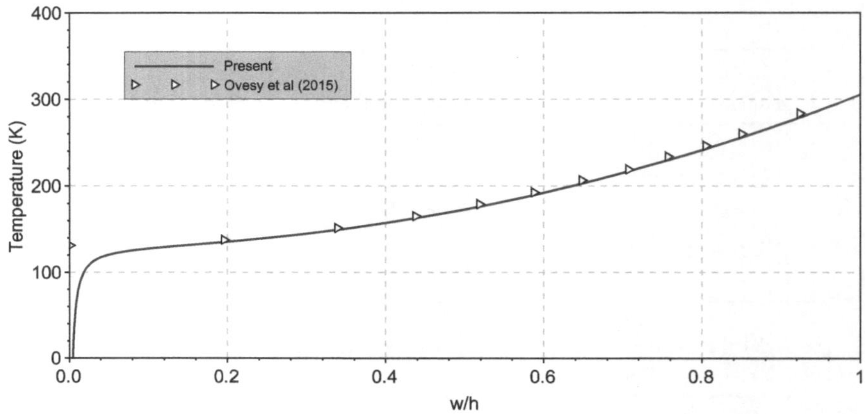

Figure 2.24 Temperature-deflection response of the simply supported isotropic rectangular plate under a uniform temperature distribution with geometrical ratios ($a/h = 100$, $a/b = 1$).

2.5.3 THERMAL POST-BUCKLING ANALYSIS OF FGM PLATES

2.5.3.1 Isotropic plates

An isotropic rectangular plate is first considered to validate the geometrically non-linear formulation. The geometrical and material properties of the plate are taken as indicated in [99] where: $a/b = 1$, $a/h = 100$ are the geometrical ratios, $E = 1GPa$ is the Young's modulus, $\nu = 0.25$ is the Poisson coefficient and $\alpha = 10^{-6}/°C$ is the thermal expansion coefficient. The plate is meshed using 32×32 four-nodes elements as shown in Fig. 2.16. Moreover, the whole of the plate is subjected to uniform temperature distribution through the thickness ($\Delta T = 500K$) and the four edges (*AB*, *BC*, *CD* and *DA*) are simply supported as assumed in [99, 293]. Figure 2.24 depicts the obtained load-displacement curve of the isotropic plate compared to the result given by Ovesy et al. [99]. A good agreement between the results is shown which demonstrates the aptitude of the present modified or improved FSDT formulation to predict the thermal post-buckling behavior with high accuracy.

2.5.3.2 FGM plates

The effectiveness and usefulness of the present formulation in the prediction of the thermal post-buckling behavior of FGM rectangular plates subjected to thermal loads is now examined. The FG plate is made of a mixture of Nickel (Ni) and Alumina (Al_2O_3) where the geometrical ratios are given as: $a/b = 1$ and $a/h = 100$. The material properties are listed in Tab. 2.21. The plate is modeled using 32×32 four-nodes elements. The boundary conditions on four edges are considered to be clamped (CCCC) as indicated in [175].

As the first thermal loading case, a uniform temperature rise through the thickness is assumed. The effects of the power law index p on the post-buckling behavior of FGM plates are examined and the non-dimensional deflection curves versus the temperature rise of the clamped FGM rectangular plate are plotted in Fig. 2.25. The obtained results match well with those presented by Na et al. [175] which outline the capacity of the present formulation to reproduce the post-buckling behavior of such structures with good accuracy. Figure 2.26 depicts the post-buckling response of the clamped FGM rectangular plate as well as its corresponding mode shapes.

The second thermal load consists of a non-linear temperature distribution across the thickness. The geometrical and material properties of $Ni-Al_2O_3$ FGM plate are taken as mentioned in [175]. The applied boundary condition is clamped in all edges. The obtained non-dimensional deflection

Table 2.21
Material properties of the Nickel (Ni) and Alumina (Al_2O_3) phases [131, 112].

Materials	Properties				
	E (GPa)	$Density(Kg/m^3)$	$Conductivity(W/mK)$	Thermal expansion $(/°C)$	v
Ni	199.5	8900	90.7	13.3×10^{-6}	0.3
Al_2O_3	393	3970	30.1	8.8×10^{-6}	0.25

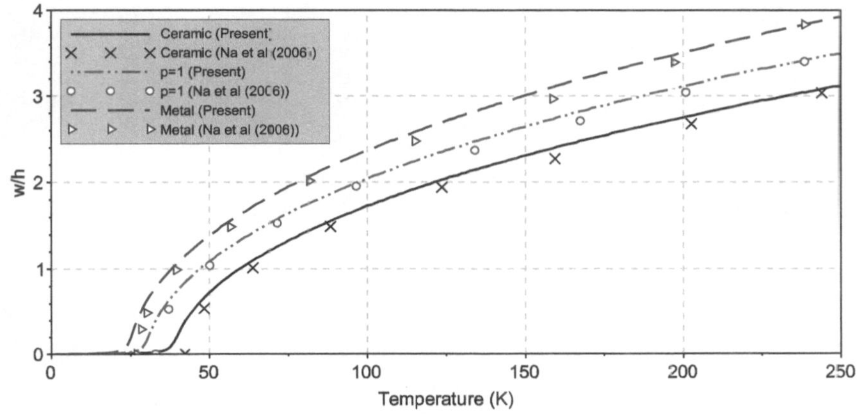

Figure 2.25 Non-dimensional deflection versus the temperature rise of a clamped FGM plate under a uniform temperature distribution with the geometrical ratios $(a/h = 100,\ a/b = 1)$.

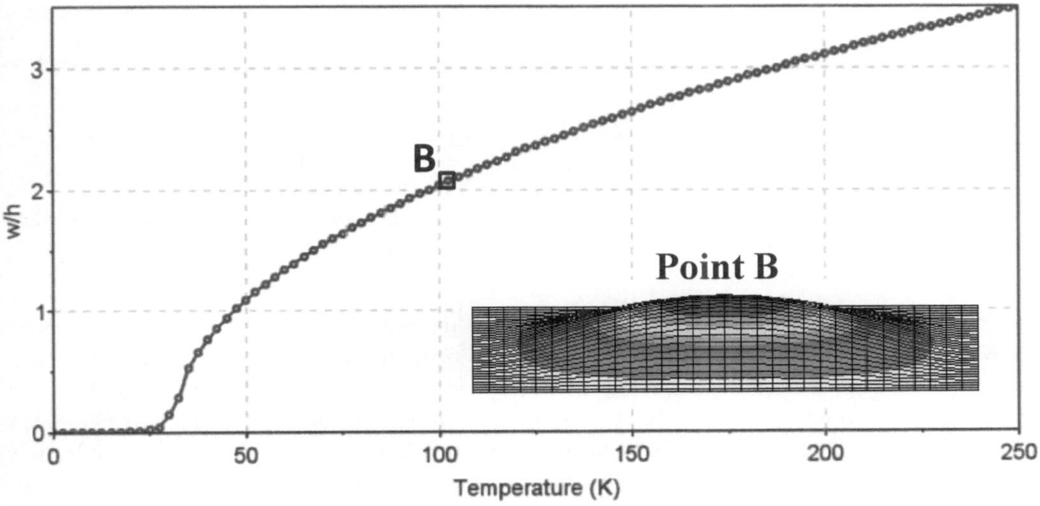

Figure 2.26 Thermal post-buckling response and its corresponding mode shape at the point B of the clamped FGM rectangular plate subjected to uniform temperature distribution with the geometrical ratios $(a/h = 100,\ a/b = 1)$ and power-law index $p = 1$.

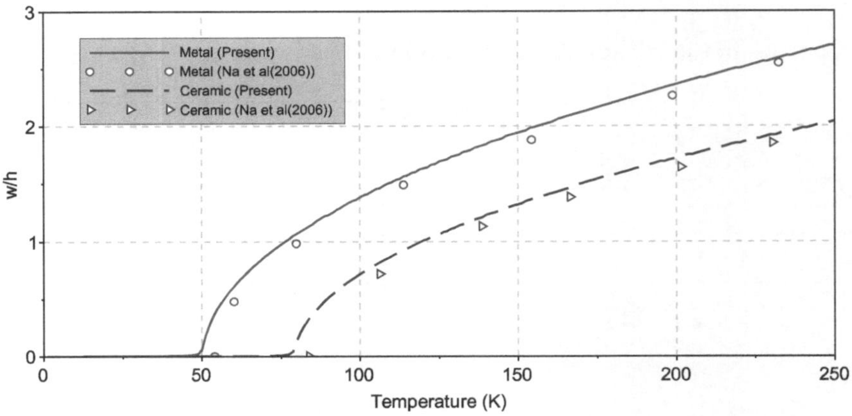

Figure 2.27 Non-dimensional deflection versus temperature rise of the clamped FGM rectangular plate under non-linear temperature distribution with the geometrical ratios ($a/h = 100$, $a/b = 1$).

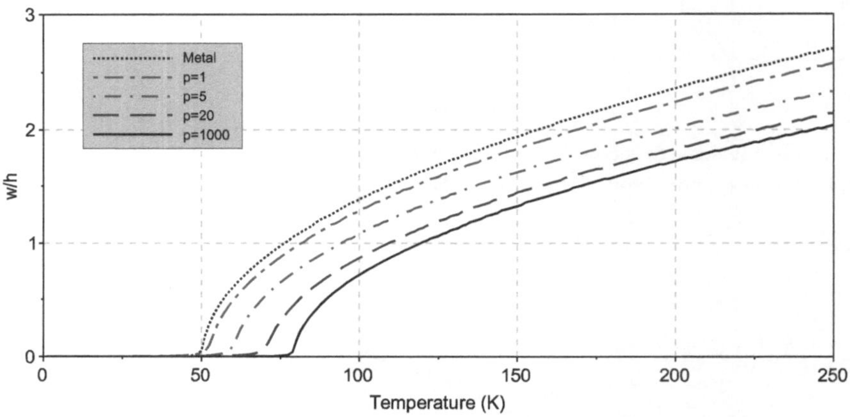

Figure 2.28 Non-dimensional deflection versus temperature rise of the clamped FGM rectangular plate under non-linear temperature distribution and different power-law index value.

curves are presented in Fig. 2.27. As can be remarked, the non-dimensional deflection responses agree well with the solution of Na et al. [175] which demonstrates the aptitude of the present model in the prediction of post-buckling behavior of FGM plates under different types of thermal loads.

Figure 2.28 shows the effects of the power-law index p on non-linear deflections of FG plates subjected to non-linear temperature distribution. It can be remarked that, the non-linear deflection increases with the temperature variation. In addition, the non-dimensional deflection of the fully metal plate is higher than that of the fully ceramic one. Furthermore, the deflection decreases as the power index p increases. In fact, for the high values of power index p, the plate becomes more stiffer due to the high flexural rigidity of the FG plate. Then, the effect of the length-to-thickness ratio (a/h) and the plate aspect ratio (a/b) on the thermal post-buckling behavior of the clamped FG plates are examined. In fact, Figs. 2.29(a) and 2.29(b) show that the decrease of the length-to-thickness ratio ($a/h = 100$ to 50) leads to a decrease in the non-dimensional deflection of the FG plate subjected to uniform or non-linear temperature distributions through the thickness. Accordingly, when the

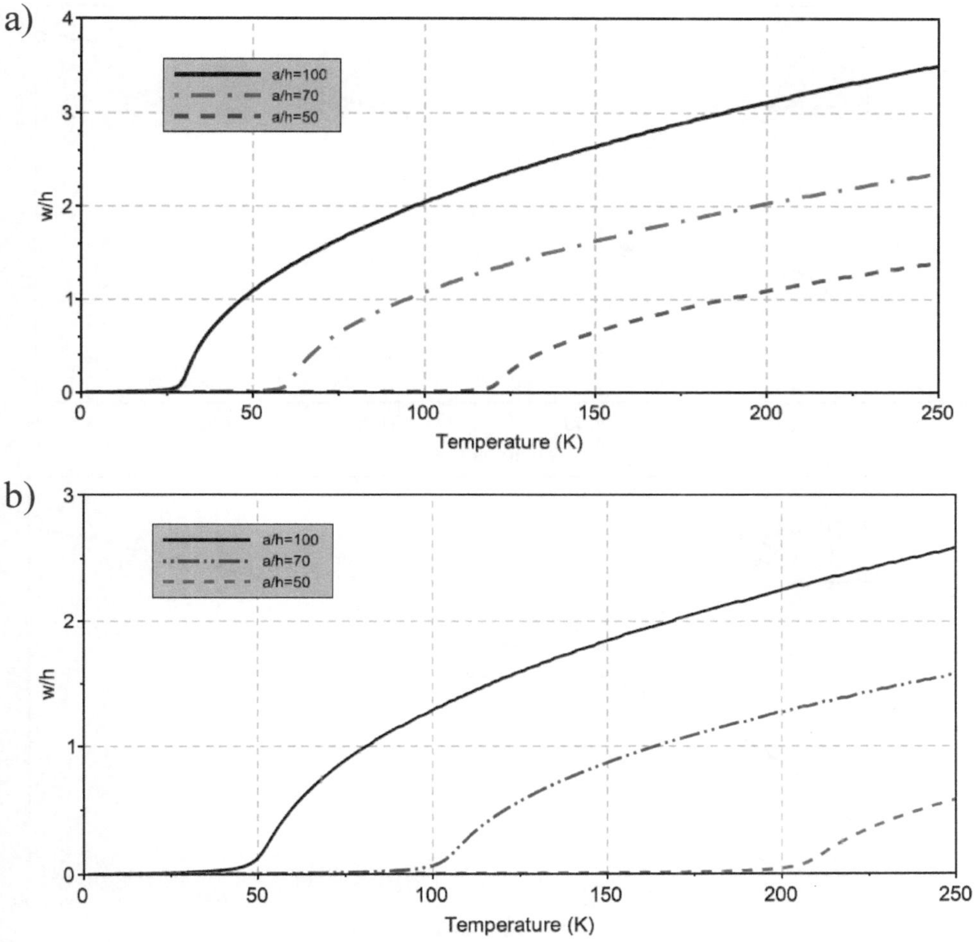

Figure 2.29 Non-dimensional deflection versus temperature rise of the clamped FGM rectangular plate subjected to: a) uniform temperature rise and b) non-linear temperature rise with different length-to-thickness ratios (a/h) and power-law index $p = 1$

length-to-thickness ratio decreases the FG plate behaves as thick plate and hence becomes more stiffer for the change of temperatures. As well as, Figs. 2.30(a) and 2.30(b) reveal the effect of the plate aspect ratio (a/b) on the non-dimensional deflection of the FGM plate under a uniform and non-linear temperature rise. Indeed, the rise of the geometrical ratio (a/b) induces an increase in the non-dimensional deflection of the FG plates. Furthermore, the thermal load increases with the non-dimensional deflection, which can be explained by the sensitivity of the stiffness to the thermal loadings.

The effects of boundary conditions on the non-linear thermal post-buckling behavior of the rectangular $(Ni - Al_2O_3)$ FG plates are also examined. A set of boundary conditions is considered: all edges are clamped (CCCC) or simply supported (SSSS), two sides (AD) and (BC) are simply supported and the others (AB) and (DC) are clamped (SCSC) or two sides (AD) and (BC) are simply supported and the others (AB) and (DC) are free (SFSF). Figures 2.31(a) and 2.31(b) illustrate the deflection-temperature curves of the square $(Ni - Al_2O_3)$ plate subjected to uniform and non-linear

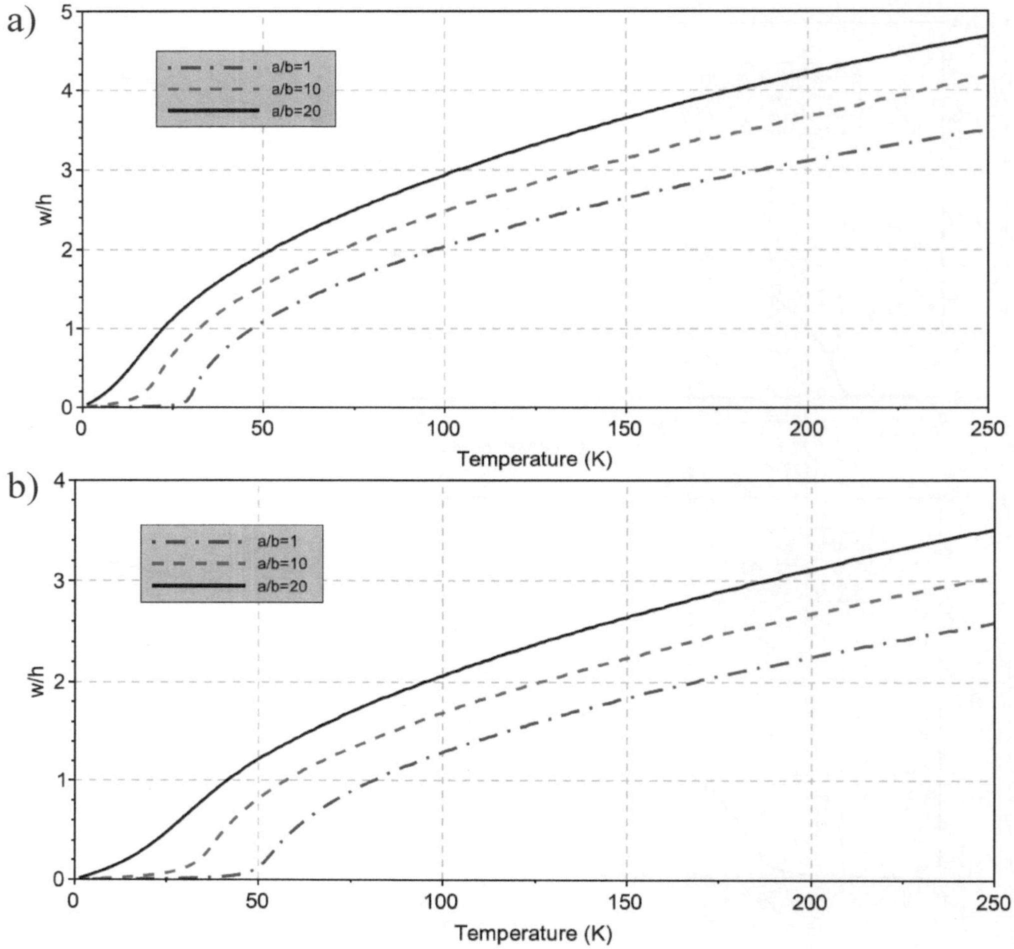

Figure 2.30 Non-dimensional deflection versus temperature rise of the clamped rectangular FGM plate subjected to: a) uniform temperature rise and b) non-linear temperature rise with different aspect ratios (a/b) and power-law index $p = 1$.

temperatures respectively and under different boundary conditions and with a power index $p = 1$ and length-to-thickness ratio $a/h = 1000$. It can be observed that the SFSF boundary condition induces more transverse displacement than the SCSC and SSSS boundary conditions. Since free boundary condition (F) imposes less constraints than simply supported (S) or clamped (C) ones, the deflection of the plate is more higher with a free boundary condition than the other ones. As expected, the fully clamped plate has the lowest deflection and this is for weak values of temperatures. However, for a high temperature, this effect is inversed and the plate bends easily. This can be explained by the fact that more constraints are imposed on the structure, more the plate becomes sensitive to the change of temperature, which affects the stiffness of the FGM structure.

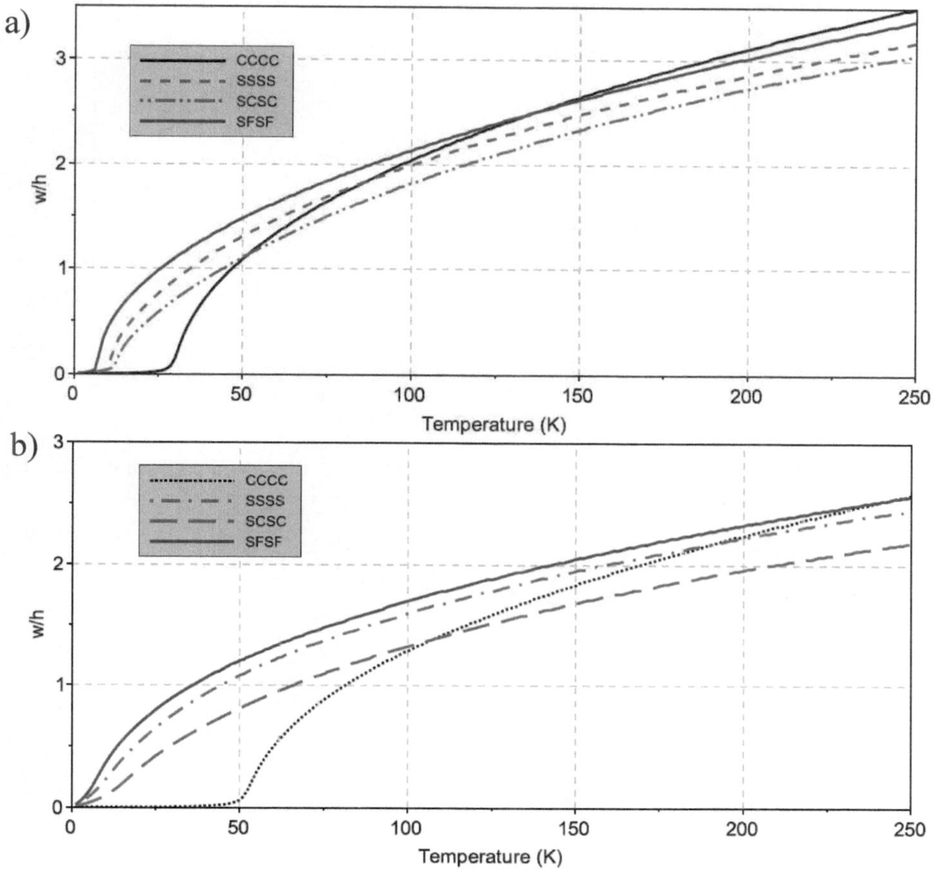

Figure 2.31 Non-dimensional deflection versus temperature rise of the rectangular FGM plate subjected to: a) uniform temperature rise and b) non-linear temperature rise for various boundary conditions and with length-to-thickness ratio ($a/h = 100$) and power-law index $p = 1$.

2.5.4 THERMAL POST-BUCKLING ANALYSIS OF FGM CYLINDRICAL SHELLS

2.5.4.1 Isotropic cylindrical shells

An isotropic cylindrical shell with radius-to-thickness ratio $R/h = 400$ and $L^2/Rh = 300$ is considered. The material properties of the cylinder are defined as follows: $E = 70GPa$, $\nu = 0.3$ and $\alpha = 23 \times 10^{-6}/°C$. The thermal post-buckling behavior of the shell is examined in the case of uniform temperature distribution across the thickness ($\Delta T = 300K$). The cylindrical shell is meshed using 128×20 four-nodes elements along the circumferential and length directions, respectively as depicted in Fig. 2.20 with two clamped boundaries (CC). As shown in Fig. 2.32, the maximum point of radial deflection of the cylindrical shell is not at the mid-length, but near to the end about $35m$. Since there are no results found in the literature matching with the thermal post-buckling behavior of isotropic clamped cylindrical shells, the maximum load-deflection curve is compared with the numerical solution of ABAQUS software, as depicted in Fig. 2.33. It is clear that the present finite element method is in good agreement with the numerical solutions given by ABAQUS which leads to the validatation of the accuracy and the efficiency of the present method.

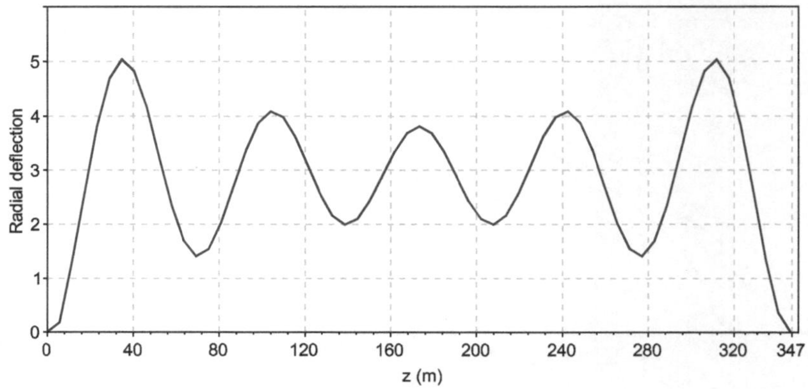

Figure 2.32 Radial deflection of the clamped isotropic cylindrical shell under a uniform temperature distribution ($T_{initial} = 0K$, $T_{final} = 300K$).

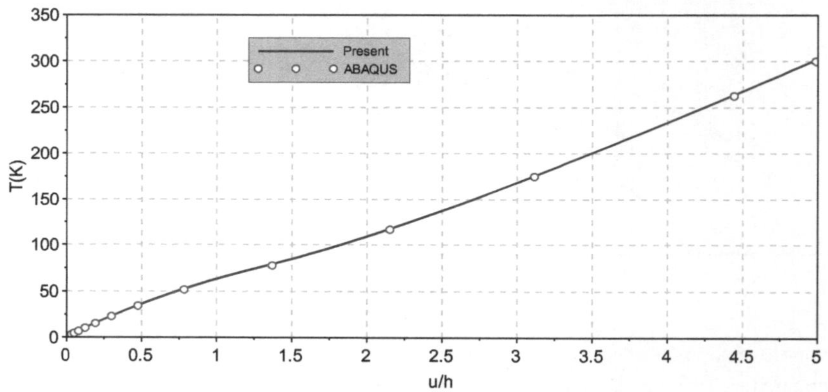

Figure 2.33 The maximum radial deflection of the clamped isotropic cylindrical shell under a uniform temperature rise.

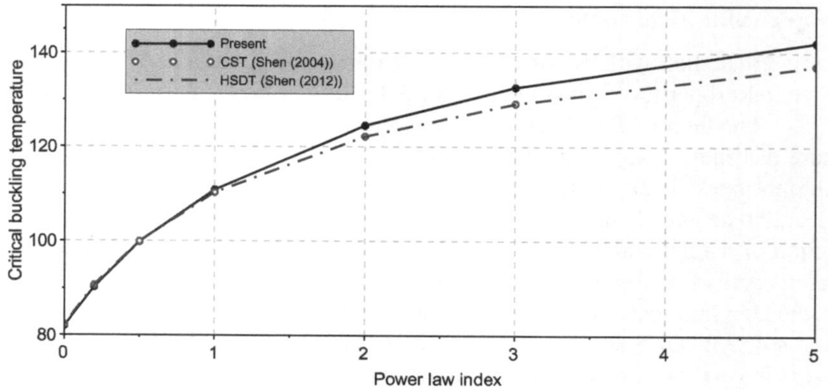

Figure 2.34 Critical buckling temperature of the FG cylindrical shell ($Si3N4/SUS304$) subjected to uniform temperature rise with TD material properties ($R/h = 400$, $L^2/Rh = 300$ and $h = 1mm$).

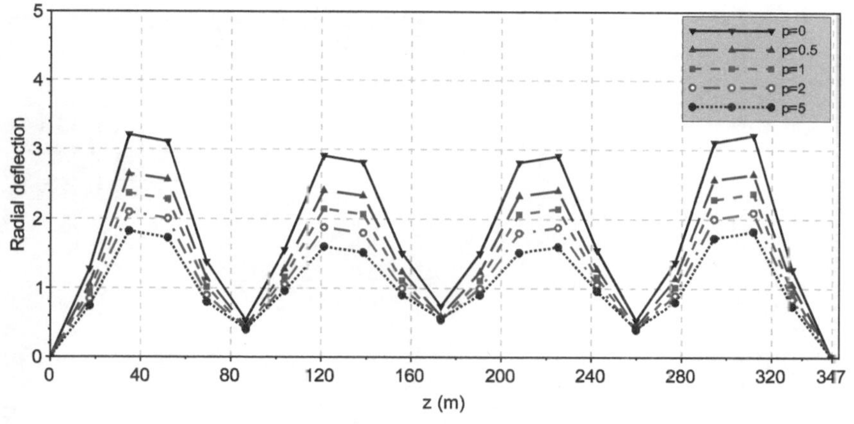

Figure 2.35 Radial deflection of the clamped FGM cylindrical shell subjected to uniform temperature distribution ($T_{initial} = 0K$, $T_{final} = 300K$).

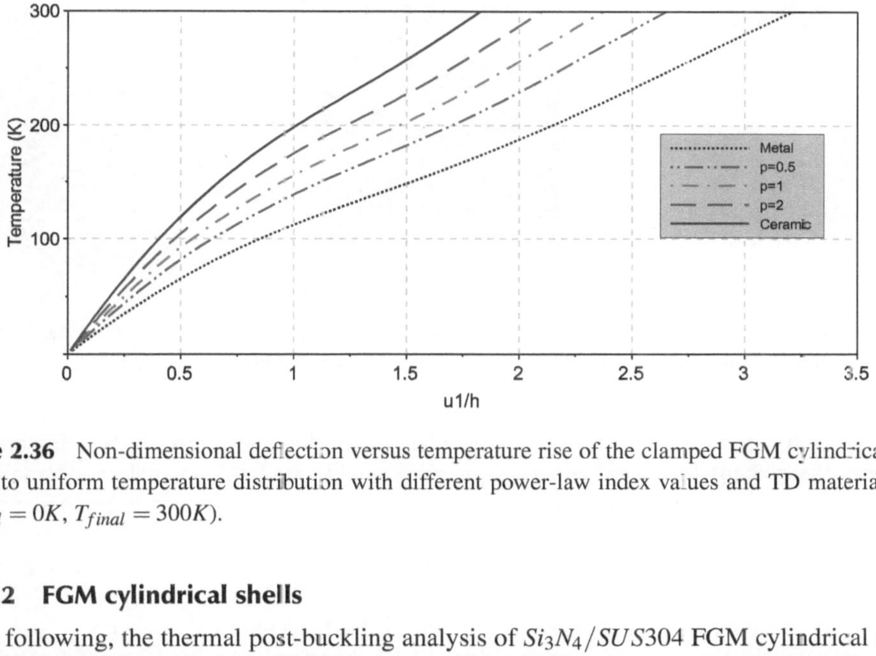

Figure 2.36 Non-dimensional deflection versus temperature rise of the clamped FGM cylindrical shell subjected to uniform temperature distribution with different power-law index values and TD material properties ($T_{initial} = 0K$, $T_{final} = 300K$).

2.5.4.2 FGM cylindrical shells

In the following, the thermal post-buckling analysis of $Si_3N_4/SUS304$ FGM cylindrical shells subjected to uniform temperature distribution through the thickness is examined. The used geometrical properties here are given as follows: $R/h = 400$ and $L^2/Rh = 300$. As well as, the corresponding material properties are listed in Tab. 2.19. The used mesh is the same as mentioned earlier in section 2.4.2. As a first study, the curves of the critical buckling temperature versus the power index p of a simply supported (SS) FGM cylindrical shell are plotted in Fig. 2.34 and compared with those presented by Shen using either the CST [222] and the HSDT theories [225]. Results given by the present finite element method are slightly higher than those of Shen [222, 225] for a power law index $p > 1$ and this is can be explained in one hand by the utilization of Von karman assumptions in both theories which includes only membrane forces and it is restricted to moderately small deformations,

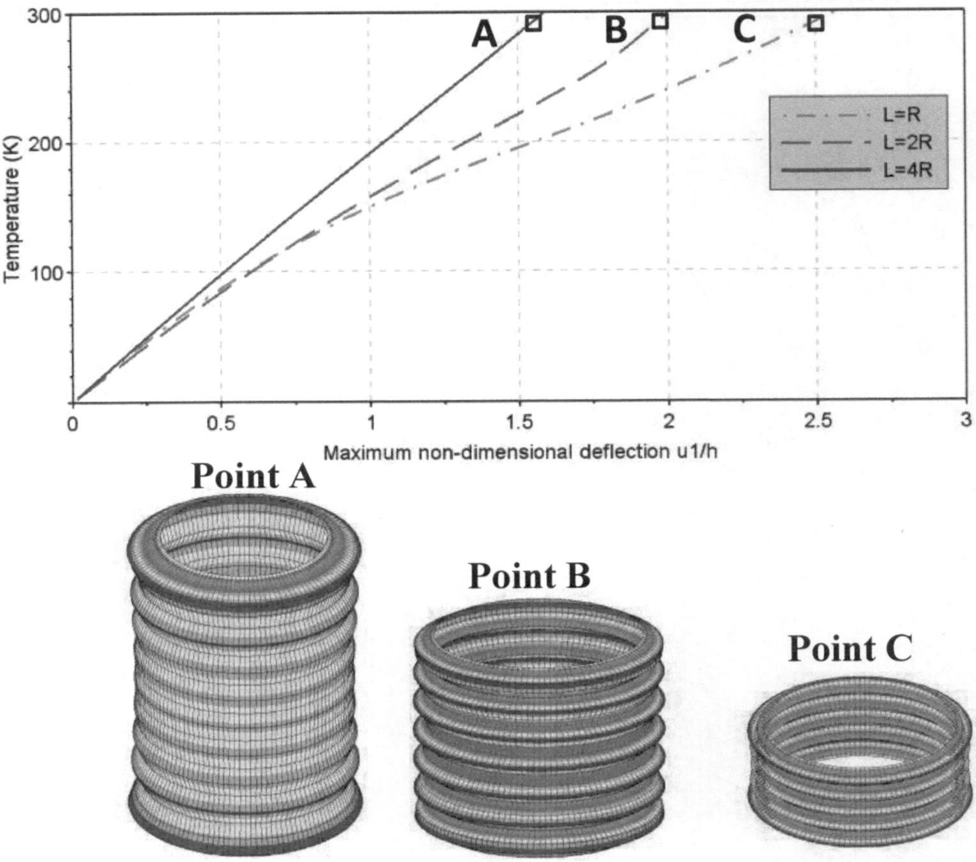

Figure 2.37 Non-dimensional deflection versus temperature rise of the clamped FGM cylindrical shell subjected to uniform temperature distribution with different (L/R) and power-law index value $p = 1$.

and on the other hand by the type of the solutions technique. In fact, in our paper the finite element method is performed using the Gauss numerical integration while in [225], an analytical approach is adopted with the boundary layer solutions. Hence, the present formulation constitutes an alternative to model large deformations with a good representation of large displacements and finite rotations. In addition, it can be seen that the increases of the power index p leads to an increase in the critical buckling temperature.

Figure 2.35 depicts the radial deflection curves of the FGM cylindrical shell under a uniform temperature distribution for different power index p with the clamped boundary condition on its two curved edges. The maximum value of the radial deflection of the FGM cylindrical shell is found closer to the applied boundary condition. Figure 2.36 presents the maximum thermal post-buckling load-deflection curves for FGM cylindrical shell versus different power law index p. It can be seen that the maximum non-dimensional deflection increases as the temperature increases. Furthermore, the thermal post-buckling path decreases as the power index p increases. This is due to the fact that increasing the power index p increases the volume fraction of ceramic which is able to withstand high temperature. Therefore, the FGM materials constitute a good way of heat resistant. On the other hand, Fig. 2.37 shows the thermal post-buckling behavior of clamped FGM cylinders with different lengths ($L = R$, $L = 2R$ and $L = 4R$) under a uniform thermal loads ($\Delta T = 300K$). Clearly, increasing lengths of the cylinder increases the stiffness and causes hence less deflection.

2.6 CONCLUSION

In this chapter, thermal free vibration, buckling and post-buckling of FGM plates and shells using the modified FSDT theory are presented. Several numerical applications are provided in order to show the applicability and efficiency of the present (I-FSDT) finite element model to predict the thermal responses of FG-based structures. These tests may serve as references for structural engineers, designers and researchers of FGM-based components subjected to severe thermal gradients.

3 Buckling and post-buckling of FGM plates and shells

In this chapter, both buckling and post-buckling analysis of FGM plates and shells under thermal and mechanical loadings are presented. For the former, the extended Kirchhoff finite shell model is used to outline its capability for predicting thermal buckling and post-buckling responses of FGM plates and shells with good accuracy, whereas for the latter the double directors shell theory (DDST) is used to analyze the buckling and post-buckling behaviors of FGM structures subjected to different compressive loads. The development steps of these finite element models are as given previously in Chapter 1. Numerical results are carried out with a check out of the formulation in each case, then parametric studies and comparisons are conducted on the effects of FG power-index, boundary conditions, thermal/or mechanical loads and geometrical parameters for buckling and post-buckling of FGM plates and shells.

3.1 INTRODUCTION

Buckling and post-buckling characteristics are one of the major design criteria for plates/panels for their optimal usage. Therefore, it is important to study the buckling and post-buckling characteristics of such structures under thermal/or mechanical loading for accurate and reliable design. Thin plates and shells are often used in many engineering applications. For instance, curved panels have gained attention in rockets and hypersonic airplane where the thermal loading is a key factor in the determination of the dynamic response of such structure. As well as, thermal stresses developed due to elevated temperature will lead to buckling failure of these slender structural members. For that, FGMs have been regarded as a convenient strategy for handling thermal buckling and post-buckling problems due to their superior mechanical and thermal properties and their lightweight compared to other conventional engineering materials and structures. On the other hand, when a shell structure is subjected to a compressive load along its edge, the shell is deformed but remains flat as a long as the edge forces are adequately small and there is no initial imperfection in geometry of the structure. As the load increases, an unstable state is reached as a result of which the shell structure bends slightly. The minimum in-plane edge compressive load required just to initiate such instability is called critical buckling load parameter. The determination of such factor has also attracted the attention of many researchers which have developed many analytical and numerical methods to analyze as well as to predict the buckling behavior of plate and shell structures.

Eslami et al. present buckling analysis of FG plates using the CPT theory [108] or the high-order shear deformation plate (HSDT) theory [216]. As well as, Najafizadeh et al. presented mechanical buckling analysis of circular plates composed of FGM materials. The governing equations are established using the Love-Kirchhoff assumptions [176] and the HSDT theory [178]. Zhao et al. investigate the mechanical and thermal buckling behavior of FG plates [291] and FG cylindrical shell panels [326] using the FSDT theory in conjunction with the element-free kp-Ritz method. The effects of the volume fraction exponent, boundary conditions and some geometrical parameters on buckling behavior of such structures are also examined. Based on the HSDT theory and Levy-type solution, Bodaghi and Saidi have proposed an analytical approach for buckling analysis of thick FG rectangular plates [32]. The effects of power law index, plate thickness, aspect ratio, loading types and boundary conditions on the critical buckling load of the FG rectangular plate are studied and discussed. Furthermore, an efficient and simple refined theory is presented for buckling analysis of FG plates by Thai and Choi [258]. For shells analysis, Bagherizadeh et al. have investigated the

DOI: 10.1201/9781003483786-3

mechanical buckling of FGM cylindrical shell which are embedded in an outer elastic medium and subjected to combined axial and radial compressive loads [16]. Theoretical formulations are presented based on a HSDT theory with consideration of the transverse shear strains. The effects of shell geometry, the volume fraction exponent and the foundation parameters on the critical buckling load are also studied. Hajlaoui et al. [92] investigated the buckling behavior of FGM structures using an enhanced solid-shell finite element and transverse shear correction. Numerical results are presented for FGM plates and cylindrical shells in order to study their mechanical buckling behavior. However, the used 3D model requires generally an additional computational time compared to 2D elements which can be prohibitive for some applications. Herein, the developed 2D finite element models are applied for both thermal and mechanical buckling and post-buckling of FGM plates shells. In the first part of this chapter, thermal buckling and post-buckling of FGM structures using the extended Kirchhoff-Love model is performed in order to trace the potential of the present finite element frameworks for the analysis of thin and thick shells. In the second part, buckling and post-buckling of FGM structures using the double directors shell (DDST) model is presented and illustrated.

3.2 BUCKLING AND POST-BUCKLING OF FGM UNDER THERMAL LOADINGS

In this section, numerical analysis of buckling and post-buckling responses of FG square, skew plates and cylindrical panels under thermal loadings are performed using the proposed Kirchhoff shell formulation. The corresponding governing equations are as developed in Chapter 1. These equations are discretized using a four-nodes shell element with five degree of freedom ($DOFs$)/node ($SQAD45$). The critical buckling temperature is determined via the resolution of the buckling eigenvalue problem as mentioned in Chapter 2. Validation study is first presented for both isotropic plates and cylindrical panels to check the accuracy of the present model in the prediction of the critical buckling temperature and the post-buckling responses of these structures subjected to different forms of temperature distributions. Then, the obtained results of FGM studied structures are performed in order to outline the capability of the present model to draw the thermal buckling and post-buckling responses of such structures with good accuracy.

3.2.1 VALIDATION STUDY

3.2.1.1 Isotropic square plate

For the verification of the thermal buckling response obtained from the present model, an isotropic square plate is first considered. The plate is meshed using 32×32 of SQAD45 finite elements and its material properties are given as: Young's modulus $E = 1GPa$, Poisson coefficient $\nu = 0.3$ and thermal expansion coefficient $\alpha = 2 \times 10^{-6}/°C$. The geometric properties of this plate are: the aspect ratio $a/b = 1$ and the length-to-thickness ratio $a/h = 100$ as shown in Fig. 3.1. It should be mentioned that the abbreviation SQAD45 refers to the used Kirchhoff finite shell element. The plate is assumed to be under two types of thermal loadings: uniform temperature rise and linear temperature distribution through the thickness. Table 3.1 presents the solutions of the critical buckling temperature ΔT_{cr} (in K) of the isotropic plates under different temperature distributions and for various boundary conditions obtained from the present formulation, as well as those provided from analytical method [89] and finite element methods [259]. The definition of the boundary condition is given as follows:

$$Simplysupported(S): \begin{cases} u = w = \Theta_1 = 0 forside(AD)and(BC), \\ v = w = \Theta_2 = 0 forside(AB)and(DC), \end{cases}$$
$$Clamped(C): u = v = w = \Theta_1 = \Theta_2 = 0.$$

The obtained results reveal a good agreement and they are more closer to the analytical solution than the one derived from the finite element procedure which affirms the capability of the present

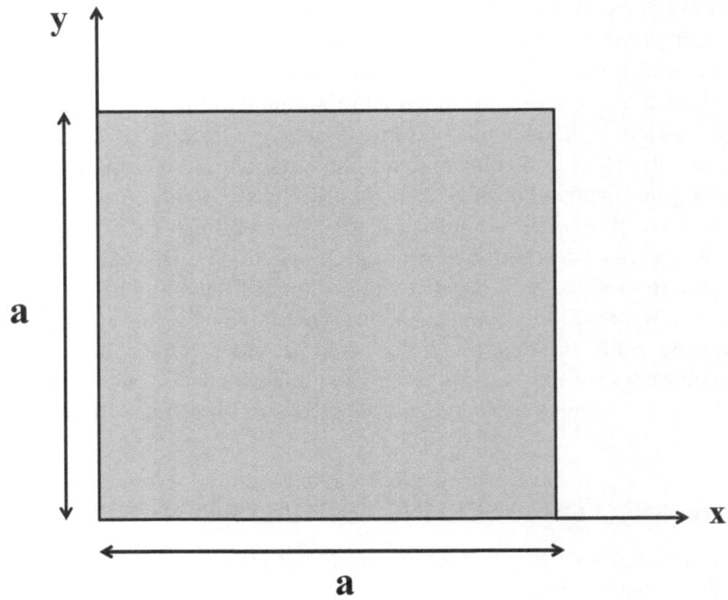

Figure 3.1 Schematic of the square plate.

Table 3.1
Comparison study of critical buckling temperature ΔT_{cr} (in K) for isotropic square plates under different temperature distributions and for clamped (CCCC) and simply supported (SSSS) boundary conditions.

BC	Results	Uniform temperature rise	Linear temperature rise
CCCC	Present	168.172	336.345
	[89]	168.71	337.42
	[259]	167.70	332.50
SSSS	Present	63.326	126.653
	[89]	63.27	126.54
	[259]	63.33	126.00

model in the prediction of thermal buckling behavior of square plates. In addition, the effect of the aspect ratio (a/b) on the critical buckling temperature of a simply supported isotropic plate under uniform temperature distribution is depicted in Fig. 3.2. The obtained results are presented in terms of non-dimensional critical buckling temperature $\Delta T_{cr}^{*} = \alpha \times \Delta T_{cr} \times 10^4$ and they are compared to those available in the literature [8]. As can be seen from Fig. 3.2, a good agreement between the present results and those given by [8] is revealed. From Fig. 3.2, it can be also remarked that the non-dimensional critical buckling temperature increases by the increase of the aspect ratio (a/b). In fact, as the aspect ratio (a/b) the width of the plate decreases and it becomes hence more sensitive to the temperature parameter.

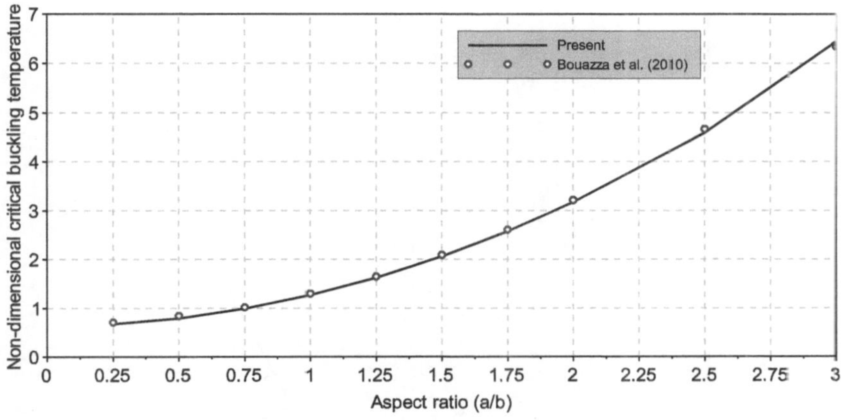

Figure 3.2 Non-dimensional critical buckling temperature ΔT^*_{cr} versus the aspect ratio a/b of the isotropic square plate under uniform temperature distribution with the geometrical ratio ($a/h = 100$).

3.2.1.2 Isotropic skew plate

The effectiveness of the present model and its ability in the prediction of thermal buckling behavior of isotropic skew plates subjected to uniform temperature rise is now examined. The geometrical characteristics of the skew plate are as depicted in Fig. 3.3 where $a/b = 1$ and $a/h = 1000$ as indicated in [198, 263]. Its material properties: the Young's modulus, the Poisson's ratio and the thermal expansion coefficient are equal to: $E = 1GPa$, $\nu = 0.3$ and $\alpha = 1 \times 10^{-6}/°C$, respectively. The definition of the boundary conditions for simply supported (S) or clamped (C) edges is given as mentioned earlier in section 5.1.1. The comparison study is carried out using the present *SQAD*45 finite element for various skew angles (α = 0°, 15°, 30° and 45°). The non-dimensional critical buckling temperatures ($\Delta T^*_{cr} = E\alpha b^2 h \Delta T_{cr}/\pi^2 D$ with $D = Eh^3/12(1-\nu^2)$) of the clamped isotropic skew plate are listed in Tab. 3.2, as well as the corresponding results given by [198] and [263]. The present results are very close to those obtained from the references [198, 263] which leads to validation of the present formulation. Therefore, the current model is convenient for thermal buckling analysis of thin skew plates.

Table 3.2
Comparison study of non-dimensional critical buckling temperature ΔT^*_{cr} for the isotropic skew plate under uniform temperature distribution.

Skew angle	Present	Ref	FSDT[263]	
	*SQAD*45	[198]	FI	SI
0°	3.724	3.71	3.714	3.710
15°	3.958	3.95	3.952	3.946
30°	4.796	4.80	4.815	4.795
45°	6.849	-	7.799	6.938

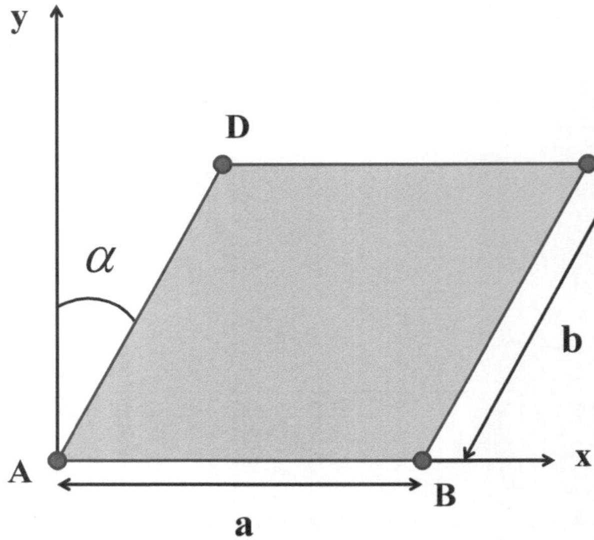

Figure 3.3 Schematic and geometrical characteristics of the skew plate.

3.2.1.3 Isotropic cylindrical panel

An isotropic cylindrical panel subjected to uniform temperature distribution is studied. The panel is meshed using 32×20 of $SQAD45$ finite elements. Its geometry is illustrated in Fig. 3.4 and the corresponding properties are given as: the thickness $h = 1mm$, the thickness ratio $A/h = 100$, the curvature ratio $R/A = 2$, the aspect ratio $A/L = 1$ and the angle $\theta = 15°$. The material properties are given as follows [28]: $E = 210GPa$, $\nu = 0.3$ and $\alpha = 12.6 \times 10^{-6}/°C$. The critical buckling temperature ΔT_{cr} (in K) of the isotropic cylindrical panel under uniform temperature distribution is computed for various thickness ratio (A/h) and for different boundary conditions and compared to those given by [28]. A sequence of letters containing "C" or "F", is used to denote fully clamped (C) or free edges (F). Three combinations of boundary conditions are considered: CCCC, CCFC and CFCF. It should be mentioned that the first and third letter indicate the boundary condition applied to curved edge while the latter letters refer to straight edge. From Fig. 3.5, it can be seen that the results obtained by the proposed model agree well with the solution of [28] which leads to verify the aptitude of the present formulation in the prediction of the thermal buckling behavior of isotropic cylindrical panels. As well as, Fig. 3.5 shows that the critical buckling temperature decreases as the thickness ratio A/h increases for all combinations of boundary conditions. In fact, more the cylindrical panel becomes thin, more the stiffness of the panel is decreased. On the other hand, the fully clamped (CCCC) or clamped in three edges and free in one edge (CCFC) cylindrical panels leads to higher values of critical buckling temperatures compared to the other boundary condition (CFCF). This means that the applied boundary conditions have a significant effect on the thermal buckling strength of such structures.

3.2.2 THERMAL BUCKLING AND POST-BUCKLING ANALYSIS OF FGM STRUCTURES

3.2.2.1 Thermal buckling responses of FGM plates and cylindrical panels

Here, the thermal buckling behavior of FGM structures is investigated. In fact, a series of analysis varying the power law index, the aspects ratios, the type of thermal loadings, boundary conditions and others geometrical parameters is carried out on FGM plates and cylindrical panels in order

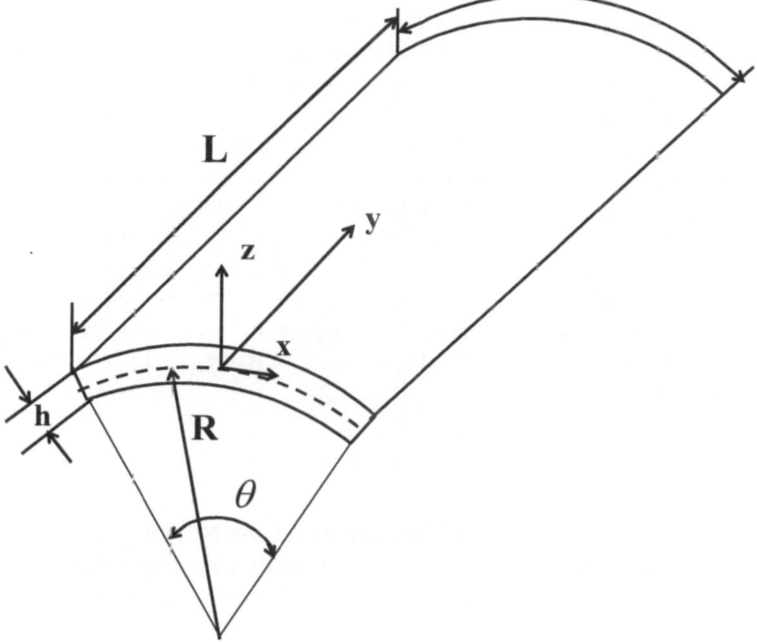

Figure 3.4 Geometry of the cylindrical panel.

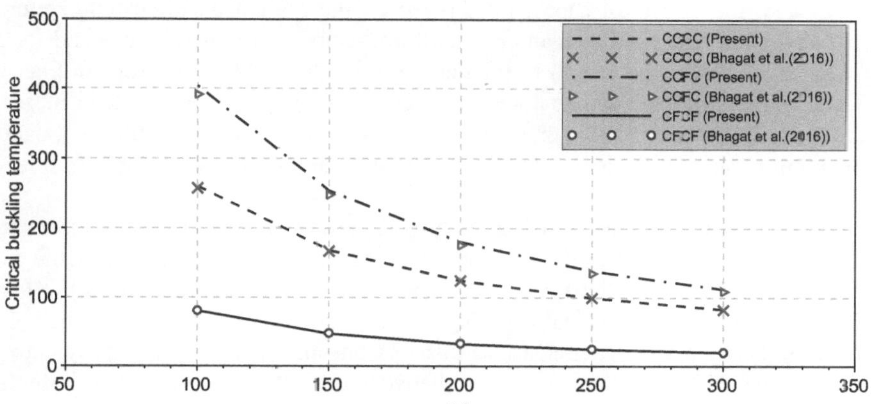

Figure 3.5 Critical buckling temperature ΔT_{cr} (in K) of the isotropic cylindrical panel under uniform temperature rise versus thickness ratio (A/h) for different boundary conditions (CCCC, CCFC, CFCF).

to show the effects of such parameters on the buckling behavior of these structures. For that, the thermal buckling of FGM square plate under uniform temperature distribution is first investigated. The plate is made of Aluminium (Al) and Alumina (Al_2O_3) phases and the adopted mesh involves

32×32 of *SQAD*45 finite elements. The material properties are listed in Tab. 3.4. The geometrical characteristics of the plate are the same as mentioned in [114]. Figure 3.6 depicts the obtained critical buckling temperatures ΔT_{cr} (in K) versus the power law index p compared to those given by [114] under uniform temperature distribution with fully clamped (CCCC) and fully simply supported (SSSS) boundary conditions. A good agreement between the results implies that the present model has the capacity to reproduce the thermal buckling responses of FGM plates. It should be mentioned that the little difference between the results in the case of simply supported boundary condition and for $p \geq 1$ can be explained by the effect of shear deformations which is neglected in our approach. For the purpose of the validation of the present formulation with analytical and finite element methods [110, 114], further numerical results of the critical buckling temperature ΔT_{cr} (in K) of the FGM (Al/Al_2O_3) plates with geometrical characteristics: $a/b = 1$ and $a/h = 100$, under uniform temperature distribution and for simply supported boundary condition are provided in Tab. 3.3. For the different power law indexes (p), the present results show an excellent agreement with those published in the literature [110, 114] which highlight again the effectiveness and reliability of the present model.

Table 3.3
Comparison study of critical buckling temperature ΔT_{cr} (in K) for FG simply supported (SSSS) square plates under uniform temperature distribution for different power law index value p.

Results	p				
	0	0.5	1	2	10
Present	17.11	10.12	9.15	9.11	8.62
[114]	17.50	9.9	8.0	7.10	-
[109]	17.09	-	7.94	7.26	7.46

On the other hand, the effects of aspect ratio (b/a) on the critical buckling temperature of a clamped (Al/Al_2O_3) plate under uniform and non-uniform temperature distributions and for different power law index p values with length-to-thickness ratio ($a/h = 100$) are shown in Figs. 3.7(a) and 3.7(b). Based on these plots, it can be noted that the critical temperature parameter increases as the aspect ratio (b/a) increases and this is for both types of temperature distributions. In addition, the thermal buckling parameter decreases as the power law index p increases which implies a rise in the stiffness of the FGM square plate.

In the second example, the thermal buckling analysis is performed for (Al/Al_2O_3) skew plate under uniform and non-uniform temperature distributions. The considered mesh is the same as mentioned earlier in section 5.1.2. The used geometrical properties are as indicated in [105]: $b/a = 1$, $b/h = 100$. Table 3.5 compares the critical buckling temperatures ΔT_{cr} (in K) of a clamped (CCCC) FGM skew plates under two sets of temperature distributions and with different power law index p. The comparison reveals an excellent agreement between the obtained results and those cited in [105] leading hence to validate again the ability of the present model to reproduce with accuracy the thermal buckling behavior of FGM skew plates. According to Tab. 3.5, the relative error between the present result and those given by [105] in the case of a skew angle ($\alpha = 0°$, $15°$ and $30°$) does not exceed 1.3% for uniform temperature rise, while it can reach 36% in the case of non-uniform temperature distribution. In fact, the solution of the critical temperature of plate under non-uniform temperature rise in [105] is achieved using the polynomial series method, however, in this study, the thermal buckling is computed using the Gauss integration method. So, the present method is more convenient for thermal buckling analysis of FGM structures under non-uniform temperature distribution. Furthermore, Tab. 3.6 illustrates the effect of boundary conditions and the power law index

Table 3.4
Material properties of Aluminium (Al), Alumina (Al_2O_3) and Zirconia (ZrO_2) phases [291].

Material	Properties			
	E (GPa)	ν	Conductivity (W/mK)	Thermal expansion (/°C)
Aluminum (Al)	70	0.3	204	23×10^{-6}
Alumina (Al_2O_3)	380	0.3	10.4	7.4×10^{-6}
Zirconia (ZrO_2)	151	0.3	2.09	10.0×10^{-6}

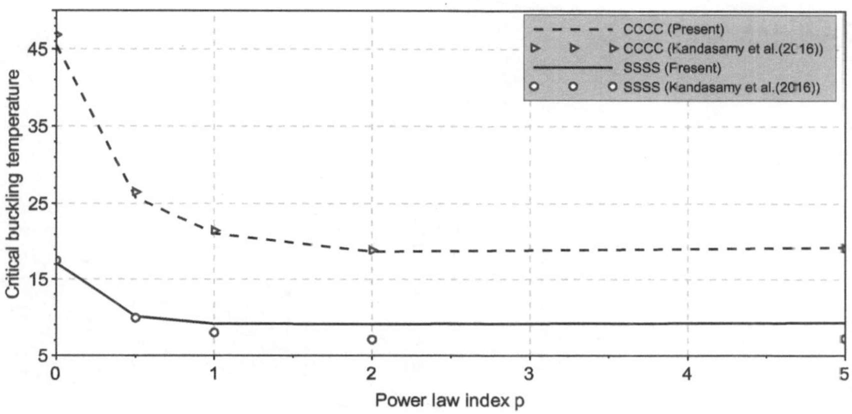

Figure 3.6 Critical buckling temperature ΔT_{cr} (in K) versus the power law index p of the rectangular plate under uniform temperature distribution for fully clamped (CCCC) and simply supported (SSSS) boundary conditions and with the geometrical ratios ($a/h = 100$, $a/b = 1$).

p on the critical buckling temperature ΔT_{cr} (in K) of the (Al/Al_2O_3) skew plate for uniform and non-uniform temperature rise. It can be seen that ΔT_{cr} increases as the power law index p decreases and this is for the two sets of temperature and for all the combination of boundary conditions. As well as, for (CCCC) and (SCSC) boundary conditions, the values of critical buckling temperature are much greater than those for the fully simply supported (SSSS) skew plates. Thus, more the plate is constrained, more its rigidity becomes higher which leads to high critical buckling temperature. In addition, the corresponding buckling mode shapes of the FGM skew plates are illustrated in Fig. 3.8.

The variation of buckling temperature versus the skew angle for the simply supported FGM skew plate under two studied thermal loads and for different power law index p is shown in Figs. 3.9(a) and 3.9(b), respectively. The results of Fig. 3.9 indicate that as the skew angle of the plate increases, the critical buckling temperature parameter increases. In fact, for a skew angle higher than 30°, the increase of the thermal buckling parameter is more pronounced. Thus, a high value of skew angles induces the concentration of stresses at the corner of the FGM plate where its rigidity becomes more sensitive to the change of temperature. This observation is similar to those for fully clamped FGM skew plate illustrated in Figs. 3.10(a) and 3.10(b). Moreover, when comparing the influence of thermal loading type on critical buckling behavior of FGM skew plate, it can be observed that the effect of non-uniform thermal load is more prominent than the uniform temperature rise. Further conclusions can be deduced from Figs. 3.11 and 3.12 which depict the effects of length-to-thickness

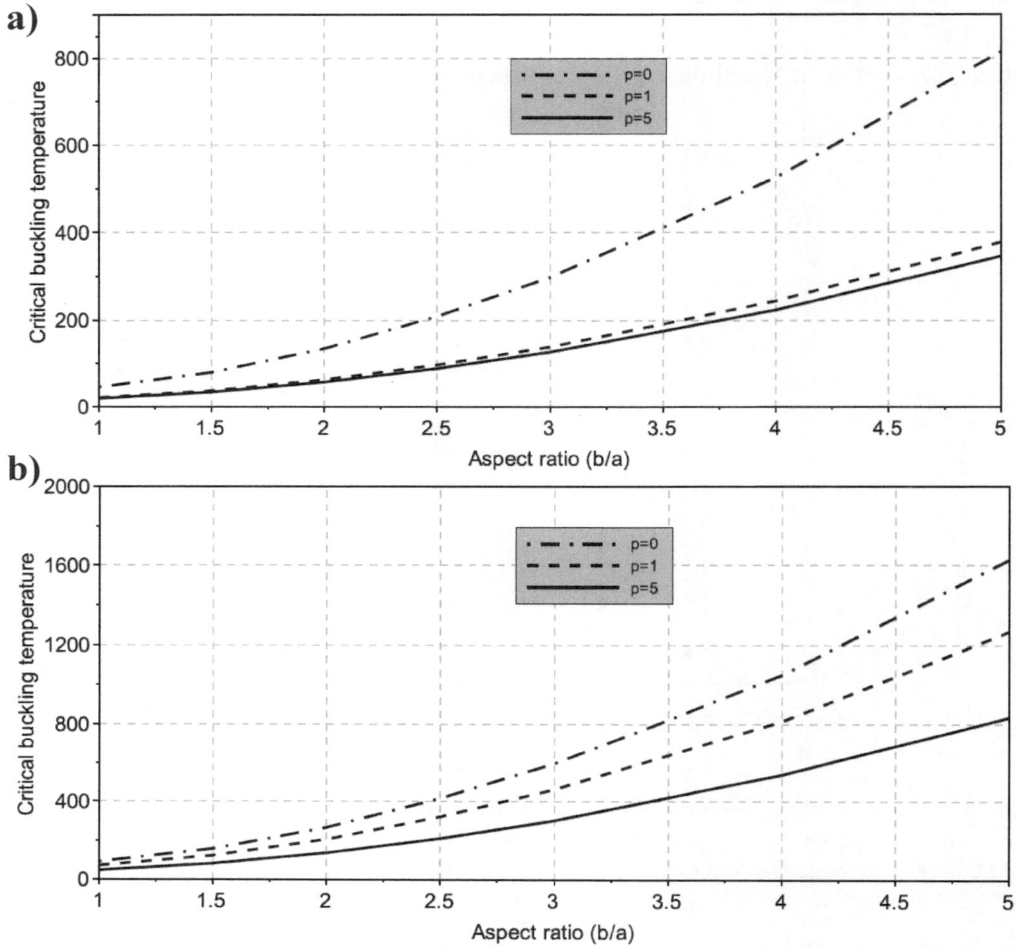

Figure 3.7 Critical buckling temperature ΔT_{cr} (in K) of the FG clamped square plate with: a) uniform temperature distribution and b) non-uniform for different power law index p versus the aspect ratio b/a with length-to-thickness ratio ($a/h = 100$).

ratio (a/h) and the aspect ratio (b/a) on thermal buckling behavior of clamped FGM skew plates under uniform temperature rise for different skew angles and with a power law index $p = 5$. Figure 3.11 shows that the increase of the length-to-thickness ratio (a/h) induces a decrease in the critical buckling temperature parameter and this is for all values of skew angles. In fact, the effect of the length-to-thickness ratio is more pronounced for FGM skew plate with a skew angle $\alpha = 60^\circ$ compared to the other values of skew angles. This reveals that the skew angle has a significant effect on thermal buckling behavior of FGM skew plates. Furthermore, it can be remarked that when the length-to-thickness ratio ($a/h \geq 50$) increases, the FGM skew plate behaves as a thin structure and hence becomes less sensitive to the change of temperature. On the other hand, Fig. 3.12 illustrates the effect of the aspect ratio (b/a) on the critical buckling temperature of the FGM skew plate. It can be seen that the critical buckling temperature parameter rises with the increase of the aspect ratio (b/a) for all values of skew angles of the clamped FGM skew plate. This observation reflects that the variation of this geometrical parameter affects the flexural rigidity of the plate and hence

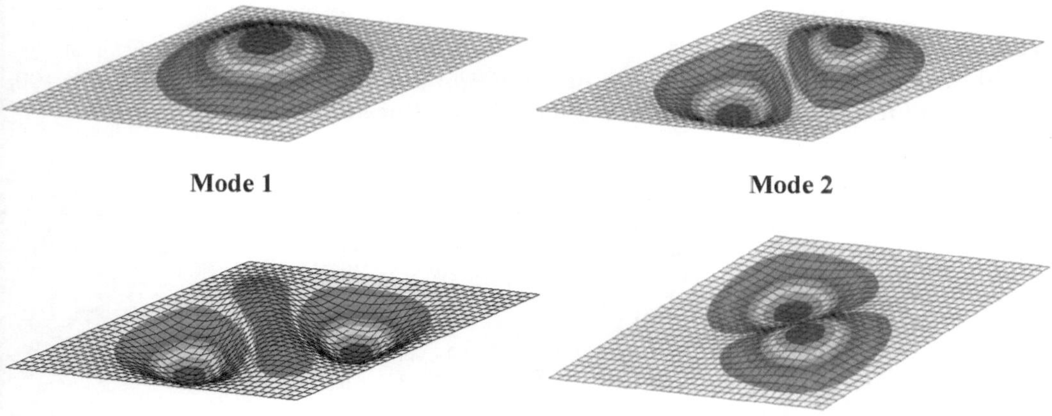

Figure 3.8 Buckling mode shapes of the FGM skew plates under uniform temperature rise with clamped edges.

Table 3.5
Critical buckling temperature rise ΔT_{cr} (in K) of clamped Al/Al_2O_3 skew plate.

Skew angle	Temperature distribution	Results	p: 0	0.5	1	2	5
0°	Uniform	*present*	45.452	25.766	21.123	18.731	19.324
		[105]	45.0471	25.6925	21.0670	18.6771	19.2688
	Non-uniform	*present*	90.904	89.862	70.497	54.941	46.442
		[105]	80.6943	72.0952	53.5903	40.14719	34.6020
15°	Uniform	*present*	48.307	27.386	22.453	19.912	20.541
		[105]	48.2540	27.3395	22.4175	19.87436	20.5040
	Non-uniform	*present*	96.615	95.514	74.937	58.405	49.369
		[105]	86.5081	77.8334	58.0947	43.6616	37.5974
30°	Uniform	*present*	58.535	33.192	27.222	24.148	24.909
		[105]	57.8237	32.7614	26.8633	23.8158	24.5703
	Non-uniform	*present*	117.07	115.768	90.858	70.835	59.870
		[105]	105.6474	96.7241	72.9233	55.2311	47.4583

the thermal buckling behavior of the FGM skew plate.

To illustrate the present approach for thermal buckling analysis of FGM shells, a clamped (Al/ZrO_2) cylindrical panel under thermal loadings is considered with the properties as listed in Tab. 3.4. The geometric properties of the FGM panel are the same as mentioned in [326]: $L = 0.2$, $R = 1$, $h = 0.002$. The considered mesh of the panel is 32×32 $SQAD45$ finite elements. Fig. 3.13 presents the buckling temperature rise ΔT_{cr} (in K) versus the power law index p of the clamped FGM cylindrical panel under uniform in comparison with [326]. It is observed that a good agreement has been achieved in this comparison study.

Figure 3.14 presents the buckling temperature rise ΔT_{cr} (in K) versus the power law index p of the Al/ZrO_2 cylindrical panel under uniform and non-uniform temperature distributions with the geometrical ratios ($A/L = 1, R/A = 2$). For both thermal loadings, the critical temperature decreases when the power law index p increases. This is due to the decreases in the stiffness of the panel

Table 3.6
Critical buckling temperature rise ΔT_{cr} (in K) of Al/Al_2O_3 skew plate ($b/a = 1$, $b/h = 100$, α=30°) for different boundary conditions and different power law index p.

p	Temperature distribution	CCCC	SCSC	SSSS
0	Uniform	58.535	61.925	44.261
	Non-uniform	117.07	123.851	88.548
0.5	Uniform	33.192	35.22	25.454
	Non-uniform	115.768	123.478	93.534
1	Uniform	27.222	29.031	21.526
	Non-uniform	90.858	97.548	93.533
5	Uniform	24.909	26.673	20.303
	Non-uniform	59.87	64.479	51.87
10	Uniform	25.592	27.273	20.226
	Non-uniform	55.303	59.204	45.817

with a rise in this exponent. In fact, the high values of the power law exponent correspond to high portion of metal (Al) in comparison with the ceramic part (ZrO_2). As well as, the panel under non-uniform temperature rise induces higher values of critical buckling temperature compared to panel under uniform temperature distribution. Therefore, the use of the numerical integration method and the dependency of the thermal load on the geometrical parameter can affected significantly the thermal buckling behavior of FGM cylindrical panels. Now, the effects of the thickness ratio (A/h) on the thermal buckling behavior of the clamped FGM cylindrical panel under the two sets of temperature distributions are examined in Figs. 3.15(a) and 3.15(b). It can be observed that the critical temperature ΔT_{cr} (in K) decreases as the thickness ratio (A/h) increases and this is for all thermal loading types. Indeed, the increase in the thickness ratio (A/h) induces a flat panel with low values of buckling strength due to the decrease in the bending stiffness of the FGM cylindrical panel. As well as, the buckling temperature decreases as the power law exponent p rises which leads to an increase in the stiffness of the panel. The two first mode shapes of the Al/ZrO_2 cylindrical panel are plotted in Fig. 3.16.

3.2.2.2 Thermal post-buckling responses of FGM plates and cylindrical panels

In order to validate the efficiency of the present formulation for thermal post-buckling analysis, two examples are considered for which the thermal post-buckling responses of isotropic square and skew plates are compared to the solutions available in the literature. First, an isotropic square plates subjected to uniform temperature rise is examined. The geometrical and material properties of the plate are taken as indicated in [231] where: $a/b = 1$, $a/h = 10$ are the geometrical ratios, $E = 1GPa$ is the Young's modulus, $\nu = 0.3$ is the Poisson's coefficient and $\alpha = 10^{-4}/°C$ is the thermal expansion coefficient. The plate is modeled using 32×32 of $SQAD45$ finite elements. The obtained results are presented in terms of dimensionless critical temperature $T = \Delta T/\Delta T_{cr}$ with $\Delta T_{cr} = 119.783$. As can be seen from Fig. 3.17, a good agreement between the obtained results and those provided by [231] is revealed which leads to verification of the effectiveness of the present model and draw hence its ability to predict the thermal post-buckling behavior of plates with good accuracy.

Secondly, the thermal post-buckling path of a clamped isotropic skew plate under uniform temperature distribution is analyzed. The material properties and the geometric parameters are: $E = 1GPa$, $\nu = 0.3$, $\alpha = 10^{-4}/°C$, $a/h = 100$ and the skew angle is taken equal to 45°. The isotropic

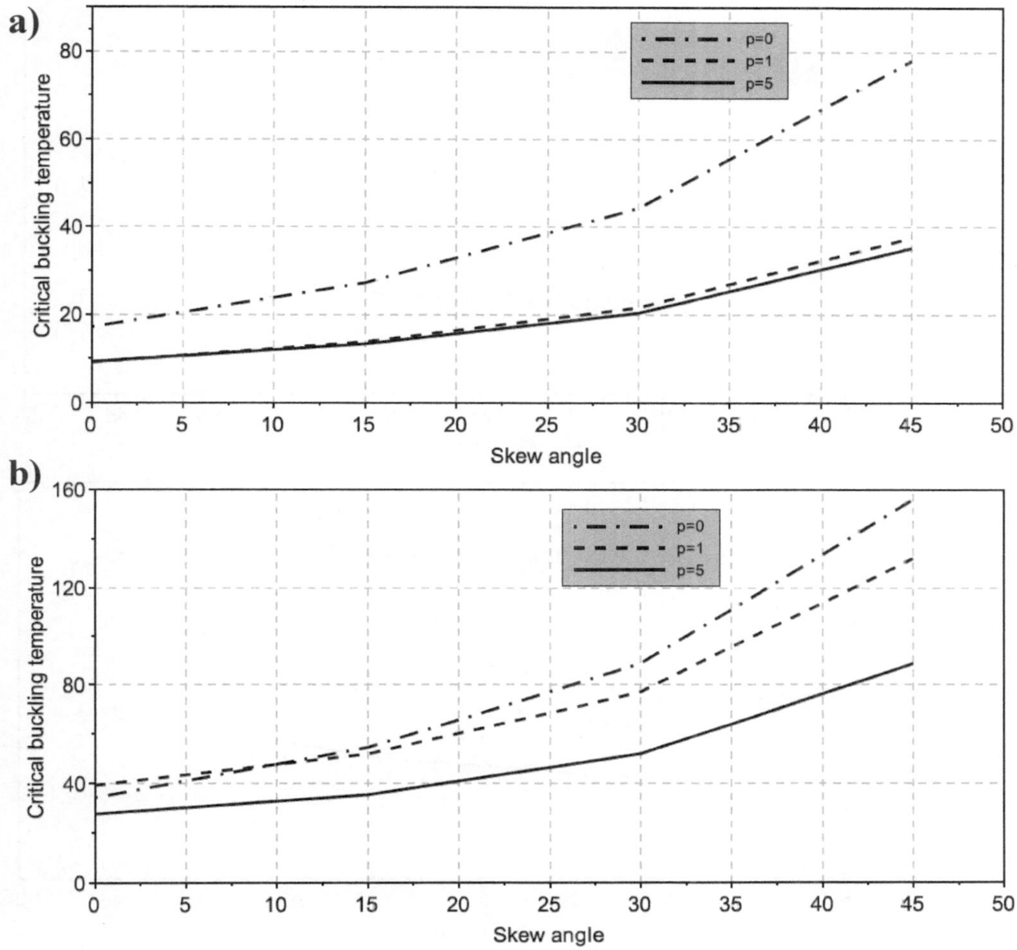

Figure 3.9 Critical buckling temperature ΔT_{cr} (in *K*) versus the skew angle (in *Deg*) of the SSSS FGM skew plate ($b/a = 1$, $b/h = 100$) subjected to: (a) uniform temperature rise and (b) non-uniform temperature rise for different power law index p.

skew plate is meshed using 32×32 of *SQAD*45 finite elements. Figure 3.18 depicts the obtained non-dimensional temperature T^* $(= E\alpha a^2 hT/\pi^2 D$, with $D = Eh^3/12(1-\nu^2))$ compared to the result given by [199]. An excellent agreement between the results is highlighted which demonstrates the aptitude of the present formulation to predict the thermal post-buckling behavior of skew plates with good accuracy. It should be mentioned that the difference between the results at $\frac{w}{h} = 0$, observed in Figs. 3.17 and 3.18, can be explained by the different theories and methods in the resolution procedure. In fact, in our paper a finite element method is elaborated in the sense of non-linear analysis with the introduction of large displacement and finite rotations while in [199] and [231] an analytical method is employed based on Van karman assumptions which include only the effect of membrane forces.

Now, the thermal post-buckling responses of FGM structures are examined for square plates, skew plates and cylindrical panels. Figure 3.19 depicts the thermal post-buckling curves of a clamped Al/Al_2O_3 FGM square plates under uniform temperature distribution for different power law co-

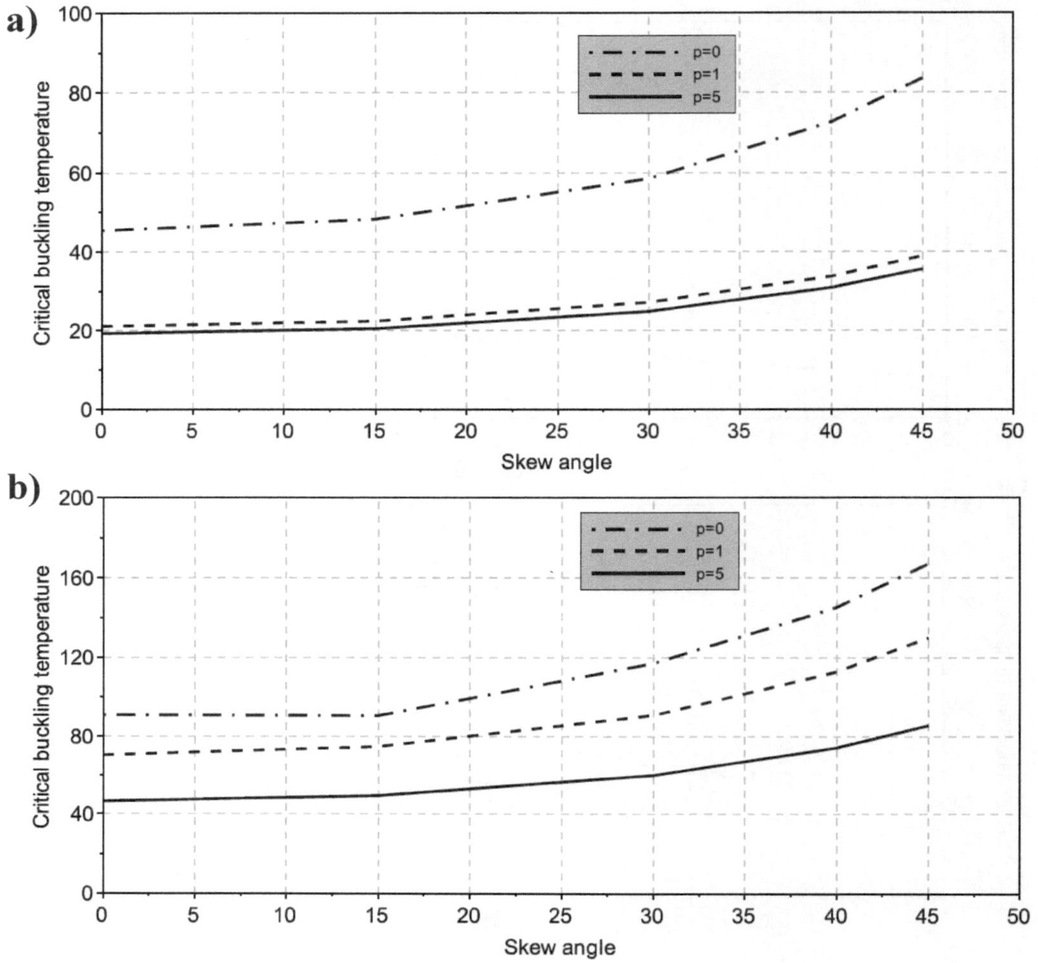

Figure 3.10 Critical buckling temperature ΔT_{cr} (in K) versus the skew angle (in *Deg*) of the CCCC FGM skew plate ($b/a = 1$, $b/h = 100$) subjected to: (a) uniform temperature rise and (b) non-uniform temperature rise for different power law index p.

efficient p. The geometrical properties and the used mesh are the same as mentioned earlier in section 5.2.1. In fact, the increases of the power index p produces less deflection and the thermal post-buckling path decreases. Figure 3.20 shows the thermal post-buckling behavior of a simply supported $Si_3N_4/SUS304$ FGM skew plates under non-uniform temperature distribution for different power law index p. The geometrical parameters are: $a/h = 100$, $a/b = 1$, Skew angle $\alpha = 45^\circ$ and the temperature dependent material properties are as listed in Tab. 3.7. It can be observed that as the power index p increases, the post-buckling response of the FG plate decreases. In fact, a high value of the power index p induces a rise in the volume fraction of ceramic portion which is able to withstand high temperature. Therefore, the FGM materials constitute a good way of heat resistant. As well as, the effect of skew angle on the thermal post-buckling behavior of a simply supported $Si_3N_4/SUS304$ skew plates under non-uniform temperature distribution is also examined as depicted in Fig. 3.21. It can be seen that with an increase in the skew angle, the post-buckling path is reduced for an applied thermal load which induces the high thermal load capacity of the FG

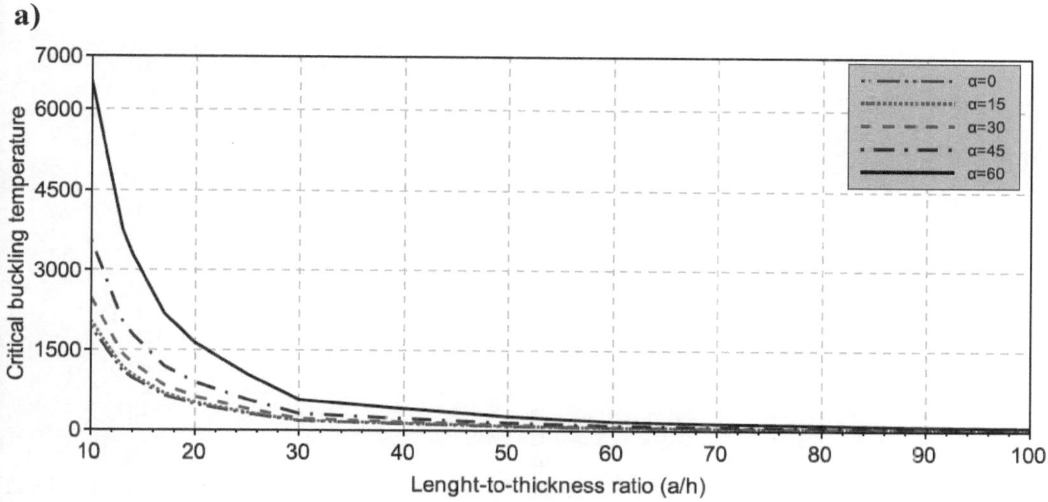

Figure 3.11 Critical buckling temperature ΔT_{cr} (in K) versus length-to-thickness ratios (a/h) of the clamped (CCCC) FGM skew plate under uniform temperature rise for various skew angles and with a power law index value $p = 5$.

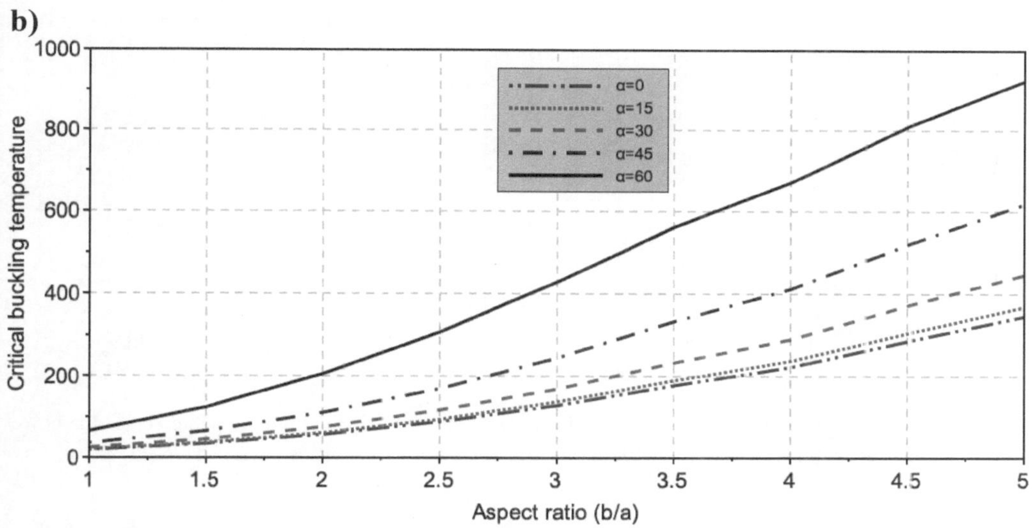

Figure 3.12 Critical buckling temperature ΔT_{cr} (in K) versus aspect ratio (b/a) of the clamped (CCCC) FGM skew plate under uniform temperature rise for various skew angles and with a power law index value $p = 5$.

skew plate with the rise of the skew angle. As a result, designer can adjust the thermal capacity of such structures with the variation of the skew angle geometrical parameter.

Figure 3.22 illustrates the post-buckling path of Al/ZrO_2 cylindrical panels under uniform temperature load for different power law index p and with geometrical properties ($L = 0.2$, $R = 1$, $h = 0.002$). As the temperature load increases, the non-dimensional deflection increases. In addition, the thermal post-buckling response increases as the power index p increases. This can be explained by the flexural rigidity of the FGM panel which is high with a fully ceramic phase ($p = 0$)

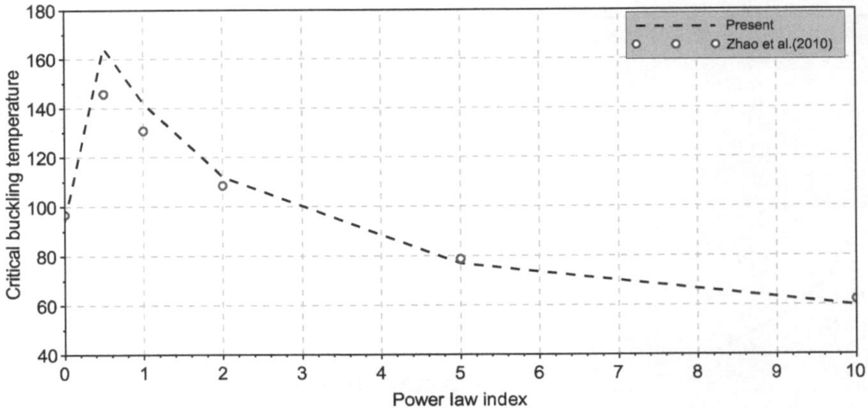

Figure 3.13 Critical buckling temperature rise for an Al/ZrO_2 panel with two sides simply-supported and two sides clamped (CSCS) under non-linear temperature distribution with the geometrical ratios ($L = 0.2$, $R = 1$, $h = 0.002$).

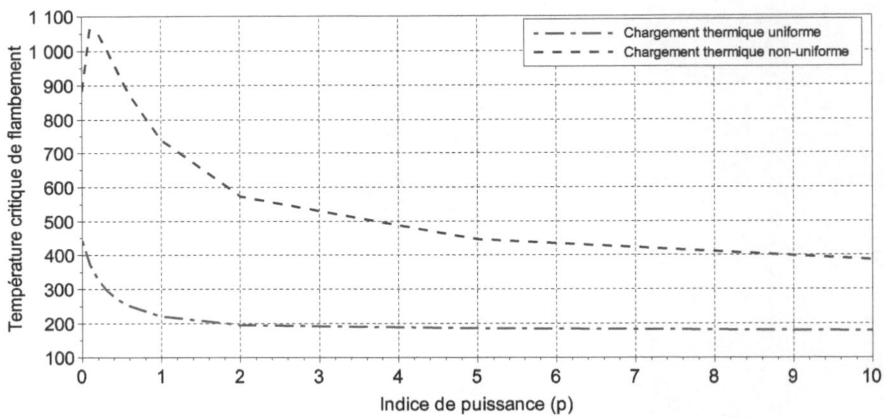

Figure 3.14 Critical buckling temperature ΔT_{cr} (in K) of a clamped FGM cylindrical panel versus the power law index p under different temperature distributions with the geometrical ratios ($A/L = 1$, $R/A = 2$).

compared to the metallic part and can hence reaching its maximum thermal load capacity. In Fig. 3.23 the thermal post-buckling curves evaluated for the clamped FGM cylindrical panel ($p = 0.5$) under uniform and non-uniform temperature distributions are highlighted. The uniform temperature distribution leads to more deflection compared to the non-uniform temperature distribution which induces the significant effect of temperature distribution profile on thermal post-buckling behavior of FGM structures. For the sake of completeness, it should be noted that the non-dimensional deflections in the case of large displacement can reach to ($w/h = 6$) as shown in Figs. 3.19, 3.22 and 3.23, whereas, in the case of small displacements the deflection does not exceed ($w/h = 1.5$) due to the inclusion of Von Karman assumptions.

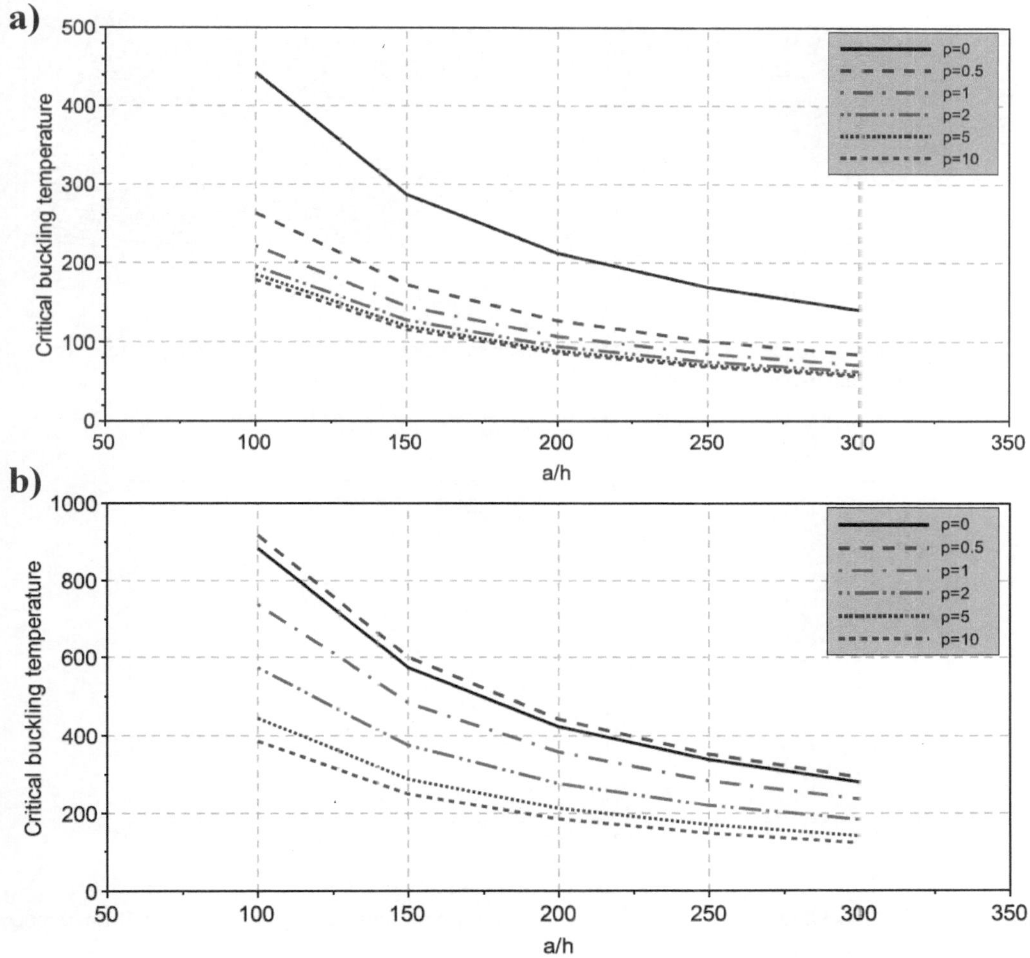

Figure 3.15 Critical buckling temperature ΔT_{cr} (in K) versus the thickness ratio (A/h) of a clamped FGM cylindrical panel subjected to: (a) uniform temperature rise and (b) non-uniform temperature rise for different power law index p and with the geometrical ratios $(A/L = 1, R/A = 2)$ versus the thickness ratio (A/h).

3.3 BUCKLING OF FGM PLATES AND SHELLS UNDER MECHANICAL LOADINGS

In this section, buckling and post-buckling responses of FGM plates and shells subjected to in-plane compressive loads are performed using the double directors shell model (DDST). Different numerical tests are conducted to show the ability of these finite element models for predicting the mechanical behaviors of such structural components with accuracy.

3.3.1 NUMERICAL RESOLUTION

The buckling eigenvalue problem is established via the construction of material and geometrical tangent parts (as mentioned in Eqs. (1.114) and (1.117), section.1.5.4, Chapter 1) which reflect the linear and non-linear contributions of the stiffness matrix of the shell structure in order to estimate the buckling loads. As a result, the buckling eigenvalue problem can be written as follows:

$$(k_m + \lambda_{cr} k_g) U_n = 0. \tag{3.1}$$

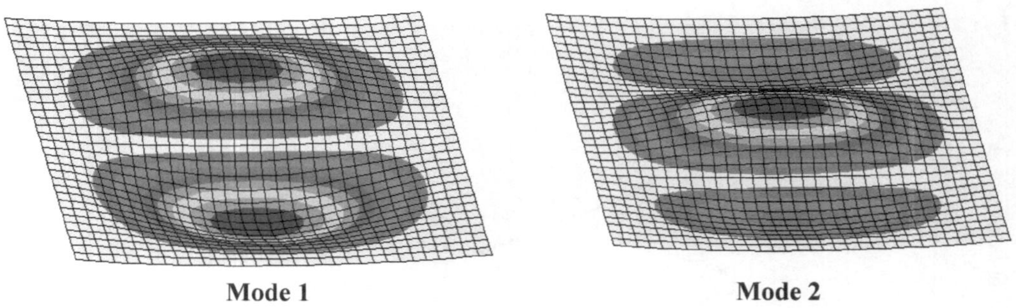

Figure 3.16 Buckling mode shapes of the FGM cylindrical panel under non-uniform temperature rise with two edges clamped and two edges simply supported.

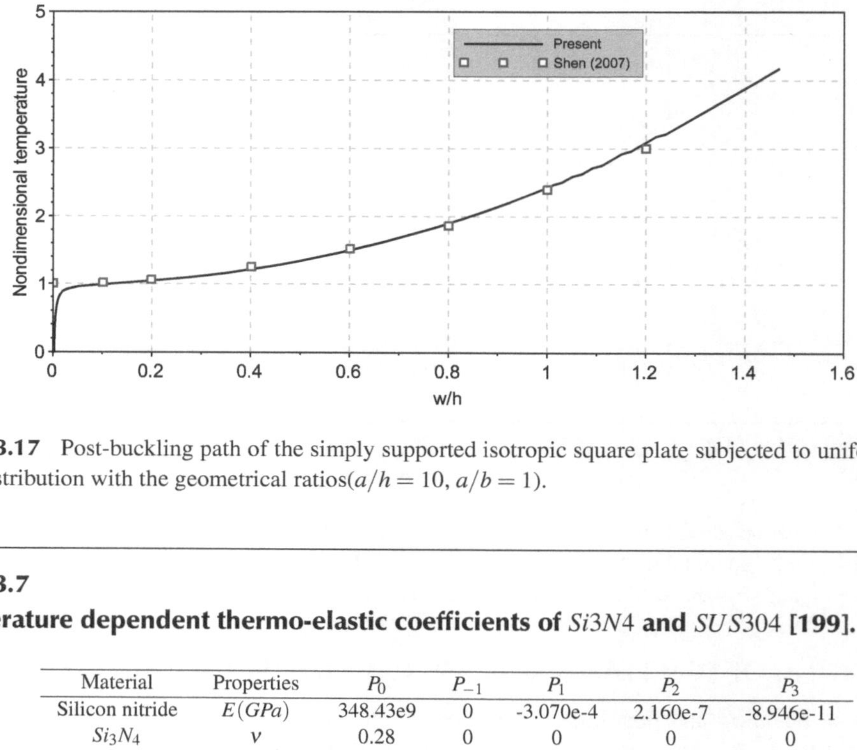

Figure 3.17 Post-buckling path of the simply supported isotropic square plate subjected to uniform temperature distribution with the geometrical ratios($a/h = 10$, $a/b = 1$).

Table 3.7
Temperature dependent thermo-elastic coefficients of $Si3N4$ and $SUS304$ [199].

Material	Properties	P_0	P_{-1}	P_1	P_2	P_3
Silicon nitride	$E(GPa)$	348.43e9	0	-3.070e-4	2.160e-7	-8.946e-11
Si_3N_4	ν	0.28	0	0	0	0
	$\alpha(K^{-1})$	5.8723e-6	0	9.095e-4	0	0
	$K(W/mK)$	9.19	0	0	0	0
	$\rho(Kg/m^3)$	2370	0	0	0	0
Stainless steel	$E(GPa)$	201.04e9	0	3.079e-4	-6.534e-7	0
$SUS304$	ν	0.28	0	0	0	0
	$\alpha(K^{-1})$	12.330e-6	0	8.086e-4	0	0
	$K(W/mK)$	12.04	0	0	0	0
	$\rho(Kg/m^3)$	8166	0	0	0	0

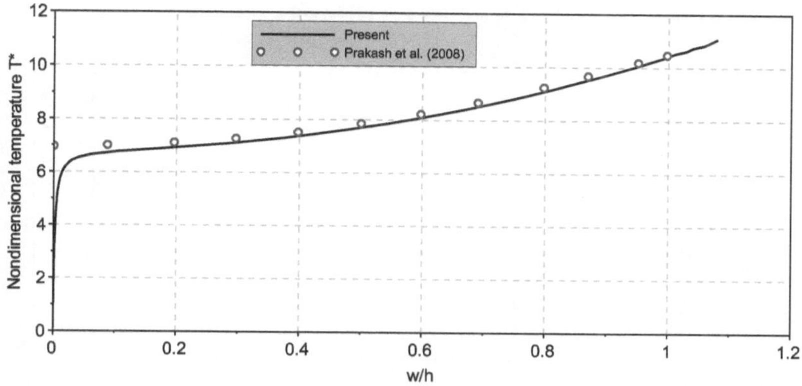

Figure 3.18 Post-buckling path of the clamped isotropic skew plate subjected to uniform temperature distribution with the geometrical ratios($a/h = 100$, $a/b = 1$, $\alpha = 45°$).

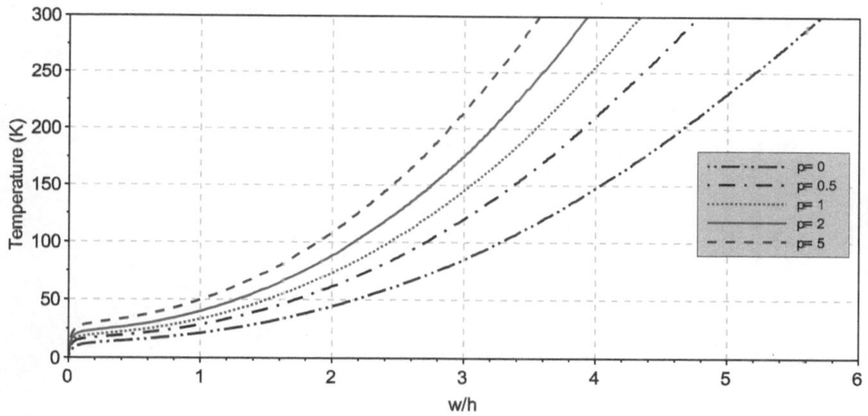

Figure 3.19 Thermal post-buckling path of a clamped Al/Al_2O_3 FGM square plates under uniform temperature distribution for different power law index p and with the geometrical parameters ($a/a = 100$, $a/b = 1$).

where k_m and k_g represent the linear stiffness matrix and the geometric stiffness matrix, respectively. λ_{cr} is the buckling load parameter and denotes the proportional increase in the load needed to reach neutral equilibrium. By applying to the structure a reference level of loading, N_{ref}, and carrying out a standard linear static analysis, the membrane stresses in the elements will be obtained. Hence, the buckling load can be written as:

$$N_{cr} = \lambda_{cr} N_{ref}. \tag{3.2}$$

3.3.2 BUCKLING OF FGM PLATES UNDER IN-PLANE COMPRESSIVE LOADS

3.3.2.1 Comparison and convergence study of isotropic plates

The first comparison is carried out for a simply supported isotropic square plate which is subjected to uniaxial compressive pressure. The geometrical characteristics of the studied plate are as follows [291, 135]: the length a, the width b and the thickness h where $a = b = 10$ in and $h = 0.2$ in. For

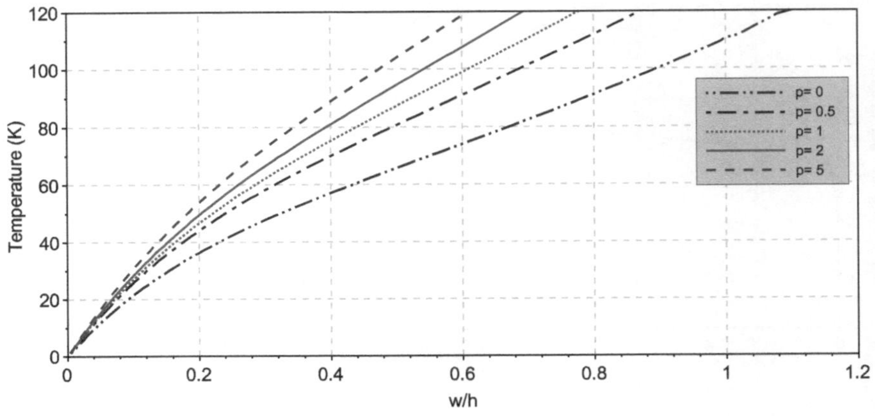

Figure 3.20 Thermal post-buckling path of a simply supported $Si_3N_4/SUS304$ FGM skew plates under non-uniform temperature distribution for different power law index p and with the geometrical parameters ($a/h = 100$, $a/b = 1$, Skew angle $\alpha = 45°$).

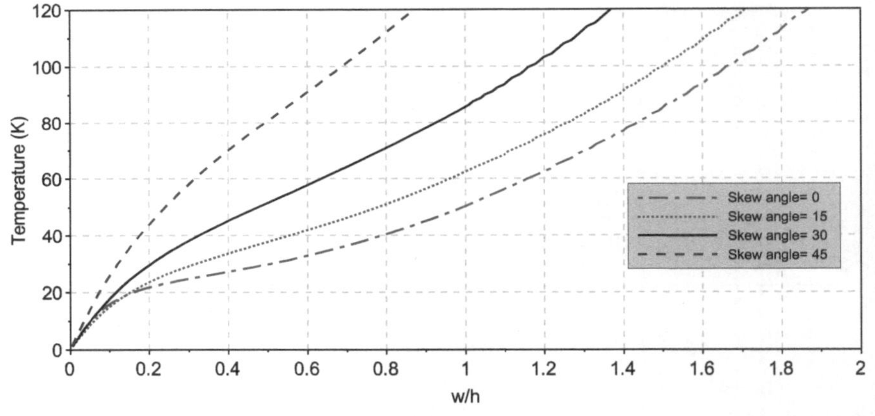

Figure 3.21 Thermal post-buckling path of a simply supported $Si_3N_4/SUS304$ FGM skew plates under non-uniform temperature distribution for different skew angles and with the geometrical parameters ($a/h = 100$, $a/b = 1$).

its material properties, the Young modulus and the Poisson ratio are equal to: $E = 3 \times 10^6 psi$ and $\nu = 0.3$, respectively. A convergence study in terms of number of used SHO4 elements is performed to verify the accuracy and efficiency of the present finite element shell model. It should be noted that the abbreviation SHO4 is used to refer the present double directors finite shell element. The obtained results are presented in terms of non-dimensional critical buckling load parameter $\overline{N}_{cr} = N_{cr}a^2/D_0$ with $D_0 = Eh^3/12(1-\nu^2)$ and they are compared to those available in the literature [291, 135, 262]. Table 3.8 lists the four buckling modes of the isotropic plate meshed using 12×12 to 18×18 SHO4 finite elements. It can be seen that the non-dimensional critical buckling load for a meshing of 18×18 SHO4 finite elements converges well with the solutions obtained either by the FSDT theory [291, 135] or the analytical method [262] which demonstrate the ability of the present model to reproduce with good accuracy the critical buckling load parameter $\overline{N}_{cr}$ of isotropic plates. It

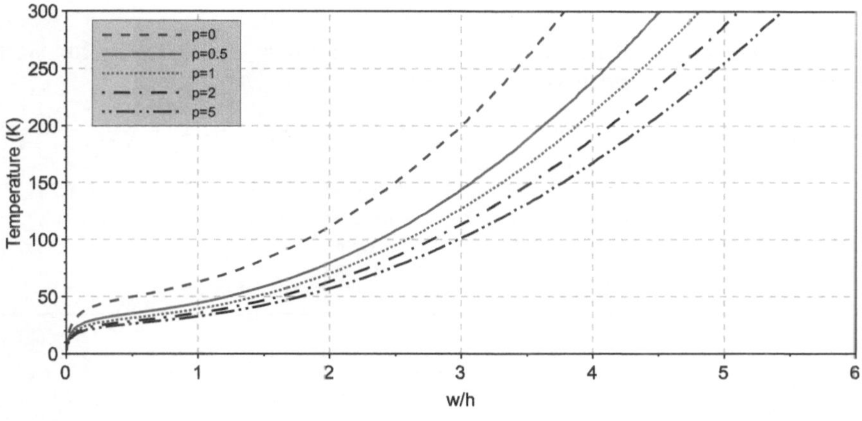

Figure 3.22 Thermal post-buckling response for an Al/ZrO_2 clamped (CCCC) panel under uniform temperature distribution with the geometrical parameters ($L = 0.2, R = 1, h = 0.002$).

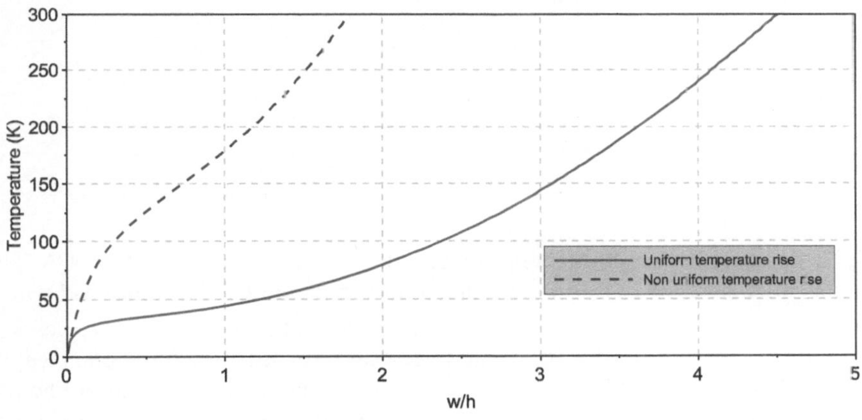

Figure 3.23 Thermal post-buckling paths for an Al/ZrO_2 clamped (CCCC) panel under uniform and non-uniform temperature rise through thickness with the geometrical ratios ($L = 0.2, R = 1, h = 0.002$).

is also worth to mention that the current results sticks well with the analytical solutions given by Timenshenko and Gere [262] more than those of Lei et al. [135] and Zhao et al. [291] which indicate the high accuracy of the present model in the restitution of numerical solutions for buckling analysis. So, to keep a general formulation, 18×18 SHO4 finite elements is adopted for meshing the plate geometry in the below numerical tests.

3.3.2.2 Mechanical buckling behavior of FGM plates

The second example consists of a functionally graded simply supported (SSSS) square plate which is made of aluminium (Al) and zirconia (ZrO_2) phases. The geometrical properties of the FGM plate are as indicated in [291] where the values of length and thickness sides are: $a = b = 0.2m$ and $h = 0.01m$, respectively. The material properties of each phase are: The Young modulus of metal phase (Al) with $E_m = 70 \times 10^9 N/m^2$, the Young modulus of ceramic phase (ZrO_2) with

Table 3.8
Convergence and comparison study of buckling load factor $\overline{N}_{cr} = N_{cr}a^2/D_0$ for isotropic square plate under uniaxial compressive pressure

Mode	Present-SHO4				FSDT.[291]	FSDT.[135]	Analytical.[262]
	12 × 12	14 × 14	16 × 16	18 × 18			
1	39.7051	39.6323	39.5813	39.5449	39.2040	39.1744	39.4761
2	62.7026	62.4078	62.2148	62.0838	60.3722	60.4874	61.6814
3	114.3688	113.0438	112.1957	111.6133	103.963	105.391	109.656
4	161.8817	160.7205	159.9707	159.4502	151.987	152.501	157.904

Table 3.9
Critical buckling load parameter $\overline{N}_{cr}^{FGM}$ of simply supported (SSSS) square Al/ZrO_2 plate subjected to uniaxial compression load in x direction for different values of power index p $(p=0,0.2,0.5)$.

Mode	power-law index p					
	0		0.2		0.5	
	Present	Ref.[291]	Present	Ref.[291]	Present	Ref.[291]
1	81.743	82.827	72.477	73.321	64.079	64.814
2	127.229	125.989	112.881	111.579	99.988	98.652
3	223.420	213.385	198.400	189.108	175.599	167.217
4	316.980	309.849	281.515	274.502	249.278	242.494

$E_c = 151 \times 10^9 N/m^2$ and the two Poisson ratios $\nu_m = \nu_c = 0.3$. The whole of the FGM plate is subjected to uniaxial compressive load in the x direction and the critical buckling load parameter $\overline{N}_{cr}^{FGM} = N_{cr}a^2/D_0$ for the first four modes varying the power index p is given in Tables 3.9 and 3.10. As can be seen, Tables 3.9 and 3.10 reveal a good agreement between the obtained results and those provided by Zhao et al. [291] which leads to the validation of the ability of the present model in the prediction of the mechanical buckling behavior of FGM plates with a good accuracy. In fact, the relative error between the obtained results and those of Zhao et al. [291] does not exceed generally 2% for the two first buckling modes while it can reach 6% for the two other modes and this is can be explained by the FSDT theory used in [291] which introduces a shear correction factors. So, the present model is more general and convenient for buckling analysis of FGM plates notably for high buckling modes. As well as, the mode shapes corresponding to first four buckling mode are illustrated in Fig. 3.24.

Then, the effects of length-to-thickness ratio a/h, aspect ratio b/a, boundary conditions, types of loadings and variations of the power law index p on non-dimensional critical buckling load $\overline{N}_{cr}^{FGM} = N_{cr}a^2/E_m h^3$ are examined. In fact, Fig. 3.25 illustrates the variation of non-dimensional critical buckling load versus the length-to-thickness ratio a/h of a simply supported (SSSS) square Al/ZrO_2 plate under different types of loadings. It can be seen that the non-dimensional critical buckling load increases by the increase of length-to-thickness ratio a/h for all types of loadings. As well as, this effect is more pronounced for thick FGM plates when $a/h \prec 10$ and such behavior can be explained by the influence of the transverse shear deformations in the FGM plate. On the other hand, for a fixed length-to-thickness ratio a/h, the non-dimensional critical buckling load increases as the power law index p decreases and this is for all types of loadings which implies that

Table 3.10

Critical buckling load parameter $\overline{N}_{cr}^{FGM}$ of simply supported (SSSS) square Al/ZrO_2 plate subjected to uniaxial compression load in x direction for different values of power index p ($p = 1, 2, 5$).

Mode	power-law index p					
	1		2		5	
	Present	Ref.[291]	Present	Ref.[291]	Present	Ref.[291]
1	57.233	57.901	52.333	53.022	48.400	49.173
2	89.287	83.105	81.605	80.606	75.254	74.670
3	156.802	149.201	142.986	136.214	131.524	126.061
4	222.744	215.495	203.196	192.971	186.818	182.606

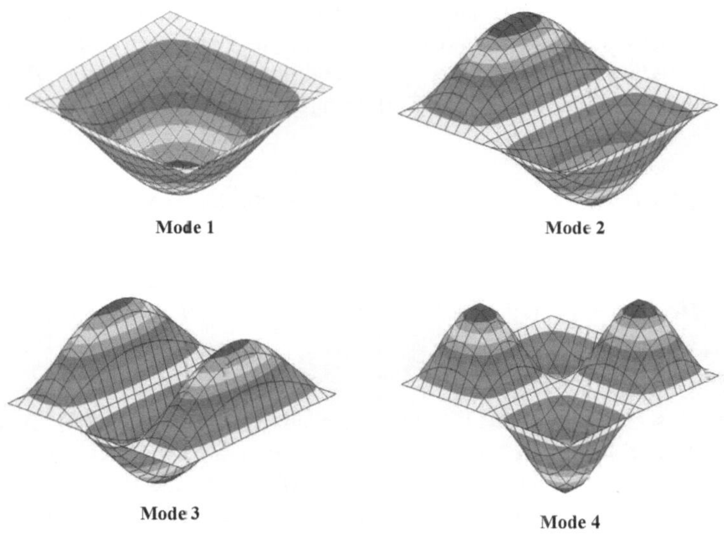

Figure 3.24 First four buckling mode shapes of the simply supported (SSSS) square Al/ZrO_2 plate.

for a fully-ceramic plate ($p = 0$), high values of critical buckling load are induced due to the high stiffness of the ceramic part compared to the metallic part. The effects of aspect ratio (b/a) on non-dimensional critical buckling load of simply supported Al/ZrO_2 plate under uniaxial compression in x direction, uniaxial compression in y direction and biaxial compression loads are shown in Figs. 3.25(a), 3.25(b) and 3.25(c), respectively. The length-to-thickness ratio (a/h) is assumed to be 100. In the case of uniaxial compression load in x direction, as depicted in Fig. 3.25(a), the non-dimensional critical buckling load decreases by the increases of the aspect ratio b/a except for the values of aspect ratios between 0.1 and 0.5 where this effect is inversed. In fact, for ($0.1 \prec b/a \prec 0.5$), the total area of the plate is so small which manifests by an increase in the critical buckling load while for the values of b/a higher than 0.5, the plate becomes more larger and hence less sensitive to the change of this parameter. For uniaxial compression load in y direction and biaxial compression load, the increases of the aspect ratio (b/a) between 0.1 and 0.5 observed in the case of uniaxial compression load in x direction is neglected for these cases. This is due to that for ($b/a \prec 0.5$), the width side is sufficiently large to support the compression load while the length side which is so small in the case of uniaxial compression load in x direction becomes more sensitive to the variation of aspect ratios in the range [0.1−0.5] which manifest by an increase in the critical buckling load. Otherwise, the variation of the aspect ratio (b/a) induces generally a decrease in the critical buckling load.

Further conclusions can be deduced from Figs. 3.27(a), 3.27(b) and 3.27(c) which demonstrate the effects of boundary conditions on non-dimensional critical buckling load of the Al/ZrO_2 plate under uniaxial compression in x direction, biaxial compression and biaxial compression and tension loads respectively. For boundary conditions of the four edges of the FGM plate, a sequence of letters containing "S", "C" or "F", is used to denote simply supported (S), fully clamped (C) or free (F), respectively. It can be remarked that a fully clamped (CCCC) FGM plate has the highest value of critical buckling load while the simply supported in two edges and free in the other edges (SFSF) FGM plate has the lowest one. The buckling curves corresponding to other boundary conditions are located between (CCCC) and (SFSF) curves. Since the constraint of clamped edge is stronger than simply supported or free edges, the values of critical buckling load for a fully clamped FGM plates are higher than the other imposed constraints and this is due to the high value of stiffness of the FGM plate corresponding to fully clamped boundary condition. Moreover, the non-dimensional critical buckling load of plate under uniaxial compression in x direction is greater than that under biaxial compression and less than that under biaxial compression and tension as shown in Fig. 3.28 which implies that the type of applied load can affect significantly the mechanical buckling behavior of such structures.

3.3.3 BUCKLING OF FGM CYLINDRICAL PANELS UNDER IN-PLANE COMPRESSIVE LOADS

Now, the mechanical buckling behavior of cylindrical panels is performed. For that, we start by a comparison study in the isotropic case, then the analysis is extended to FGM cylindrical panels in order to show the effectiveness and the usefulness of the (DDST) model.

3.3.3.1 Isotropic cylindrical panels

A simply supported isotropic cylindrical panel is considered where its geometrical and material properties are as follows: the length side $L = 10in$, the span angle $\theta_0 = 0.2in$, the radius $R = 50in$, the thickness $h = 0.1in$, the Young modulus $E = 3 \times 10^6 psi$ and the Poisson ratio $\nu = 0.3$. The panel is subjected to uniform compression load along its curved sides (AB) and (DC) and its geometry is illustrated in Fig. 3.29. The panel is meshed using 32×20 SHO4 finite elements. The non-dimensional critical buckling parameter $\overline{N}_{cr} = N_{cr}R/Eh^2$ is computed for the first four buckling modes and compared with those available in the literature using the FSDT theory [326] and the analytical method [262]. The obtained results are reported in Tab. 3.11 which reveals a good agreement between the results. It can be also seen, that the value of the current critical buckling parameter for the first buckling mode is more closer to the analytical solution than the one derived from the FSDT theory [326] which affirms the efficiency of the present model in the prediction of mechanical buckling behavior of cylindrical panels.

3.3.3.2 FGM cylindrical panels

In this section, the mechanical buckling analysis of FGM cylindrical panels is also provided. In fact, a simply supported (SSSS) cylindrical panel made of Aluminium (Al) and Zirconia (ZrO_2) constituents is considered with a length side $L = 0.1m$, a span angle $\theta_0 = 0.2rad$, a radius $R = 0.5m$ and a thickness $h = 0.001m$. The material properties of the cylindrical panel are the same as mentioned earlier in section 3.3.2.2. The obtained results for non-dimensional critical buckling load parameter $\overline{N}_{cr}^{FGM} = N_{cr}R/E_m h^2$ for the first four buckling modes varying the power-law index p are given in Tabs. 3.12 and 3.13, respectively. It can be observed that the present results are in good

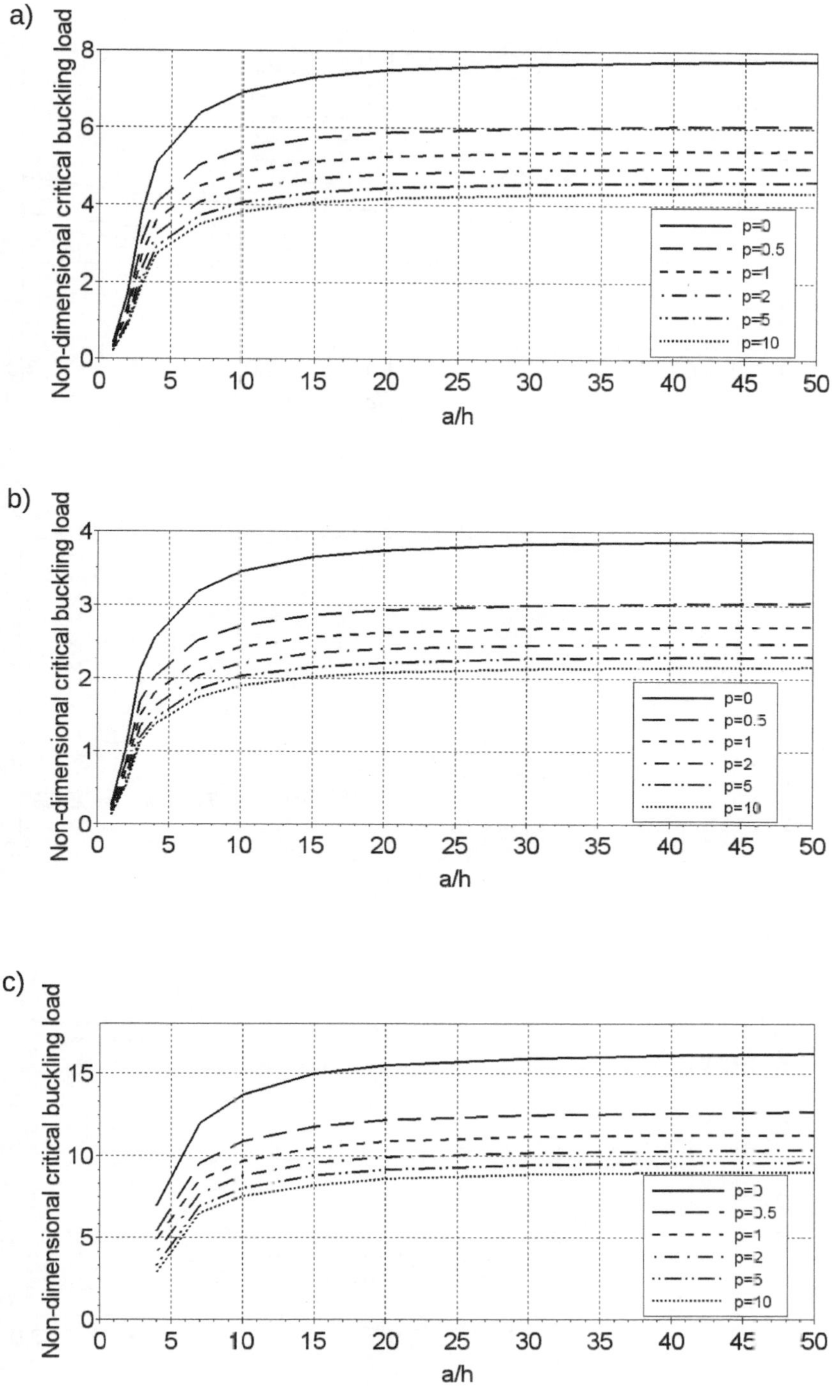

Figure 3.25 The effects of length-to-thickness ratio a/h on the non-dimensional critical buckling load of a simply supported (SSSS) square Al/ZrO_2 plate under (a) uniaxial compression in x direction (b) biaxial compression and (c) biaxial compression and tension loads.

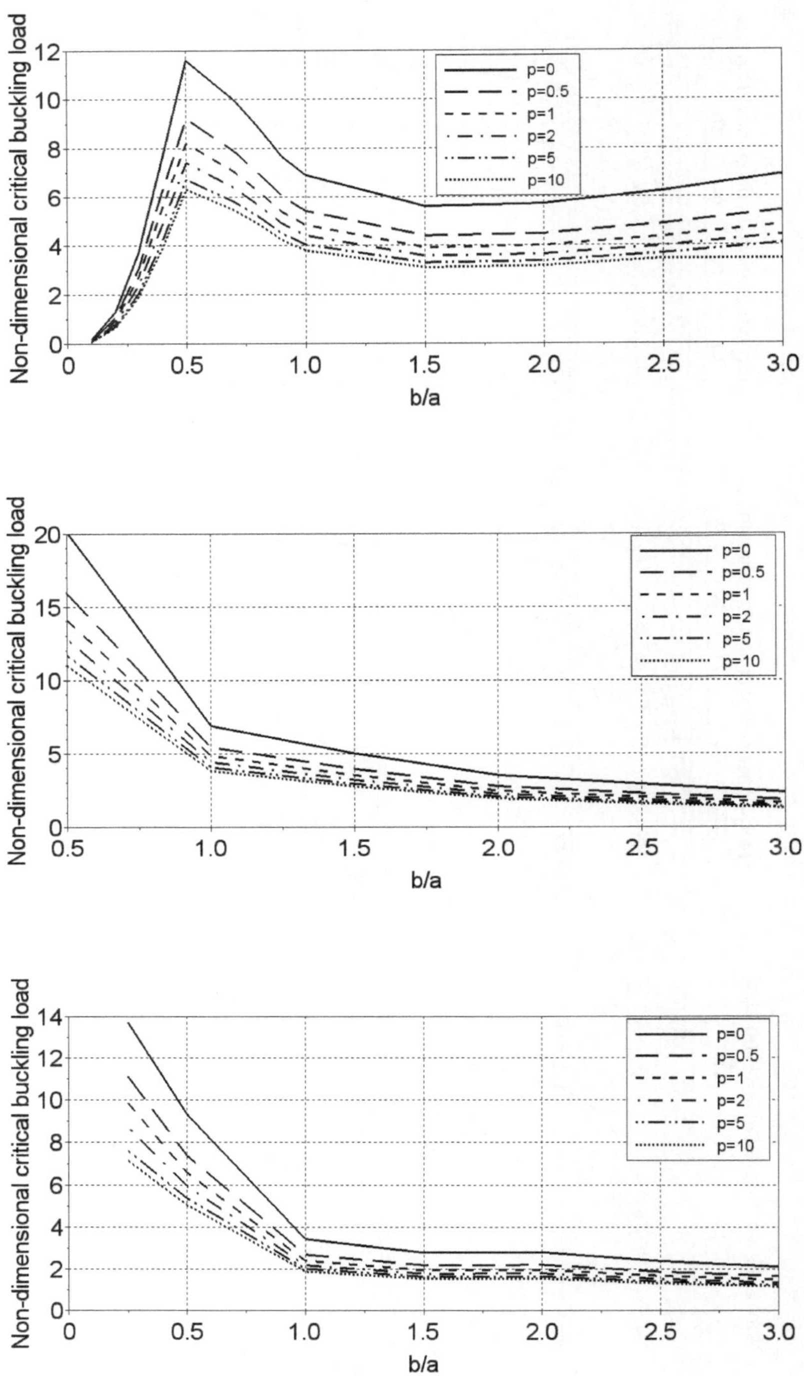

Figure 3.26 The effects of aspect ratio b/a on the non-dimensional critical buckling load of a simply supported (SSSS) Al/ZrO_2 plate under (a) uniaxial compression in x direction (b) uniaxial compression in y direction and (c) biaxial compression loads.

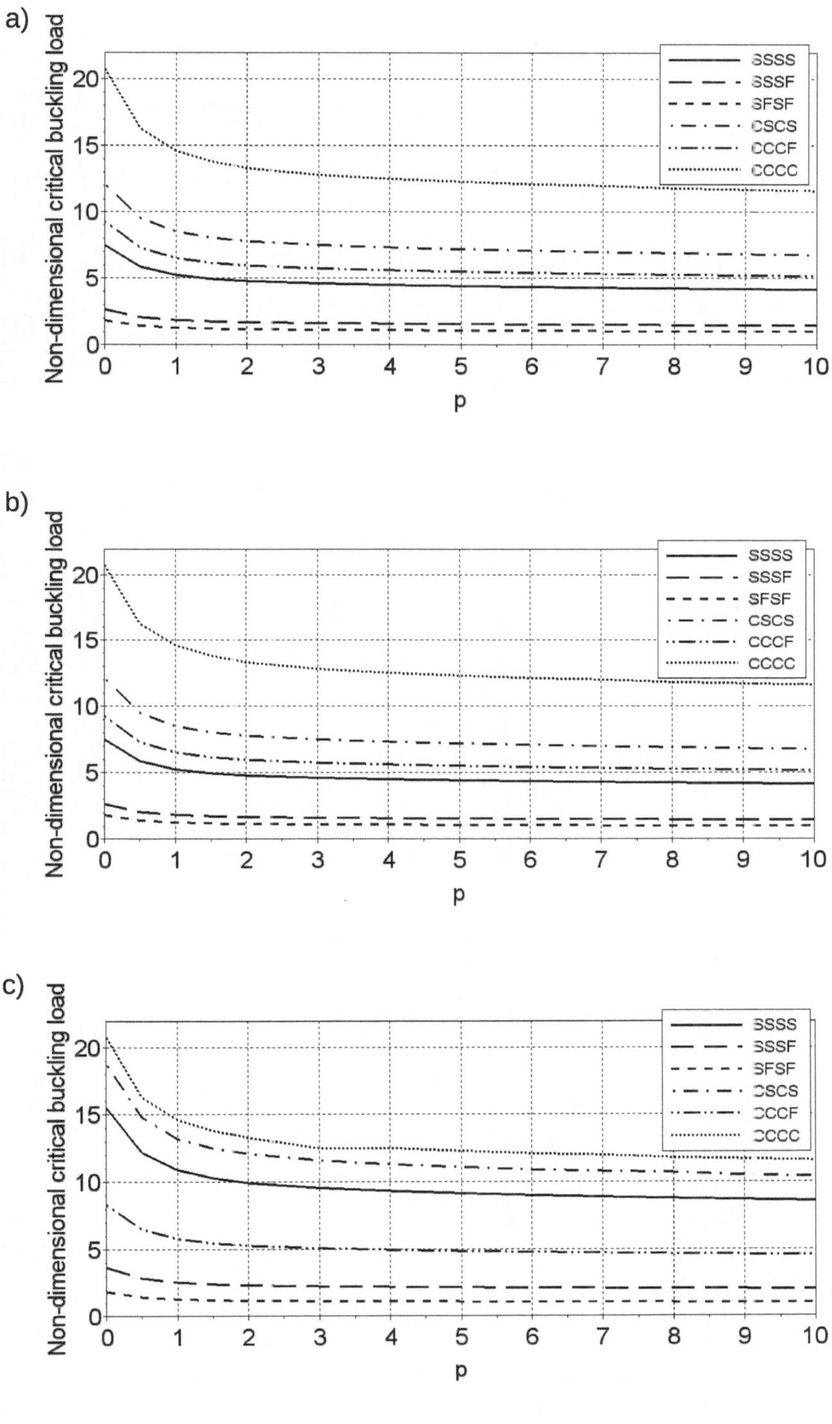

Figure 3.27 The effects of boundary conditions on the non-dimensional critical buckling load of the Al/ZrO_2 plate under (a) uniaxial compression in x direction (b) biaxial compression and (c) biaxial compression and tension loads.

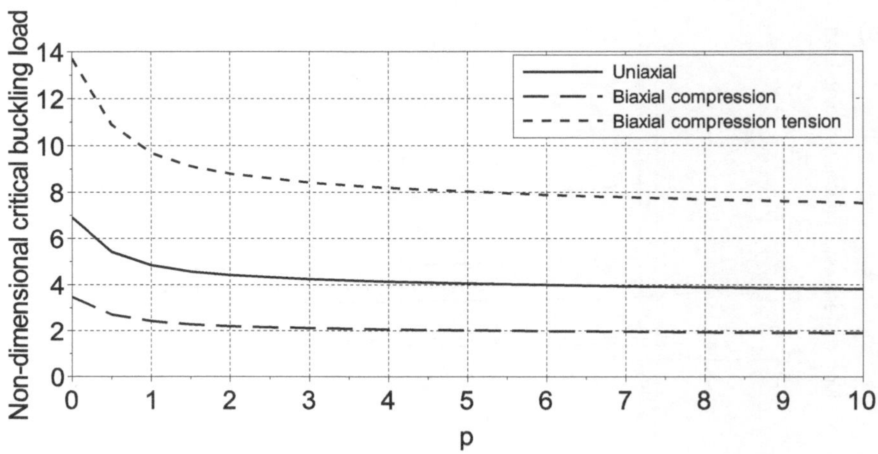

Figure 3.28 The effects of types of loadings on the non-dimensional critical buckling load of the simply supported (SSSS) square Al/ZrO_2 plate.

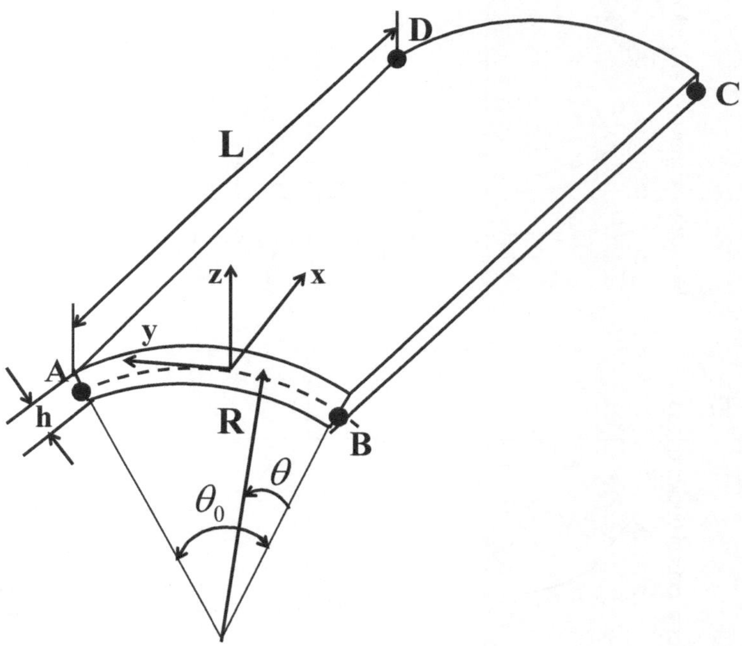

Figure 3.29 Geometry of the cylindrical panel.

agreement with those given by Zhao et al. [326] except for some high buckling modes where the relative error between the results can reach up to 7%. In fact, the FSDT theory used in [326] present less accuracy for high buckling modes due its introduction of shear correction factors which outline the necessity of high-order models for accurate solutions.

Now, the effects of length-to-thickness, radius-to-thickness ratios, span angle θ and boundary conditions on mechanical buckling behavior of FGM cylindrical panels are examined. The panel

Table 3.11

Non-dimensional critical buckling parameter $\overline{N}_{cr}$ for a simply supported (SSSS) isotropic cylindrical panel under uniform compression load.

Mode	Present-SHO4	FSDT.[326]	Analytical.[262]
1	0.6000	0.5960	0.6052
2	0.6740	0.6535	
3	0.7160	0.7053	
4	0.9240	0.8485	

Table 3.12

Critical buckling load parameter $\overline{N}_{cr}^{FGM}$ of simply supported (SSSS) Al/ZrO_2 cylindrical panel under uniform compression load for different values of power-law index p ($p = 0, 0.5, 1$).

Mode	power-law index p					
	0		0.5		1	
	Present	Ref.[326]	Present	Ref.[326]	Present	Ref.[326]
1	1.2900	1.2768	1.0200	1.0390	0.9020	0.9313
2	1.4600	1.4098	1.1300	1.1314	1.000	1.0149
3	1.5500	1.5149	1.2200	1.2509	1.0700	1.1241
4	1.9900	1.8434	1.5400	1.4683	1.3700	1.3175

Table 3.13

Critical buckling load parameter $\overline{N}_{cr}^{FGM}$ of simply supported (SSSS) Al/ZrO_2 cylindrical panel under uniform compression load for different values of power-law index p ($p = 2, 5, 10, 20$).

Mode	power-law index p							
	2		5		10		20	
	Present	Ref.[326]	Present	Ref.[326]	Present	Ref.[326]	Present	Ref.[326]
1	0.8060	0.8366	0.7250	0.7464	0.6820	0.6933	0.6490	0.6525
2	0.9090	0.9217	0.8320	0.8377	0.7840	0.7812	0.7430	0.7324
3	0.9510	1.0043	0.8470	0.8828	0.7980	0.8153	0.7640	0.7677
4	1.2500	1.2042	1.1500	1.1016	1.0900	1.0334	1.0300	0.9666

is subjected to uniaxial compression load along its curved edges (AB) and (DC) while its straight sides (AD) and (BC) are unloaded. Figure 3.30 shows the variation of non-dimensional critical buckling load versus the length-to-thickness ratio L/h of the simply supported Al/ZrO_2 cylindrical panel under uniform compression load. It can be observed that the non-dimensional critical buckling load decreases as the length-to-thickness ratio L/h increases for the different values of the power law index p. In fact, when the length-to-thickness ratio L/h increases, the FGM cylindrical panel behaves as thin shell structure which buckles more easily and this is due to the decreases in the stiffness of the shell structure. Figure 3.31 illustrates the effect of radius-to-thickness ratio R/h on non-dimensional critical buckling load of the simply supported Al/ZrO_2 cylindrical panel for different power law index p. It can be seen that the critical buckling load decreases as the radius-to-thickness ratio R/h increases while it becomes constant after reaching $R/h = 500$ and this is for each value of power law index p. From Fig. 3.32, it can be also remarked that for a value of span angle $0.1 \prec \theta \prec 0.2$, the non-dimensional critical buckling load decreases slightly but it stabilizes after reaching $\theta = 0.2$ which demonstrates that the span angle geometrical parameter does not have a significant effect on the critical buckling load parameter of the simply supported Al/ZrO_2 cylindrical panel. Figure 3.50 demonstrates the effect of boundary conditions on non-dimensional mechanical critical buckling load of the Al/ZrO_2 cylindrical panel under uniaxial compression load along its curved sides versus the power law index p. It can be seen that non-dimensional critical buckling load decreases as the power law index increases, and the variation of the non-dimensional critical buckling load is considerable when the power law index p is small. This is due to the fact that higher values of power law index correspond to high portion of metal in comparison with the ceramic part. As well as, different combinations of boundary conditions are considered where a fully clamped edge is denoted by (C), a fully simply supported edge S and free F. For SSSS, CSCS and CCCC boundary conditions, the values of critical buckling load are higher than those for SSSF, SFSF and CCCF boundary conditions and this is due to the more constraints imposed by the clamped boundary condition compared to other boundary conditions: simply supported S or free F. Therefore, the type of imposed boundary conditions can affected significantly the mechanical buckling behavior of FGM cylindrical panels under a compressive load.

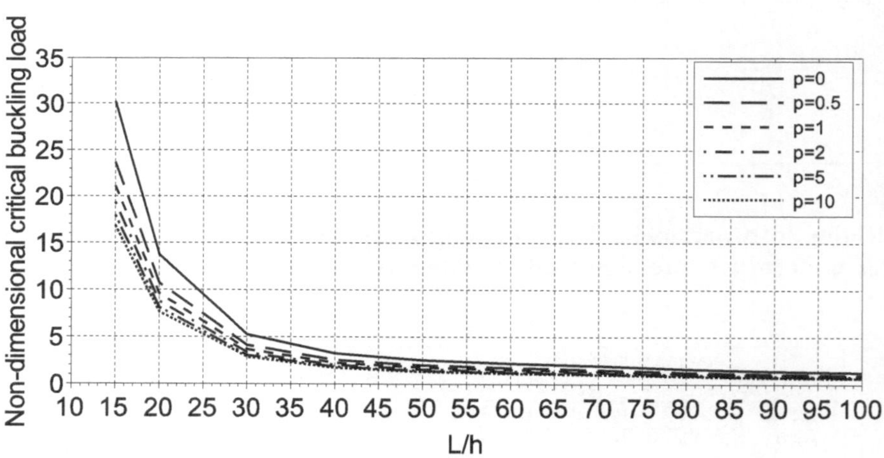

Figure 3.30 Non-dimensional critical buckling load parameter $\overline{N}_{cr}^{FGM}$ versus the length-to-thickness ratios L/h for a simply supported SSSS Al/ZrO_2 cylindrical panel under uniform compression load along its curved edges.

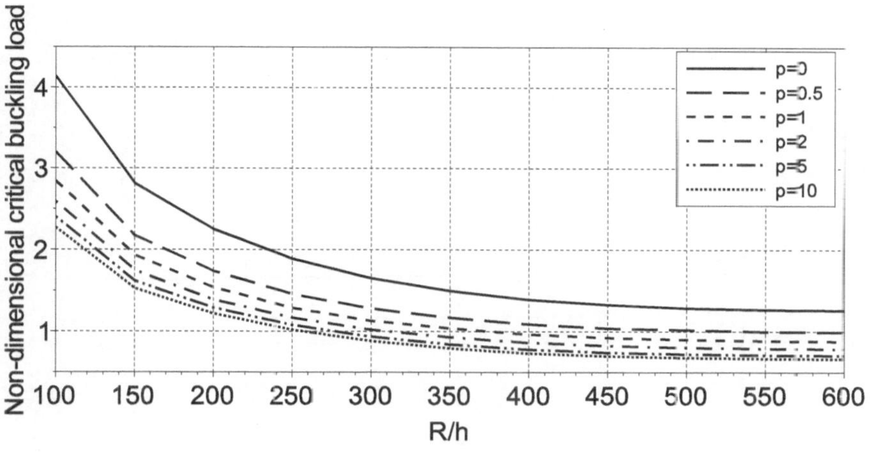

Figure 3.31 Non-dimensional critical buckling load parameter $\overline{N}_{cr}^{FGM}$ versus the radius-to-thickness ratios R/h of a simply supported (SSSS) Al/ZrO_2 cylindrical panel under uniform compression load along its curved edges.

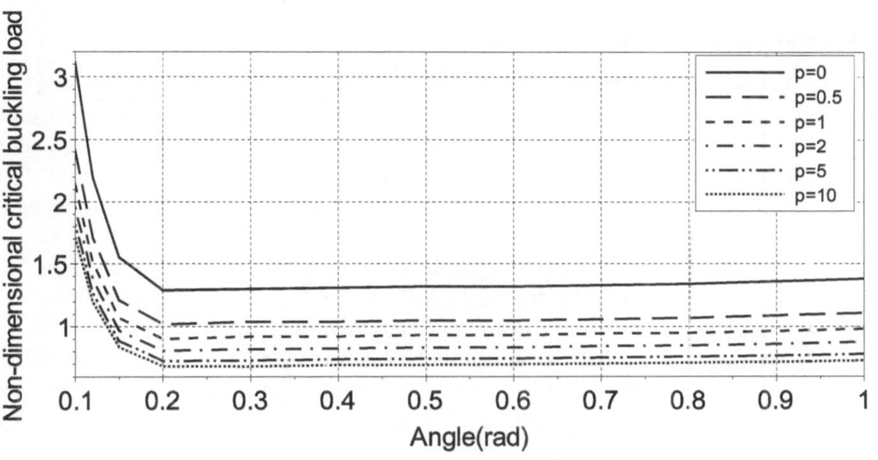

Figure 3.32 Non-dimensional critical buckling load parameter $\overline{N}_{cr}^{FGM}$ versus the span angle θ of a simply supported (SSSS) Al/ZrO_2 cylindrical panel under uniform compression load along its curved edges.

3.4 POST-BUCKLING OF FGM PLATES AND SHELLS UNDER MECHANICAL LOADINGS

For post-buckling analysis and with an initial perturbation of the external load, the equilibrium responses of studied structures are obtained using the iterative and incremental Newton-Raphson method combined with the arc length method where its steps are as highlighted in Fig. 3.34.

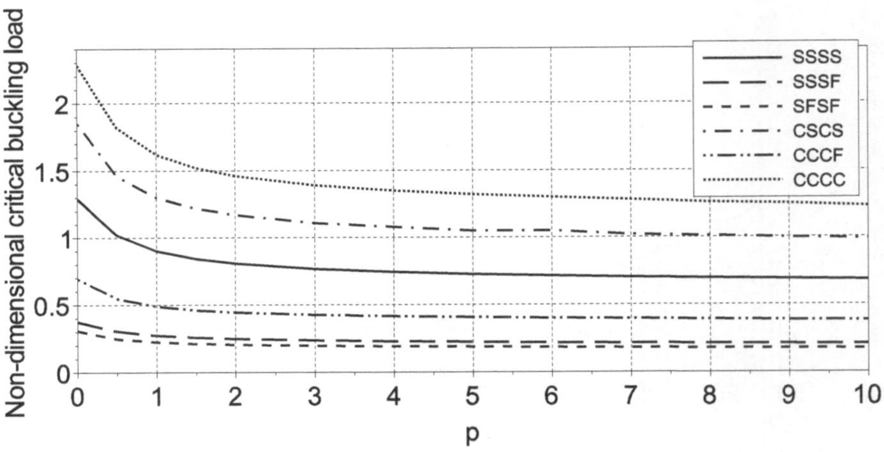

Figure 3.33 Effects of boundary conditions on non-dimensional mechanical critical buckling load parameter of Al/ZrO_2 cylindrical panel under uniform compression load along its curved edges.

3.4.1 MECHANICAL POST-BUCKLING ANALYSIS OF PLATES

3.4.1.1 Verification of the present model for isotropic plates

In the first step of this work, the post-buckling behavior of an isotropic square plate subjected to uniaxial compression load is examined where the present solutions are compared to those of Yamaki [295, 296] and Lee et al. [134], respectively. For the sake of comparison, the geometrical characteristics of the plate are set equal to $a/b = 1$ and $a/h = 120$. The Poisson's ratio is equal to $\nu = 1/3$. The applied boundary condition is of type (CCSS) where the loaded edges are clamped and unloaded ones are simply supported as shown in Fig. 3.35(a). Results are performed in terms of non-dimensional parameters where the non-dimensional load factor $\overline{N}$ is computed with the following expression: $\overline{N} = N_x a^2/\pi^2 E h^3$; a and h are the length and thickness of the plate, respectively; E refers to Young's modulus and N_x is the uniformly distributed load along the x direction.

Comparison is provided in Fig. 3.38 which outlines the aptitude of the present model in the restitution of the post-buckling behavior of isotropic plates with good accuracy. In fact, the present post-buckling curve agrees well with the analytical solutions provided by Yamaki [295, 296] for plates without (Yamaki-01) and with (Yamaki-02) initial deflections and with the finite element solution given by Lee et al. [134] using the FSDT theory which affirms the high accuracy of the present model. It should be mentioned that the present results are obtained using 18×18 SHO4 of the developed finite elements and it will be adopted in the following post-buckling analysis of plates since it provides accurate results compared to the literature.

3.4.1.2 Post-buckling responses of FG plates under mechanical loads

After performing the comparison study in the case of isotropic plates, novel numerical results from the present model for FG plates subjected to various types of mechanical loadings are provided. To this end, a FGM square $(a = b)$ plate composed of Aluminium (Al) and Zirconia (ZrO_2) phases is considered with different length-to-thickness ratios $(a/h = 40, 100, 200)$. The type of applied boundary is (CCSS) as defined in Fig. 3.35(a). Three different types of applied mechanical loads, namely uniaxial compression in the x direction, biaxial compression and biaxial compression and tension loads are analyzed. The definition of each loading type is given in Fig. 3.37. Accordingly to the open literature, the post-buckling behavior of plates under uniaxial compression load in the x direction is the frequently studied case, so to provide more general analysis about post-buckling behavior of

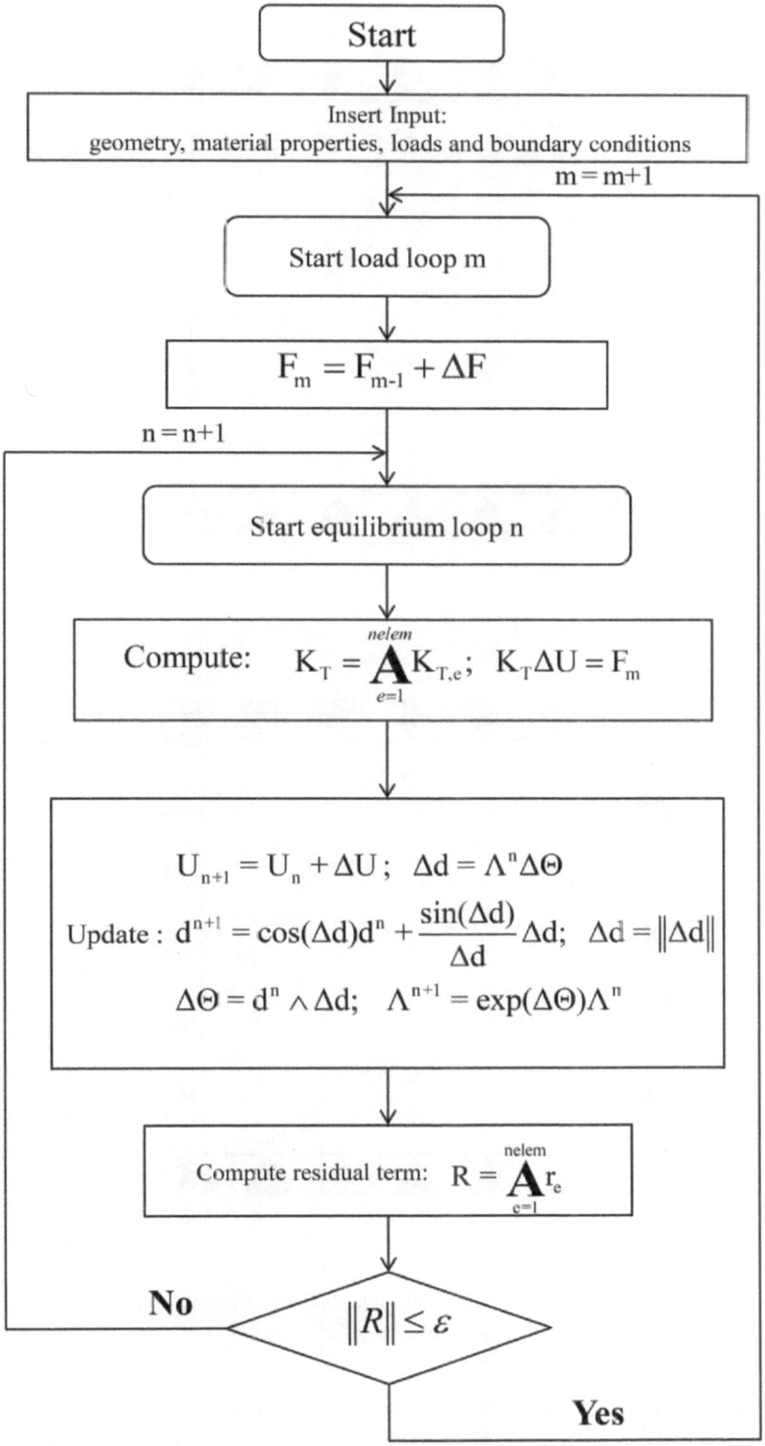

Figure 3.34 Numerical resolution flowchart of the proposed method for post-buckling analysis of FG and nanocomposite structures.

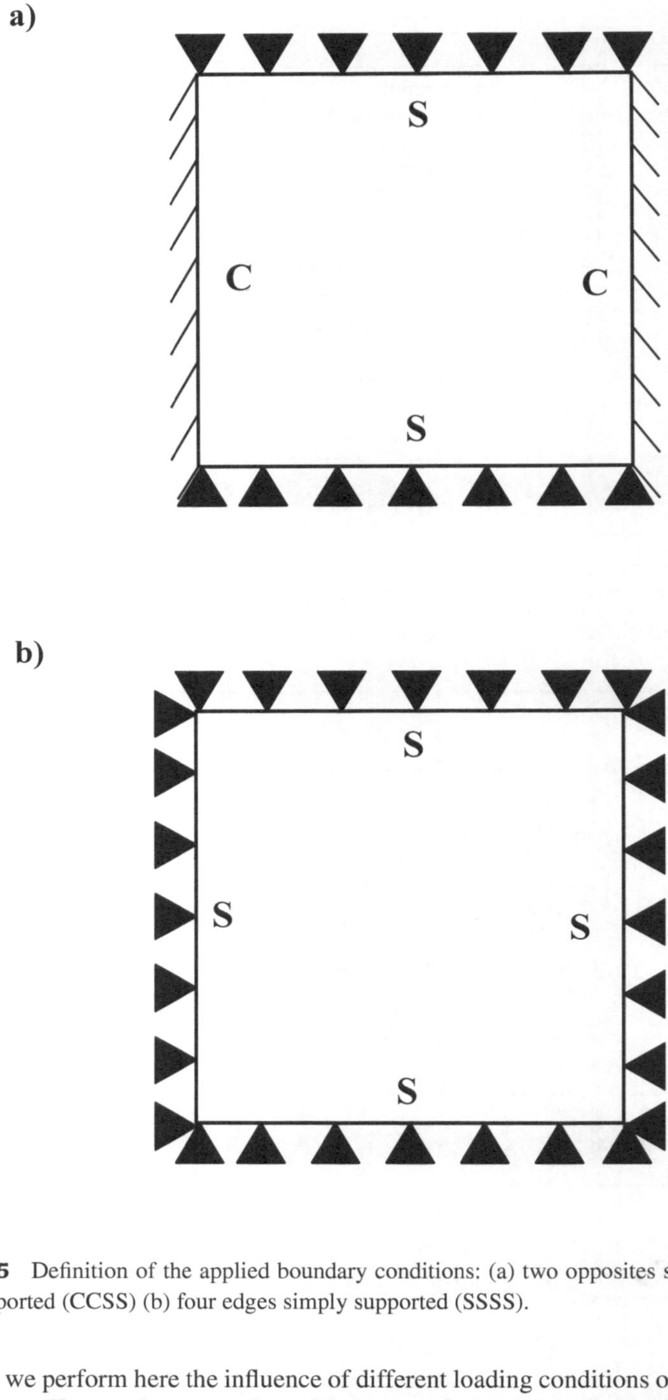

Figure 3.35 Definition of the applied boundary conditions: (a) two opposites sides clamped and the others simply supported (CCSS) (b) four edges simply supported (SSSS).

FG plates, we perform here the influence of different loading conditions on post-buckling responses of FG plates. The results are evaluated in terms of non-dimensional load factor $\overline{N} = Na^2/E_m h^3$ where E_m and h refer to Young's modulus of metal phase and the total thickness of the plate, respectively. N is the uniformly distributed load in the appropriate direction according to the type of applied mechanical load as defined in Fig. 3.37. The non-dimensional deflection is computed in the

c)

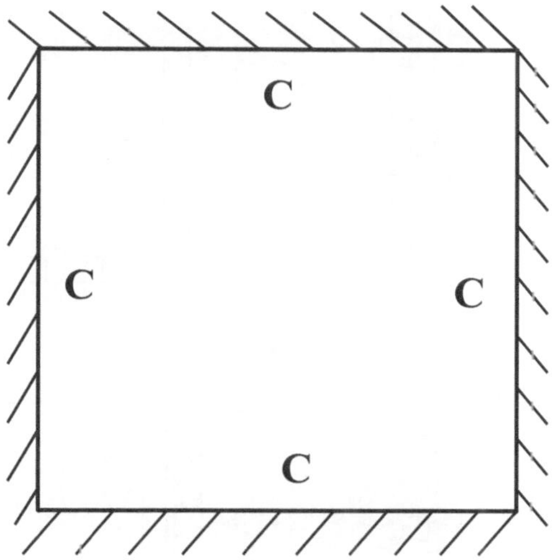

d)

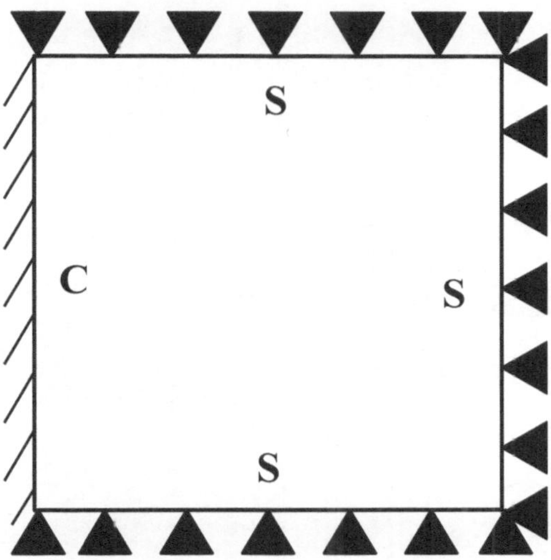

Figure 3.36 Definition of the applied boundary conditions: (c) four edges clamped (CCCC) (d) one edge clamped and the others simply supported (CSSS).

center of the plate as: $\overline{W} = \frac{1}{h}W(\frac{a}{2};\frac{a}{2})$. The material properties of metal and ceramic constituents including Young's moduli and Poisson's ratio are taken as indicted in [134]: $E_m = 70 \times 10^9 N/m^2$; $E_c = 151 \times 10^9 N/m^2$ and $\nu_m = \nu_c = 0.3$. Figures 3.39, 3.40 and 3.41 show the mechanical post-buckling responses of Al/ZrO_2 plates (CCSS) subjected to uniaxial compression in the x direction; biaxial compression and biaxial compression and tension loads with three sets of length-to-

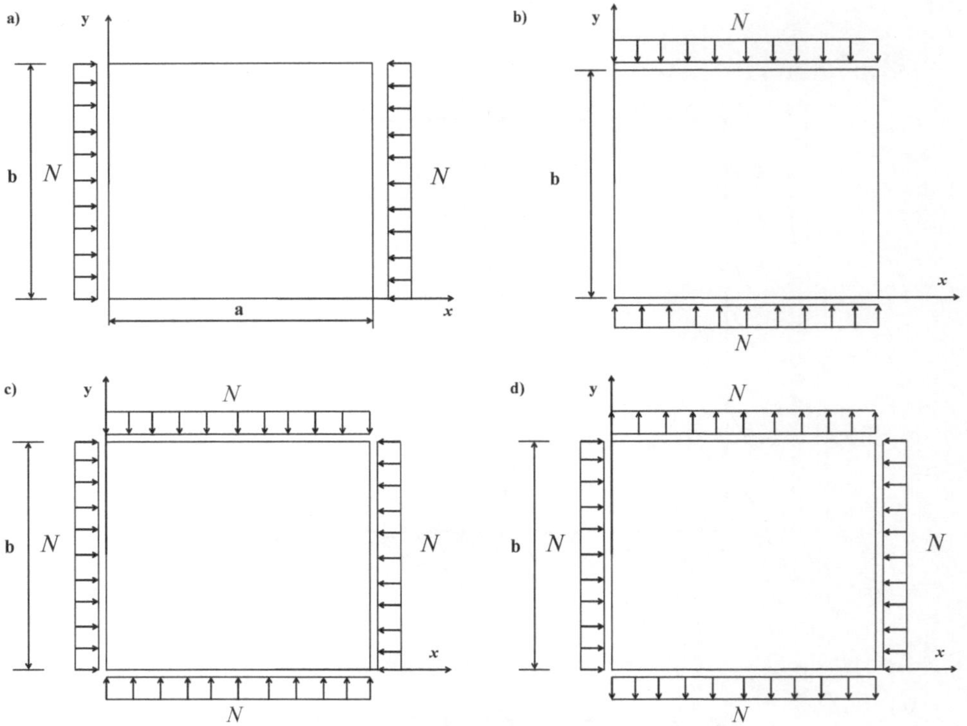

Figure 3.37 Geometry of the plate with (a) uniaxial compression in x direction (b) uniaxial compression in y direction (c) biaxial compression and (d) biaxial compression and tension loads.

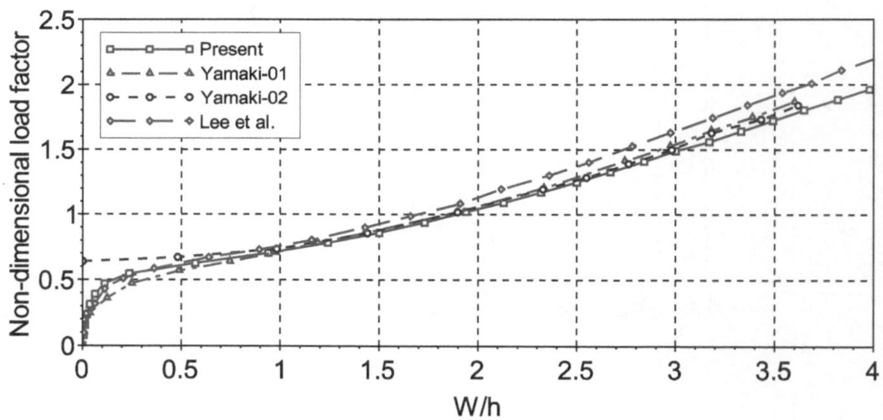

Figure 3.38 Post-buckling response of isotropic square plate (CCSS) subjected to uniaxial compression load.

thickness ratios ($a/h = 40, 100, 200$), respectively. Different power-law index are taken into account ($p = 0, 0.5, 1, 2, \infty$) to explore the effects of variation of this parameter on post-buckling responses of such structures. Firstly, and for the purpose of validation of the present model for FG plates, Figure 3.40(a) depicts the post-buckling equilibrium path obtained from the present model and those

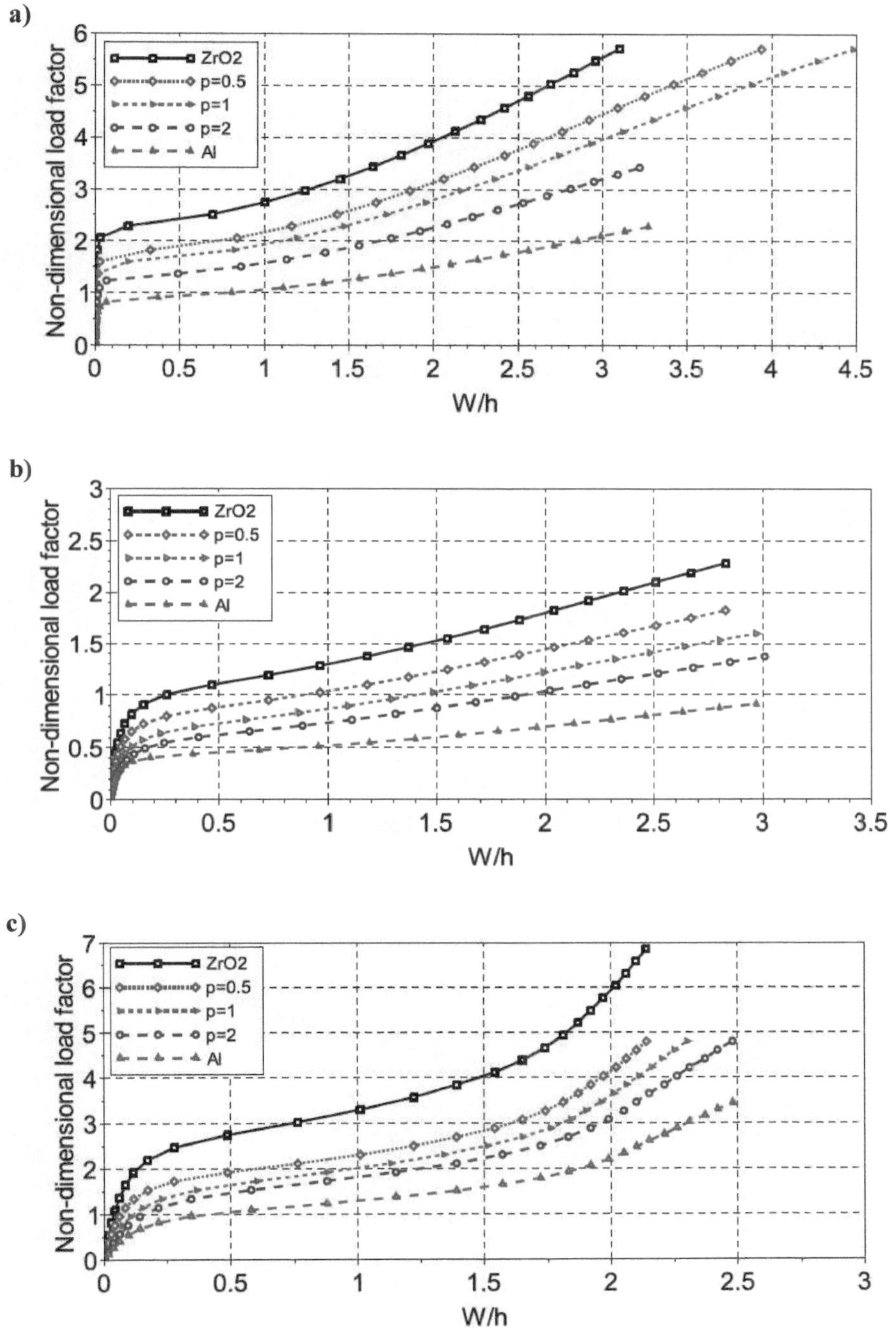

Figure 3.39 Post-buckling responses of Al/ZrO_2 square plate subjected to (a) uniaxial in x direction, (b) biaxial compression and (c) biaxial compression and tension loads (CCSS, $a/h = 40$).

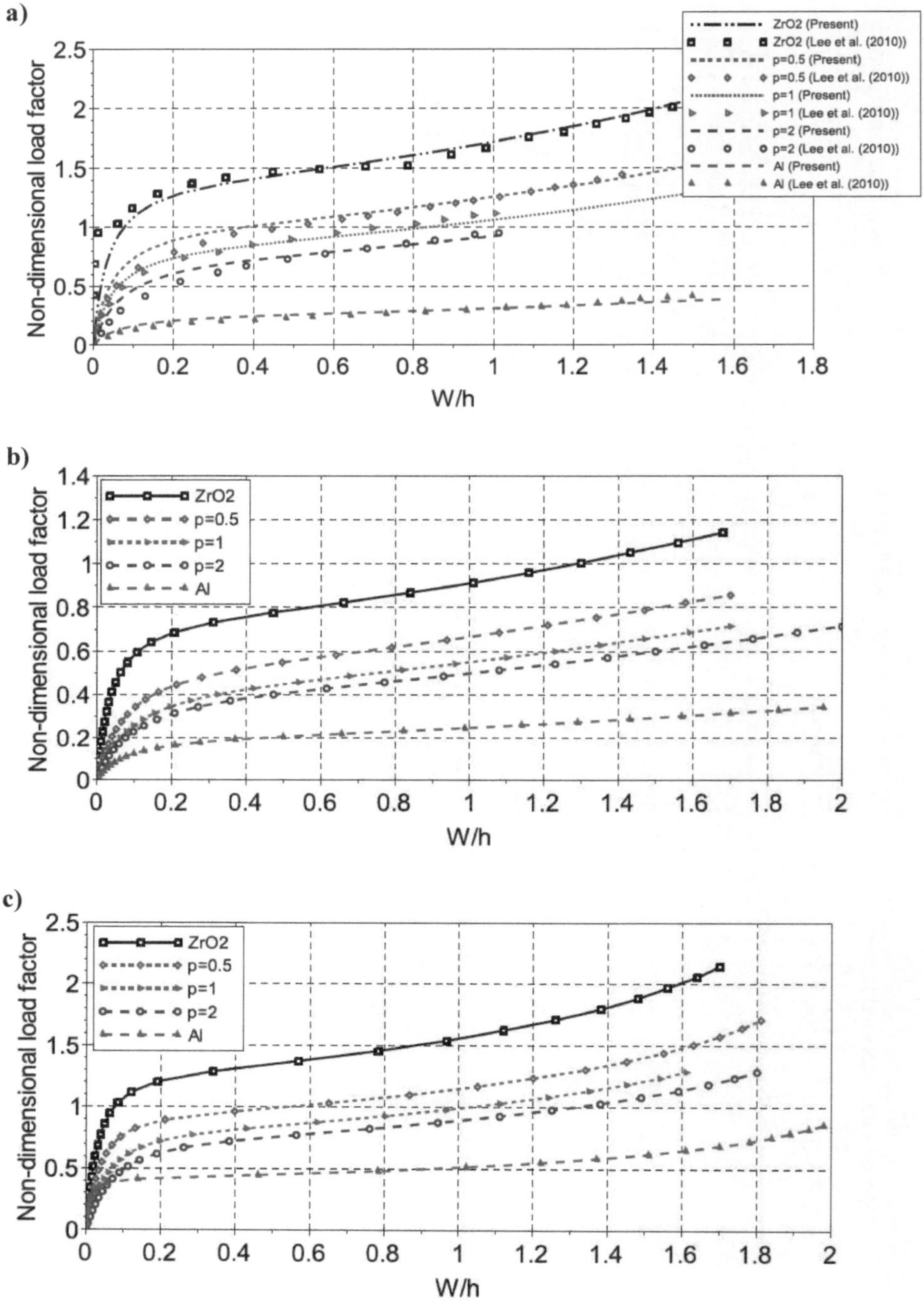

Figure 3.40 Post-buckling responses of Al/ZrO_2 square plate subjected to (a) uniaxial in x direction, (b) biaxial compression and (c) biaxial compression and tension loads (CCSS, $a/h = 100$).

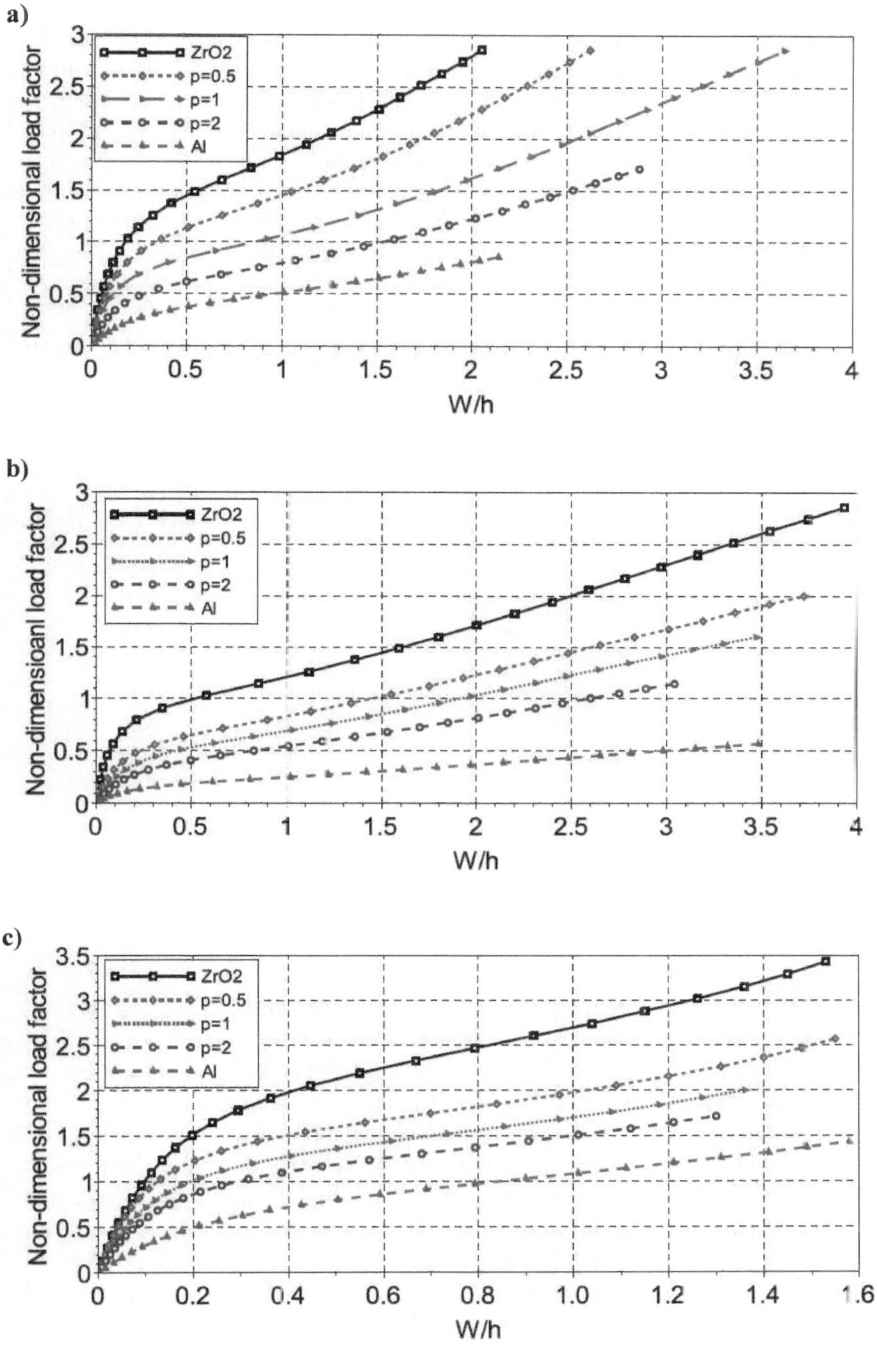

Figure 3.41 Post-buckling responses of Al/ZrO_2 square plate subjected to (a) uniaxial in x direction, (b) biaxial compression and (c) biaxial compression and tension loads (CCSS, $a/h = 200$).

given by Lee et al. [134] of the Al/ZrO_2 plate (CCSS, $a/h = 100$) in the case of uniaxial compression in the x direction and this is for different power-law index. It is seen that the results are in close agreement which proves again the accuracy and the effectiveness of the present formulation for post-buckling analysis of FG plates. As well as, it can be observed from Figs. 3.39, 3.40 and 3.41 that the pure Zirconia (ZrO_2) and Aluminium (Al) plates induce the highest and lowest load factors, respectively and this is expected since the stiffness of the ceramic phase is more higher than the metal one. The post-buckling curves of the FG plates with ($p = 0.5, 1, 2$) are located between those of the pure Zirconia and Aluminium plates. In fact, the post-buckling strength of the FG plates decreases as the power-law index increases and this is for all applied mechanical loadings. Additionally, it is revealed that for a lower length-to-thickness ratio ($a/h = 40$), the post-buckling paths of the Al/ZrO_2 plate are achieved with high load factors compared to the other cases ($a/h = 100, 200$). Indeed, when the length-to-thickness ratio decreases, the FG plate behaves as thick structure and requires hence high load levels to generate the post-buckling behavior due its high stiffness. Figure

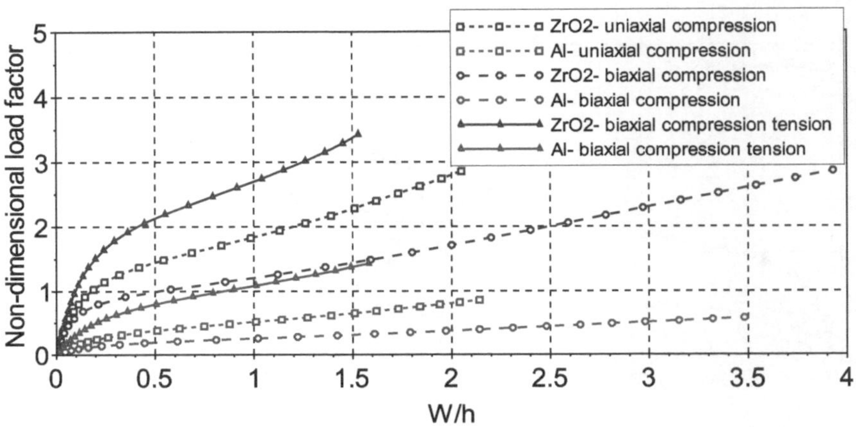

Figure 3.42 The influence of different types of mechanical loadings on post-buckling responses of FGM square plates (CCSS, $a/h = 200$).

3.42 provides the influence of different types of loads on post-buckling behavior of Al/ZrO_2 plate (CCSS, $a/h = 200$). For the sake of brevity, only pure ceramic and metal phases are considered. For both phases, it can be seen that the biaxial compression load induces the maximum post-buckling deflection while the biaxial compression and tension load presents the lowest one. In fact, the biaxial compression load stimulates more the buckling phenomenon by the rise in the edges compression forces and the FG plate bends hence easily while in the biaxial compression and tension load, the tension effort compensates the compression ones and leads to high flexural rigidity of the FG plate which manifest by less post-buckling deflection. Consequently, the type of applied load has a significant effect on post-buckling responses of FG plates which must be considered in the analysis. Next, the effect of plate aspect ratio (b/a) on post-buckling responses of Al/ZrO_2 square plate subjected to uniaxial compression load in x direction with ($a/h = 100$) and (CCSS) type of boundary condition is examined for two power-law index ($p = 0$) and ($p = 2$) through Figs. 3.43(a) and 3.43(b), respectively. In these figures, four magnitudes of aspect ratio are taken into account ($b/a = 1, 1.5, 2, 3$). It can be remarked that, as the aspect ratio (b/a) increases, as the post-buckling deflection of FG plates is enhanced. Indeed, when the aspect ratio rises ($b/a \succ 1$), the plate becomes larger to support more compression forces which enhances the post-buckling deflection due to the decreases in the flexural rigidity of the FG plates. Therefore, more the aspect ratio is increased, more the post-buckling deflection is pronounced.

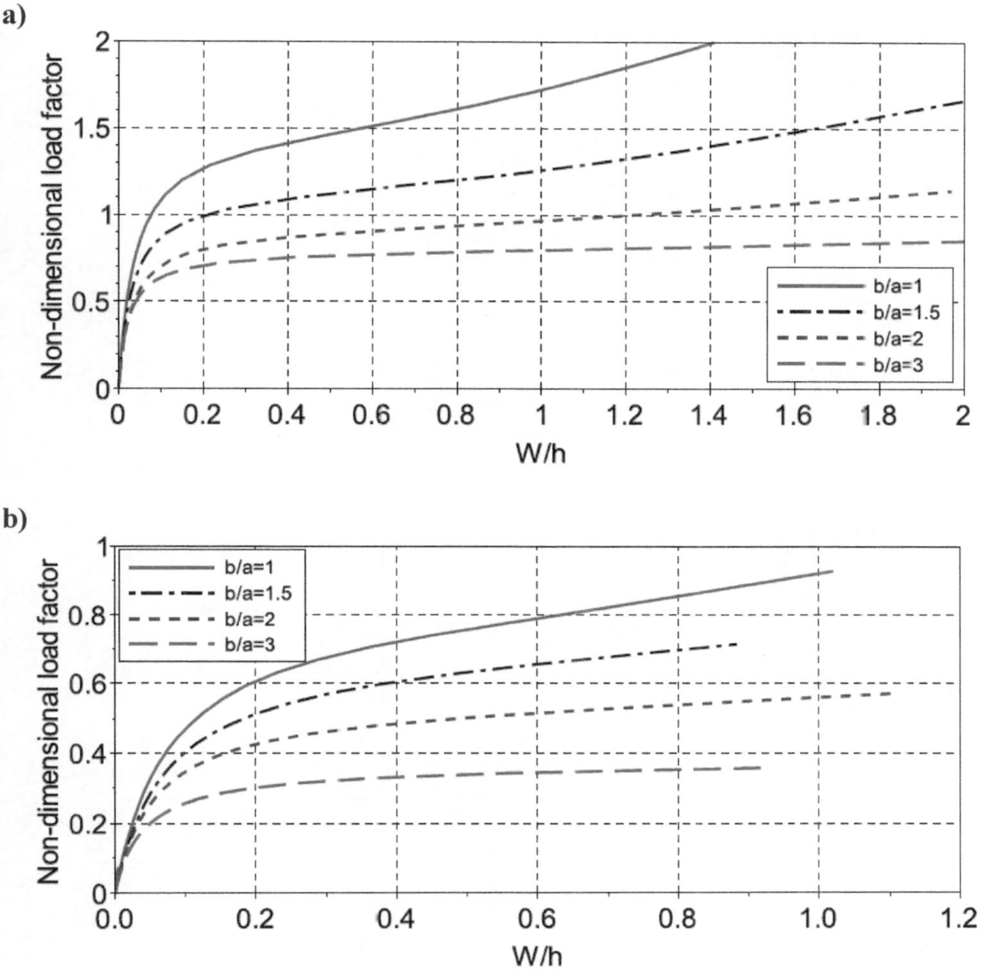

Figure 3.43 The influence of aspect ratio on post-buckling responses of Al/ZrO_2 plates subjected to uniaxial compression load in x direction for two power-law index (a) $p = 0$ and (b) $p = 2$, respectively with (CCSS, $a/h = 100$).

Figures 3.44(a), 3.44(b) and 3.44(c) depict the post-buckling equilibrium path of Al/ZrO_2 square plate subjected to different types of boundary conditions and mechanical loadings with ($a/h = 100$, $p = 1$). Four sets of applied boundary conditions, namely CCSS, CCCC, SSSS and CSSS are analyzed where the definition of each boundary condition is as given in Figs. 3.35 and 3.36. For all types of mechanical loadings, Figure 3.44 demonstrates the effects of boundary conditions on post-buckling behavior of FG plates. In fact, the post-buckling curves induced by CCCC FG plates are higher than CCSS, CSSS and SSSS plates. Thus, they can be explained by the rise in the stiffness of the structure via the more imposed constraints of a clamped edge in comparison to a simply supported edge. Therefore, more constraints are applied to the structure, more the post-buckling strength of the FG plate is enhanced.

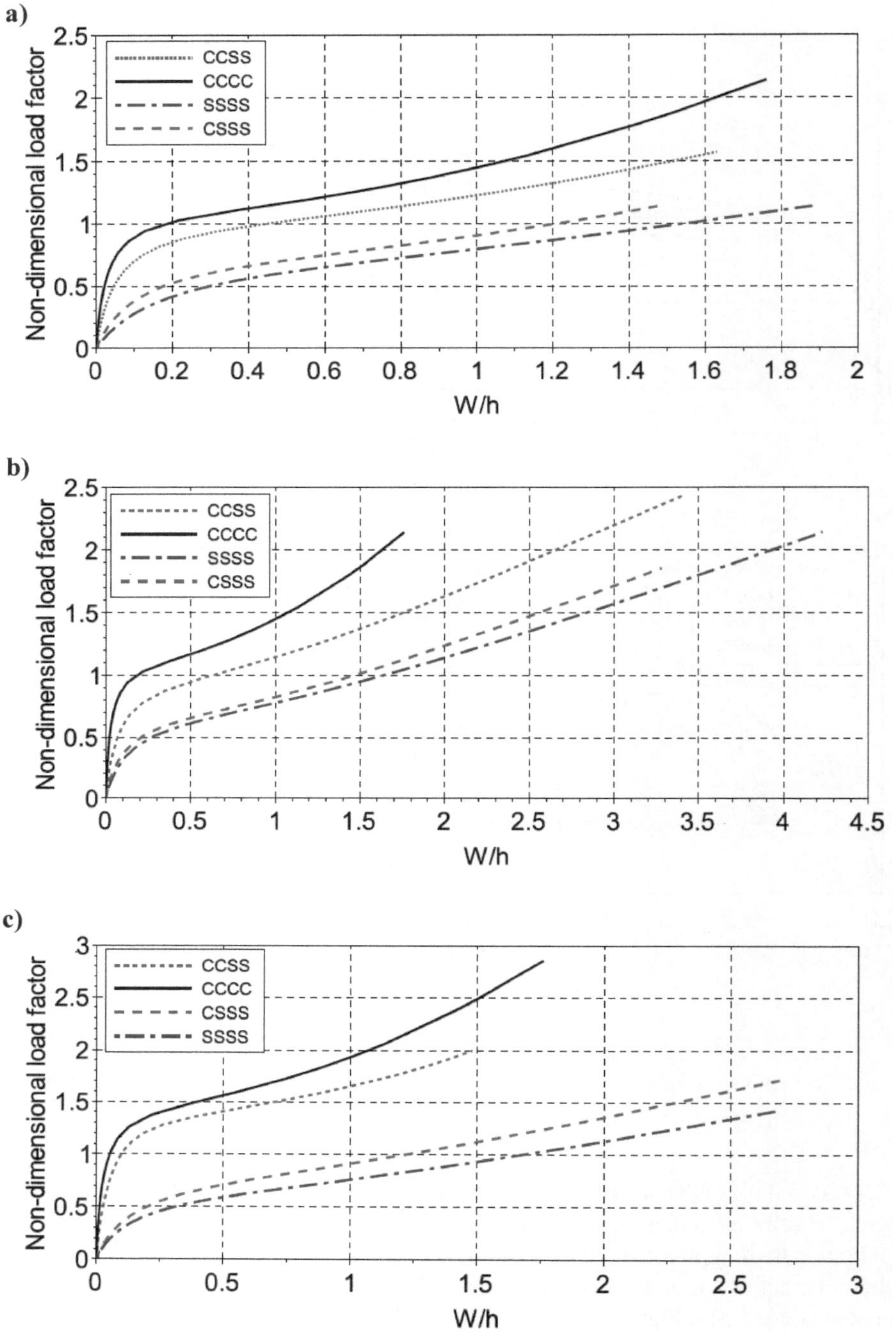

Figure 3.44 The influence of different boundary conditions on post-buckling responses of Al/ZrO_2 square plate subjected to (a) uniaxial in x direction, (b) biaxial compression and (c) biaxial compression and tension loads with ($a/h = 100$, $p = 1$).

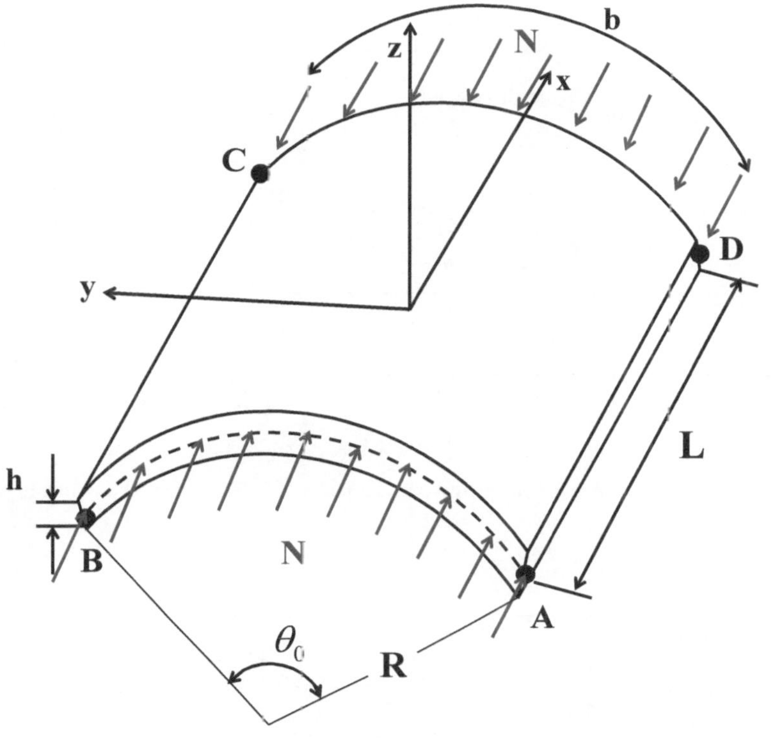

Figure 3.45 Geometry of the cylindrical panel (CCSS) subjected to uniaxial compression load.

3.4.2 MECHANICAL POST-BUCKLING RESPONSES OF CYLINDRICAL PANELS

In this section, mechanical post-buckling responses of isotropic, FGM and nanocomposite cylindrical panels are presented. In fact, a validation study is carried out first in the isotropic case. Then, the results are performed for FGM cylindrical panels in order to show the capacity of the present formulation in the prediction of mechanical post-buckling behavior of such curved panels.

3.4.2.1 Validation for isotropic cylindrical panels

To validate the proposed formulation for curved shells, an isotropic cylindrical panel made of Aluminium (Al) is considered. The geometrical and material properties of the panel are taken as [147, 144]: the length $L = 14.75in$, the width $b = 14.5in$, the radius $R = 60in$, the total thickness $h = 0.13in$, the Young's modulus $E = 10Msi$ and the Poisson's ratio $\nu = 0.33$. The panel is subjected to uniaxial compression load ($P_b = Nb$) along its curved edges (AB) and (CD) as illustrated in Fig. 3.45. Furthermore, the curved edges are clamped (CC) and the straight ones are simply supported (SS). The used mesh involves 10×10 SHO4 finite elements. The post-buckling response of the isotropic cylindrical panel evaluated in its center is shown in Fig. 3.46. The present response is compared to those of Thornburgh and Hilburger [261] and Liew et al. [147]. It can be seen that before the critical buckling point, the present curve is in good agreement with those derived from the literature, especially the one given by Liew et al. [147] in which the effect of shear deformations is considered while this effect is neglected in [261]. Beyond this point, some discrepancies are observed and this is can be explained by the different shell theories used in the literature. In fact, in [261], the authors have used a finite element method based on Kirchhoff-Love thin shell theory

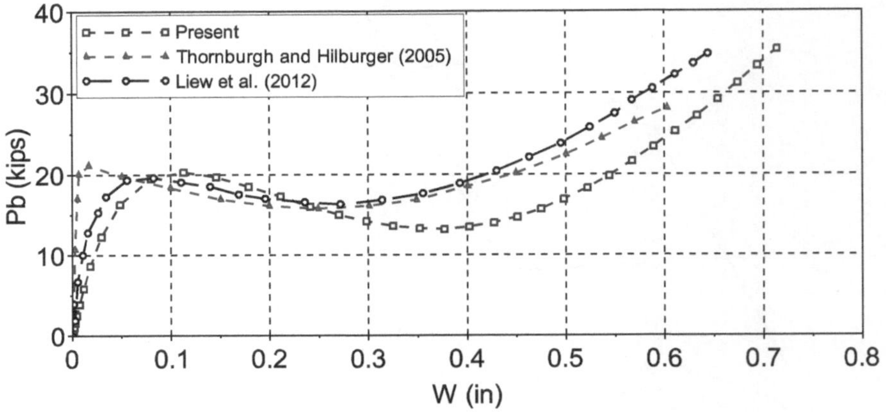

Figure 3.46 Post-buckling response of the isotropic cylindrical panel (CCSS) subjected to uniaxial compression load.

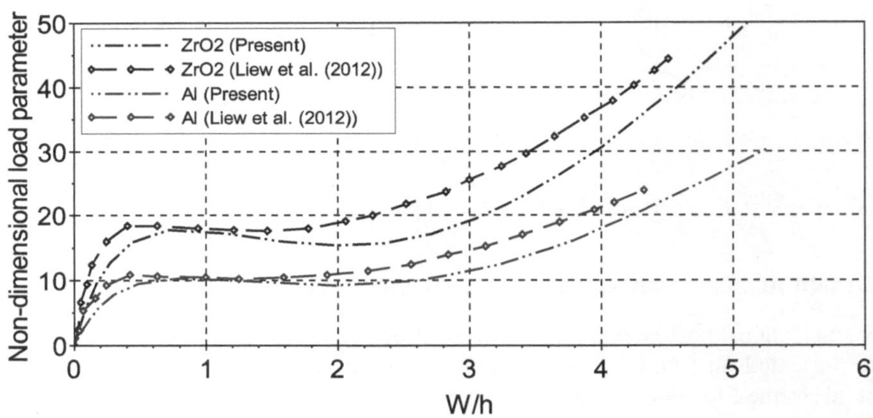

Figure 3.47 Post-buckling responses of the (Al/ZrO_2) cylindrical panel (CCSS) subjected to uniaxial compression load.

while in [147], a meshless Ritz method based on the FSDT theory is used. In both theories, the Von-Karman kinematic assumptions are introduced which include only small strains and moderate rotations, whereas, the present model is formulated via a high-order shell theory with large deflections and rotations. Since different shell theories are adopted, a little difference is revealed after the instability point. Otherwise, the obtained curves are visually correlated.

3.4.2.2 Mechanical post-buckling responses of FG cylindrical panels

This section investigates the mechanical post-buckling behavior of FG cylindrical panels subjected to uniaxial compression load as shown in Fig. 3.45. The panel is made of Aluminium (Al) and Zirconia (ZrO_2) phases where its material characteristics are as mentioned in section 3.3.3.2. The geometrical parameters of the panel are as follows [147]: the length $L = 0.2m$, the span angle $\theta_0 = 0.2rad$, the radius $R = 1m$ and the total thickness $h = 0.002m$. The applied boundary

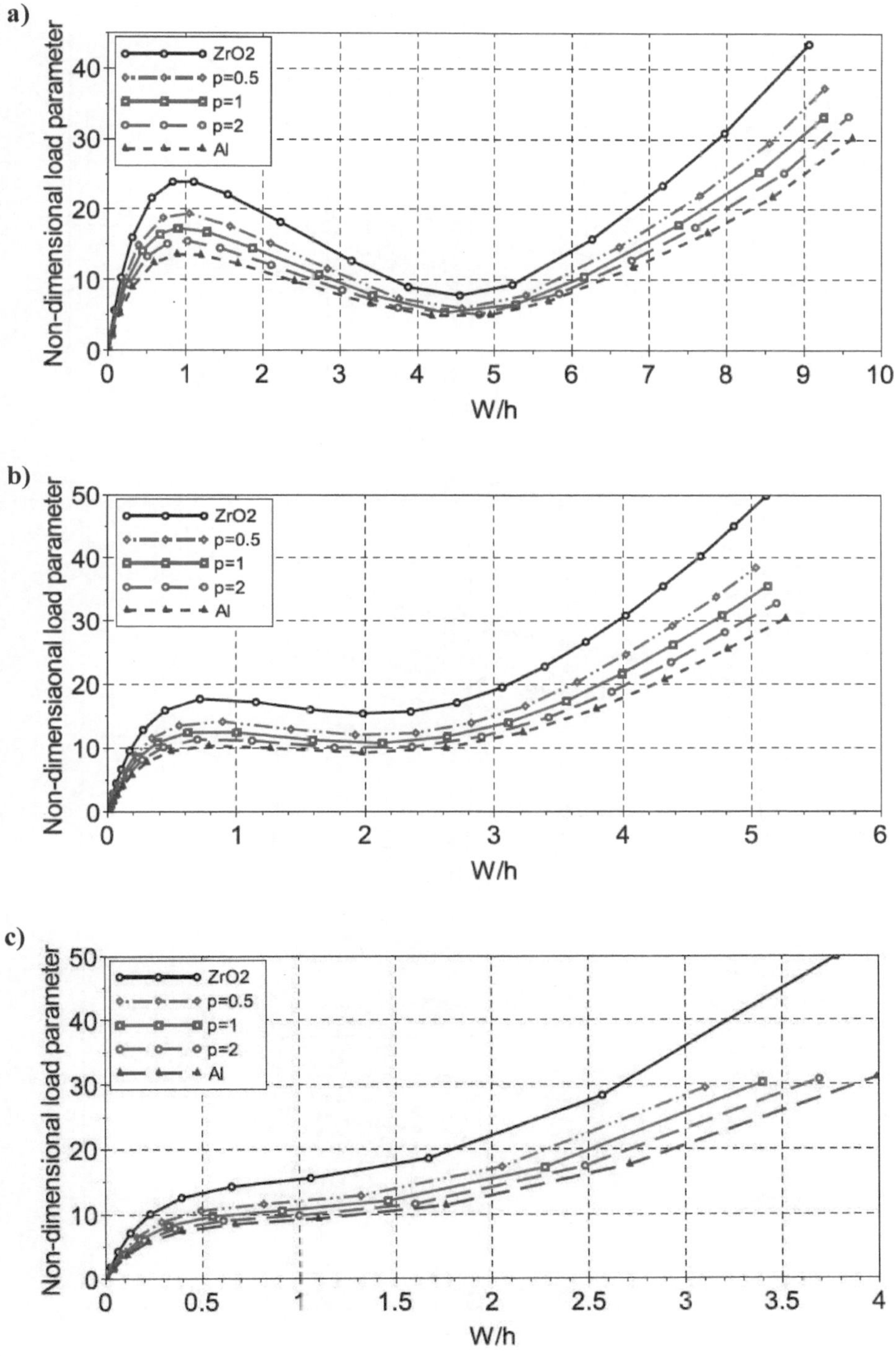

Figure 3.48 Influence of the thickness parameter on the post-buckling responses of the (Al/ZrO_2) cylindrical panel (CCSS,$R = 1$) under uniaxial compression load and with different power-law index and thickness parameter (a) $h = 0.001$, (b) $h = 0\ 002$ and (c) $h = 0.004$.

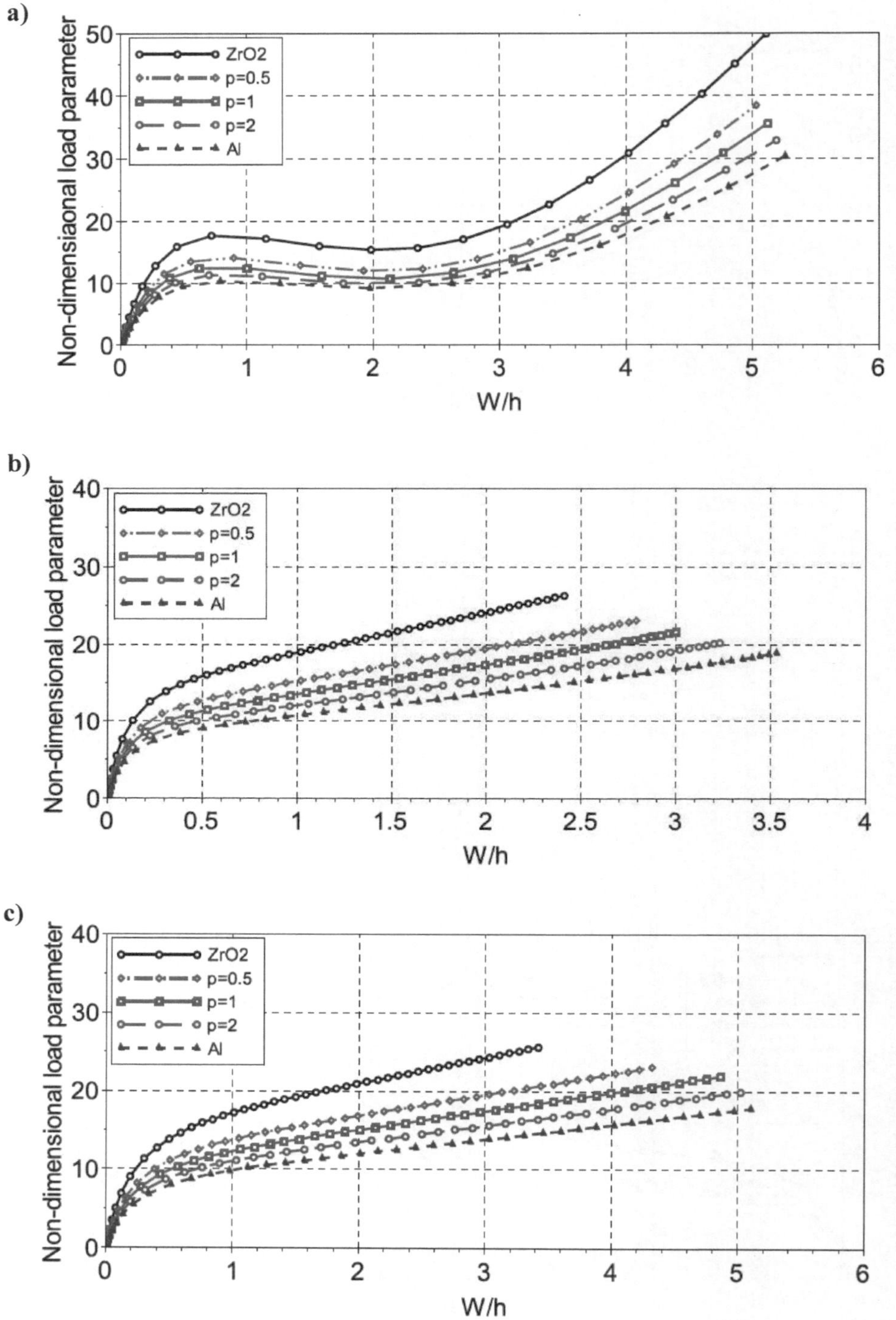

Figure 3.49 Influence of the radius parameter on the post-buckling responses of the (Al/ZrO_2) cylindrical panel (CCSS, $h = 0.002$) under uniaxial compression load and with different power-law index and radius parameter (a) $R = 1$, (b) $R = 2$ and (c) $R = 3$.

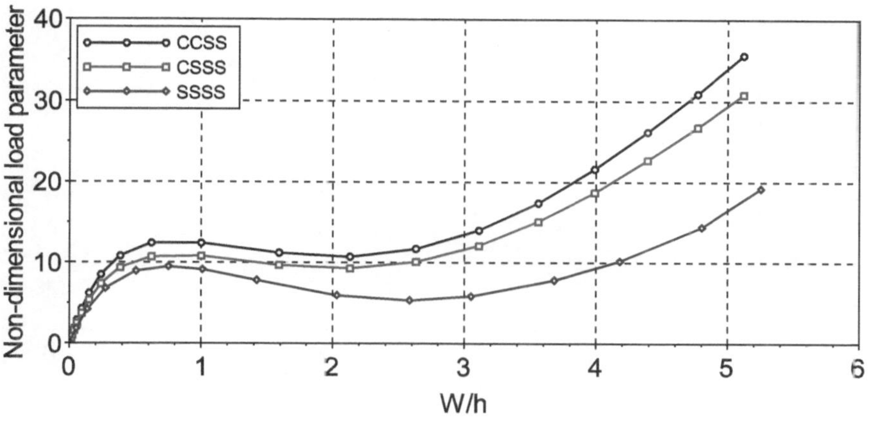

Figure 3.50 Post-buckling responses of the (Al/ZrO_2) cylindrical panel ($R = 1$, $h = 0.002$) subjected to different boundary conditions ($p = 1$).

condition is of type (CCSS) as defined earlier in section 5.2.1. The whole of the panel is meshed using 10×10 SHO4 finite elements. The results are evaluated in terms of non-dimensional parameters including the central deflection $\overline{W} = \frac{W}{h}$ and the load parameter $\overline{N} = NL^2/E_m h^3$ with N as the uniaxial applied load. Firstly, the comparison between the present results and those given by Liew et al. [147] is provided in Fig. 3.47. A good agreement is revealed and the little difference is due to the different theories. In fact, Liew et al. used the element-free kp-Ritz method which is based on the FSDT theory with Van-Karman assumptions while in the present model, a high-order shear deformable model with large deflections and rotations is provided. Hence, the present model constitutes a more general efficient model to predict the post-buckling behavior of FG cylindrical panels under uniaxial compression load. Next, we provide some new results within the proposed double directors finite element shell model. The influence of thickness geometrical parameter (h) on post-buckling responses of the (Al/ZrO_2) cylindrical panel under uniaxial compression load and with different power-law index is provided in Fig. 3.48. For that, three sets of thickness parameter ($h = 0.001$, $h = 0.002$ and $h = 0.004$) are considered. It can be seen that, as the thickness parameter decreases, as the post-buckling behavior of the FG cylindrical panel becomes more prominent. In fact, for low values of the thickness parameter (thin panel, $h = 0.001$), the critical buckling point reaches its maximum and the slope of the curve becomes deeper while this effect is attenuated for thick panel ($h = 0.001$). Therefore, the variation of the thickness parameter has a significant effect on post-buckling responses of FG cylindrical panels which must be considered in the design phase. In addition, from Fig. 3.48, it can be remarked that the post buckling strength of the (Al/ZrO_2) cylindrical panel decreases as the power-law index (p) increases. Indeed, the maximum value of the critical buckling point is reached for Zirconia (ZrO_2) phase, whereas, the minimum one occurs for the Aluminium (Al) phase and this can be explained by the high stiffness of the ceramic phase compared to the metallic one. Equally, the effect of the variation of the radius (R) parameter on the post-buckling of FG cylindrical panels are also examined. Figures 3.49(a), 3.49(b) and 3.49(c) show the post-buckling responses of the (Al/ZrO_2) cylindrical panel (CCSS) under uniaxial compression load and with different values of the power law-index (p) and radius parameter ($R = 1$, $R = 2$ and $R = 3$), respectively. In this case, the thickness parameter is equal to $h = 0.002$. Figure 3.49 reveals that the increase of the radius parameter (from $R = 1$ to $R = 3$) induces a decrease in the post-buckling strength of the (Al/ZrO_2) cylindrical panel and this is for all power-law index. In fact, with the rise of the radius parameter, the slope of the curve vanishes after the critical buckling

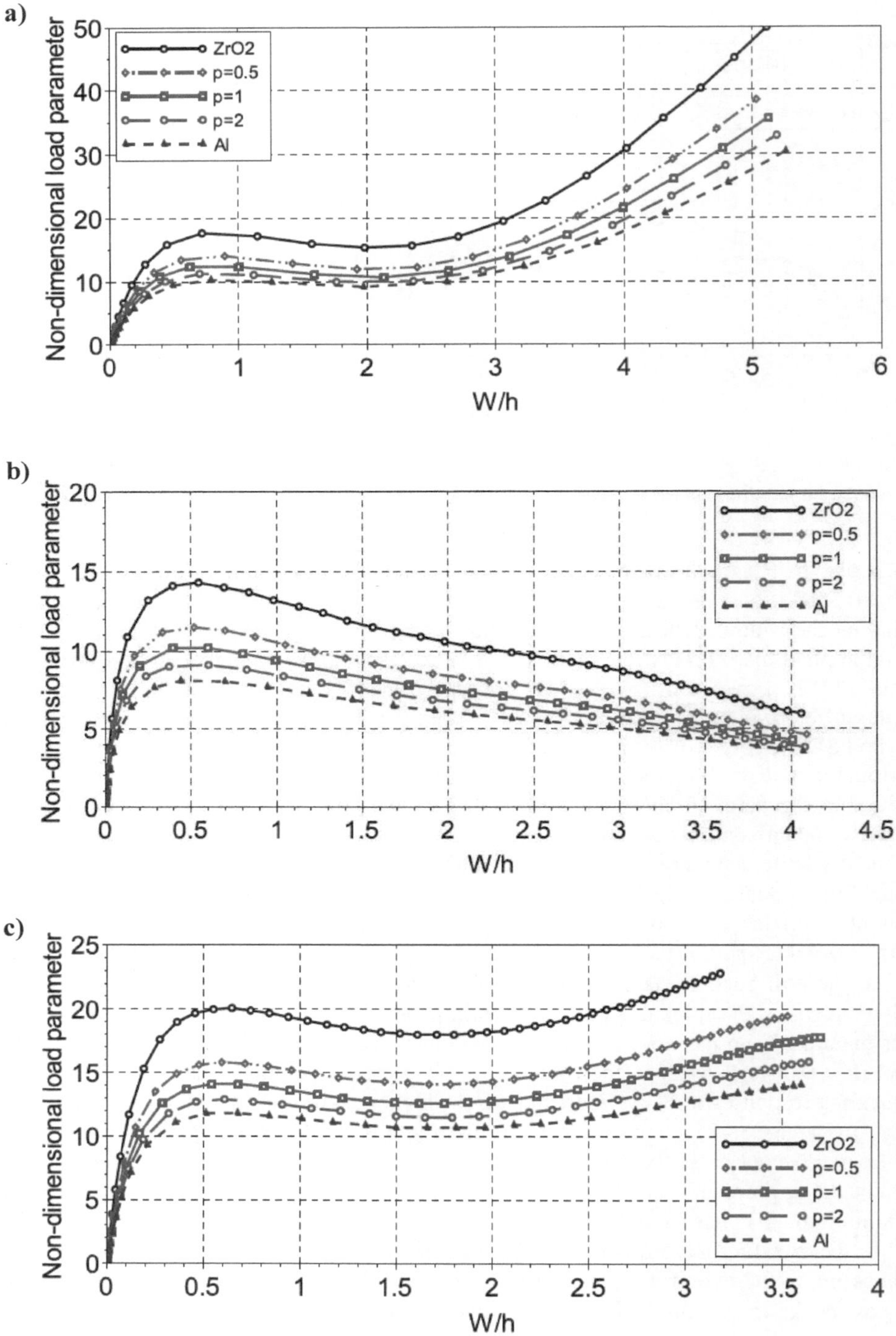

Figure 3.51 Influence of the span angle (θ_0) on the post-buckling responses of the (Al/ZrO_2) cylindrical panel (CCSS, $L = 0.2m$, $R = 1$, $h = 0.002$) under uniaxial compression load and with different power-law index and span angle (a) $\theta_0 = 0.2rad$, (b) $\theta_0 = 0.3rad$ and (c) $\theta_0 = 0.4rad$.

point and this can be explained by the diminution of the curvature effect and the panel becomes hence flat.

Continuously, the effect of boundary conditions on the post-buckling behavior of the (Al/ZrO_2) cylindrical panel under uniaxial compression load, subjected to three types of boundary conditions, namely, (CCSS, CSSS, SSSS) and with a power-law index ($p = 1$) is provided in Fig. 3.50. The FG panel has the same geometrical parameters as mentioned at the beginning of this section with ($h = 0.002$). For (CSSS) panel, the curved edge (AB) is clamped and the other one are simply supported while for (SSSS) one, all the edges of the panel are simply supported. As can be seen, the (CCSS) boundary condition induces the highest strength of the post-buckling response in contrast to the (SSSS) one which presents the lowest one. In fact, when more constraints are imposed on the panel, it becomes more stiffer due to its high flexural rigidity which manifest by a high value of the critical buckling point limit and a less deflection. As well as, the effect of variation of the span angle θ_0 on post-buckling responses of the (Al/ZrO_2) cylindrical panel ($L = 0.2m$, $R = 1$, $h = 0.002$) under uniaxial compression load is examined. Figures 3.51(a), 3.51(b) and 3.51(c) depict the post-buckling responses of the (Al/ZrO_2) cylindrical panel (CCSS, $L = 0.2m$, $R = 1$, $h = 0.002$) under uniaxial compression load and with different power-law index and for three sets of span angle ($\theta_0 = 0.2, 0.3, 0.4rad$), respectively. It can be seen, that the variation of the span angle from $\theta_0 = 0.2rad$ to $\theta_0 = 0.3rad$ induces a decrease in the post-buckling curves (for lower buckling loads) after the critical buckling point limit and this is for all power law-index. Then, the curves start to increase again in this zone (beyond the critical buckling point limit) with a span angle equal to $\theta_0 = 0.4rad$. This outlines that the FG panels are so sensitive to the variation of this parameter.

3.5 CONCLUSION

Thermal and mechanical buckling and post-buckling analysis of FGM structures using the extended Kirchhoff-Love model for thermal analysis and the double directors high-order shear deformation model (DD-HSDT) for mechanical analysis are presented in this chapter. Firstly, buckling and post-buckling responses of FGM plates and shells under thermal loadings are presented and discussed. The effects of several geometrical and material characteristics including the power-law index, aspect ratios, boundary conditions, thermal loadings and others' parameters are explored and examined. It is revealed that the functionally graded property enhances the mechanical and thermal behavior of structural components. Furthermore, buckling and post-buckling analysis of FGM plates and shells subjected to different edges compression loads are investigated with the variation of several parameters. The efficiency and reliability of these models in the prediction of the thermal and mechanical behaviors of FGM structures is highlighted via several numerical simulations. Comparative studies are also conducted in order to show the validity of these models and to outline their high accuracy in the analysis. The applicability of these models will be extended to remaining types of composite materials, namely functionally graded carbon nanotubes reinforced composites (FG-CNTRC) which constitute a new class of advanced materials that attracted the attention of several researchers, in the last decades. The effective material properties of these (FG-CNTRC) structures, as well as the performed analysis on this new issue will be presented and highlighted in the next chapter.

4 Mechanical analysis of FG-CNTRC structures

This chapter deals with mechanical analysis of functionally graded carbon nanotubes reinforced composite (FG-CNTRC) plates and shells using different kinematic models, which are developed in Chapter 1. Herein, static bending, stresses, free vibration and dynamic applications of FG-CNTRC plates and shells are performed to show the effects of carbon nanotubes volume fraction, types of distributions, namely uniform (UD) and functionally graded distributions (FG-V, FG-A, FG-O and FG-X) on mechanical responses of such structural components. The effective materials properties of FG-CNTRC structures are determined using an extended rule of mixture to capture the scale dependency of carbon nanotubes. Different applications to FG-CNTRC square plates, skew plates and curved panels are presented to illustrate the efficiency and reliability of developed finite element models in the prediction of mechanical behavior of composites structures reinforced by carbon nanotubes .

4.1 INTRODUCTION

The carbon nanotubes (CNTs) were discovered by Iijima [104] in 1991 and they consist of rolled graphene sheets with a cylindrical shape. They could be either single-walled (SWCNTs) or multi-walled (MWCNTs) in nanosized scale. In addition, these CNTs are considered the first advanced material of the twenty-first century due to their unique morphology, novel physico-chemical properties, exceptional mechanical, electrical, and thermal properties and, also, their versatile applications. In fact, many experimental and theoretical results [277, 168, 74] have been focused on their material properties. For instance Esawi and Farag [72] have outlined that the Young's moduli of CNTs are greater than 1 TPa and tensile strength exceeds that of steel by over one-order of magnitude. Such kinds of properties make them an alternative of reinforcements in composite structures [132]. Hence, the incorporation of CNTs into composite structures leads to a novel class of structural members labeled (CNTRCs). These CNTRCs can be embedded in beam, plate or shell as structural components with nonuniform distribution across the whole of structure. Indeed, according to the traditional approach of manufacturing of nanocomposites, these CNTRCs can be distributed either uniformly or randomly such that the resulting mechanical, thermal, or physical properties do not vary spatially at the macroscopic level.

To account for this distribution, the concept of functionally graded materials (FGMs) [146] can be combined with carbon nanotubes reinforcements formed hence the so-called functionally graded carbon nanotube reinforced composites (FG-CNTRCs). These new trends of materials have attracted many industries such as aerospace, automobile, energy, structural and chemical industry. For instance, in the electronic field, these FG-CNTRCs are used to increase the conductivity of electrodes in lead-acid batteries due to their superconducting properties at low temperature. On the other hand, the good damping properties of FG-CNTRCs and thermal and chemical resistance influenced the aerospace industry to manufacture various parts, which include rotor blades, stabilizers and fuselage portions. Furthermore, regarding the weight reduction and good structural properties of CNTs, the FG-CNTRCs are the best replacements for composites. As well as, the lightweight metallic, ceramic, and polymeric composites for armor and armaments and barrier materials for chemical-biological protection in military applications focus on utilization of FG-CNTRCs. In the academic field, this combination was first proposed by Shen [223] and the results demonstrated that non-linear bending behavior can be considerably improved through the use of a functionally graded

DOI: 10.1201/9781003483786-4

Table 4.1
The volume fraction of CNTs for various types of CNT distribution [224]

CNTs distribution	V_{CNT}
UD	V^*_{CNT}
FG-V	$(1+2\frac{z}{h})V^*_{CNT}$
FG-O	$2(1-2\frac{\lvert z \rvert}{h})V^*_{CNT}$
FG-X	$4\frac{\lvert z \rvert}{h}V^*_{CNT}$
FG-Λ	$(1-2\frac{z}{h})V^*_{CNT}$

distribution of carbon nanotubes in the matrix. Then, following his work, various investigations were reported on the mechanics behavior of FG-CNTRC structures and most of them are recorded in a recent review elaborated by Liew et al.[145]. In this review, the authors show that many aspects of mechanical behavior analysis of FG-CNTRCs were carried out covering static, free vibration, buckling, post-buckling and non-linear analysis. Even several works are done on this interesting thematic, in the last decades, our attention will be on linear and non-linear static bending analysis, as well as free vibration, dynamic, buckling and post-buckling analysis of FG-CNTRC plates and shells within the developed theoretical frameworks in Chapter 1. Our contributions through these works consists of the implementation of improved and enhanced finite element models, unlike the standard (CPT), (FSDT) or (HSDT) theories presented in the literature. Furthermore, some analysis conducted on dynamic, buckling ans post-buckling responses of (FG-CNTRC) structures constitutes the first new results on this area. Therefore, the proposed models are recommended to investigate a wide range of analysis, especially for complex geometries where the fast computational time is required with a high accuracy. In this chapter, the effective material properties of (FG-CNTRC) structures is first presented. Then, a series of different numerical applications are presented and discussed including comparison studies and parametric analysis to show the effects of carbon nanotubes (CNTs) and its forms of distributions on the mechanical responses of such structures.

4.2 MATERIAL PROPERTIES OF FG-CNTRC SHELLS

The considered CNTRC shell is made from a mixture of single-walled carbon nanotubes (SWCNTs) which are functionally graded in the thickness direction of isotropic matrix. Uniform distribution (UD) and three types of functionally graded distributions of CNTs are used and denoted by FG-V, FG-O and FG-X, in which the top, the mid-plane, and both top and bottom surfaces of the shell are rich of CNTs, respectively. According to the investigations of Shen et al. [223, 225, 228], the volume fractions of CNTs for each type of distribution, can be expressed as indicated in Tab. 8.29. The profiles of these distributions are illustrated in Fig. 4.1. From Tab. 4.1, it can be remarked that both uniform and functionally graded profiles of CNTRC shells will have the same total volume fraction of CNTs which is denoted V^*_{CNT}. The expression of V^*_{CNT} is given, as a function of mass density of CNTs, ρ^{CNT}, mass density of matrix ρ^m and mass fraction of CNTs w_{CNT} as follows:

$$V^*_{CNT} = \frac{w_{CNT}}{w_{CNT} + \frac{\rho^{CNT}}{\rho^m} - \frac{\rho^{CNT}}{\rho^m} w_{CNT}}. \tag{4.1}$$

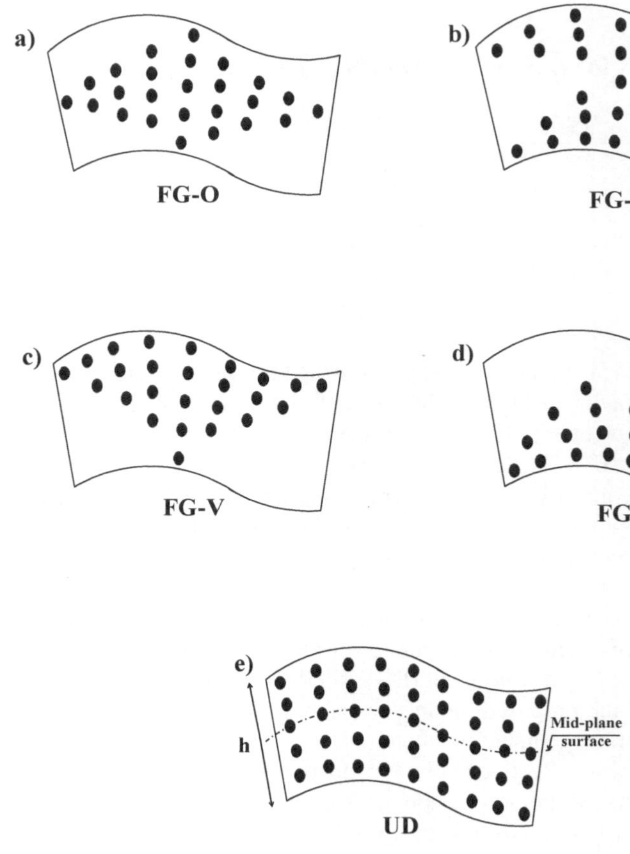

Figure 4.1 The different profiles of distributions of CNTRC shells.

The effective material properties of the CNTRC shell are estimated using the Modified Rule of Mixtures (MROM) as:

$$\begin{aligned} E_{11} &= \eta_1 V_{CNT} E_{11}^{CNT} + V_m E_m, \\ \frac{\eta_2}{E_{22}} &= \frac{V_{CNT}}{E_{22}^{CNT}} + \frac{V_m}{E_m}, \\ \frac{\eta_3}{G_{12}} &= \frac{V_{CNT}}{G_{12}^{CNT}} + \frac{V_m}{G_m}, \end{aligned} \tag{4.2}$$

where E_{11}^{CNT}, E_{22}^{CNT} and E_m are the Young's modulus of CNTs and isotropic matrix respectively. G_{12}^{CNT} and G_m are the shear modulus of CNTs and matrix, respectively. In addition, η_i (i = 1,2,3), are the CNT efficiency parameters which can be determined by matching the effective parameters of CNTRC obtained from the molecular dynamic (MD) simulations [95]. V_{CNT} and V_m are the volume fractions of the CNTs and the isotropic matrix respectively, which are related by the following expression:

$$V_m + V_{CNT} = 1. \tag{4.3}$$

The expressions of the Poisson ratio ν_{12} and the mass density ρ are given by:

$$\nu_{12} = V_{CNT}^{*} \nu_{12}^{CNT} + V_m \nu_m, \tag{4.4}$$

Table 4.2

Material properties of PmPV matrix at room temperature (300K) and (10,10) single-walled carbon nanotube (SWCNTs) [223].

(10,10) SWCNTs	PmPV matrix
$E_{11}^{CNT} = 5.6466TPa$	$E_m = 2.1GPa$
$E_{22}^{CNT} = 7.0800TPa$	$\rho^m = 1150Kg/m^3$
$G_{12}^{CNT} = 1.9445TPa$	$v_m = 0.34$
$\rho^{CNT} = 1400Kg/m^3$	
$v_{12}^{CNT} = 0.175$	

Table 4.3

The values of CNT efficiency parameters for three different values of CNTs volume fractions [223].

V_{CNT}^*	η_1	η_2	η_3
0.11	0.149	0.934	0.934
0.14	0.150	0.941	0.941
0.17	0.149	1.381	1.381

and

$$\rho = V_{CNT}\rho^{CNT} + V_m\rho^m, \tag{4.5}$$

where as mentioned earlier, ρ^{CNT} and ρ^m are the mass density of the CNT and matrix phases, respectively, and v_{12}^{CNT} and v_m are the Poisson ratio of the CNT and the matrix, respectively.

4.3 LINEAR STATIC AND STRESSES RESPONSES OF FG-CNTRC PLATES

In this section, linear static and stresses responses of FG-CNTRC structures are presented. The present results are investigated using the double directors high-order shear deformable shell model (DD-HSDT). In the presented works, poly(m-phenylenevinylene)-co-[(2,5-dioctoxy-p-phenylene) vinylene], referred to as (PmPV) is considered as the matrix and the used CNTs are assumed to be of the type of armchair (10,10) single-walled carbon nanotube (SWCNTs). The material properties of both matrix and CNTs are given according to molecular dynamic (MD) simulation as shown in Tab. 4.2. Furthermore, the CNT efficiency parameters η_i (i $= 1,2,3$) introduced in Eq. (4.2), can be estimated by matching Young's and shear moduli of CNTRC given by the modified rule of mixture to those obtained by the molecular dynamics simulations [95]. The values of these parameters for three different values of CNT volume fractions are given in Tab. 4.3. Moreover, we assume that $G_{12} = G_{13} = G_{23}$.

4.3.1 FG-CNTRC SQUARE PLATE

4.3.1.1 Static deflections of FG-CNTRC plates

The static analysis of FG-CNTRC square plate subjected to uniformly distributed transverse load equal to $q_0 = -1.0 \times 10^5 N/m^2$ are performed, where the geometry of the plate is as illustrated in Fig. 4.2. Various distributions of CNTs are considered: UD, FG-O, FG-X and FG-V with different sets

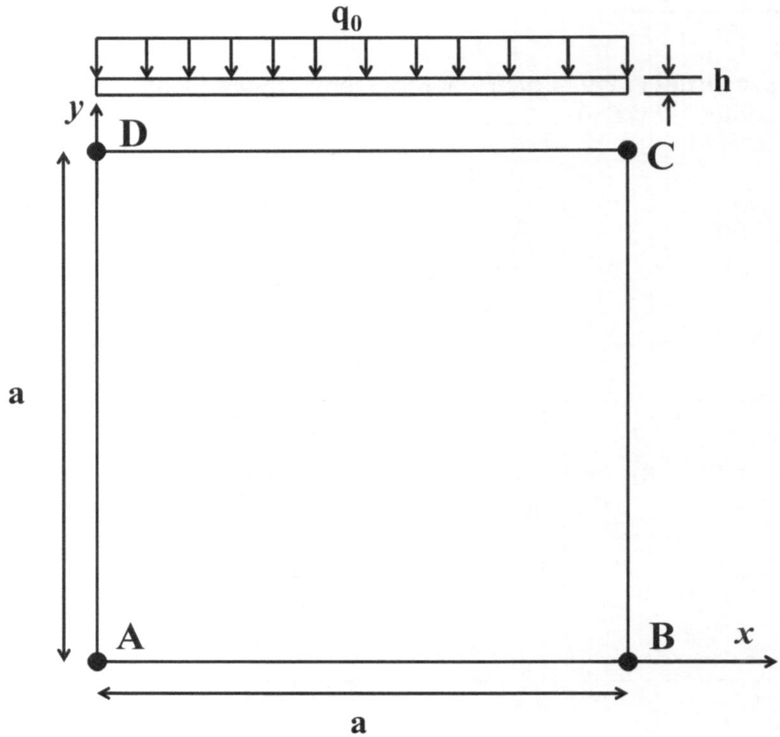

Figure 4.2 Geometry of FG-CNTRC square plate.

of length-to-thickness ratios ($a/h = 10, 20, 50$) and boundary conditions. The used non-dimensional parameters are listed in Tab. 4.4. The comparison of the central non-dimensional transverse deflection $\overline{w}$ with those of published literature [328] is illustrated in Tabs. 4.5 and 4.6. As can be seen, a good agreement among the results is shown which draw the correctness and reliability of the present (DD-HSDT) finite shell model in the analysis of the bending behavior of FG-CNTRC plates. In fact, the relative error between the obtained non-dimensional transverse deflection and those given by Zhu et al.[328] does not exceed generally 3% for various length-to-thickness ratios, types of CNTs distributions and boundary conditions except for the case where the length-to-thickness ratio is equal to 10, this error can reach 13% for some distributions (for instance, this error is about 13% for the simply supported FG-V-CNTRC plate with different volume fractions of CNTs V_{CNT}^{*}). The difference between results, in this case can be explained by the limitation of the applicability of the FSDT theory used in [328]. So, regarding the length-to-thickness ratio (10), which indicates a thick plate, the used FSDT theory in literature can generate a significant difference due its introduction of shear correction coefficients. Also, according to the tables, when the length-to-thickness ratio increases the FG-CNTRC plate behaves as thin plate and the reference results [328] converges well with the results of the present formulation. Consequently, the proposed double directors shell formulation is more general and convenient for all types (thin, moderately thick and thick) of FG-CNTRC plates.

Table 4.4
Non-dimensional used parameters.

Non-dimensional parameter	Expression
Central deflection	$\overline{w} = \frac{1}{h} w(\frac{a}{2}, \frac{a}{2})$
Central axial stress	$\overline{\sigma}_{xx} = \frac{h^2}{\lvert q_0 \rvert a^2} \sigma_{xx}(\frac{a}{2}, \frac{a}{2})$
In-plane shear stress	$\overline{\sigma}_{xy} = \frac{h^2}{\lvert q_0 \rvert a^2} \sigma_{xy}(0,0)$
Shear transverse stresses	$\overline{\sigma}_{xz} = \frac{h^2}{\lvert q_0 \rvert a^2} \sigma_{xz}(0, \frac{a}{2})$
	$\overline{\sigma}_{yz} = \frac{h^2}{\lvert q_0 \rvert a^2} \sigma_{yz}(\frac{a}{2}, 0)$
Thickness coordinates	$\overline{z} = \frac{z}{h}$

4.3.1.2 Stresses responses of FG-CNTRC plates

Secondly, the non-dimensional central axial stresses $\overline{\sigma}_{xx}$ versus the thickness direction of the FG-CNTRC square plate are depicted in Fig. 4.3 and 4.4 for various boundary conditions with the volume faction ($V_{CNT}^* = 0.17$) and length-to-thickness ratio ($a/h = 50$). The obtained results are similar to those given in [328] which affirms the validity of the present formulation. Figs. 4.3 and 4.4 show that the symmetry of UD, FG-O and FG-X distributions implies an anti-symmetry on the central axial stresses along the thickness direction. However, for FG-V distribution, the central axial stress is close to zero at the bottom surface where the concentration of CNTs vanishes. Furthermore, the values of the axial stresses corresponding to the clamped FG-CNTRC plate in all edges are much smaller than those derived from the other boundary conditions whereas when the FG-CNTRC plate presents an identical boundary condition along the longitudinal direction (SSSS, SCSC and SFSF), the values of $\overline{\sigma}_{xx}$ are very close and reaches its maximum.

On the other hand, and since the FG-O-CNTRC square plate presents the maximum of non-dimensional deflection, the effect of CNTs' volume fraction on axial stress $\overline{\sigma}_{xx}$, in plane shear stress $\overline{\sigma}_{xy}$ and transverse shear stresses $\overline{\sigma}_{xz}$ and $\overline{\sigma}_{yz}$ for this distribution are examined for a thin ($a/h = 50$) simply supported FG-CNTRC square plate. Indeed, Figs. 4.5(a) and 4.5(b) show that the variation of volume fraction of CNTs V_{CNT}^* does not have a significant effect on axial $\overline{\sigma}_{xx}$ and in-plane shear $\overline{\sigma}_{xy}$ stresses while this effect is more pronounced for transverse shear stresses $\overline{\sigma}_{xz}$ and $\overline{\sigma}_{yz}$. Figures 4.6(c) and 4.6(d) show that the increase of V_{CNT}^* leads to a decrease in the amplitude of axial and transverse shear stresses which reaches its minimum in the mid-surface (z = 0) where the concentration of CNTs is maximal. In top ($z = +h/2$) and bottom ($z = -h/2$) surfaces, the amplitudes of $\overline{\sigma}_{xz}$ and $\overline{\sigma}_{yz}$ are equal to zero due to the deformation kinematic assumptions which are based on the satisfaction of the zero condition of the transverse shear stress in both top and bottom surfaces. Therefore, since the volume fraction V_{CNT} of CNTs is graded through the thickness of the CNTRC plate, the shear parts of stresses are those influenced by the variation of V_{CNT} while the membrane parts remains intact.

Further conclusions can be derived from Fig. 4.7 where the effect of boundary conditions for a thick FG-X-CNTRC square plate ($a/h = 10$) and CNTs with a volume fraction $V_{CNT}^* = 0.11$ is performed. As can be seen from Fig. 4.7, the type of introduced boundary conditions can affect the amplitude of non-dimensional axial stress $\overline{\sigma}_{xx}$. Indeed, since the constraint of clamped edge is stronger than simply supported or free edges, the amplitude of $\overline{\sigma}_{xx}$ with two edges simply supported and others free (SFSF) is higher than those with four edges simply supported (SSSS), clamped (CCCC) or two opposite edges simply supported and the other two clamped (SCSC). Hence, the type of imposed boundary conditions is pronounced for a thick FG-CNTRC square plate.

Table 4.5

Non-dimensional central deflection $\overline{w}$ for FG-CNTRC square plate with different volume fractions of CNTs, length-to-thickness ratio and boundary conditions (SSSS, CCCC).

V^*_{CNT}	a/h		SSSS		CCCC	
			Present	Ref.[328]	Present	Ref.[328]
0.11	10	UD	3.725×10^{-3}	3.739×10^{-3}	1.982×10^{-3}	2.228×10^{-3}
		FG-V	3.869×10^{-3}	4.466×10^{-3}	2.118×10^{-3}	2.351×10^{-3}
		FG-O	5.411×10^{-3}	5.230×10^{-3}	2.546×10^{-3}	2.512×10^{-3}
		FG-X	3.145×10^{-3}	3.177×10^{-3}	1.816×10^{-3}	2.109×10^{-3}
	20	UD	3.633×10^{-2}	3.628×10^{-2}	1.292×10^{-2}	1.339×10^{-2}
		FG-V	3.885×10^{-2}	4.879×10^{-2}	1.548×10^{-2}	1.593×10^{-2}
		FG-O	6.235×10^{-2}	6.155×10^{-2}	1.928×10^{-2}	1.860×10^{-2}
		FG-X	2.699×10^{-2}	2.701×10^{-2}	1.088×10^{-2}	1.150×10^{-2}
	50	UD	1.156	1.155	0.262	0.2618
		FG-V	1.257	1.653	0.3650	0.3649
		FG-O	2.162	2.157	0.4796	0.4719
		FG-X	0.791	0.790	0.1888	0.1894
0.14	10	UD	3.280×10^{-3}	3.306×10^{-3}	1.818×10^{-3}	2.087×10^{-3}
		FG-V	3.393×10^{-3}	3.894×10^{-3}	1.925×10^{-3}	2.182×10^{-3}
		FG-O	4.669×10^{-3}	4.525×10^{-3}	2.306×10^{-3}	2.313×10^{-3}
		FG-X	2.830×10^{-3}	2.844×10^{-3}	1.685×10^{-3}	1.979×10^{-3}
	20	UD	2.995×10^{-2}	3.001×10^{-2}	1.133×10^{-2}	1.188×10^{-2}
		FG-V	3.199×10^{-2}	4.025×10^{-2}	1.338×10^{-2}	1.390×10^{-2}
		FG-O	5.123×10^{-2}	5.070×10^{-2}	1.651×10^{-2}	1.604×10^{-2}
		FG-X	2.258×10^{-2}	2.256×10^{-2}	1.000×10^{-2}	1.036×10^{-2}
	50	UD	0.9159	0.9175	0.2123	0.2131
		FG-V	0.9976	0.9976	0.2942	0.2955
		FG-O	1.736	1.738	0.3852	0.3805
		FG-X	0.6265	0.6271	0.1553	0.1560
0.17	10	UD	2.386×10^{-3}	2.394×10^{-3}	1.259×10^{-3}	1.412×10^{-3}
		FG-V	2.475×10^{-3}	2.864×10^{-3}	1.346×10^{-3}	1.486×10^{-3}
		FG-O	3.456×10^{-3}	3.378×10^{-3}	1.593×10^{-3}	1.595×10^{-3}
		FG-X	2.033×10^{-3}	2.012×10^{-3}	1.169×10^{-3}	1.318×10^{-3}
	20	UD	2.352×10^{-2}	2.348×10^{-2}	8.284×10^{-3}	8.561×10^{-3}
		FG-V	2.517×10^{-2}	3.173×10^{-2}	9.970×10^{-3}	1.021×10^{-2}
		FG-O	4.053×10^{-2}	4.020×10^{-2}	1.228×10^{-2}	1.198×10^{-2}
		FG-X	1.754×10^{-2}	1.737×10^{-2}	7.061×10^{-2}	7.290×10^{-2}
	50	UD	0.7535	0.7515	0.1703	0.1698
		FG-V	0.820	1.082	0.2387	0.2384
		FG-O	1.417	1.416	0.3125	0.3085
		FG-X	0.5158	0.5132	0.1231	0.1223

4.3.2 FG-CNTRC CYLINDRICAL PANEL

4.3.2.1 Static deflections of FG-CNTRC panels

Now, a FG-CNTRC cylindrical panel of length L, radius R, span angle θ_0 and thickness h is considered. The values of the geometrical parameters are assumed as: $\theta_0 = 0.1rad$, $h = 0.002m$, $h/R = 0.002$ and $L/R = 0.1$. Moreover, the CNTs are assumed to be uniaxially aligned in axial direction and functionally graded in thickness direction of the cylindrical panel, as shown in Fig. 5.41. To validate the static analysis in this case, the obtained results for FG-CNTRC cylindrical shell panel under uniformly distributed load $q_0 = 0.1MPa$ are compared with those given by Zhang et al. [316]. In the present analysis, a discretization with 32×20 SHO4 finite elements of the FG-CNTRC

Table 4.6

Non-dimensional central deflection $\overline{w}$ for FG-CNTRC square plate with different volume fractions of CNTs , length-to-thickness ratio and boundary conditions (SCSC, SFSF).

V_{CNT}^{*}	a/h		SCSC		SFSF	
			Present	Ref.[328]	Present	Ref.[328]
0.11	10	UD	3.323×10^{-3}	3.325×10^{-3}	3.425×10^{-3}	3.444×10^{-3}
		FG-V	3.426×10^{-3}	3.853×10^{-3}	3.568×10^{-3}	4.186×10^{-3}
		FG-O	4.558×10^{-3}	4.433×10^{-3}	5.152×10^{-3}	4.967×10^{-3}
		FG-X	2.851×10^{-3}	2.867×10^{-3}	2.868×10^{-3}	2.905×10^{-3}
	20	UD	3.406×10^{-2}	3.628×10^{-2}	3.339×10^{-2}	3.341×10^{-2}
		FG-V	3.610×10^{-2}	4.433×10^{-2}	3.574×10^{-2}	4.544×10^{-2}
		FG-O	5.454×10^{-2}	5.389×10^{-2}	5.867×10^{-2}	5.797×10^{-2}
		FG-X	2.593×10^{-2}	2.587×10^{-2}	2.477×10^{-2}	2.484×10^{-2}
	50	UD	1.104	1.099	1.067	1.063
		FG-V	1.188	1.504	1.160	1.540
		FG-O	1.9162	1.909	2.031	2.030
		FG-X	0.7765	0.7728	0.7337	0.7338
0.14	10	UD	2.961×10^{-3}	2.974×10^{-3}	3.000×10^{-3}	3.029×10^{-3}
		FG-V	3.043×10^{-3}	3.412×10^{-3}	3.111×10^{-3}	3.615×10^{-3}
		FG-O	4.018×10^{-3}	3.916×10^{-3}	4.378×10^{-3}	4.236×10^{-3}
		FG-X	2.581×10^{-3}	2.583×10^{-3}	2.574×10^{-3}	2.593×10^{-3}
	20	UD	2.853×10^{-2}	2.852×10^{-2}	2.749×10^{-2}	2.760×10^{-2}
		FG-V	3.023×10^{-2}	3.689×10^{-2}	2.936×10^{-2}	3.722×10^{-2}
		FG-O	4.603×10^{-2}	4.557×10^{-2}	4.761×10^{-2}	4.719×10^{-2}
		FG-X	2.191×10^{-2}	2.184×10^{-2}	2.076×10^{-2}	2.078×10^{-2}
	50	UD	0.8896	0.8890	0.8468	0.8505
		FG-V	0.960	1.234	0.920	1.229
		FG-O	1.583	1.583	1.613	1.619
		FG-X	0.6217	0.6206	0.5837	0.5854
0.17	10	UD	2.124×10^{-3}	2.124×10^{-3}	2.195×10^{-3}	2.207×10^{-3}
		FG-V	2.184×10^{-3}	2.461×10^{-3}	2.287×10^{-3}	2.691×10^{-3}
		FG-O	2.915×10^{-3}	2.865×10^{-3}	3.289×10^{-3}	3.206×10^{-3}
		FG-X	1.827×10^{-3}	1.804×10^{-3}	1.858×10^{-3}	1.842×10^{-3}
	20	UD	2.199×10^{-2}	2.190×10^{-2}	2.161×10^{-2}	2.162×10^{-2}
		FG-V	2.328×10^{-2}	2.834×10^{-2}	2.317×10^{-2}	2.961×10^{-2}
		FG-O	3.543×10^{-2}	3.518×10^{-2}	3.816×10^{-2}	3.787×10^{-2}
		FG-X	1.673×10^{-2}	1.653×10^{-2}	1.608×10^{-2}	1.596×10^{-2}
	50	UD	0.7169	0.7135	0.6952	0.6950
		FG-V	0.7717	0.9790	0.7569	1.010
		FG-O	1.253	1.252	1.332	1.333
		FG-X	0.502	0.4990	0.4770	0.4758

cylindrical panel is established, then the non-dimensional central deflection $\overline{w}$ are derived for different volume fractions ($V_{CNT}^{*} = 0.11, 0.14, 0.17$) and two types of boundary conditions: all sides (AB), (BC), (CD) and (DA) are simply supported (SSSS) or clamped (CCCC). In this example, only the matrix is changed which is considered here of Poly (methyl methacrylate), referred as PMMA with young modulus $E_m = 2.5GPa$. The others material properties are the same used for the FG-CNTRC square plate.

Table 4.7 lists the obtained non-dimensional central deflection $\overline{w}$ compared to those given in [316] for different volume fractions V_{CNT}^{*}. Table 4.7 reveals a good agreement between the results leading hence to validate again the capability of the present formulation based on double directors shell element to reproduce with accuracy the central deflections of FG-CNTRC cylindrical panel.

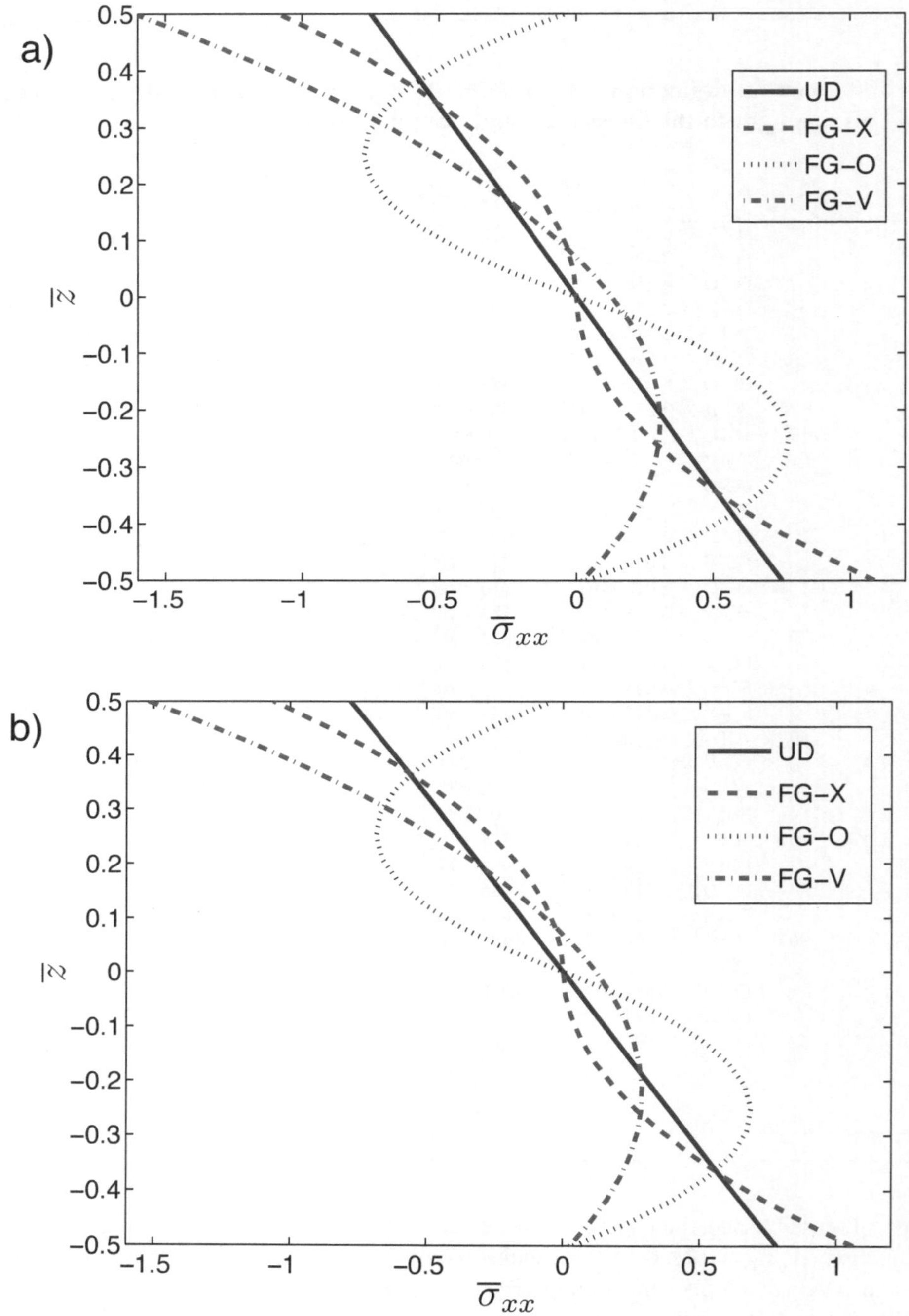

Figure 4.3 Variation of non-dimensional central axial stresses $\overline{\sigma}_{xx}$ through the thickness for FG-CNTRC square plate subjected to uniform load $q_0 = -0.1MPa$ and various boundary conditions: a) SSSS b) SCSC with $(a/h = 50)$ and $(V^*_{CNT} = 0.17)$.

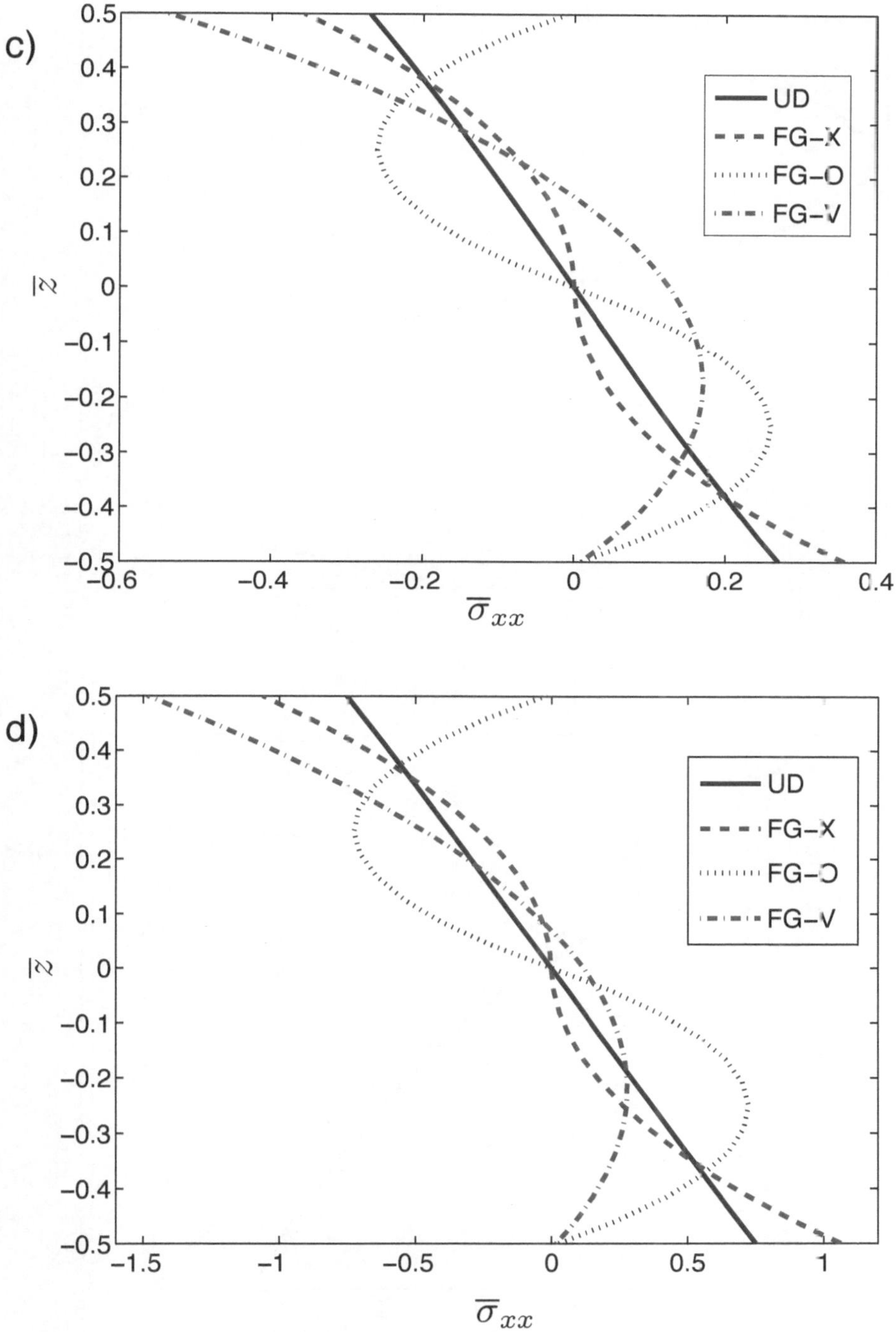

Figure 4.4 Variation of non-dimensional central axial stresses $\overline{\sigma}_{xx}$ through the thickness for FG-CNTRC square plate subjected to uniform load $q_0 = -0.1MPa$ and various boundary conditions: c) CCCC d) SFSF with $(a/h = 50)$ and $(V^*_{CNT} = 0.17)$.

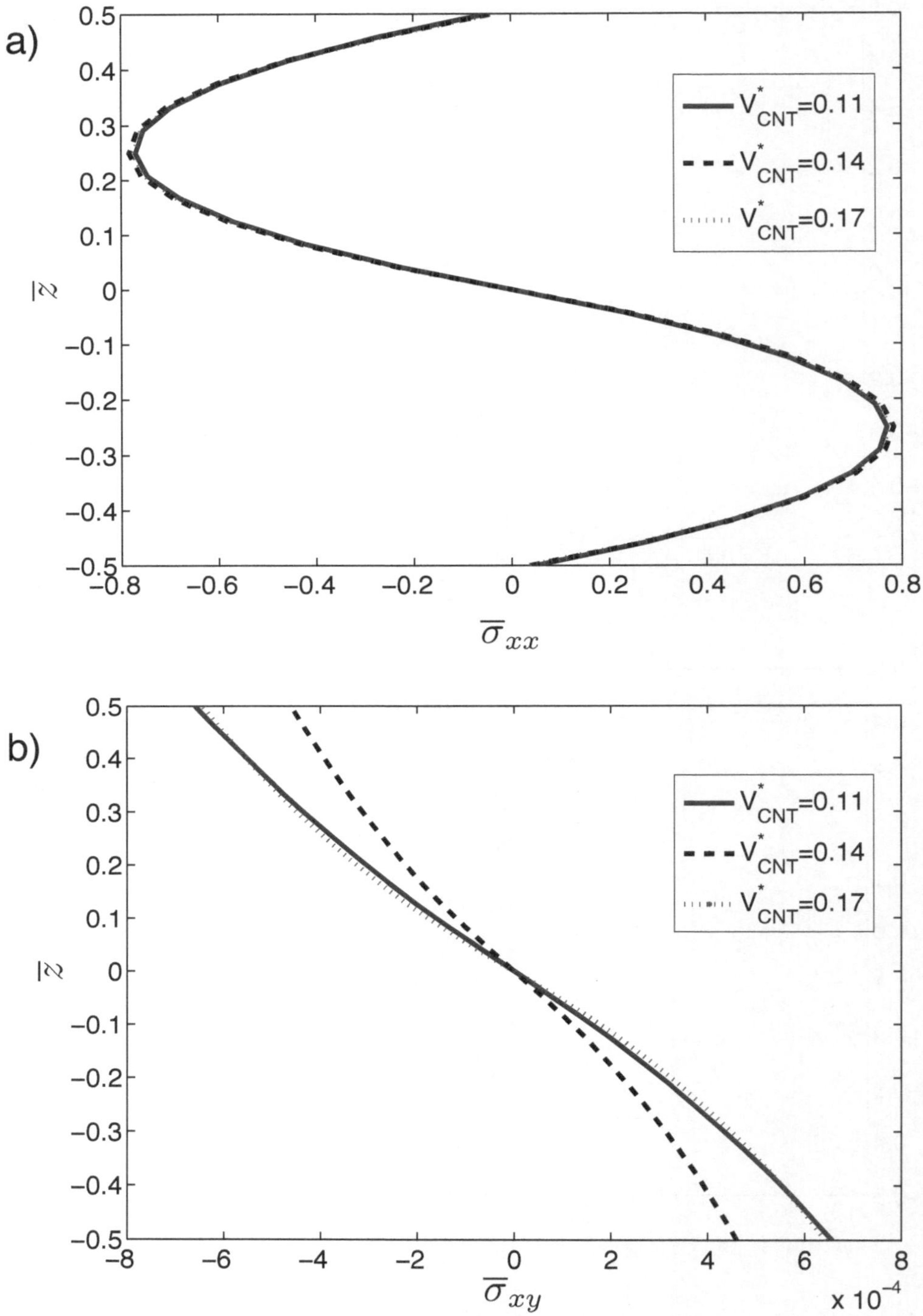

Figure 4.5 Effect of variation of CNTs volume fraction V^*_{CNT} on a) non-dimensional axial stress $\overline{\sigma}_{xx}$ and b) non-dimensional in plane stress $\overline{\sigma}_{xy}$ for thin $(a/h = 50)$ simply supported FG-O-CNTRC square plate under a uniform load $q_0 = -0.1MPa$.

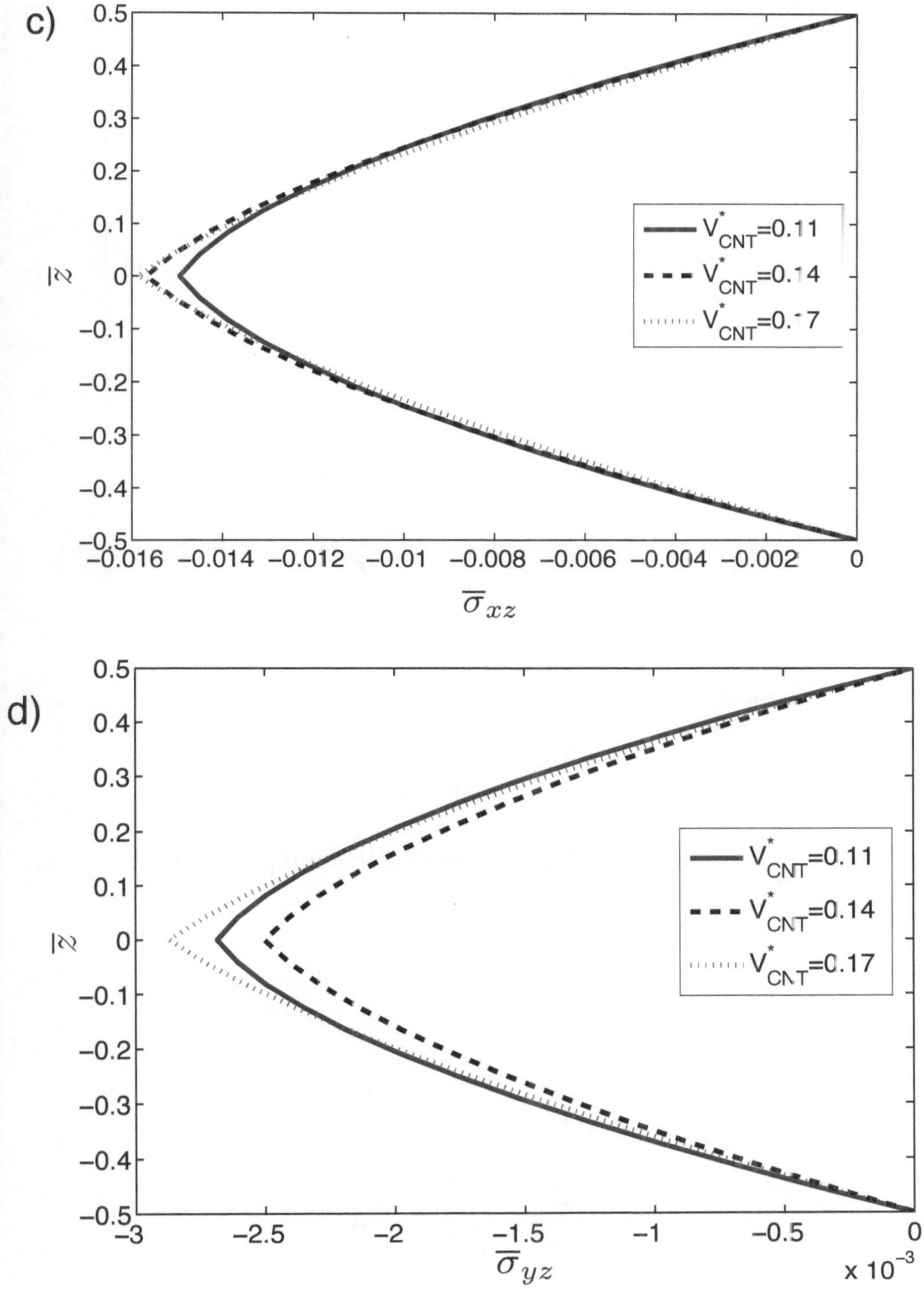

Figure 4.6 Effect of variation of CNTs volume fraction V_{CNT}^{*} on c) non-dimensional axial shear stress $\overline{\sigma}_{xz}$ and d) non-dimensional transverse shear stress $\overline{\sigma}_{yz}$ for thin $(a/h = 50)$ simply supported FG-O-CNTRC square plate under a uniform load $q_0 = -0.1MPa$.

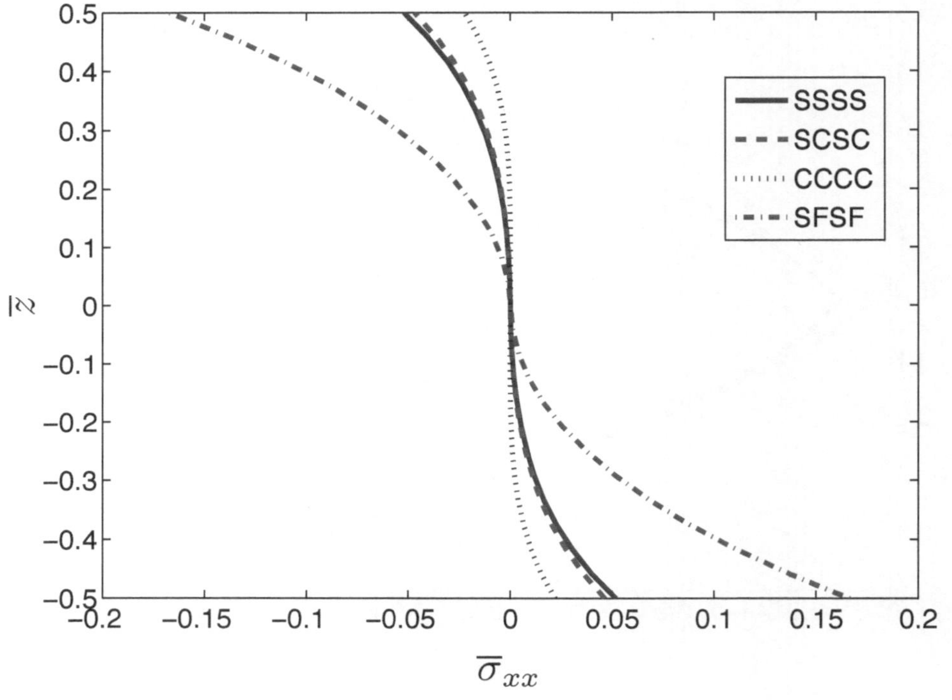

Figure 4.7 Effect of boundary conditions on the non-dimensional axial stress $\overline{\sigma}_{xx}$ for FG-X-CNTRC square plate under a uniform load $q_0 = -0.1MPa$ ($a/h = 10$, $V_{CNT}^* = 0.11$).

Table 4.7

Non-dimensional central deflection $\overline{w}$ for FG-CNTRC cylindrical panel with different volume fraction of CNTs.

V_{CNT}^*		SSSS		CCCC	
		Present	Ref.[316]	Present	Ref.[316]
0.11	UD	1.1279	1.1156	0.2508	0.2500
	FG-V	1.6002	1.5515	0.3468	0.3482
	FG-O	2.0648	2.0221	0.4538	0.4477
	FG-X	0.7765	0.7708	0.1799	0.1800
0.14	UD	0.8966	0.8888	0.2028	0.2024
	FG-V	1.2856	1.2536	0.2804	0.2815
	FG-O	1.6713	1.6421	0.3664	0.3613
	FG-X	0.6155	0.6114	0.1473	0.1472
0.17	UD	0.7354	0.7270	0.1631	0.1625
	FG-V	1.0486	1.0171	0.2272	0.2281
	FG-O	1.3557	1.3287	0.2965	0.2935
	FG-X	0.5059	0.5009	0.1173	0.1165

In addition, Table 4.8 compares non-dimensional central deflection of the FG-CNTRC cylindrical panel with different edge-to-radius ratios (L/R) and for a volume fraction $V_{CNT}^* = 0.11$. The comparison shows an excellent agreement between the obtained non-dimensional central deflection and

Table 4.8

Non-dimensional central deflection $\overline{w}$ for FG-CNTRC cylindrical panel with different edge-to-radius ratios (L/R) and volume fraction $V_{CNT}^* = 0.11$ of CNTs.

L/R		SSSS		CCCC	
		Present	Ref.[316]	Present	Ref.[316]
0.1	UD	1.1279	1.1156	0.2508	0.2500
	FG-V	1.6002	1.5515	0.3468	0.3482
	FG-O	2.0648	2.0221	0.4538	0.4477
	FG-X	0.7765	0.7708	0.1799	0.1800
0.15	UD	4.9940	4.8888	1.0939	1.0878
	FG-V	5.0721	6.1967	1.4376	1.4453
	FG-O	8.0749	7.8261	1.7947	1.7831
	FG-X	3.6060	3.5468	0.7946	0.7908
0.2	UD	11.905	11.585	2.5154	2.4977
	FG-V	14.686	13.391	3.0179	3.0325
	FG-O	16.827	16.278	3.5338	3.5187
	FG-X	9.1526	8.9446	1.9488	1.9349
0.25	UD	19.835	19.275	3.9046	3.8725
	FG-V	22.801	20.761	4.3196	4.3310
	FG-O	25.086	24.350	4.8034	4.7848
	FG-X	16.210	15.797	3.2460	3.2184
0.3	UD	26.698	25.985	4.8547	4.8105
	FG-V	29.0877	26.752	5.0785	5.0786
	FG-O	31.2656	30.502	5.4588	5.4368
	FG-X	22.8983	22.318	4.2667	4.2268

those cited in [316] for different edge-to-radius ratios (L/R). Similar concordance is revealed from Tab. 4.9 of the normalized central deflection when moderately thicker FG-CNTRC cylindrical panels are considered ($h = 0.004m$). Hence, the developed SHO4 discrete double directors shell element has the aptitude to draw the static behavior of FG-CNTRC shell structures with a good accuracy.

4.3.2.2 Stresses analysis of FG-CNTRC panels

In the following discussion, the effect of different types of CNTs profiles on the non-dimensional central deflection $\overline{w}$ of FG-CNTRC cylindrical panel along center-line $(x, \theta_0/2)$ is presented in Fig. 4.9. According to Fig. 4.9, the FG-O form has the highest value of non-dimensional central deflection $\overline{w}$ while the FG-X form has the lowest one. Thus, the CNTs distributed near the top and bottom surfaces are more stiffer than those distributed near the mid-surface. Since, the FG-O-CNTRC has the highest non-dimensional central deflection $\overline{w}$, the effect of volume fraction of CNTs is also examined. Indeed, Fig. 4.10 shows that the increase of volume fraction of CNTs V_{CNT}^* leads to a decrease in non-dimensional central deflection $\overline{w}$ for FG-O-CNTRC cylindrical panel. This affirms that the reinforcement of CNTs leads to an increase in the stiffness of the FG-CNTRC cylindrical panel which highlight the performance of such reinforcement in the enhancement of mechanical behavior of composite structures.

Furthermore, the effect of volume fraction of CNTs V_{CNT}^* on non-dimensional axial stress $\overline{\sigma}_{xx}$, non-dimensional in plane shear stress $\overline{\sigma}_{xy}$ and non-dimensional shear stresses $\overline{\sigma}_{xz}$ and $\overline{\sigma}_{yz}$ for FG-O-CNTRC cylindrical panel is investigated. Figures 4.11(a) and 4.11(b) show that the volume fraction of CNTs has not a significant effect in membrane stresses $\overline{\sigma}_{xx}$ and $\overline{\sigma}_{xy}$ of FG-O-CNTRC cylindrical panel while this effect is more pronounced with the shear stresses $\overline{\sigma}_{xz}$ and $\overline{\sigma}_{xy}$ (See Fig. 4.12). Indeed, the magnitude of non-dimensional axial and transverse shear stresses decreases with the increase of volume fraction $V_{CNT}^* = 0.17$ in mid-surface (z = 0) of FG-O-CNTRC cylindrical panel

Table 4.9

Non-dimensional central deflection $\overline{w}$ for FG-CNTRC cylindrical panel with different edge to radius ratios (L/R) and for volume fraction $V_{CNT}^{*} = 0.11$ of CNTs and thickness $h = 0.004m$.

L/R		SSSS		CCCC	
		Present	Ref.[316]	Present	Ref.[316]
0.1	UD	0.0788	0.0787	0.0231	0.0236
	FG-V	0.1083	0.1073	0.0292	0.0297
	FG-O	0.1388	0.1345	0.0369	0.0361
	FG-X	0.0566	0.0567	0.0185	0.0191
0.15	UD	0.3322	0.3307	0.0852	0.0854
	FG-V	0.4356	0.4234	0.1075	0.1081
	FG-O	0.5329	0.5293	0.1327	0.1310
	FG-X	0.2437	0.2428	0.0659	0.0665
0.2	UD	0.7815	0.7764	0.1877	0.1868
	FG-V	0.9530	0.9166	0.2236	0.2240
	FG-O	1.1070	1.1022	0.2634	0.2614
	FG-X	0.6036	0.5995	0.1488	0.1485
0.25	UD	1.2923	1.2834	0.2941	0.2917
	FG-V	1.4698	1.4130	0.3280	0.3274
	FG-O	1.6404	1.6374	0.3694	0.3674
	FG-X	1.0565	1.0481	0.2442	0.2423
0.3	UD	1.7267	1.7157	0.3738	0.3699
	FG-V	1.8635	1.8001	0.3955	0.3939
	FG-O	2.0278	2.0278	0.4315	0.4294
	FG-X	1.4797	1.4679	0.3245	0.3210

where the concentration of CNTs is maximal. Thereby, the shear parts of stresses are those influenced by the variation of volume fraction of CNTs which can be explained by the combination of the parabolic distribution of shear stresses with the linear distribution of volume fraction of CNTs .

4.3.3 FG-CNTRC SKEW PLATE

4.3.3.1 Deflection responses of FG-CNTRC skew plate

After the validation of the present formulation based on double directors shell element using comparative studies of the obtained results with those available in the open literature for the two first examples, the static analysis is performed here for FG-CNTRC skew plate using the developed formulation. Indeed, according to the existing bibliography, the only reference in static analysis for FG-CNTRC skew plate is provided by Gracia-Macias et al. [84]. However, in their study, they use the FSDT theory with a linearly isotropic transverse constitutive equations which is a particular case of the orthotropic behavior of CNTRC. Consequently, to keep a general formulation, the present double directors shell formulation is adopted to illustrate the static behavior of FG-CNTRC skew plate. Figure 4.13 shows the geometrical properties of the studied FG-CNTRC skew plate which are the length side a, the skew angle α and the thickness h. For this example, two types of boundary conditions are considered: all sides (AB), (BC), (AD) and (DC) are simply supported (SSSS) or clamped (CCCC). Moreover, UD and four types of FG skew plates are considered: FG-O, FG-X, FG-V and FG-Λ to investigate the bending analysis of FG-CNTRC skew plate under uniform transverse load $q_0 = -0.1MPa$. The mechanical properties of the matrix and the CNT reinforcements are the as in section 4.2. In addition, a set of length-to-thickness ratios is considered $(a/h = 10, 50)$ with various skew angles $(\alpha = 90^o, 60^o, 45^o, 30^o)$ and a thickness $h = 0.002m$. The whole of the FG-CNTRC skew plate is meshed using 32×32 SHO4 finite elements.

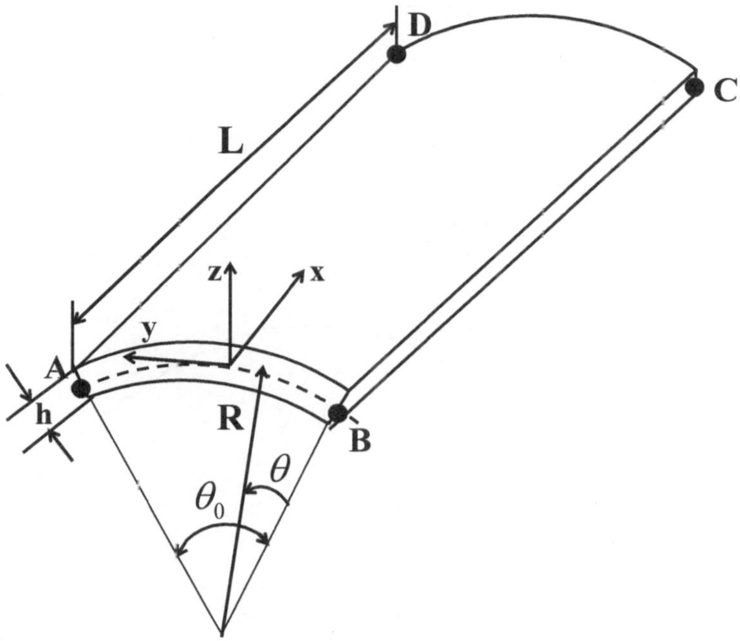

Figure 4.8 Geometry of FG-CNTRC cylindrical panel.

Tables 4.10 and 4.11 list the non-dimensional deflection $\overline{w}$ for the five types of FG-CNTRC skew plates subjected to uniform transverse load $q_0 = -0.1MPa$ with different volume fractions of CNTs, length-to-thickness ratios $(a/h = 10, 50)$ and various skew angles for (SSSS) and (CCCC) boundary conditions. As can be seen, the effect of skew angles on non-dimensional deflection $\overline{w}$ is prominent. Indeed, when the skew angle increases and reaches ($\alpha = 30^o$), the central deflection $\overline{w}$ decreases. Moreover, the FG-O form has the highest values of central deflection while the FG-X form has the lowest ones. Also, FG-V and FG-Λ forms have the same values of central deflection. This is due to the symmetric distribution of these two profiles via mid-surface: when the top surface is rich on CNTs for FG-V form, the bottom surface is rich on CNTs for FG-Λ. Hence, these two forms behave identically in the reproduction of central deflection. As well as, the increase in volume fraction of CNTs (V^*_{CNT}) leads to a significant decrease in central deflection $\overline{w}$ while the increase of length-to-thickness ratio leads to an increase in central deflection. Hence, the reinforcements of thick $(a/h = 10)$ FG-CNTRC skew plates is more stiffer than thin $(a/h = 50)$ FG-CNTRC skew plates. This outlines the advantage of FG materials as good reinforcements which affects the stiffness of the shell structure. Indeed, more these reinforcements are close to the top and bottom surfaces, more the FG-CNTRC skew plate is stiffer. Consequently, these reinforcements plays a significant role in the adjustment of the stiffness of the whole FG-CNTRC structure.

4.3.3.2 Stresses analysis of FG-CNTRC skew plates

Figures 4.14(a) and 4.14(b) shows that the variation of skew angle has a significant effect on non-dimensional central deflection $\overline{w}$ for various types of distribution which indicates the stiffening effect of higher values of skew angle α and affirms the above conclusions. Since the (CCCC) skew plate having more kinematic boundary constraints than the (SSSS) one, it is expected that the amplitudes

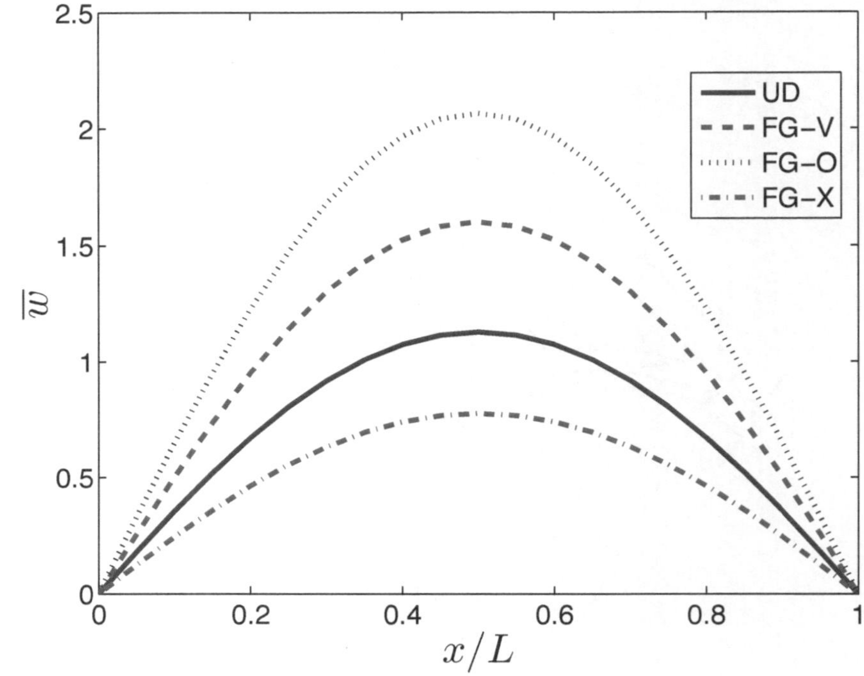

Figure 4.9 Non-dimensional central deflection $\overline{w}$ of FG-CNTRC cylindrical panel along center-line $(x, \theta_0/2)$.

of non-dimensional central axial stress $\overline{\sigma}_{xx}$ for clamped FG-CNTRC skew plate are more smaller than those of simply supported FG-CNTRC skew plate as depicted in Fig. 4.15. Furthermore, and since the FG-O-CNTRC has the highest value of central deflection and the skew angle $(\alpha = 30^o)$ has the most significant effect on central deflection, the effect of variation of volume fraction of CNTs V^*_{CNT} on central axial stress $\overline{\sigma}_{xx}$, in plane stress $\overline{\sigma}_{xy}$, axial shear stress $\overline{\sigma}_{xz}$ and transverse shear stress $\overline{\sigma}_{yz}$ for simply supported FG-O-CNTRC skew plate under uniform load $q_0 = -0.1MPa$ and with $(a/h = 50)$ and skew angle $(\alpha = 30^o)$ is also investigated. Indeed, Figs. 4.16(a) and 4.16(b) show that the volume fraction V^*_{CNT} has an effect on central axial $\overline{\sigma}_{xx}$ stress and in-plane shear stress $\overline{\sigma}_{xy}$ notably in half of the top and bottom surfaces $(z = \pm h/4)$ with an increase of the peak of the amplitude of the corresponding stresses. This effect can be possibly because of the concentration of CNTs which moves from the mid-surface $(z = 0)$ to the half of top and bottom surface $(z = \pm h/4)$ when the area of the skew plate becomes reduced with a pronounced skew angle $(\alpha = 30^o)$. As well as, from Figs. 4.17(c) and 4.17(d), it can be remarked that the increase of V^*_{CNT} induces a little effect on shear stresses which generates an increase in the amplitude of the peak of transverse shear stress $\overline{\sigma}_{xz}$ and a decrease in the amplitude of the peak of transverse shear stress $\overline{\sigma}_{yz}$. This is different to the observations related to shear stresses for the simply supported FG-O-CNTRC square plate under the same conditions. This effect may be related essentially to the skew angle which reaches its maximum $(\alpha = 30^o)$ for this case compared to the case of FG-O-CNTRC square plate where the skew angle is equal to $(\alpha = 90^o)$. It is worth to noticed that the values of non-dimensional central deflection $\overline{w}$ for FG-CNTRC skew plate with a skew angle $(\alpha = 90^o)$ converges to those obtained for FG-CNTRC square plate which validates the efficiency of the present formulation to investigate the static behavior of FG-CNTRC shell structures. Hence, for FG-CNTRC skew plates, and among all parameters: profile of CNTs, volume fraction of CNTs V^*_{CNT}, length-to-thickness ratios (a/h) and

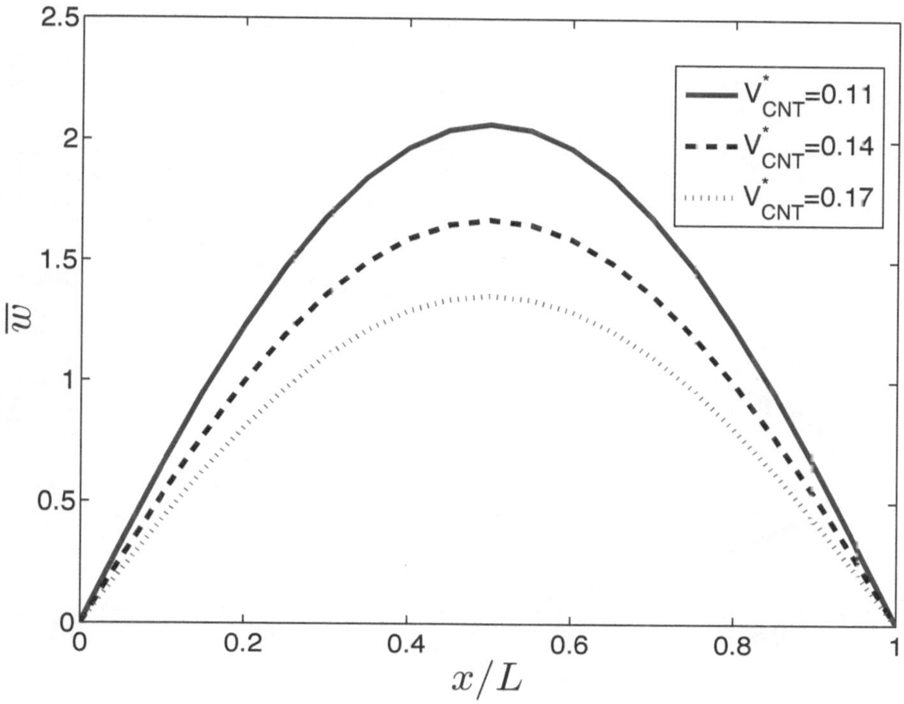

Figure 4.10 Effect of variation of V^*_{CNT} on non-dimensional central deflection $\overline{w}$ for FG-O-CNTRC cylindrical panel along center-line $(x, \theta_0/2)$.

skew angle (α), the static behavior of such structures is more sensitive to skew angle than others parameters. Therefore, it is also worthwhile to examine the effect of skew angles on stresses of FG-CNTRC skew plates.

Figs. 4.18 and 4.19 (a)–(d) demonstrate the significant effect of variation of skew angles on non-dimensional axial stress $\overline{\sigma}_{xx}$, non-dimensional in plane shear stress $\overline{\sigma}_{xy}$, non-dimensional transverse shear stresses $\overline{\sigma}_{xz}$ and $\overline{\sigma}_{yz}$ for a simply supported FG-O-CNTRC skew plate under a uniform load $q_0 = -0.1MPa$ and with $(a/h = 50)$ and $(V^*_{CNT} = 0.17)$. Indeed, for central axial stress $\overline{\sigma}_{xx}$ and in plane stress $\overline{\sigma}_{xy}$, the increase of skew angle leads to a decrease in the amplitude of stresses curves. In addition, the effect of skew angle is more prominent with transverse shear stress $\overline{\sigma}_{xz}$ which manifest by an inversion of the sign of the amplitude of transverse shear stress $\overline{\sigma}_{xz}$ for skew angle ($\alpha = 90^o$) compared to the other angles ($\alpha = 30^o, 45^o, 60^o$). This can be explained by the trigonometrical effect of skew angle which implies three different zones: the first one vary from 30° to 45° where the maximum of amplitude of axial shear stress (at mid-plane surface) increases. The second one vary from 45° to 60° where the maximum of amplitude of axial shear stress decreases. The third zone vary from 60° to 90° where the maximum of amplitude of axial shear stress decreases continuously until the point of inversion of its sign which reaches its maximum with 90°. Therefore, the maximum of amplitude of axial shear stress may behave as a trigonometric function versus the skew angle. Moreover, the transverse shear stress $\overline{\sigma}_{yz}$ is less influenced by the variation of skew angles even a decrease in the peak of the amplitude of the transverse shear stress curve which can be observed. Thereby, these phenomena occur as the skew angle increases and reaches its maximum with ($\alpha = 30^o$). So, the increase of skew angle affects both membrane and shear parts of stresses due to the shear flexibility of FG-CNTRC skew plate to this variation. Therefore, the skew angle has

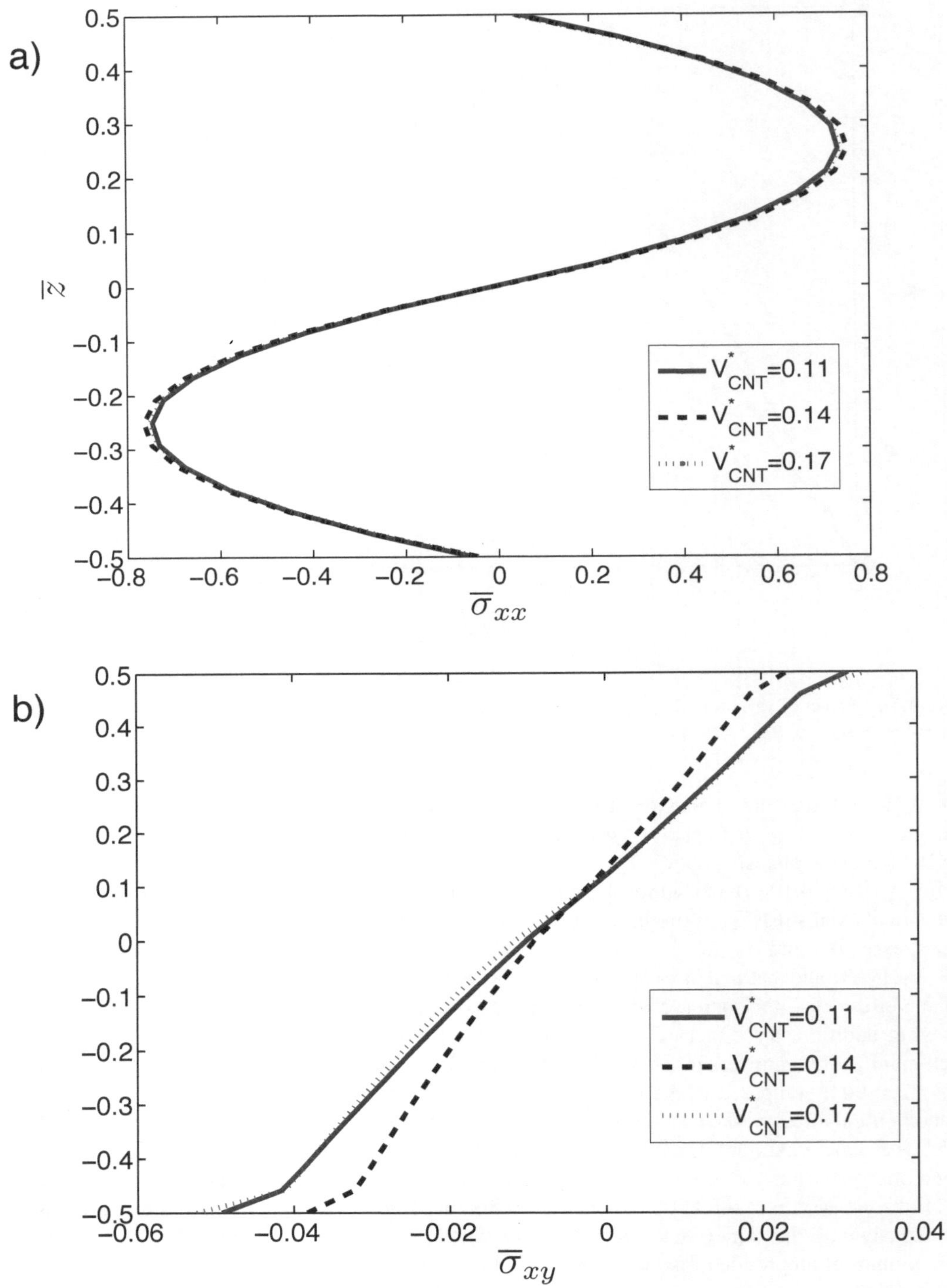

Figure 4.11 Effect of CNTs volume fraction V^*_{CNT} on a) non-dimensional axial stress $\overline{\sigma}_{xx}$ and b) non-dimensional in plane stress $\overline{\sigma}_{xy}$ for simply supported FG-O-CNTRC cylindrical panel under a uniform load $q_0 = 0.1MPa$ $(h = 0.002m,\ L/R = 0.1)$.

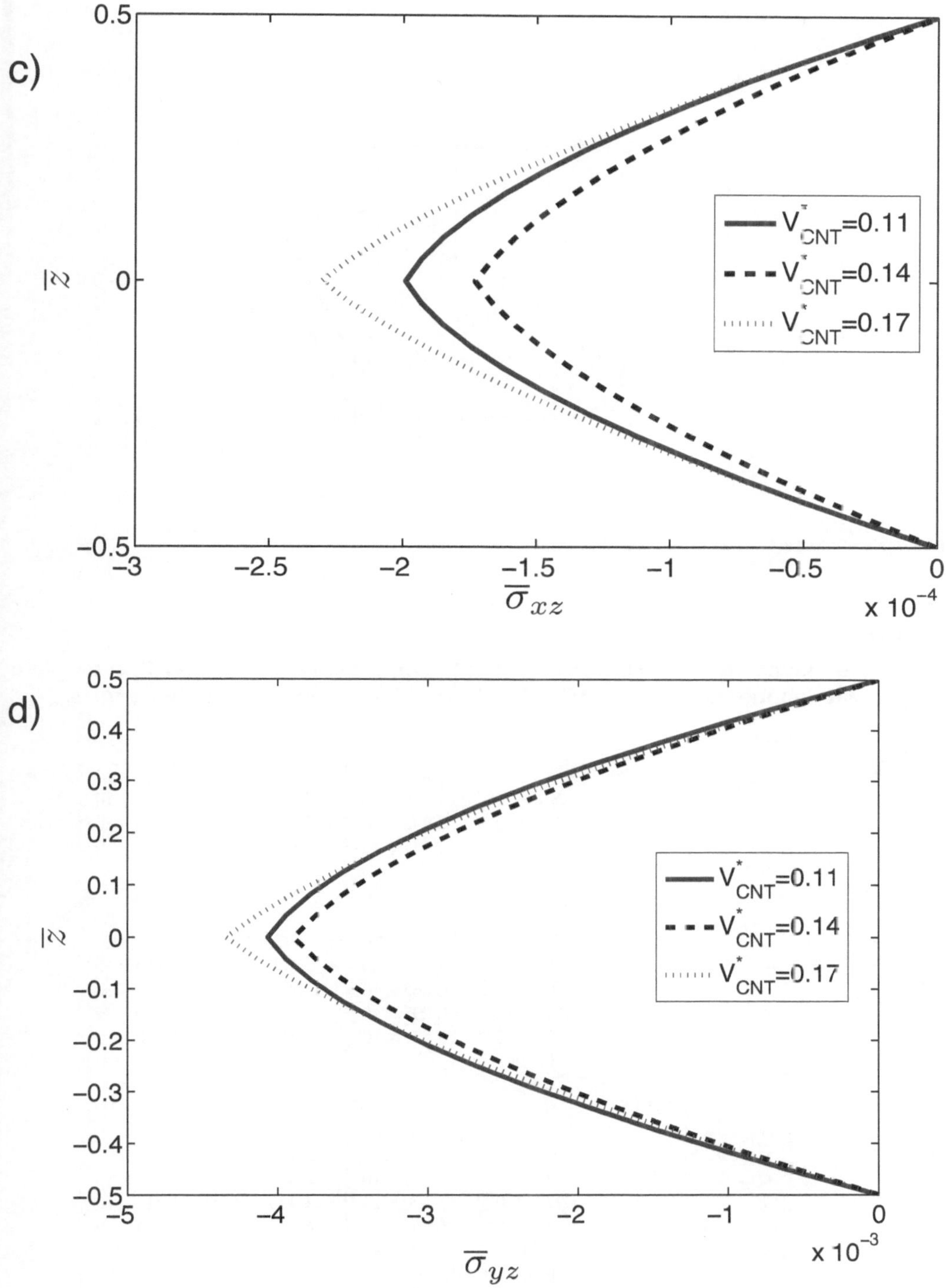

Figure 4.12 Effect of CNTs volume fraction V_{CNT}^{*} on c) non-dimensional axial shear stress $\overline{\sigma}_{xz}$ and d) non-dimensional transverse shear stress $\overline{\sigma}_{yz}$ for simply supported FG-O-CNTRC cylindrical panel under a uniform load $q_0 = 0.1MPa$ $(h = 0.002m,\ L/R = 0.1)$.

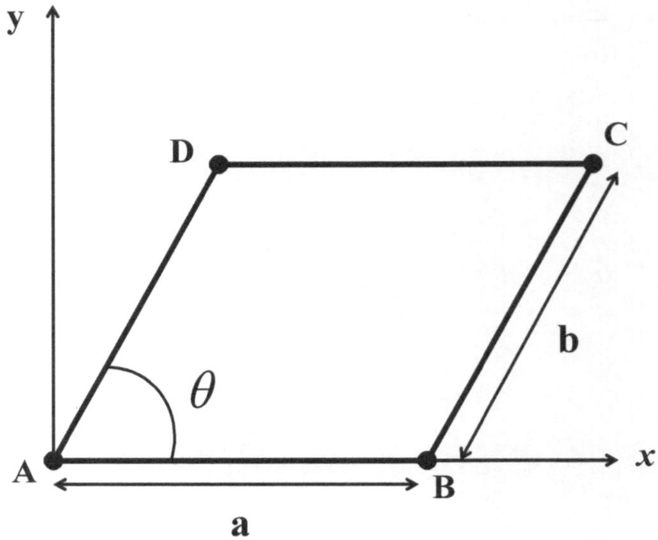

Figure 4.13 Geometry of FG-CNTRC skew plate.

Table 4.10

Non-dimensional central deflection, $\overline{w}$, for simply supported (SSSS) FG-CNTRC skew plate with different volume fractions of CNTs , length-to-thickness ratios and skew angles.

V^*_{CNT}	a/h	$\alpha(^o)$	UD	FG-X	FG-V	FG-Λ	FG-O
0.11	10	90	3.719×10^{-3}	3.139×10^{-3}	3.864×10^{-3}	3.864×10^{-3}	5.405×10^{-3}
		60	3.111×10^{-3}	2.694×10^{-3}	3.215×10^{-3}	3.215×10^{-3}	4.200×10^{-3}
		45	2.186×10^{-3}	1.956×10^{-3}	2.242×10^{-3}	2.242×10^{-3}	2.713×10^{-3}
		30	1.015×10^{-3}	9.481×10^{-4}	1.027×10^{-3}	1.027×10^{-3}	1.141×10^{-3}
	50	90	1.1566	0.7916	1.2581	1.2581	2.1619
		60	0.9835	0.6648	0.9769	0.9769	1.3982
		45	0.5836	0.4702	0.6426	0.6426	0.8077
		30	0.2631	0.2257	0.2875	0.2875	0.3313
0.14	10	90	3.274×10^{-3}	2.824×10^{-3}	3.388×10^{-3}	3.388×10^{-3}	4.663×10^{-3}
		60	2.784×10^{-3}	2.694×10^{-3}	2.865×10^{-3}	2.865×10^{-3}	3.717×10^{-3}
		45	1.994×10^{-3}	1.795×10^{-3}	2.037×10^{-3}	2.037×10^{-3}	2.470×10^{-3}
		30	9.490×10^{-4}	8.838×10^{-4}	9.570×10^{-4}	9.570×10^{-4}	1.070×10^{-3}
	50	90	0.9158	0.6266	0.9982	0.9982	1.7361
		60	0.7402	0.5454	0.8089	0.8089	1.1868
		45	0.5053	0.4021	0.5548	0.5548	0.7121
		30	0.2367	0.2008	0.2581	0.2581	0.3016
0.17	10	90	2.382×10^{-3}	2.029×10^{-3}	2.472×10^{-3}	2.472×10^{-3}	3.452×10^{-3}
		60	1.987×10^{-3}	1.725×10^{-3}	2.047×10^{-3}	2.047×10^{-3}	2.681×10^{-3}
		45	1.391×10^{-3}	1.239×10^{-3}	1.420×10^{-3}	1.420×10^{-3}	1.728×10^{-3}
		30	6.436×10^{-4}	5.940×10^{-4}	6.467×10^{-4}	6.467×10^{-4}	7.244×10^{-4}
	50	90	0.7534	0.5158	0.8206	0.8206	1.4171
		60	0.5775	0.4290	0.6313	0.6313	0.9031
		45	0.3745	0.2994	0.4117	0.4117	0.5196
		30	0.1678	0.1412	0.1825	0.1825	0.2136

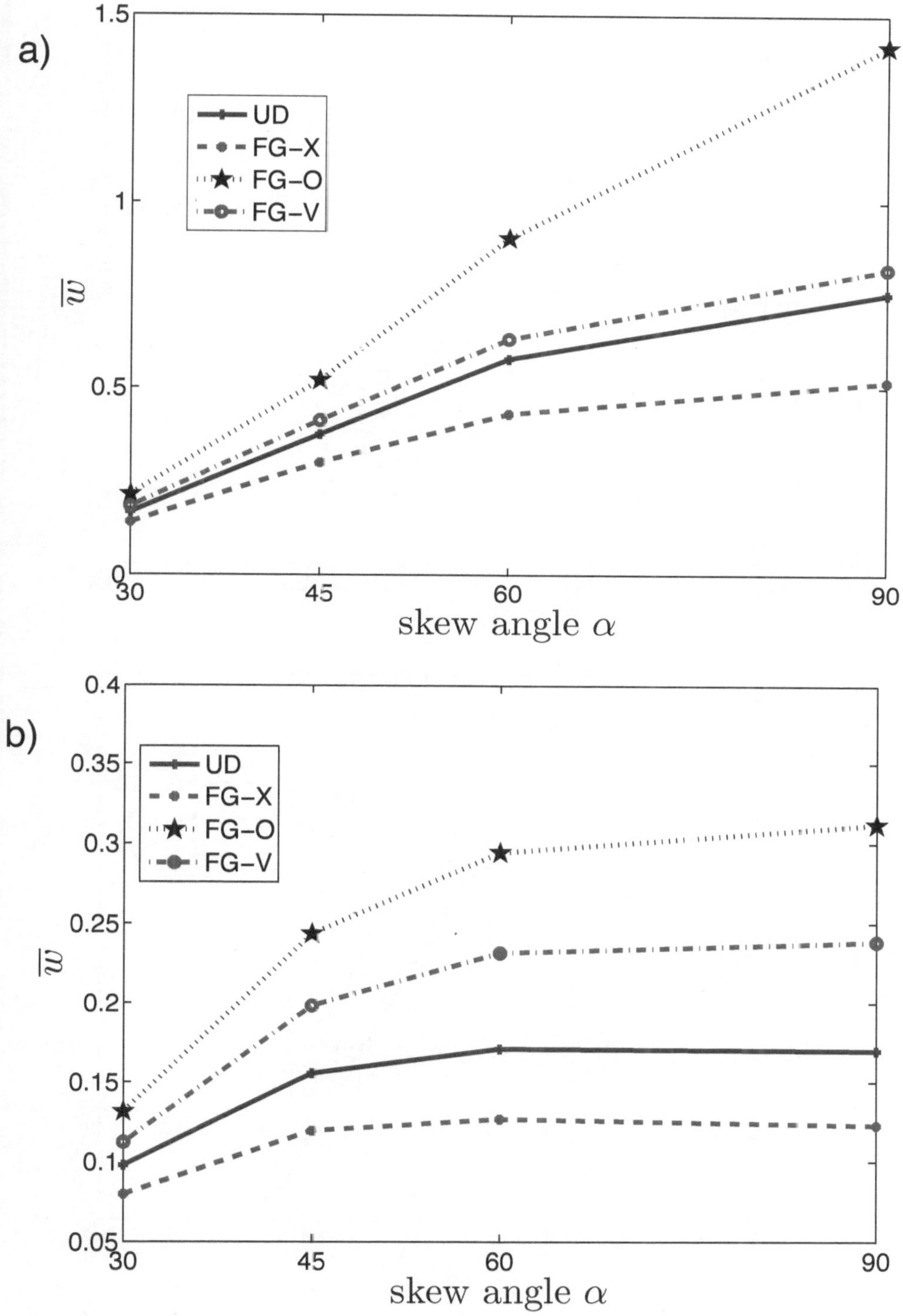

Figure 4.14 Effect of skew angle α on non-dimensional central deflection $\overline{w}$ for a) simply supported (SSSS) and b) clamped (CCCC) FG-CNTRC skew plates under uniform load $q_0 = -0.1MPa$ and with $V^*_{CNT} = 0.17$ and $a/h = 50$.

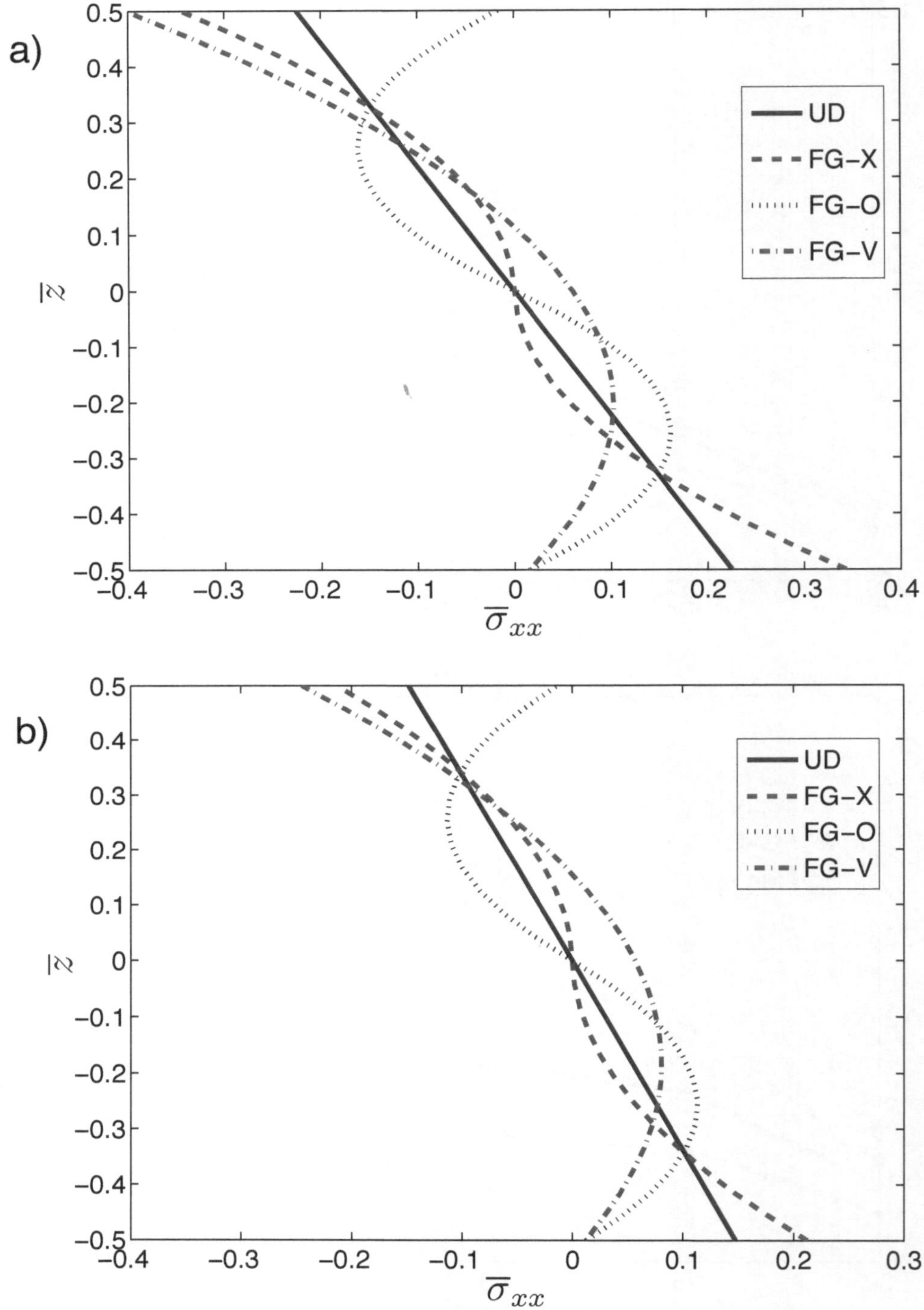

Figure 4.15 Non-dimensional central axial stress $\overline{\sigma}_{xx}$ of a) simply supported (SSSS) and b) clamped (CCCC) FG-CNTRC skew plate under uniform load $q_0 = -0.1MPa$ with $V^*_{CNT} = 0.17$, $a/h = 50$ and skew angle ($\alpha = 30^o$).

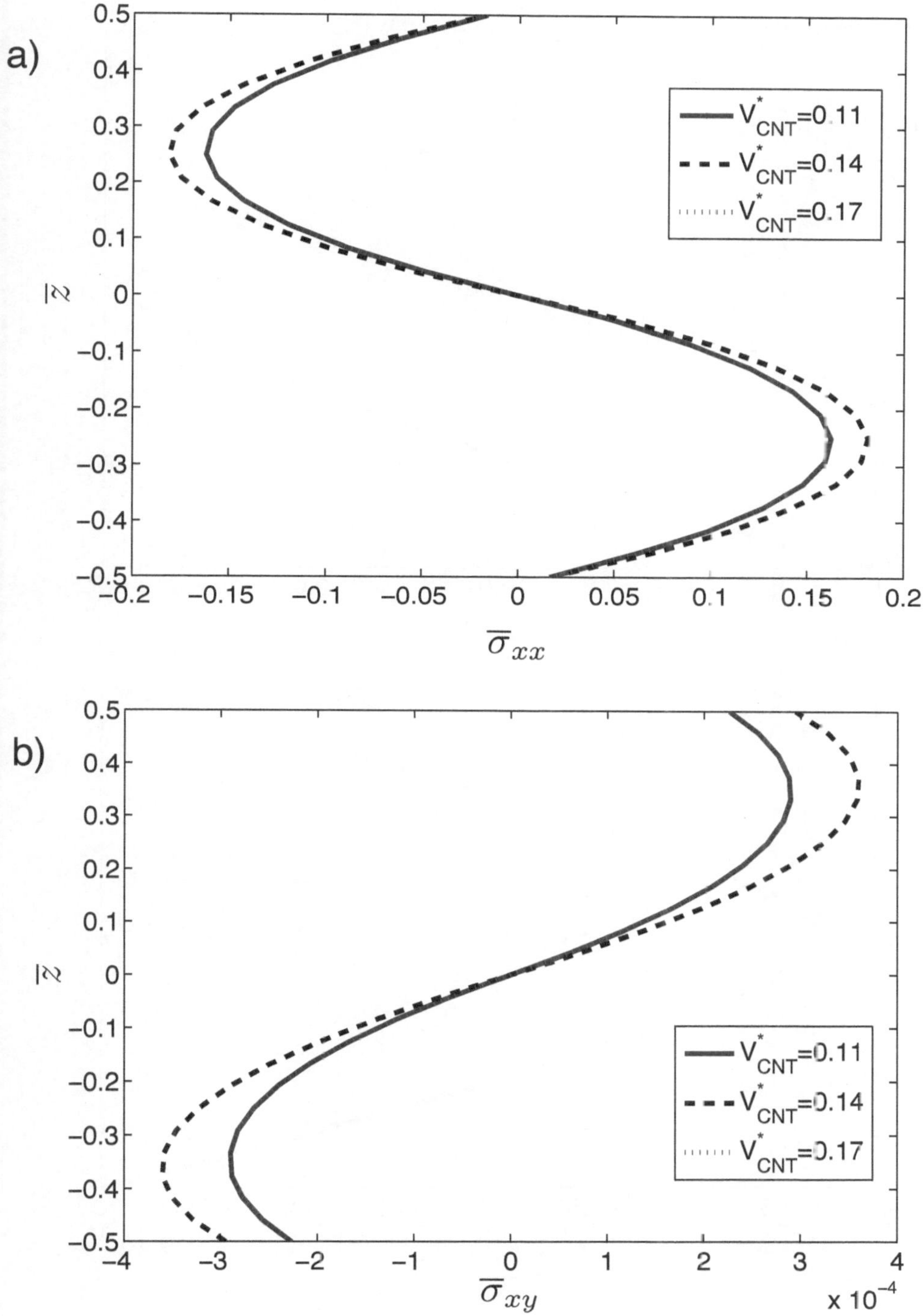

Figure 4.16 Effect of variation of CNTs volume fraction V^*_{CNT} on a) non-dimensional axial stress $\overline{\sigma}_{xx}$ and b) non-dimensional in plane stress $\overline{\sigma}_{xy}$ for a simply supported FG-O-CNTRC skew plate under a uniform load $q_0 = -0.1MPa$ with $a/h = 50$, $\alpha = 30^o$.

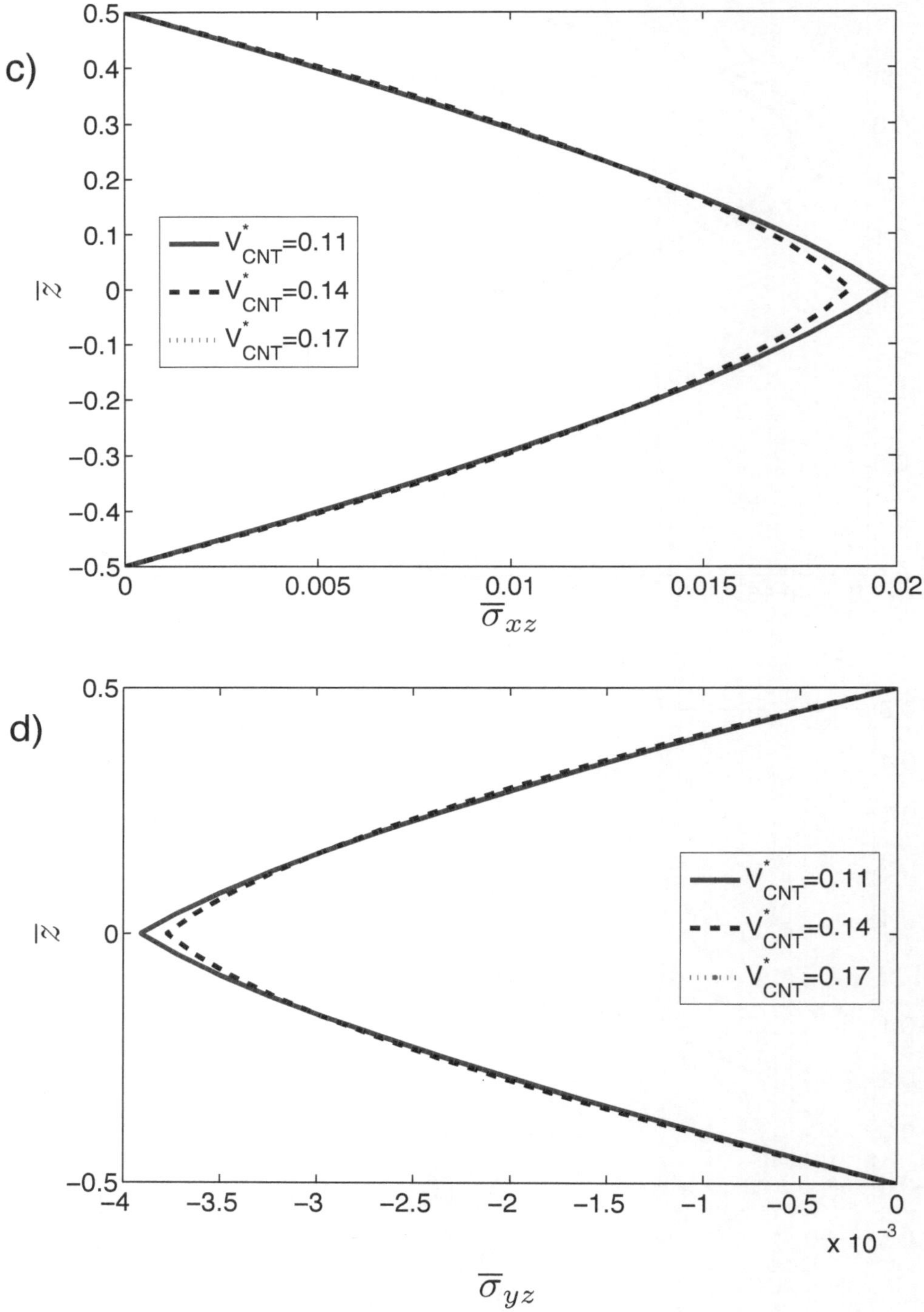

Figure 4.17 Effect of variation of CNTs volume fraction V_{CNT}^{*} on c) non-dimensional axial shear stress $\overline{\sigma}_{xz}$ and d) non-dimensional transverse shear stress $\overline{\sigma}_{yz}$ for a simply supported FG-O-CNTRC skew plate under a uniform load $q_0 = -0.1MPa$ with $a/h = 50$, $\alpha = 30^o$.

Table 4.11

Non-dimensional central deflection, $\overline{w}$, for clamped (CCCC) FG-CNTRC skew plate with different volume fractions of CNTs , length-to-thickness ratios and skew angles.

V^*_{CNT}	a/h	$\alpha(^o)$	UD	FG-X	FG-V	FG-Λ	FG-O
0.11	10	90	1.966×10^{-3}	1.803×10^{-3}	2.103×10^{-3}	2.103×10^{-3}	2.521×10^{-3}
		60	1.757×10^{-3}	1.629×10^{-3}	1.853×10^{-3}	1.853×10^{-3}	2.151×10^{-3}
		45	1.339×10^{-3}	1.275×10^{-3}	1.387×10^{-3}	1.337×10^{-3}	1.550×10^{-3}
		30	6.757×10^{-4}	6.471×10^{-3}	6.848×10^{-4}	6.848×10^{-4}	7.327×10^{-3}
	50	90	0.2616	0.1883	0.3648	0.3648	0.4783
		60	0.2642	0.1958	0.3560	0.3560	0.4521
		45	0.2414	0.1865	0.3074	0.3074	0.3746
		30	0.1527	0.1267	0.1760	0.1760	0.2024
0.14	10	90	1.803×10^{-3}	1.674×10^{-3}	1.911×10^{-3}	1.911×10^{-3}	2.283×10^{-3}
		60	1.626×10^{-3}	1.518×10^{-3}	1.701×10^{-3}	1.701×10^{-3}	1.973×10^{-3}
		45	1.251×10^{-3}	1.178×10^{-3}	1.288×10^{-3}	1.288×10^{-3}	1.443×10^{-3}
		30	6.392×10^{-4}	6.106×10^{-4}	6.452×10^{-4}	6.452×10^{-4}	5.930×10^{-3}
	50	90	0.2118	0.1549	0.2941	0.2941	0.3843
		60	0.2172	0.1629	0.2922	0.2922	0.3706
		45	0.2034	0.1579	0.2596	0.2596	0.3176
		30	0.1347	0.1110	0.1562	0.1562	0.1815
0.17	10	90	1.249×10^{-3}	1.161×10^{-3}	1.336×10^{-3}	1.336×10^{-3}	1.577×10^{-3}
		60	1.115×10^{-3}	1.040×10^{-3}	1.173×10^{-3}	1.173×10^{-3}	1.349×10^{-3}
		45	8.485×10^{-4}	7.952×10^{-4}	8.751×10^{-4}	8.751×10^{-4}	9.763×10^{-4}
		30	4.273×10^{-4}	4.057×10^{-4}	4.302×10^{-4}	4.302×10^{-4}	4.617×10^{-4}
	50	90	0.1698	0.1227	0.2386	0.2386	0.3117
		60	0.1711	0.1267	0.2317	0.2317	0.2944
		45	0.1557	0.1194	0.1984	0.1984	0.2437
		30	9.773×10^{-2}	7.954×10^{-2}	0.1121	0.1121	0.1313

a significant effect on bending behavior of FG-CNTRC skew plates and the type of reinforcement plays an important role in the adjustment of the stiffness of the whole structure.

4.4 FREE VIBRATION ANALYSIS OF FG-CNTRC STRUCTURES

In this section, free vibration and dynamic responses of FG-CNTRC structures are performed using the (DD-HSDT) shell model.

4.4.1 VIBRATIONAL PROBLEM FOR FG-CNTRC STRUCTURES

The determination of the frequencies of FG-CNTRC shell structure is achieved via the resolution of the following vibrational eigenvalue problem:

$$(\mathbf{K}-\omega^2\mathbf{M})\cdot\{\Lambda\}=0. \tag{4.6}$$

where ω is the eigen-frequency of the structure and $\mathbf{K}$ and $\mathbf{M}$ are the global stiffness and mass matrices of the FG-CNTRC shell, respectively. The mass matrix $\mathbf{M}$ is derived from the variation of the kinetic energy and can be expressed as [308]:

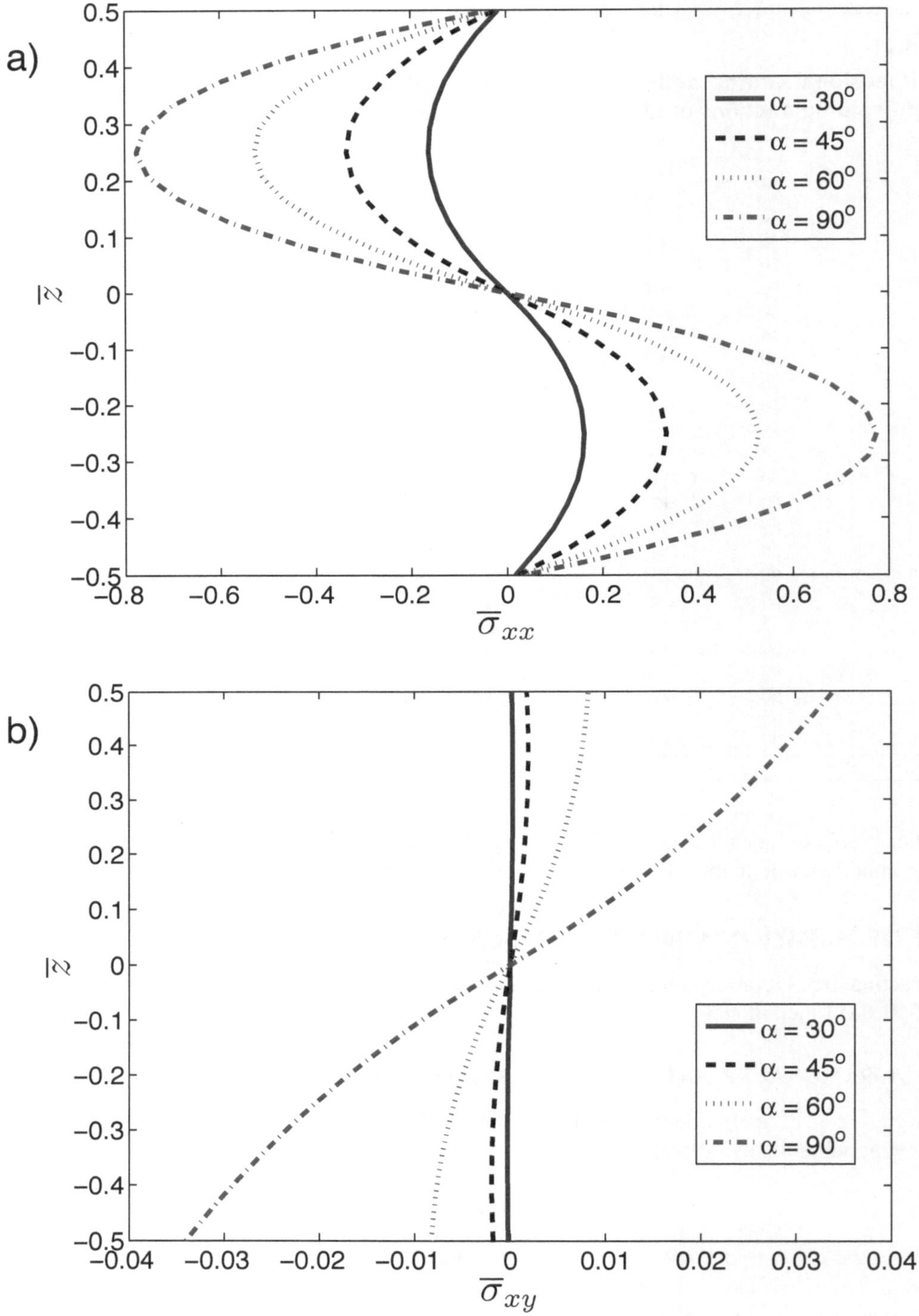

Figure 4.18 Effect of skew angles on a) non-dimensional central axial stress $\overline{\sigma}_{xx}$ and b) non-dimensional in plane stress $\overline{\sigma}_{xy}$ for FG-O-CNTRC skew plate under a uniform load $q_0 = -0.1MPa$ and with $(a/h = 50)$ and $(V^*_{CNT} = 0.17)$.

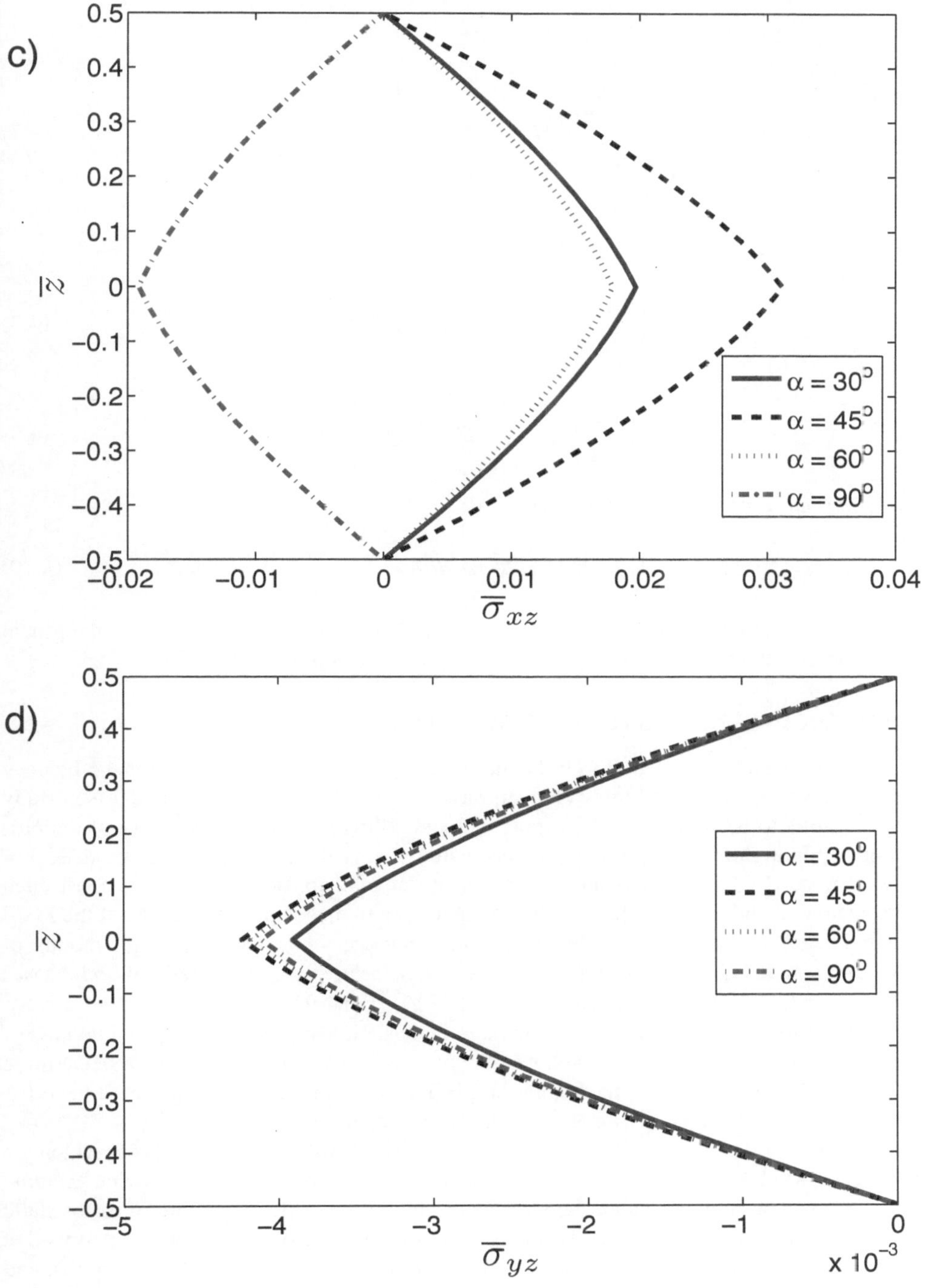

Figure 4.19 Effect of skew angles on c) non-dimensional axial shear stress $\overline{\sigma}_{xz}$ and d) non-dimensional transverse shear stress $\overline{\sigma}_{yz}$ for FG-O-CNTRC skew plate under a uniform load $q_0 = -0.1MPa$ and with $(a/h = 50)$ and $(V^*_{CNT} = 0.17)$.

$$M_{II} = \begin{bmatrix} M_{1I} & & & & & & & \\ & M_{1I} & & & & 0 & & \\ & & M_{1I} & & & & & \\ & & & M_{2I} & & & & \\ & 0 & & & M_{2I} & & & \\ & & & & & M_{3I} & & \\ & & & & & & M_{3I} \end{bmatrix}, I = 1,..,4 \tag{4.7}$$

where

$$M_{kI} = \frac{\int_A \overline{\rho}_{kk} N^I N^I dA \int_A \overline{\rho}_{kk} dA}{\int_A (\Sigma \overline{\rho}_{kk} N^I N^I) dA}, \quad k = 1,2,3, \quad I = 1,..,4. \tag{4.8}$$

and

$$\overline{\rho} = \int_{-h/2}^{h/2} \rho \begin{bmatrix} 1 & f_1 & f_2 \\ f_1 & f_1^2 & f_1 f_2 \\ f_2 & f_1 f_2 & f_2^2 \end{bmatrix} dz, \tag{4.9}$$

For the stiffness matrix, its expression can be read as:

$$K = \int_A B^T H_T B dA \tag{4.10}$$

where H_T and B are the elastic modulus of the FG-CNTRC structure (Eq. (1.104)) and the global vector which contains the strain-displacement matrices (as indicated in Eq. (1.113)).

4.4.2 FREQUENCY RESPONSES OF FG-CNTRC PLATES

Herein, free vibration analysis of FG-CNTRC plates is provided. First, the computational efficiency of the present double directors shell finite element model is performed through a comparison study of the present results to those of ABAQUS finite element software. Then, the stability of the proposed method is highlighted through a convergence study using different sets of mesh sizes. For that, a fully clamped (CCCC) functionally graded square plate is considered, which is constituted by two isotropic metal and ceramic phases. The geometrical and mechanical properties of the FGM plate are as indicated in [265]: length side $a = 1m$, total thickness $h = 0.05m$, Young's moduli of metal and ceramic phases $E_m = 70GPa$ and $E_c = 168GPa$, respectively, Poisson ratio of both phases $\nu_m = \nu_c = 0.3$, metal and ceramic densities $\rho_m = 2707Kg/m^3$ and $\rho_c = 5700Kg/m^3$, respectively. Furthermore, the volume fraction of the ceramic material has a linear profile along the thickness of the structure as mentioned in [265]. The whole of the plate is meshed using 100×100 SHO4 finite elements. It is worth to mention that the denotation SHO4 is used to identify the present linear discrete double directors shell element. The obtained first ten frequencies are listed in Tab. 4.12. As can be seen, the present results are in good agreement with those of ABAQUS solutions [265] which are obtained using 10,000 S4 elements. Hence, the present model has the aptitude to provide accurate numerical solutions. On the other hand, the stability of the present model was improved for static analysis via refining the mesh sizes of several numerical examples and one can be referred to Wali et al. [282] for more details. For free vibration analysis, a set of mesh sizes: (4×4, 8×8, 16×16 and 32×32) is used to draw the convergence property of the normalized frequency parameter relative to SHO4 reference with 100×100 elements. Figure 4.20 shows that as the number of elements per side increases, as the normalized frequency parameter converges well with the reference solution for the three first modes of vibration. As can be seen from Fig. 4.20, a 16×16 of elements are sufficient to reach accurate vibration results. Therefore, this mesh size is adopted to perform free vibrations analysis of some examples of FG-CNTRC structures. So, within this comparison and convergence

Table 4.12
First ten natural frequencies of fully clamped FGM square plate.

Mode	Natural frequency [Hz]	
	Present-SHO4	ABAQUS [265]
1	436.320	436.237
2	873.110	872.240
3	873.110	872.240
4	1265.600	1263.240
5	1525.300	1522.000
6	1534.700	1531.550
7	1855.300	1879.600
8	1855.300	1879.600
9	2369.200	2361.220
10	2369.200	2361.220

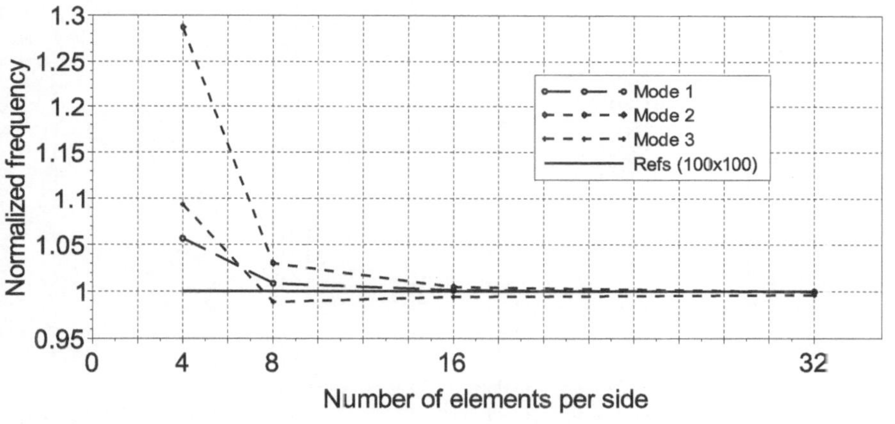

Figure 4.20 Convergence property of three first modes of fully clamped square FGM plate.

study, the present double directors finite element model has drawn its high accuracy and efficiency in the prediction of vibrational behavior of such structures.

The non-dimensional frequency parameter $\bar{\omega} = \omega(a^2/h)\sqrt{\rho^m/E_m}$ of FG-CNTRC plate with side length a and thickness $h = 0.002m$ is performed under uniform transverse load ($q_0 = -0.1MPa$) and the plate is meshed using 16×16 SHO4 finite elements. Furthermore, the free vibration behavior of the FG-CNTRC square plate is analyzed using different length-to-thickness ratios ($a/h = 10, 20, 50$) and two types of boundary conditions: four edges ((AB), (BC), (CD) and (DA)) are simply supported (SSSS) or clamped (CCCC).

Tables 4.13, 4.14, 4.15, 4.16 and 4.17 and 4.18 list the non-dimensional frequency parameter for different distributions of CNT, such that (UD), FG-V, FG-O and FG-X are uniform and the three functionally graded distributions of CNT, different volume fractions of CNT and various length-to-thickness ratios of the simply supported and clamped FG-CNTRC plates, respectively. The obtained results in terms of non-dimensional frequency parameter are compared to those of Zhu et al. [328] in order to validate the present formulation. As can be seen, the present model has the ability to restitute the natural frequency of the FG-CNTC plate with good accuracy in comparison with those

Table 4.13

Non-dimensional natural frequency $\bar{\omega} = \omega(a^2/h)\sqrt{\rho^m/E_m}$ of UD and FG-V-CNTRC square plates with different length-to-thickness ratio, two boundary conditions (SSSS, CCCC) and volume fraction of CNT $V^*_{CNT} = 0.11$.

BC	a/h	Modes (m,n)	UD		FG-V	
			Present	Ref.[328]	Present	Ref.[328]
SSSS	10	(1,1)	13.52	13.53	12.43	12.45
		(1,2)	17.58	17.70	16.94	17.06
		(1,3)	19.40	19.44	19.45	19.49
		(1,4)	19.40	19.44	19.45	19.49
		(2,1)	27.09	27.56	26.87	27.34
		(2,2)	32.86	32.56	31.63	31.41
	20	(1,1)	17.26	17.35	15.03	15.11
		(1,2)	21.23	21.51	19.66	19.90
		(1,3)	31.58	32.39	30.79	31.56
		(1,4)	38.79	38.89	38.89	38.99
		(2,1)	38.79	38.89	38.89	38.99
		(2,2)	48.06	50.19	47.26	47.73
	50	(1,1)	19.10	19.22	16.16	16.25
		(1,2)	23.06	23.40	20.85	21.14
		(1,3)	33.68	34.66	32.45	33.35
		(1,4)	51.50	54.04	50.98	53.43
		(2,1)	69.78	70.81	59.40	60.18
		(2,2)	71.21	72.90	61.42	62.78
CCCC	10	(1,1)	18.61	17.62	18.08	17.21
		(1,2)	23.62	23.04	23.31	22.96
		(1,3)	33.42	33.59	33.33	33.07
		(1,4)	36.76	33.72	35.94	33.55
		(2,1)	37.10	37.01	37.22	36.52
		(2,2)	39.48	37.31	38.86	37.43
	20	(1,1)	28.71	28.40	26.51	26.30
		(1,2)	33.05	33.11	31.35	31.49
		(1,3)	43.56	44.55	42.53	43.58
		(1,4)	60.14	59.19	57.23	56.24
		(2,1)	60.43	61.85	59.64	59.22
		(2,2)	62.50	63.04	59.67	62.60
	50	(1,1)	39.38	39.73	33.91	34.16
		(1,2)	43.05	43.87	38.35	39.04
		(1,3)	52.86	54.76	49.45	51.20
		(1,4)	70.28	74.48	68.16	72.20
		(2,1)	94.97	98.29	85.06	86.29
		(2,2)	96.70	100.53	86.93	89.05

derived from [328]. In fact, the relative error between the results does not exceed generally 3% except in some cases for some distributions and volume fractions (for instance, this error is about 10% for the mode (2,1) of clamped FG-X-CNTRC plate with different volume fraction of CNTs V^*_{CNT}). The difference between results, in these cases, can be explained by the FSDT theory used in [328] which introduces shear correction coefficients. In fact, many methods exist to evaluate the shear correction coefficients which can influence notably the values of high modes of vibrations. Otherwise, the results derived from [328] converges well with those of the proposed double directors shell element. Therefore, the present double directors shell model is more general and convenient

Table 4.14

Non-dimensional natural frequency $\bar{\omega} = \omega(a^2/h)\sqrt{\rho^m/E_m}$ of FG-O and FG-X-CNTRC square plates with different length-to-thickness ratio, two boundary conditions (SSSS, CCCC) and volume fraction of CNT $V^*_{CNT} = 0.11$.

BC	a/h	Modes (m,n)	FG-O		FG-X	
			Present	Ref.[328]	Present	Ref.[328]
SSSS	10	(1.1)	11.33	11.55	14.64	14.61
		(1.2)	16.04	16.26	18.51	18.64
		(1.3)	19.47	19.49	19.43	19.49
		(1.4)	19.47	19.49	19.43	19.49
		(2.1)	26.07	26.51	27.93	28.51
		(2.2)	29.14	30.28	34.19	33.59
	20	(1,1)	13.37	13.52	19.84	19.93
		(1,2)	18.21	18.48	23.46	23.77
		(1,3)	29.43	30.16	33.47	34.38
		(1,4)	38.94	38.99	38.87	38.99
		(2,1)	38.94	38.99	38.87	38.99
		(2,2)	42.60	43.94	49.92	52.26
	50	(1,1)	14.20	14.30	22.83	22.98
		(1,2)	19.10	19.37	26.36	26.78
		(1,3)	30.77	31.61	36.48	37.59
		(1,4)	49.03	51.37	54.19	56.94
		(2,1)	52.03	53.03	78.75	83.15
		(2,2)	54.33	55.82	81.99	84.89
CCCC	10	(1,1)	16.58	16.70	19.38	18.08
		(1,2)	22.06	22.25	24.35	23.60
		(1,3)	32.33	32.37	34.13	34.33
		(1,4)	32.37	32.85	37.43	34.46
		(2,1)	35.66	35.80	38.31	37.44
		(2,2)	37.28	37.44	40.93	37.78
	20	(1,1)	23.96	24.48	31.03	30.42
		(1,2)	29.06	29.79	35.20	35.03
		(1,3)	40.50	41.89	45.56	46.48
		(1,4)	51.86	53.55	62.12	61.98
		(2,1)	54.57	56.61	64.06	64.56
		(2,2)	57.71	60.71	65.97	65.17
	50	(1,1)	29.85	30.30	45.85	46.16
		(1,2)	34.61	35.44	49.07	49.93
		(1,3)	46.09	47.87	58.19	60.22
		(1,4)	64.90	68.84	75.06	79.53
		(2,1)	75.18	77.46	99.60	108.69
		(2,2)	77.40	80.46	109.59	110.92

for all types (thin, moderately thick and thick) of FG-CNTRC plates. In addition, it can be remarked that the values of natural frequencies associated to FG-X distribution are the highest ones while those associated to FG-O distribution are the lowest ones. This tendency of high values is more pronounced with the volume fraction $V^*_{CNT} = 0.17$, (for instance the value of the non-dimensional natural frequency reaches its maximum 138.48 with the last mode (2,2) of vibration for $V^*_{CNT} = 0.17$ in the clamped case of boundary conditions). Hence, more the CNT reinforcements are close to top and bottom surfaces, more the FG-CNTRC shell structure is stiffer and this implies high values of natural frequencies.

Table 4.15
Non-dimensional natural frequency $\bar{\omega} = \omega(a^2/h)\sqrt{\rho^m/E_m}$ of UD and FG-V-CNTRC square plates with different length-to-thickness ratio, two boundary conditions (SSSS, CCCC) and volume fraction of CNT $V_{CNT}^* = 0.14$.

BC	a/h	Modes (m,n)	UD		FG-V	
			Present	Ref.[328]	Present	Ref.[328]
SSSS	10	(1,1)	14.32	14.30	13.26	13.25
		(1,2)	18.26	18.36	17.63	17.73
		(1,3)	19.74	19.79	19.83	19.87
		(1,4)	19.74	19.79	19.83	19.87
		(2,1)	27.76	28.23	27.55	28.02
		(2,2)	34.15	33.64	33.06	32.67
	20	(1,1)	18.85	18.92	16.45	16.51
		(1,2)	22.60	22.86	20.85	21.08
		(1,3)	32.75	33.57	31.84	32.61
		(1,4)	39.49	39.58	39.66	39.75
		(2,1)	39.49	39.58	39.66	39.75
		(2,2)	49.26	51.42	48.96	51.07
	50	(1,1)	21.25	21.35	17.92	17.99
		(1,2)	24.95	25.29	22.36	22.64
		(1,3)	35.26	36.26	33.74	34.66
		(1,4)	53.01	55.60	52.34	54.83
		(2,1)	77.11	78.11	65.79	66.55
		(2,2)	77.44	80.01	67.55	68.94
CCCC	10	(1,1)	19.34	18.12	18.86	17.79
		(1,2)	24.33	23.57	24.06	23.41
		(1,3)	34.20	34.25	34.15	34.10
		(1,4)	37.76	34.65	37.59	34.27
		(2,1)	38.33	37.92	37.95	37.53
		(2,2)	40.98	37.97	40.42	38.15
	20	(1,1)	30.41	29.91	28.27	27.92
		(1,2)	34.61	34.51	32.94	32.97
		(1,3)	45.01	45.89	44.01	44.98
		(1,4)	61.64	61.62	60.39	58.95
		(2,1)	63.40	64.19	61.22	61.81
		(2,2)	65.36	64.49	62.66	64.13
	50	(1,1)	43.30	43.58	37.36	37.56
		(1,2)	46.68	47.47	41.49	42.17
		(1,3)	56.04	57.96	52.19	53.96
		(1,4)	73.11	77.3	70.67	74.78
		(2,1)	97.73	106.37	92.89	94.02
		(2,2)	104.97	106.48	94.49	96.57

On the other hand, Fig. 4.21 shows the effect of variation of volume fraction V_{CNT}^* on the natural frequency $\bar{\omega}$ with different length-to-thickness ratios and boundary conditions. It can be seen that the values of natural frequencies for different distributions of FG-CNTRC plates with clamped (CCCC) boundary condition are higher than those with simply supported (SSSS) boundary conditions. In fact, the clamped boundary condition imposed more constraints to the FG-CNTRC plate which affects the stiffness of the structure and implies hence an increase in the natural frequency of the FG-CNTRC plates. Moreover, the variation of the length-to-thickness ratio is more prominent with clamped FG-CNTRC plates compared to the simply supported FG-CNTRC plates which affirms

Table 4.16
Non-dimensional natural frequency $\bar{\omega} = \omega(a^2/h)\sqrt{\rho^m/E_m}$ of FG-O and FG-X-CNTRC square plates with different length-to-thickness ratio, two boundary conditions (SSSS, CCCC) and volume fraction of CNT $V^*_{CNT} = 0.14$.

BC	a/h	Modes (m,n)	FG-O		FG-X	
			Present	Ref.[328]	Present	Ref.[328]
SSSS	10	(1,1)	12.13	12.33	15.35	15.36
		(1,2)	16.63	16.84	19.21	19.38
		(1,3)	19.87	19.87	19.81	19.87
		(1,4)	19.87	19.87	19.81	19.87
		(2,1)	26.59	27.00	28.74	29.39
		(2,2)	30.61	31.63	35.27	34.63
	20	(1,1)	14.64	14.78	21.52	21.64
		(1,2)	19.19	19.46	25.00	25.35
		(1,3)	30.17	30.90	34.96	35.93
		(1,4)	39.73	39.75	39.63	39.75
		(2,1)	39.73	39.75	39.63	39.75
		(2,2)	45.99	47.34	51.60	54.06
	50	(1,1)	15.71	15.80	25.42	25.55
		(1,2)	20.30	20.56	28.76	29.19
		(1,3)	31.65	32.50	38.67	39.83
		(1,4)	49.83	52.18	56.47	59.33
		(2,1)	57.73	58.74	81.46	82.81
		(2,2)	59.73	61.27	90.01	91.29
CCCC	10	(1,1)	17.33	17.31	20.03	18.59
		(1,2)	22.72	22.78	25.08	24.24
		(1,3)	33.00	33.41	35.02	35.22
		(1,4)	33.85	33.44	38.17	35.41
		(2,1)	37.03	36.78	39.79	38.16
		(2,2)	38.03	38.16	42.40	38.78
	20	(1,1)	25.67	26.12	32.53	31.85
		(1,2)	30.51	31.18	36.70	36.48
		(1,3)	41.68	43.03	47.17	48.08
		(1,4)	55.00	56.40	64.00	64.33
		(2,1)	57.51	59.27	66.65	66.91
		(2,2)	58.83	61.79	68.53	67.14
	50	(1,1)	32.92	33.36	50.08	50.40
		(1,2)	37.30	38.14	53.12	54.02
		(1,3)	48.23	50.05	61.98	64.11
		(1,4)	66.65	70.64	78.75	83.39
		(2,1)	82.46	84.79	103.49	112.89
		(2,2)	84.35	87.51	117.72	119.13

that more the FG-CNTRC plate have imposed constraints, more it becomes sensitive to the change of the length-to-thickness ratio. As well as, it should be noted that the obtained results are visually in unison with those provided in [328]. Consequently, the present double directors shell model has the capacity to reproduce the dynamic behavior of FG-CNTRC plates with good accuracy.

4.4.3 FREQUENCY RESPONSES OF FG-CNTRC SKEW PLATES

The second example consists of a FG-CNTRC skew plate with length side a, width side b, skew angle θ and thickness $h = 0.2$ as depicted in Fig. 4.13. As the first example, the FG-CNTRC skew

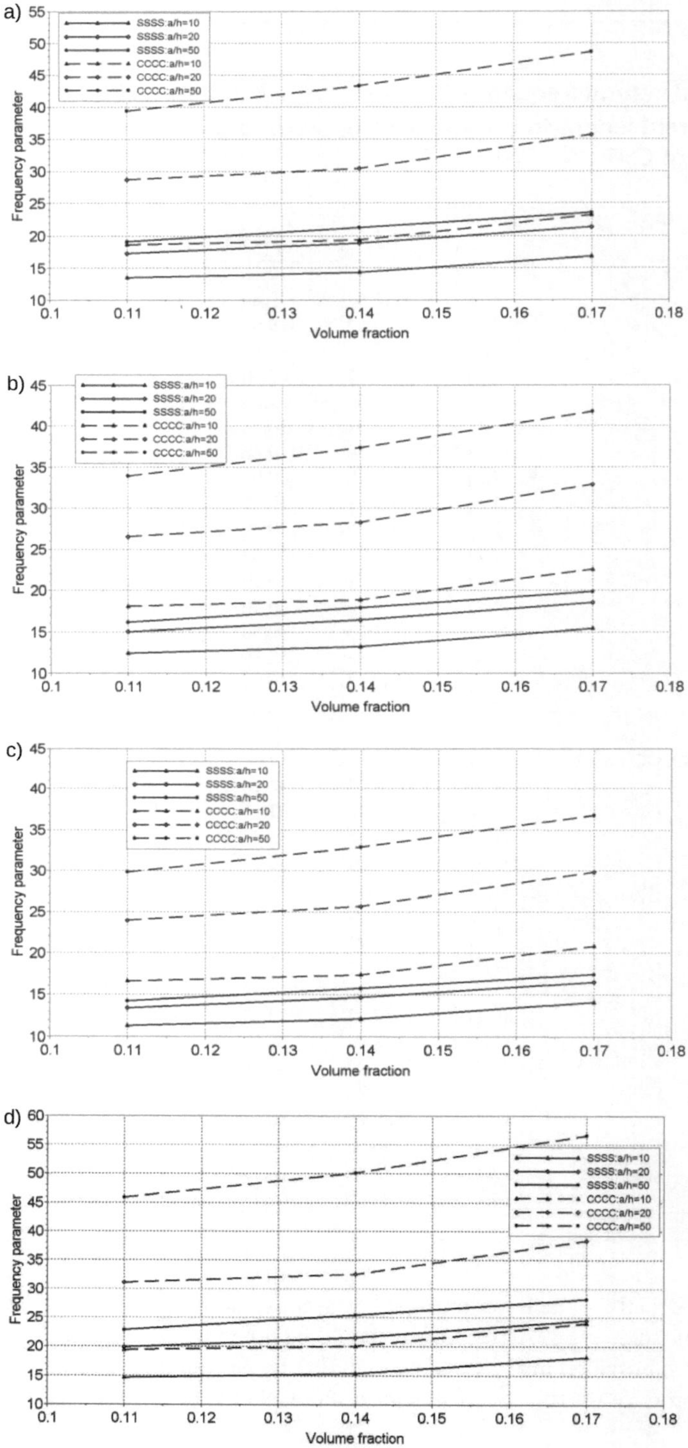

Figure 4.21 Variation of the fundamental frequency parameter $\bar{\omega}$ with the CNT volume fraction V^*_{CNT} for different profiles of CNTRC square plates. (a): UD; (b): FG-V; (c): FG-O; (d): FG-X distributions.

Table 4.17

Non-dimensional natural frequency $\bar{\omega} = \omega(a^2/h)\sqrt{\rho^m/E_m}$ of UD and FG-V-CNTRC square plates with different length-to-thickness ratio, two boundary conditions (SSSS, CCCC) and volume fraction of CNT $V^*_{CNT} = 0.17$.

BC	a/h	Modes (m,n)	UD		FG-V	
			Present	Ref.[328]	Present	Ref.[328]
SSSS	10	(1,1)	16.79	16.81	15.42	15.46
		(1,2)	21.91	22.06	21.15	21.30
		(1,3)	24.27	24.33	24.44	24.51
		(1,4)	24.27	24.33	24.44	24.51
		(2,1)	33.84	34.44	33.67	34.27
		(2,2)	40.96	40.63	39.42	39.26
	20	(1,1)	21.33	21.45	18.54	18.63
		(1,2)	26.36	26.70	24.44	24.73
		(1,3)	39.38	40.40	38.52	39.47
		(1,4)	48.54	48.67	48.88	49.02
		(2,1)	48.54	48.67	48.88	49.02
		(2,2)	60.05	62.72	58.52	59.19
	50	(1,1)	23.54	23.69	19.87	19.98
		(1,2)	28.55	28.98	25.85	26.20
		(1,3)	41.93	43.16	40.54	41.64
		(1,4)	64.29	67.47	63.90	66.94
		(2,1)	86.10	87.38	73.04	74.03
		(2,2)	87.92	90.03	75.64	77.34
CCCC	10	(1,1)	23.20	22.01	22.54	21.54
		(1,2)	29.49	28.80	29.16	28.61
		(1,3)	41.78	42.01	41.78	41.43
		(1,4)	45.81	42.13	44.82	42.11
		(2,1)	46.41	46.25	46.78	45.79
		(2,2)	49.24	46.69	48.52	47.05
	20	(1,1)	35.68	35.31	32.88	32.68
		(1,2)	41.15	41.25	39.04	39.27
		(1,3)	54.35	55.62	53.20	54.56
		(1,4)	75.14	73.76	71.15	70.14
		(2,1)	75.21	77.10	74.24	73.92
		(2,2)	77.83	78.80	74.81	78.52
	50	(1,1)	48.63	49.07	41.74	42.07
		(1,2)	53.28	54.32	47.43	48.30
		(1,3)	65.68	68.06	61.57	63.75
		(1,4)	87.60	92.86	85.25	90.29
		(2,1)	118.59	121.66	104.90	106.51
		(2,2)	119.64	124.51	107.34	110.05

plate is subjected to an uniform transverse load $q_0 = -0.1$ MPa while only two distributions of CNT are considered here which are uniform (UD) and FG-X distributions in order to analyze the vibrational behavior of FG-CNTRC skew plates. In fact, the variation of a set of parameters as the skew angle, the volume faction of CNT, thickness-to-width ratios and the plate aspect ratios are examined for different combinations of boundary conditions. For the purpose of validation of the present formulation, the material properties of FG-CNTRC skew plate are selected as Zhang et al. [314], where Poly (methyl methacrylate), refereed to PMMA matrix, is chosen for matrix with Young modulus $E_m = 2.5GPa$ and (10,10) SWCNTs are selected for reinforcements. The mechanical properties of

Table 4.18

Non-dimensional natural frequency $\bar{\omega} = \omega(a^2/h)\sqrt{\rho^m/E_m}$ of FG-O and FG-X-CNTRC square plates with different length-to-thickness ratio, two boundary conditions (SSSS, CCCC) and volume fraction of CNT $V_{CNT}^* = 0.17$.

BC	a/h	Modes (m,n)	FG-O		FG-X	
			Present	Ref.[328]	Present	Ref.[328]
SSSS	10	(1,1)	14.09	14.28	18.11	18.27
		(1,2)	19.90	20.09	23.19	23.54
		(1,3)	24.50	24.51	24.42	24.51
		(1,4)	24.50	24.51	24.42	24.51
		(2,1)	32.33	32.76	35.29	36.24
		(2,2)	36.72	37.76	42.24	42.15
	20	(1,1)	16.48	16.62	24.51	24.76
		(1,2)	22.45	22.73	29.28	29.81
		(1,3)	36.30	37.13	42.30	43.61
		(1,4)	49.01	49.02	48.84	49.02
		(2,1)	49.01	49.02	48.84	49.02
		(2,2)	53.02	54.36	63.43	66.61
	50	(1,1)	17.43	17.54	28.19	28.41
		(1,2)	23.47	23.78	32.85	33.43
		(1,3)	37.85	38.85	46.07	47.54
		(1,4)	60.35	63.17	68.98	72.57
		(2,1)	64.02	65.15	100.53	102.93
		(2,2)	66.86	68.57	100.53	105.33
CCCC	10	(1,1)	20.85	20.83	24.02	22.74
		(1,2)	27.58	27.65	30.48	29.87
		(1,3)	40.33	40.50	42.96	43.29
		(1,4)	40.86	40.78	47.23	43.58
		(2,1)	44.88	44.69	47.35	47.07
		(2,2)	46.91	47.07	50.80	47.60
	20	(1,1)	29.85	30.32	38.38	38.06
		(1,2)	36.11	36.84	43.89	44.10
		(1,3)	50.19	51.75	57.37	58.92
		(1,4)	65.09	66.65	78.63	77.64
		(2,1)	68.38	70.40	79.13	81.04
		(2,2)	71.51	75.01	81.71	82.93
	50	(1,1)	36.75	37.24	56.59	57.24
		(1,2)	42.62	43.57	60.91	62.23
		(1,3)	56.77	58.89	72.93	75.74
		(1,4)	79.96	84.71	94.92	100.85
		(2,1)	92.96	95.46	126.60	137.91
		(2,2)	95.70	99.14	135.22	138.48

CNTs are the same as mentioned for the first example while three others volume fractions of CNTs are considered such that 12%, 17% and 28% are introduced as rates for reinforcements. As a result, the efficiency parameters η_i (i = 1,2,3) becomes as listed in Tab. 4.19. We further assume that, $G_{12} = G_{13}$ and $G_{23} = 1.2G_{12}$. For this example, the used non-dimensional frequency parameter is $\bar{\omega} = \omega(b^2/\pi^2)\sqrt{\rho^m h/D}$ where ω is the angular frequency, ρ^m is the mass density of the matrix and $D = E_m h^3/(12(1-\nu_m^2))$ is the flexural rigidity of the plate as [314]. The FG-CNTRC skew plate is meshed using 32×32 SHO4 finite elements.

Tables 4.20 and 4.21 list the non-dimensional frequency parameter $\bar{\omega}$ of fully clamped UD and FG-X-CNTRC skew plates with length-to-width and thickness-to-width ratios ratio as $a/b = 1$ and

Table 4.19

The values of CNT efficiency parameters for the set of volume fractions (12%, 17% and 28%) of CNTs [314].

V^*_{CNT}	η_1	η_2	η_3
0.12	0.137	1.022	0.715
0.17	0.142	1.626	1.138
0.28	0.141	1.585	1.110

Table 4.20

Non-dimensional frequency parameter ($\bar{\omega} = \omega(b^2/\pi^2)\sqrt{\rho^m h/D}$) of fully clamped (CCCC) uniform distributed UD-CNTRC skew plate.

Skew angle $\theta°$	Modes	12%		17%		28%	
		Present	Ref.[314]	Present	Ref.[314]	Present	Ref.[314]
90	1	13.28	13.05	16.00	15.79	19.98	19.74
	2	14.71	14.38	17.98	17.71	21.73	21.47
	3	18.38	17.85	22.93	22.59	26.32	26.01
	4	24.75	23.17	31.38	31.43	34.56	34.65
	5	33.74	31.89	43.16	41.80	46.43	45.05
	6	36.00	35.74	43.25	43.76	54.33	47.70
	7	36.73	36.46	44.29	44.38	55.18	54.98
	8	38.61	38.23	46.90	44.97	57.42	56.08
60	1	13.55	13.44	16.42	16.26	20.21	20.15
	2	16.07	15.67	19.84	19.27	23.35	22.93
	3	21.61	20.49	27.19	25.69	30.50	29.13
	4	30.19	28.09	38.34	35.59	41.93	39.19
	5	35.71	36.44	43.17	43.92	53.39	52.35
	6	37.30	37.35	45.40	45.78	55.17	54.90
	7	41.04	38.45	50.65	48.29	57.05	56.80
	8	41.11	41.44	52.01	50.78	59.93	60.97
45	1	14.55	14.39	17.81	17.55	21.39	21.32
	2	19.30	18.64	24.12	23.21	27.51	26.80
	3	27.88	26.71	35.23	33.67	39.01	37.56
	4	36.62	37.40	44.68	45.69	53.66	53.00
	5	39.12	38.44	49.30	48.11	55.15	56.75
	6	40.74	41.68	50.29	51.11	59.16	61.32
	7	49.44	49.70	62.18	62.01	69.92	71.10
	8	51.16	50.53	64.39	63.54	71.86	71.35
30	1	18.82	18.70	23.49	23.38	26.89	27.14
	2	28.83	28.58	36.33	36.17	40.49	40.59
	3	41.64	41.96	52.30	53.28	58.69	60.23
	4	44.08	46.43	54.86	58.02	63.22	68.43
	5	55.98	57.75	70.23	73.15	79.04	83.24
	6	57.03	57.98	72.03	74.95	79.97	84.36
	7	71.83	75.39	90.00	95.77	101.61	109.33
	8	75.89	78.41	95.71	100.69	106.22	112.48

Table 4.21
Non-dimensional frequency parameter ($\bar{\omega} = \omega(b^2/\pi^2)\sqrt{\rho^m h/D}$) of fully clamped (CCCC) FG-X-CNTRC skew plate.

Skew angle θ°	Modes	12%		17%		28%	
		Present	Ref.[314]	Present	Ref.[314]	Present	Ref.[314]
90	1	16.08	15.87	19.39	19.15	24.34	24.05
	2	17.35	17.02	21.22	20.80	26.15	25.70
	3	20.75	20.30	26.00	25.41	31.05	30.44
	4	26.93	26.62	34.47	34.07	40.03	39.65
	5	35.90	35.09	46.58	45.52	53.18	52.07
	6	43.77	36.47	52.66	47.36	66.29	54.09
	7	44.38	44.08	53.57	53.07	67.13	66.83
	8	46.00	44.74	55.92	54.02	69.41	67.77
60	1	16.15	16.19	19.63	19.60	24.42	24.50
	2	18.48	18.24	22.91	22.48	27.76	27.45
	3	23.84	22.87	30.27	28.87	35.55	34.16
	4	32.52	30.49	41.90	39.13	48.27	45.33
	5	42.55	40.61	51.68	52.37	64.42	60.27
	6	43.54	44.17	53.66	53.30	64.93	66.93
	7	44.44	45.55	56.96	55.37	66.74	68.84
	8	47.34	48.59	58.58	59.64	71.22	73.15
45	1	16.96	17.04	20.85	20.81	25.57	25.71
	2	21.55	21.14	27.16	26.46	32.23	31.68
	3	30.30	29.31	38.83	37.39	45.08	43.65
	4	41.96	41.26	52.32	52.80	62.57	61.34
	5	43.20	45.50	54.57	55.33	65.06	68.77
	6	46.46	48.86	57.75	60.00	69.90	73.58
	7	54.43	55.45	69.06	70.44	81.35	82.53
	8	55.79	56.31	71.34	70.8	83.01	84.37
30	1	21.06	21.31	26.50	26.60	31.51	31.94
	2	31.45	31.58	40.20	40.16	46.84	47.11
	3	45.59	46.44	58.15	59.04	67.92	69.25
	4	49.36	53.65	61.94	66.42	74.02	80.60
	5	61.38	64.29	78.17	81.43	91.49	95.93
	6	61.95	65.10	79.26	81.96	92.28	97.36
	7	78.89	84.10	100.34	106.41	117.63	125.51
	8	82.13	86.21	105.40	109.98	122.09	128.41

$h/b = 0.001$, respectively. The obtained results are in good agreement with those given by Zhang el al. [314] for different skew angles. In fact, the relative error between the results does not exceed generally 3% except in the case where the skew angle is equal to ($\theta^\circ = 30$). For this case, the relative error can reach more than 5% for both UD and FG-X-CNTRC skew plates and this can be explained by the limitation of the FSDT theory used in [314] notably for high values of skew angles. Indeed, when the skew angle increases and becomes ($\theta^\circ = 30$), the stiffness of the FG-CNTRC skew plate also increases and affects notably the high modes of vibration. Therefore, a high-order model is required to estimate accurately the high modes of vibration which becomes sensitive to the increases of skew angle. For that, the proposed discrete double directors shell element is more convenient and general for such cases.

On the other hand, Tabs. 4.22 and 4.23 show the effect of boundary conditions on the vibrational behavior of UD and FG-X-CNTRC skew plates which have the same geometrical characteristics as

Table 4.22
Non-dimensional frequency parameter ($\bar{\omega} = \omega(b^2/\pi^2)\sqrt{\rho^m h/D}$) of a UD-CNTRC skew plate for various boundary conditions and with volume fraction of CNT, $V_{CNT}^* = 0.12$.

Skew angle θ°	results	Fundamental frequency parameters					
		SSSS	CCCC	SCSC	SFSF	CCCF	SSSF
90	Present.	5.96	13.28	6.33	6.33	13.01	5.76
	Ref.[314]	5.92	13.05	6.19	5.82	12.79	5.36
60	Present	12.37	13.55	12.71	12.71	13.00	12.12
	Ref.[314]	13.02	13.447	13.31	12.57	13.13	12.64
45	Present	13.39	14.55	14.11	14.11	13.25	12.76
	Ref.[314]	13.63	14.39	14.29	12.91	13.53	13.04
30	Present	16.66	18.82	18.66	18.66	14.66	14.42
	Ref.[314]	15.51	18.83	17.88	13.59	18.01	15.41

Table 4.23
Non-dimensional frequency parameter ($\bar{\omega} = \omega(b^2/\pi^2)\sqrt{\rho^m h/D}$) of a FG-X-CNTRC skew plate for various boundary conditions and with volume fraction of CNT, $V_{CNT}^* = 0.12$.

Skew angle θ°	results	Fundamental frequency parameters					
		SSSS	CCCC	SCSC	SFSF	CCCF	SSSF
90	Present	7.18	16.08	7.51	7.51	15.85	7.01
	Ref.[314]	7.108	15.87	7.37	7.02	15.55	7.07
60	Present	14.54	16.15	14.85	14.85	15.67	14.32
	Ref.[314]	15.69	16.19	15.98	15.14	15.91	15.23
45	Present	15.63	16.96	16.30	16.30	15.78	15.06
	Ref.[314]	16.30	17.04	16.91	15.58	16.27	15.70
30	Present	18.87	21.06	20.83	20.83	17.08	16.76
	Ref.[314]	17.76	21.31	20.23	16.87	21.05	17.01

mentioned in the previous results. From the current results, it can be remarked that more kinematic boundary constraints are imposed on the FG-CNTRC skew plate, more the values of fundamental frequencies are higher. This effect is slightly outlined for increased skew angle ($\theta^\circ = 30$) where the fundamental frequency parameter reaches its maximum. Hence, for an indicated volume fraction of CNTs, the variation of both boundary conditions and skew angles affects the vibrational behavior of such FG-CNTRC shell structures. In addition, the obtained results agree with the cited reference [314] except when the skew angle increases and becomes ($\theta^\circ = 30$). For example, the relative error between the results is more than 10% for skew angle ($\theta^\circ = 30$) of FG-CNTRC skew plate under SFSF and CCCF boundary conditions and a volume faction ($V_{CNT}^* = 0.12$). This observation can be explained notably by the shear correction coefficients which can take different values according to the used method. Besides, the studied example is of moderately thick FG-CNTRC skew plate while the FSDT theory is restricted to thin structures. So, to provide accurate results for all types (thin, moderately thick or thick) of studied FG-CNTRC structures, the present model constitutes an alternative for handling such structures. Moreover, results provided by Zhang et al. [314] about the influences of the aspect ratios (a/b) and thickness-to-width ratios (h/b) on the non-dimensional frequency parameter $\bar{\omega}$ are also compared with the proposed model in Tabs. 4.24 and 4.25, respectively. It is worth to mention that the frequency parameters decrease with an increase of (a/b) or (h/b) ratios which indicate the effect of introduction of transverse shear deformations notably for high modes of vibration. This comparison study leads to verification of the ability of the present model to draw the dynamic behavior of FG-CNTRC skew plates with a good accuracy. It is also worth to noticed that the proposed formulation has a relevant potential to estimate the natural frequencies of high modes of vibration in comparison to the used (FSDT) theory in [314]. In fact, this formulation is based on three generic kinematic assumptions notably the one which imposes a zero shear transverse deformation in both top and bottom surfaces with a parabolic distribution of these deformations.

4.4.4 FREQUENCY RESPONSES OF FG-CNTRC CYLINDRICAL PANELS

The third studied example is of FG-CNTRC cylindrical panel where its geometry is as shown in Fig. 5.41. This panel is assumed to be thin, of length L, radius R, span angle θ_0 and thickness h. These parameters are considered as follows: $\theta_0 = 0.1rad$, $h = 0.002m$, $h/R = 0.002$ and $L/R = 0.1$. As shown in Fig. 5.41, the CNTs are assumed to be uniaxially aligned in axial direction and functionally graded in thickness direction of the cylindrical panel. In this example, the mechanical properties of CNTs are the same as used for FG-CNTRC square plate in section 5.1 and for the matrix are the same as used in section 5.2 (PMMA matrix with young modulus $E_m = 2.5GPa$). As well as, the used mesh involves 32×20 SHO4 finite elements. Uniform (UD) and three types of graded distributions are considered here which are FG-V, FG-X and FG-O with three volume fractions of CNTs, $V^*_{CNT} = 0.11, 0.14, 0.17$ and two types of boundary conditions: all sides (AB), (BC), (CD) and (DA) are simply supported (SSSS) or clamped (CCCC). To perform the dynamic behavior of FG-CNTRC cylindrical panel, the used non-dimensional frequency parameter is: $\bar{\omega} = \omega(L^2/h)\sqrt{\rho^m/E_m}$. A comparison study with the results of Zhang el al. [316] is carried out in order to outline the applicability of the proposed method for various forms of FG-CNTRC shell structures.

Figures 4.22 and 4.23 show the effect of volume fraction of CNTs on the frequency parameter of FG-CNTRC cylindrical panel with two boundary conditions : simply supported (SSSS) and clamped (CCCC). It can be seen that when the volume fraction of CNTs V^*_{CNT} increases, the non-dimensional frequency parameter also increases and this is due to the stiffness of the FG-CNTRC cylindrical panel which becomes higher with the CNT reinforcements. As a result, the CNTs enhance the dynamic behavior of such functionally graded shell structures. Nevertheless, the variation of edge-to-thickness ratio induces a rapid decrease in the non-dimensional frequency parameter as depicted in Figs. 4.24 and 4.25. These results are similar to those given in [316]. Table 4.26 lists the first six non-dimensional frequencies of different distributions of CNTs with two boundary conditions. From this table, it can be remarked that the values of frequency parameter for a clamped FG-CNTRC cylindrical panel are higher than those for simply supported FG-CNTRC cylindrical panel. This is because the constraint of boundary condition clamped (C) is stronger than the boundary condition supported (S). Furthermore, it can be seen from Tab. 4.26 that the FG-X distribution has the highest value of non-dimensional frequency parameter while the (FG-O) distribution has the lowest one. This observation is in concordance with those of FG-CNTRC square plate which indicated that the distributions close to top and bottom surfaces enhances the dynamic behavior of FG-CNTRC shell structures. Also, the obtained results match very well with those of Zhang el al. [316]. Therefore, the proposed formulation improves its accuracy and efficiency in the prediction of the dynamic behavior of functionally graded composite shell structures reinforced by CNTs with different shapes.

Table 4.24

Non-dimensional frequency parameters ($\bar{\omega} = \omega(b^2/\pi^2)\sqrt{\rho^m h/D}$) of a clamped (CCCC) UD-CNTRC skew plate for various aspect ratios, ($V^*_{CNT} = 0.12$).

$\theta°$	Mode	Frequency parameters									
		$a/b=1$		$a/b=1.5$		$a/b=2$		$a/b=2.5$		$a/b=3$	
		Present	Ref.[314]	Present	Ref.[314]	Present	Ref.[314]	Present	Ref.[314]	Present	Ref.[314]
90	1	13.28	13.05	6.29	6.08	4.04	3.86	3.14	2.95	2.74	2.54
	2	14.71	14.38	8.75	8.25	7.22	6.69	6.29	6.15	4.69	4.56
	3	18.38	17.85	13.89	12.93	9.36	9.19	6.71	6.16	6.51	5.94
	4	24.75	23.17	16.21	15.91	11.30	10.95	8.80	8.43	7.66	7.25
	5	33.74	31.89	17.52	17.13	12.90	11.91	11.60	11.48	8.26	8.21
	6	36.00	35.74	20.83	20.07	15.76	15.04	12.58	11.62	10.38	10.23
	7	36.73	36.46	21.47	20.19	17.85	17.71	13.29	13.12	12.45	11.42
	8	38.61	38.23	26.77	25.78	19.09	18.91	13.95	13.20	13.19	12.33
60	1	13.55	13.44	6.74	6.57	4.60	4.37	3.78	3.49	3.42	3.10
	2	16.07	15.67	10.48	9.77	8.91	8.21	6.67	6.56	5.18	5.05
	3	21.61	20.49	16.52	15.50	9.98	9.87	8.77	7.92	8.45	7.65
	4	30.19	28.09	16.95	16.60	12.64	12.12	10.31	9.76	8.55	8.31
	5	35.71	36.44	19.48	18.96	16.94	15.23	12.13	12.35	9.70	9.37
	6	37.30	37.35	23.18	21.82	18.20	17.21	14.31	13.94	11.68	11.40
	7	41.04	38.45	28.34	25.60	18.30	18.59	16.59	14.70	13.74	14.33
	8	41.11	41.44	31.51	28.75	21.11	20.94	17.94	16.73	15.55	14.48
45	1	14.55	14.39	7.86	7.49	5.85	5.37	5.13	4.57	4.84	4.24
	2	19.30	18.64	13.82	13.03	10.41	10.10	7.71	7.39	6.33	5.95
	3	27.88	26.71	17.89	17.78	13.39	12.57	12.18	11.84	9.40	9.23
	4	36.62	37.40	20.53	19.85	15.47	14.81	12.79	11.84	12.58	11.71
	5	39.12	38.44	26.28	25.01	19.57	19.63	14.80	14.31	13.51	12.81
	6	40.74	41.68	28.28	27.68	21.45	20.92	16.76	16.26	13.54	13.34
	7	49.44	49.70	32.66	33.31	24.83	22.96	20.61	20.81	16.10	15.95
	8	51.16	50.53	36.63	36.78	27.47	26.53	22.06	21.69	17.76	17.58
30	1	18.82	18.83	11.80	10.98	9.96	8.69	9.38	8.00	9.15	7.53
	2	28.83	28.79	19.87	19.24	14.02	13.23	11.39	10.31	10.22	8.85
	3	41.64	42.53	27.18	25.43	20.80	20.35	15.63	15.05	12.77	11.87
	4	44.08	47.00	28.54	27.99	24.97	22.30	21.43	21.03	16.82	16.2
	5	55.98	58.53	38.53	38.64	28.30	27.01	24.56	21.84	21.86	21.49
	6	57.03	59.74	39.13	39.34	28.61	27.86	25.72	23.37	24.30	22.64
	7	71.83	76.74	48.52	47.23	36.45	36.56	28.14	27.69	25.06	23.54
	8	75.89	79.99	50.36	50.92	37.20	36.97	29.40	28.00	26.21	25.03

4.5 DYNAMIC ANALYSIS OF FG-CNTRC STRUCTURES

In this part, the dynamic behavior of functionally graded carbon nanotubes-reinforced composite shell structures (FG-CNTRC) via forced vibration analysis is presented. The governing equations of motion are developed using the linear discrete double directors (DD-HSDT) finite element model. The elaborated model is based on high-order-distribution of displacement field and uses a cubic variation of the vector position along the thickness direction. A zero transverse shear stress at top and bottom surfaces is also imposed. Four types of distributions of carbon nanotubes (CNTs) such that uniformly and three functionally graded distributions are considered. The extended rule of mixture is used to estimate the effective material properties of carbon nanotube-reinforced composite

Table 4.25
Non-dimensional frequency parameters ($\bar{\omega} = \omega(b^2/\pi^2)\sqrt{\rho^m h/D}$) of a clamped (CCCC) UD-CNTRC skew plate for various thickness-to-width ratios, ($V^*_{CNT} = 0.12$).

$\theta°$	Mode	Frequency parameters									
		$h/b = 0.001$		$h/b = 0.01$		$h/b = 0.05$		$h/b = 0.1$		$h/b = 0.2$	
		Present	Ref.[314]	Present	Ref.[314]	Present	Ref.[314]	Present	Ref.[314]	Present	Ref.[314]
90	1	13.28	13.05	12.90	12.43	8.62	7.76	5.59	4.74	3.53	2.74
	2	14.71	14.38	14.35	13.66	10.35	9.27	7.50	6.41	5.09	4.08
	3	18.38	17.85	18.04	16.97	14.26	12.77	10.93	8.91	6.19	4.80
	4	24.75	23.17	24.41	22.92	17.99	15.78	10.99	9.43	6.80	5.68
	5	33.74	31.89	33.33	31.36	18.96	16.65	12.06	9.94	7.43	6.00
	6	36.00	35.74	33.78	32.57	20.16	18.07	12.38	12.14	7.73	7.06
	7	36.73	36.46	34.52	33.22	21.44	18.92	14.53	13.26	7.85	7.19
	8	38.61	38.23	36.44	34.99	24.76	22.93	15.51	13.59	9.43	7.67
60	1	13.55	13.44	13.18	12.94	9.02	8.27	6.06	5.23	3.93	3.11
	2	16.07	15.67	15.71	15.04	11.75	10.56	8.68	7.42	5.81	4.61
	3	21.61	20.49	21.23	19.64	16.78	14.81	11.61	9.71	7.19	5.44
	4	30.19	28.09	29.69	26.83	18.49	16.69	12.23	10.32	7.40	6.06
	5	35.71	36.44	33.64	33.54	20.71	18.50	14.22	11.88	7.92	7.15
	6	37.30	37.35	35.27	34.82	22.47	19.79	14.39	13.01	9.08	7.47
	7	41.04	38.45	39.15	36.22	26.13	22.72	15.56	14.72	9.27	7.92
	8	41.12	41.44	40.08	38.03	28.12	24.67	18.08	15.47	9.95	8.78
45	1	14.55	14.39	14.19	13.78	10.13	9.21	7.13	6.12	4.72	3.65
	2	19.30	18.64	18.91	18.02	14.51	12.95	10.60	9.00	6.89	5.24
	3	27.88	26.71	27.33	25.87	20.40	18.23	13.56	11.38	8.75	6.47
	4	36.62	37.40	34.72	34.67	20.41	18.41	14.13	11.90	8.83	6.67
	5	39.12	38.44	38.03	36.59	25.47	22.75	17.64	14.72	8.97	8.06
	6	40.74	41.68	38.86	38.49	26.29	23.44	17.66	15.00	10.95	8.27
	7	49.44	49.70	47.70	46.66	31.93	28.54	17.93	17.05	11.11	9.26
	8	51.16	50.53	49.38	47.95	32.97	29.51	20.69	17.71	11.38	9.57
30	1	18.82	18.83	18.43	17.37	14.05	10.40	10.28	6.93	6.67	4.07
	2	28.83	28.79	28.17	26.19	20.86	15.98	14.55	10.24	9.12	5.77
	3	41.64	42.53	40.33	38.07	27.49	20.18	18.36	12.68	11.35	7.01
	4	44.08	47.00	42.30	41.67	28.49	21.23	19.79	12.96	12.28	7.13
	5	55.98	58.53	53.77	51.60	34.34	23.63	22.22	13.18	12.43	7.48
	6	57.03	59.74	55.10	52.68	37.64	26.84	24.86	15.53	13.72	8.29
	7	71.83	76.74	68.35	66.57	41.02	29.59	25.14	17.19	14.94	9.24
	8	75.89	79.99	72.66	69.89	45.42	29.63	26.12	17.87	15.33	9.56

(CNTRC) shell. The applicability and the performance of the present model are illustrated by three numerical examples of FG-CNTRC square plates, spherical caps and annular ring plates. The transient center deflections of the studied shell structures are computed and depicted for different volume fractions and profiles of CNTs, various boundary conditions and other geometrical parameters in order to show the effect of these parameters on dynamic behavior of FG-CNTRC shells.

4.5.1 INTEGRATION OF VIBRATION PROBLEMS IN TIME DOMAIN

The determination of the transient responses of FG-CNTRC structures is achieved via the integration of the vibrational problem, in time domain, using the Newmark's algorithm. The governing equation

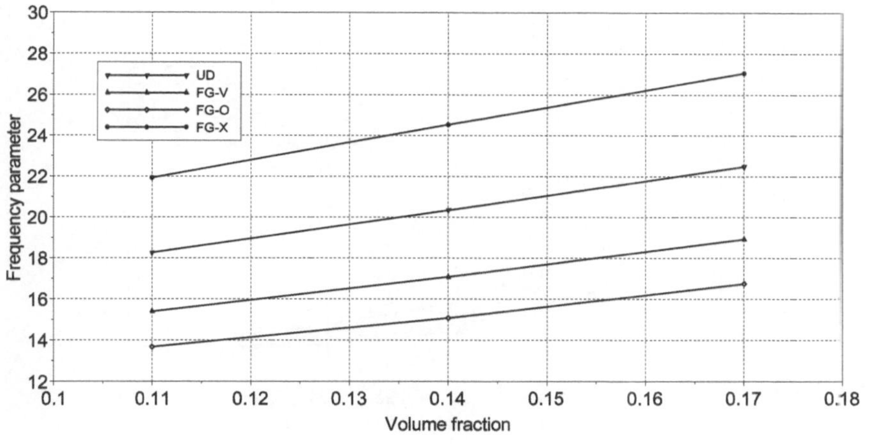

Figure 4.22 Effect of volume fraction of CNTs on frequency parameters of simply supported (SSSS) FG-CNTRC panel.

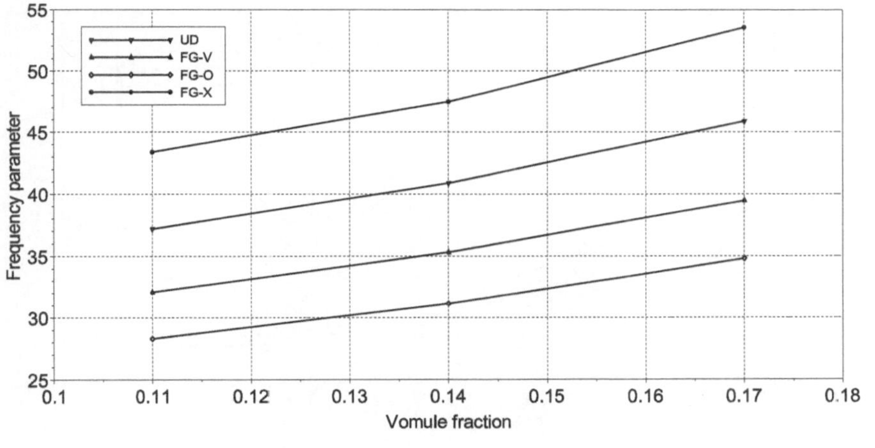

Figure 4.23 Effect of volume fraction of CNTs on frequency parameters of fully clamped (CCCC) FG-CNTRC panel.

of motion for forced vibration problem of FG-CNTRC shell structure can be written as:

$$\mathbf{M}\ddot{\mathbf{U}}_n + \mathbf{K}\mathbf{U}_n = \mathbf{F}. \tag{4.11}$$

where **F** is the external load vector. The steps of the used Newmark's algorithm can be summarized in the algorithm presented in Tab. 4.27.

4.5.2 DYNAMIC ANALYSIS OF FG-CNTRC PLATES

Herein, dynamic analysis is performed for FG-CNTRC square plate of length side $a = 25cm$, thickness $h = 5cm$ and simply supported on all its edges. As usual, UD, FG-O, FG-X and FG-V-CNTRC forms are considered where their material properties and efficiency parameters η_i (i = 1,2,3) are as

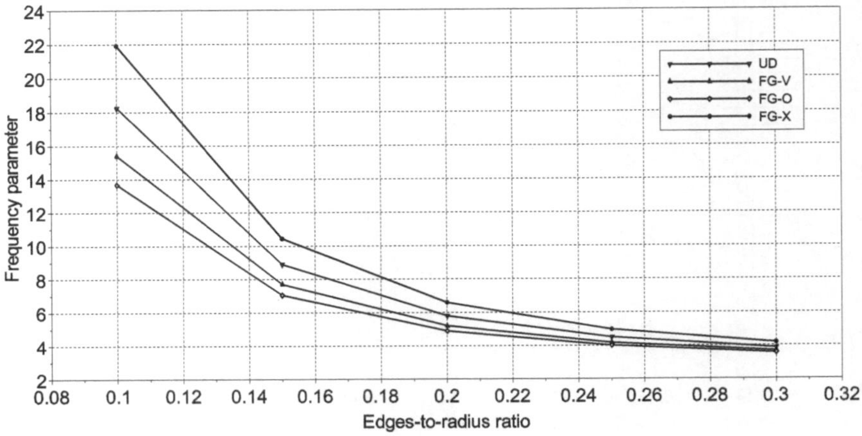

Figure 4.24 Effect of edge-to-radius ratio (L/R) on frequency parameters of simply supported (SSSS) FG-CNTRC panel ($V^*_{CNT} = 0.11$).

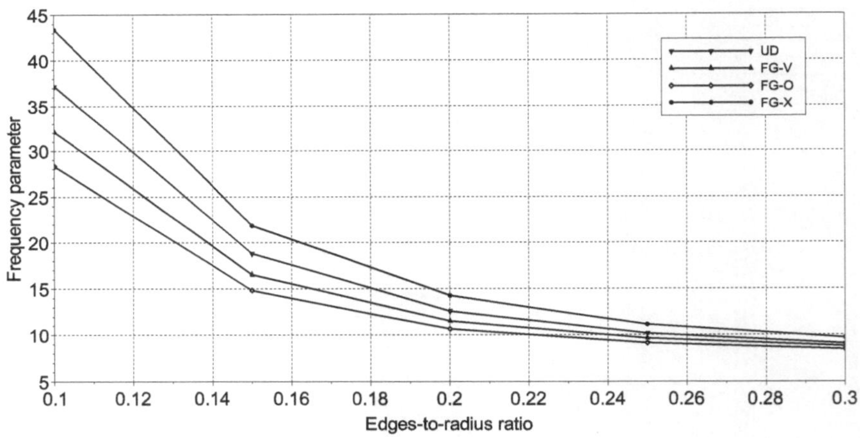

Figure 4.25 Effect of edge-to-radius ratio (L/R) on frequency parameters of fully clamped (CCCC) FG-CNTRC anel ($V^*_{CNT} = 0.11$).

mentioned Tabs. 4.2 and 4.3, respectively. This plate is subjected to a transverse sudden dynamic load as shown in Fig. 4.26. Its value is equal to $q_0 = 1 \times 10^5 N/m^2$. The used mesh involves 16×16 SHO4 finite elements. It should be mentioned that the abbreviation SHO4 is used to identify the proposed linear discrete double directors shell element.

The obtained results in terms of time-central vertical displacement compared to those given by Lei et al. [137] for different distributions of FG-CNTRC plates and different volume fractions V^*_{CNT} are illustrated in Fig. 4.27. A good agreement between the results is revealed which outlines the capacity of the present model in the prediction of temporal responses of FG-CNTRC plates with good accuracy. As well as, it can be seen that the amplitude of the central vertical displacement $w(t)$ decreases with the increase of the CNT volume fraction V^*_{CNT} for UD and three functionally

Table 4.26
Non-dimensional frequency parameter ($\bar{\omega} = \omega(L^2/h)\sqrt{\rho^m/E_m}$) of various distributions of FG-CNTRC cylindrical panel with simply supported (SSSS) and clamped (CCCC) boundary conditions, ($V^*_{CNT} = 0.11$).

	Mode	CNT distributions							
		UD		FG-V		FG-O		FG-X	
		Present	Ref.[316]	Present	Ref.[316]	Present	Ref.[316]	Present	Ref.[316]
SSSS	1	18.26	17.85	15.40	15.27	13.67	13.44	21.93	21.24
	2	22.46	22.07	20.29	20.18	18.75	18.48	25.73	25.09
	3	33.56	33.28	32.35	32.25	30.81	30.58	36.37	35.93
	4	51.84	51.77	51.34	51.41	49.48	48.70	54.64	54.53
	5	70.00	65.12	58.30	55.30	50.65	49.43	79.86	76.75
	6	72.09	67.26	60.83	58.00	53.51	51.50	85.06	78.55
CCCC	1	37.18	36.84	32.09	31.69	28.30	28.17	43.38	42.93
	2	41.20	40.92	36.76	36.56	33.33	33.21	47.01	46.64
	3	51.73	51.82	48.54	48.69	45.44	45.59	56.92	56.94
	4	69.68	70.63	67.69	68.67	64.63	65.50	74.43	75.39
	5	91.72	91.44	80.29	79.80	70.85	71.41	99.32	101.72
	6	93.91	93.61	82.93	82.58	73.86	74.35	104.68	104.00

Table 4.27
Newmark's algorithm.

- $[\overline{\mathbf{K}}] = [\mathbf{K}] + \frac{1}{\beta \Delta t^2}[\mathbf{M}] + \frac{\gamma}{\beta \Delta t}[\mathbf{C}]$
- $\{R_{t+\Delta t}\} = \{F_{t+\Delta t}\} + [\mathbf{M}]\left(\frac{1}{\beta \Delta t^2}\{U_t\} + \frac{1}{\beta \Delta t}\{\dot{U}_t\} + \left(\frac{1}{2\beta} - 1\right)\{\ddot{U}_t\}\right) +$ $[\mathbf{C}]\left(\frac{\gamma}{\beta \Delta t}\{U_t\} + \left(\frac{\gamma}{\beta} - 1\right)\{\dot{U}_t\} + \Delta t\left(\frac{\gamma}{2\beta} - 1\right)\{\ddot{U}_t\}\right)$
- Update $[\overline{\mathbf{K}}]$
- Resolution of Eq. (8.11) for $\{U_{t+\Delta t}\}$
- Computation of $\{\ddot{U}_{t+\Delta t}\}$ $\{\ddot{U}_{t+\Delta t}\} = \frac{1}{\beta}\left[\frac{1}{\Delta t^2}(\{U_{t+\Delta t}\} - \{U_t\} - \Delta t\{\dot{U}_t\}) - (\frac{1}{2} - \beta)\{\bar{U}_t\}\right]$
- Computation of $\{\dot{U}_{t+\Delta t}\}$ $\{\dot{U}_{t+\Delta t}\} = \{\dot{U}_t\} + \Delta t(1-\gamma)\{\ddot{U}_t\} + \gamma \Delta t\{\ddot{U}_{t+\Delta t}\}$

graded distributions and this can be explained by the enhancement of the stiffness with the CNTs reinforcements. Furthermore, from Fig. 4.27c and Fig. 4.27d, it is clear that the FG-X-CNTRC has the lowest vertical displacement while the FG-O-CNTRC has the highest one. Hence, the type of CNT distribution can affect the dynamic behavior of FG-CNTRC plates since the stiffness of the plate is larger with CNTs distributions close to top and bottom surfaces more than those distributed near the mid-plane surface.

In the same context of improvement of the present formulation, the dynamic behavior of FG-CNTRC square plate is examined here by comparing the present temporal results with those of Zhang et al. [312] which consider different regular polygonal FG-CNTRC plates. The comparison is carried out for the case of square plate varying the apothem-to-thickness ratio ($r = a/2$) which corresponds to the half length of the square plate in our study. The same geometry and

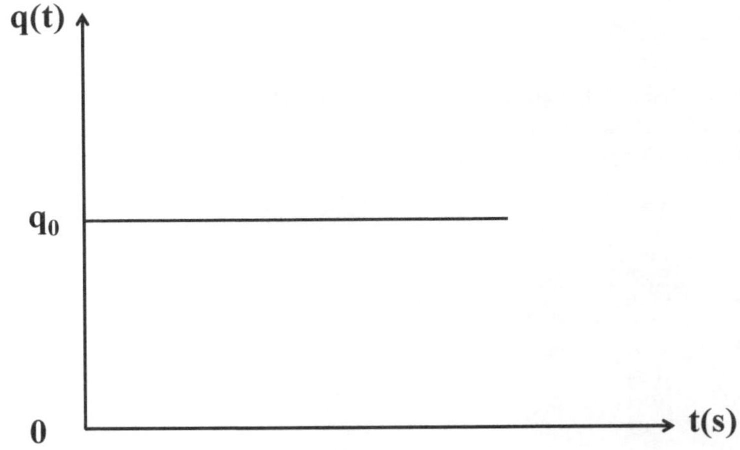

Figure 4.26 The applied transverse sudden dynamic load on the FG-CNTR structures.

mechanical properties of FG-CNTRC plates are considered as mentioned earlier at the beginning of section 4.2. Fig. 4.28 represents the time-displacement curves of UD and three CNTs distributions of simply supported FG-CNTRC square plates subjected to the same transverse sudden load $q_0 = 1 \times 10^5 N/m^2$ and for different apothem-to-thickness ratio $(r/h = 5, 10, 20)$ with a CNT volume fraction $(V^*_{CNT} = 0.11)$. Firstly, it can be observed that the obtained temporal responses agree well with those given by Zhang et al. [312] which lead to prove again the efficiency and the accuracy of the present model. From Figs. 4.28(a), 4.28(b) and 4.28(c), it is found that the maximum value and period of vertical displacement occurs for FG-O-CNTRC plates while the minimum occurs for FG-X-CNTRC plates. In addition, UD and FG-V-CNTRC plates undergo medium vertical displacement between FG-O and FG-X-CNTRC plates. Therefore, it can be concluded that the CNTs distributions close to top and bottom surfaces are more efficient in increasing the stiffness of the plate than those distributed near the mid-surface. Moreover, the increase of apothem-to-thickness ratio manifests rapidly by an increase in the amplitude and period of the central vertical displacement in the studied time range $[0-200] \times 10^{-6} s$. Indeed, when the apothem-to-thickness ratio increases, the FG-CNTRC square plate behaves as thin structure and becomes hence more sensitive to the variation of this parameter. Now, the effect of boundary conditions on dynamic behavior of FG-CNTRC plates using the present model is examined. Figs. 4.29(a), 4.29(b), 4.29(c) and 4.29(d) illustrate the vertical central displacement versus time curves of UD, FG-V, FG-O and FG-X CNTRC square plates, respectively with different boundary conditions and under a volume fraction $(V^*_{CNT} = 0.11)$. A combination of clamped (C) and simply supported (S) boundary conditions is considered and the apothem-to-thickness ratio is chosen as $(r/h = 5)$. The used boundary conditions are defined as follows:

$$\text{Simply supported (S):} \quad \begin{array}{l} v = w = \Theta_1^1 = \Theta_1^2 = 0 \text{ for sides (AD) and (BC),} \\ u = w = \Theta_2^1 = \Theta_2^2 = 0 \text{ for sides (AB) and (DC),} \end{array}$$

$$\text{Clamped (C): } u = v = w = \Theta_1^1 = \Theta_2^1 = \Theta_1^2 = \Theta_2^2 = 0,$$

From Fig. 4.29, it is shown that the amplitude of the vertical central displacement of the simply supported FG-CNTRC plate is higher than the other curves containing clamped edges which indicate that more constraints are imposed on the structure, more the amplitude of the vertical central displacement becomes small. Hence, the type of imposed boundary condition can affect the dynamic

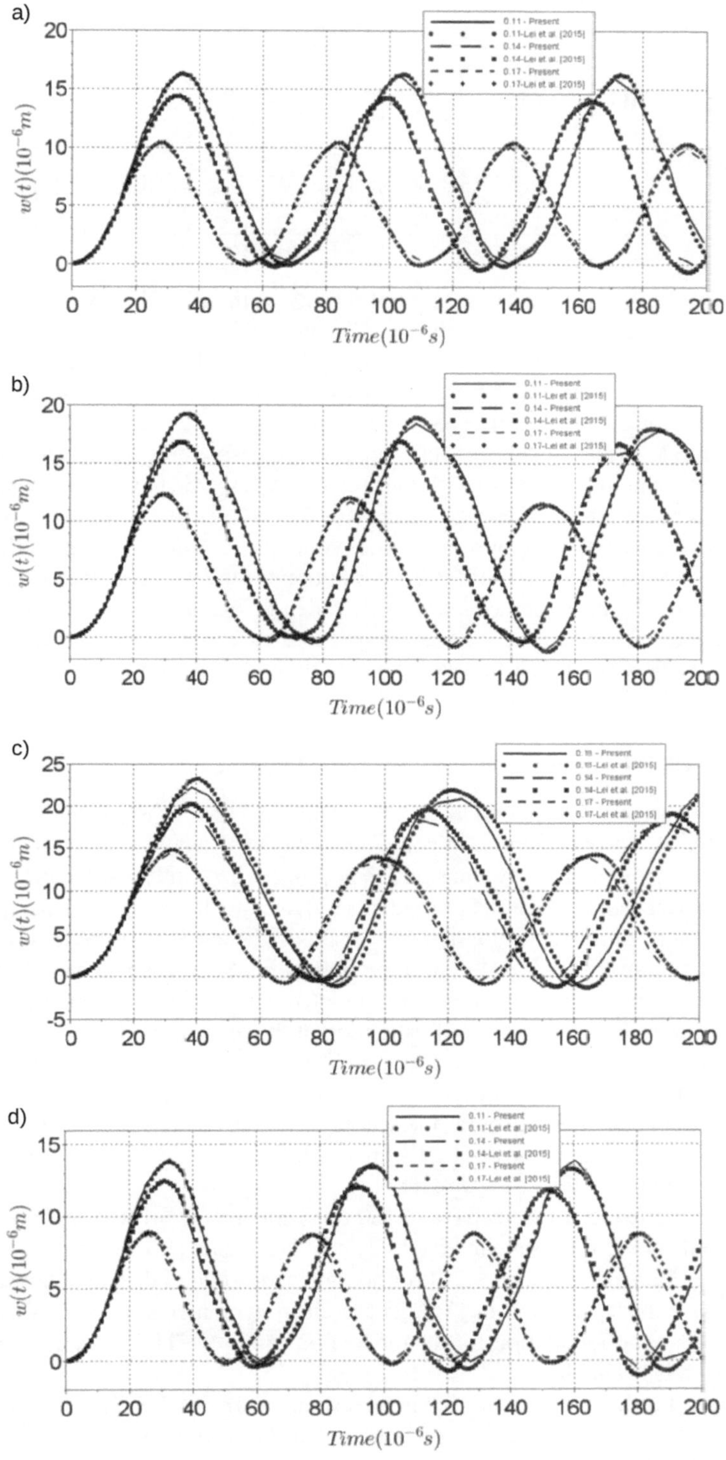

Figure 4.27 Temporal responses of (a) UD, (b) FG-V, (c) FG-O, (d) FG-X CNTRC plate with different CNTs volume fractions.

Table 4.28
Parameters of spherical cap [76].

Parameter	R	θ	E	ν	h	H	ρ	F
Value	4.76	10.9°	10^7	0.3	0.01576	0.0859	0.000245	100

response of FG-CNTRC plates due to the flexural rigidity of the FG-CNTRC plate which becomes more sensitive to the variation of the boundary conditions.

The dynamic response of the FG-CNTRC square plate with UD and three types of CNTs reinforcements (FG-V, FG-O and FG-X) under a harmonic loading is also examined here. In fact, the applied loading is of the form $q(t) = q_0 \cos(\omega t)$ where $q_0 = 40N$ and $\omega = 9 \times 10^4\ rad/s$. The harmonic time-central vertical displacement responses corresponding to various profiles of FG-CNTRC square plates and under a CNT volume fraction ($V^*_{CNT} = 0.17$) are presented in Fig. 4.30. Stable results are observed in the interval of time $[0-2] \times 10^{-4} s$. It can be also observed that the various types of CNT distributions contribute to the pronounced changes in amplitude. As expected, the amplitude value of the FG-X type of CNT distributions is the smallest one while the amplitude of the FG-O type is the highest one. This observation can be explained by the load capacity of the FG-X-CNTRC square plate which is the highest due to the distribution of CNTs near the top and bottom surfaces, in this case which makes the flexural rigidity of the FG-X-CNTRC plate more stronger than the others distributions.

4.5.3 DYNAMIC RESPONSES OF FG-CNTRC SPHERICAL CAPS

Secondly, we perform the dynamic behavior of a clamped FG-CNTRC spherical cap. Since, no results are available in the open literature concerning dynamic analysis of FG-CNTRC spherical shells except the one presented by Thomas and Roy [260] which induces damping characteristics of FG-CNTRC spherical shell. Even the presented results show the damping capacity of such shell structures which is an essential factor for dynamic analysis, the used eight-nodes finite element may generate further computational time in the solution procedure. So to keep a general formulation, the proposed double directors finite element model is adopted here to perform dynamic analysis of FG-CNTRC spherical shells. For the purpose of validation of the present model, numerical calculations are carried out first for the isotropic spherical cap. This numerical example is also examined for the isotropic case in [76, 93]. The geometrical and material characteristics of the isotropic spherical shell are given in Tab. 4.28. Furthermore, the whole of the structure is subjected to concentrated apex load (see Fig. 4.31) with a mesh of 217 nodes and 192 elements. As well as, one layer of SHO4 elements is used in the thickness direction.

The value of the applied load between 0 and 280 μs is equal to $F = 100$. The normalized vertical displacement at shell apex is illustrated in Fig. 4.32 and it is compared to the linear dynamic response provided by Duarte Filho and Armando [76] for homogeneous material. The obtained results are in good agreement with those reported in [76]. These close agreements serve to verify the present approach and establish the foundation for its application to FG-CNTRC spherical shells where no comparative solution is available.

Now, we present the first results on dynamic responses of FG-CNTRC spherical shells. The FG-CNTRC spherical cap is subjected to sudden applied uniform pressure loading of value $q_0 = -1 \times 10^6\ N/m^2$ as depicted in Fig. 4.26 and the applied boundary condition is the clamped type. The effect of CNTs profiles distributions is first examined for different CNTs volume fractions. Figures 4.33, 4.34 and 4.35 represent the vertical central deflection versus the time for uniform (UD) and

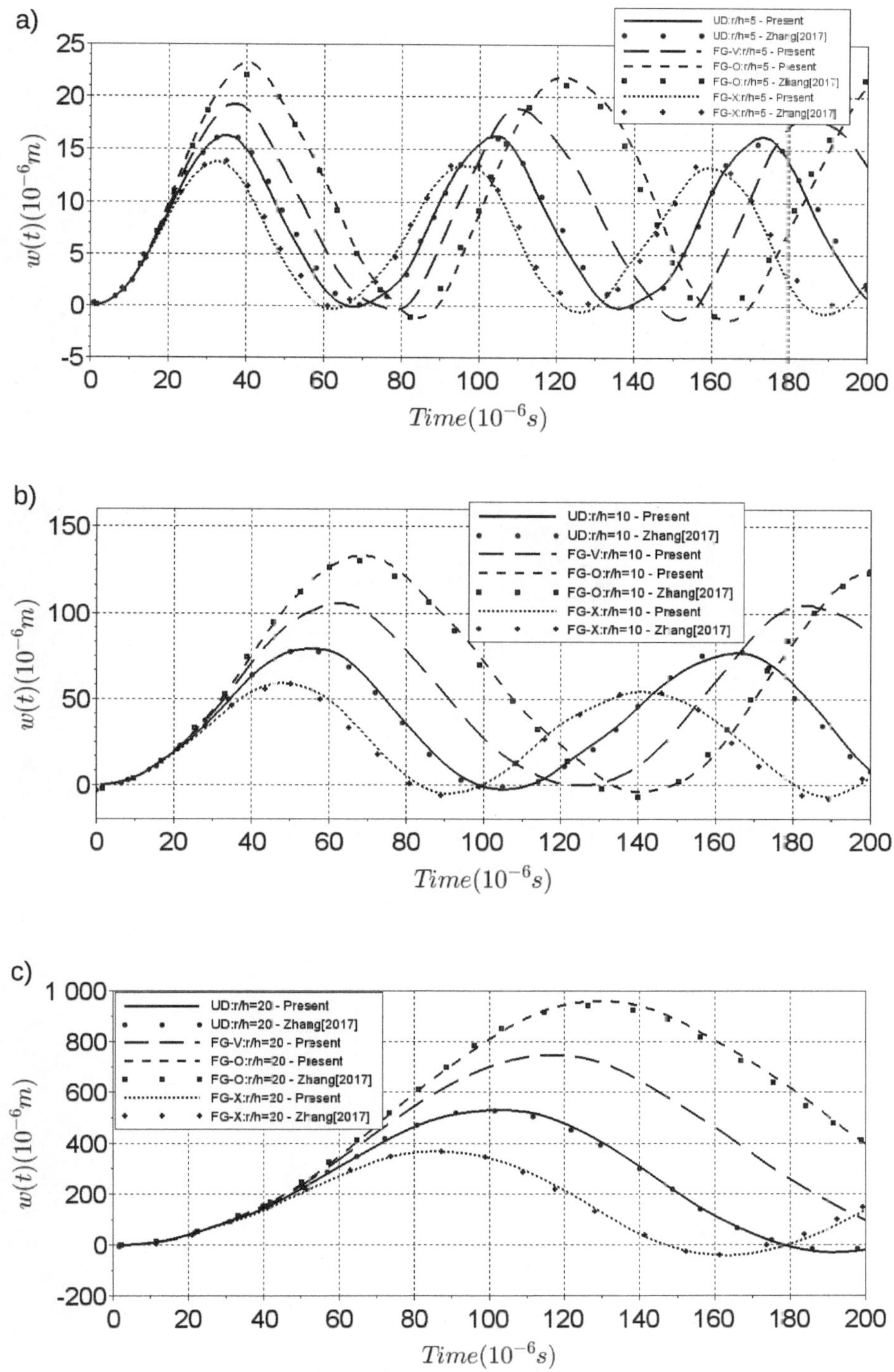

Figure 4.28 Central vertical displacement time-history for different distributions of FG-CNTRC square plate and with (a) $r/h = 5$, (b) $r/h = 10$ and (c) $r/h = 20$ apothem-to-thickness ratios, $V_{CNT}^* = 0.11$.

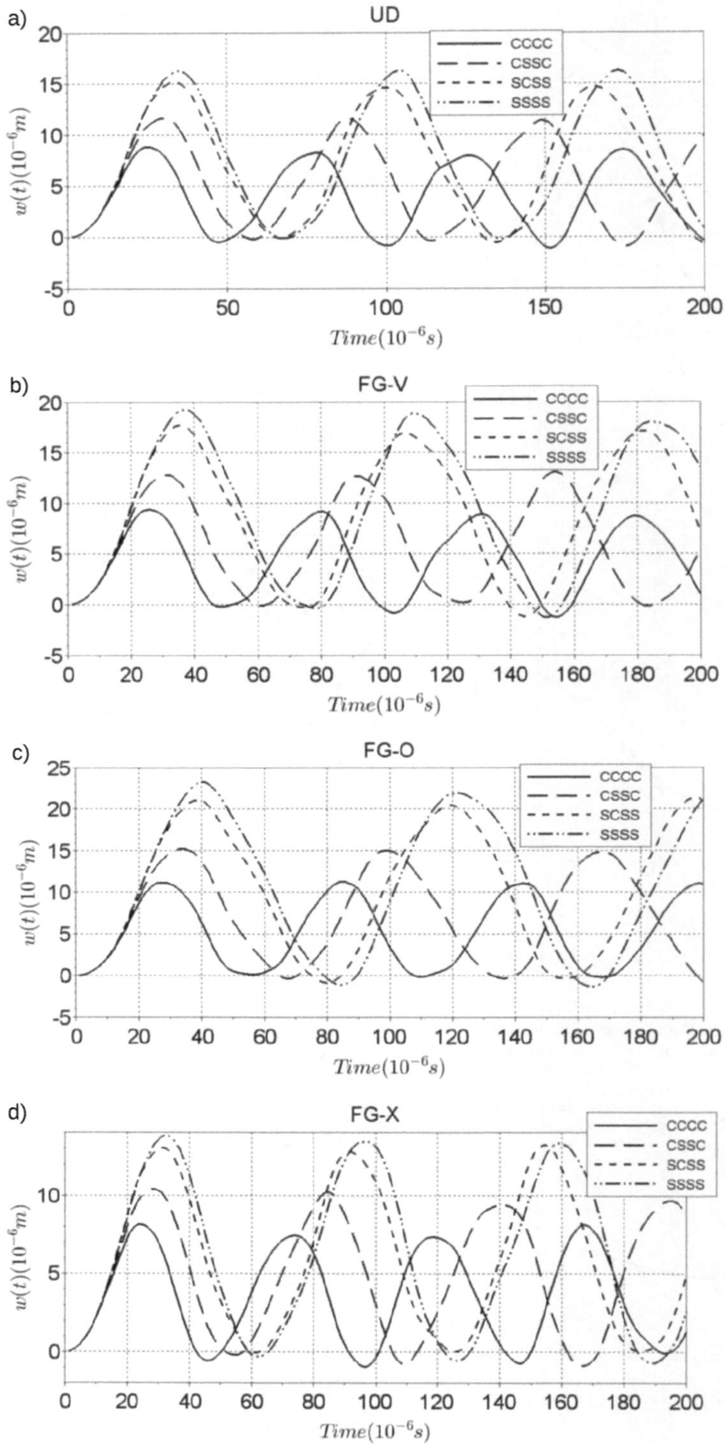

Figure 4.29 Vertical central deflection versus time curves for (a) UD, (b) FG-V, (c) FG-O and (c) FG-X CNTRC square plate with different boundary conditions and volume fraction $V_{CNT}^{*} = 0.11$.

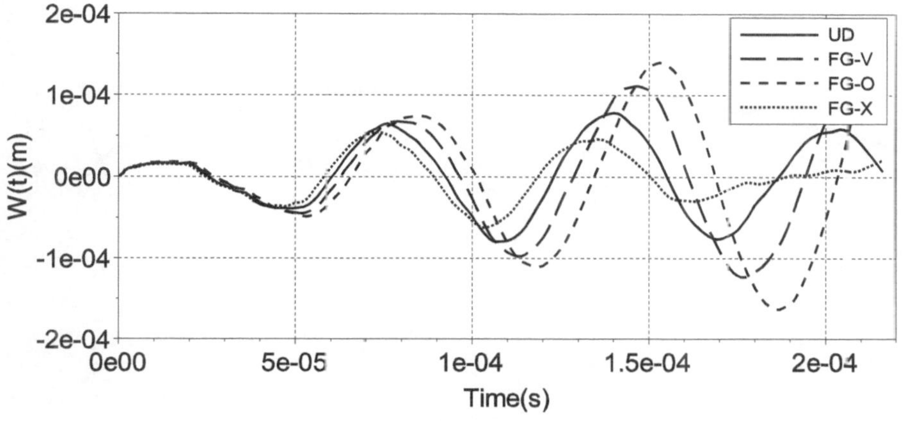

Figure 4.30 Vertical central deflection versus time curves for various CNTs profiles of FG-CNTRC square plates under a harmonic loading and a CNT volume fraction ($V^*_{CNT} = 0.17$).

three profiles of CNT distributions: FG-V, FG-O and FG-X-CNTRC spherical caps and under three values of volume fractions ($V^*_{CNT} = 0.11, 0.14, 0.17$) respectively. It is clear that the value of vertical central deflection of FG-X distribution is the lowest one while the value of FG-O distribution is the highest one. As well as, the value of period deceases from FG-O curve to FG-X curve. Thereby, the CNTs distributions close to top and bottom surfaces are more efficient in the enhancement of the stiffness of the FG-CNTRC shell structure than to those near the mid-surface or uniform distributed. Moreover, when the volume fraction of CNTs increases, the value of the vertical central deflection decreases for each type of distribution, for instance the value of central deflection of FG-X distribution deceases about 40% from $V^*_{CNT} = 0.11$ to $V^*_{CNT} = 0.17$. Hence, more the FG-CNTRC spherical shell is reinforced by CNTs, more the stiffness of the structure becomes higher which implies a decrease in the amplitude of vibrations. Therefore, the CNTs reinforcements constitute an efficient method to control the vibration phenomenon for structural engineering applications. It should be noted that the oscillations of the spherical cap shell are well stable around its equilibrium configuration and this is for different CNTs distributions. Hence, no snap-through phenomenon is occurred in this case since the present study is performed below the critical load point of instability which ensures the efficiency of the present model.

On the other hand, the effect of variation of radius-to-thickness ratio (R/H) is examined here for the FG-CNTRC spherical cap. In fact, Figs. 4.36, 4.37 and 4.38 illustrate the temporal responses of the clamped FG-CNTRC spherical shell for different profiles of CNTs and with a volume fraction $V^*_{CNT} = 0.11$. It can be remarked that for ($R/H = 20$), the amplitude of vibration does not exceed ($6 \times 10^{-6} m$) for the different profiles of CNTs while this amplitude reaches ($1000 \times 10^{-6} m$) with ($R/H = 300$) radius-to-thickness ratio which indicates the significant effect of the increase of this parameter on forced vibrations of FG-CNTRC spherical shells. Moreover, the FG-X form has the lowest amplitude while the FG-O has the highest one and this is for different radius-to-thickness ratio which affirms that the CNTs distributed near to top and bottom surface are more efficient than those distributed near the mid-surface. Therefore, the CNTs reinforcements can be a good solution for passive vibrations and one can choose a special kind of CNTs distributions to adjust the stiffness of the structure and hence the vibration amplitude of the studied shell structure. Similar to the previous example, the harmonic response of the FG-CNTRC spherical cap for different CNTs distributions is also examined. Fig. 4.39 shows the harmonic responses for UD and three graded CNTs distributions of the FG-CNTRC spherical cap which is subjected to a harmonic loading of

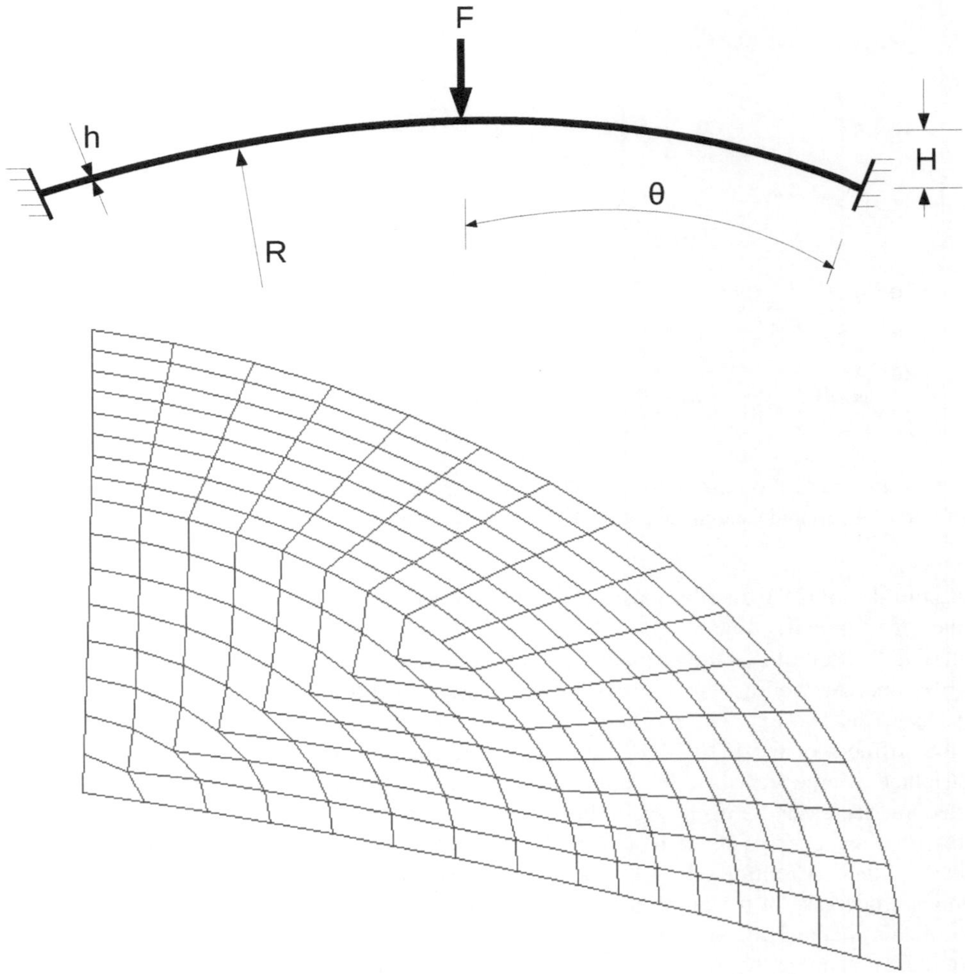

Figure 4.31 Geometry of the spherical cap.

amplitude $q_0 = 40N$ and under a pulsation excitation equal to $\omega = 1500rad/s$. As can be seen, the oscillations are periodic around the equilibrium position of the structure and the UD and FG-V distributions are localized between FG-X and FG-O curves which reflects the effect of the type of CNT distribution in the enhancement of the dynamic response of such structures.

4.5.4 DYNAMIC ANALYSIS OF FG-CNTRC ANNULAR RING PLATES

Since no results are available for comparison with dynamic analysis of annular ring plates in both isotropic and FG-CNT reinforcements, we provide here, the first known temporal responses for annular ring plates. This example was used frequently by many authors [39, 37] to test finite element shell formulations with geometrical non-linearity. Thus, it motivate us to perform dynamic analysis of such structure with linear isotropic and FG-CNT material laws. The geometry of the example consists of an annular plate with a slit cut along the line A-B; with internal radius $r = 6m$, external radius $R = 10m$ and constant shell thickness $h = 0.03m$ as shown in Fig. 4.40(a). The elastic material

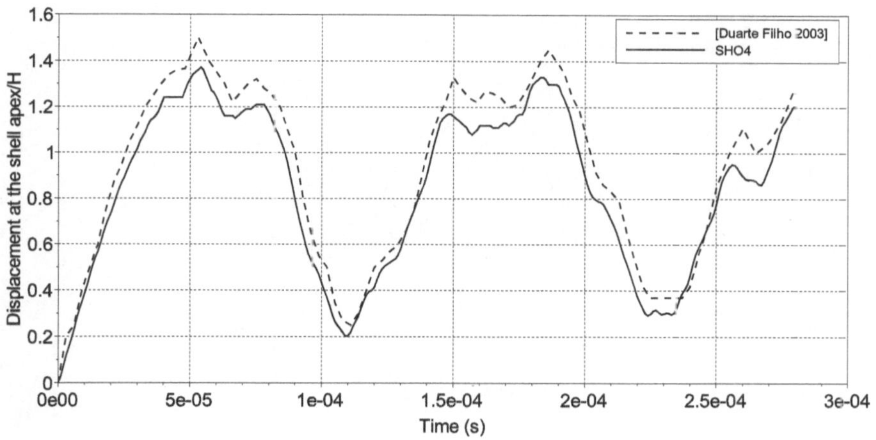

Figure 4.32 Transient center deflection of isotropic spherical cap.

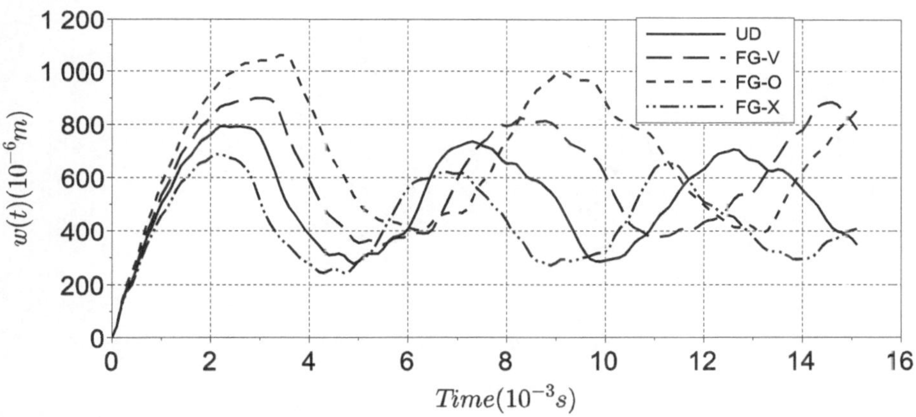

Figure 4.33 Temporal responses of clamped FG-CNTRC spherical cap for different distributions of CNTs and under a volume fraction, $V_{CNT}^{*} = 0.11$.

properties are: $E = 70 \times 10^9 N/m^2$, $\rho = 2700 Kg/m^3$ and $\nu = 0.3$. The ring plate is loaded at its free edge, the other edge is fully clamped. The maximum applied load, as shown in Fig. 4.26, at the free edge is fixed to $p = 3N/m$. The corresponding static deformed shape is given in Fig. 4.40(b). Moreover, the ring plate is discretized using 8×48 SHO4 finite shell elements. The normalized static and dynamic responses of the isotropic ring plate are illustrated in Fig. 4.41. It can be seen that the oscillations are well stable over the period of time $[0-150]s$ and the steady state response is well established around the static response which may serve as reference solution for dynamic analysis of isotropic ring plates. Then, the effect of introduction of CNT reinforcements in the structure is examined. The material properties of the used matrix and the CNTs are as reported earlier in Tabs. 4.2 and 4.3, respectively. As can be seen from Fig. 4.42, which illustrates the transient deflection response of the FG-CNTRC ring plate for various CNTs profiles distributions under a volume fraction $V_{CNT}^{*} = 0.11$, the FG-X form presents the lowest value of deflection while the FG-O

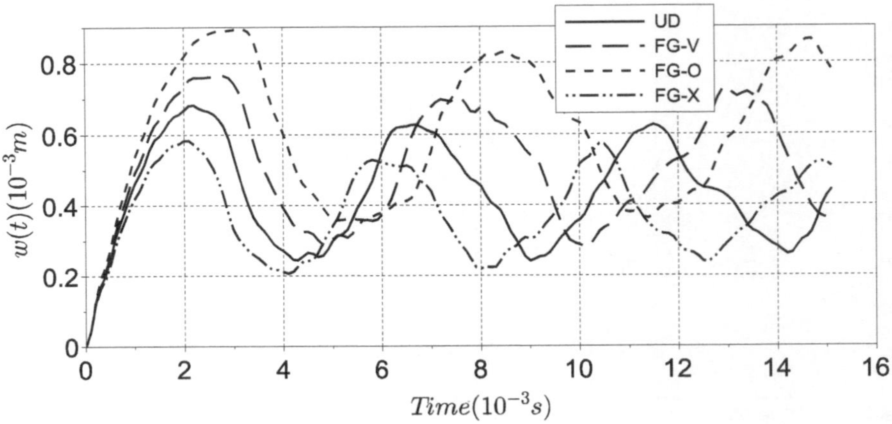

Figure 4.34 Temporal responses of clamped FG-CNTRC spherical cap for different distributions of CNTs and under a volume fraction, $V_{CNT}^* = 0.14$.

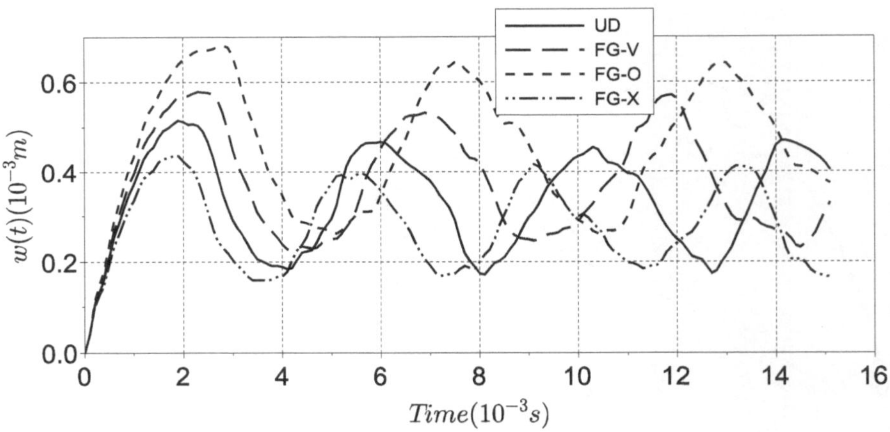

Figure 4.35 Temporal responses of clamped FG-CNTRC spherical cap for different distributions of CNTs and under a volume fraction, $V_{CNT}^* = 0.17$.

form has the highest one. As well as, the UD and FG-V forms are between FG-X and FG-O curves. Hence, more the CNTs reinforcements are closer to top and bottom surfaces, more the stiffness of the structure becomes higher which induces a decrease in transient response of the FG-CNTRC ring plates.

Since the FG-X form has the lowest amplitude value, the effect of CNTs volume fractions on dynamic response of FG-X-CNTRC ring plate is also examined. Figure 4.43 shows the temporal responses of FG-X-CNTRC ring plate varying the CNTs volume fraction from $V_{CNT}^* = 0.11$ to $V_{CNT}^* = 0.17$. It can be seen that, as the CNTs volume fraction V_{CNT}^* increases, the temporal response of FG-CNTRC ring plate decreases which can be explained by the increase of the stiffness of the structure with the percentage of added CNTs. Therefore, more the FG-CNTRC ring plate is reinforced by CNTs, more the structure becomes stiffer which enables the designer to adjust the stiffness of the structure by the addition of the appropriate percentage of CNTs. Moreover, the effect of thickness parameter is highlighted via Fig. 4.44 which depicts the temporal response of

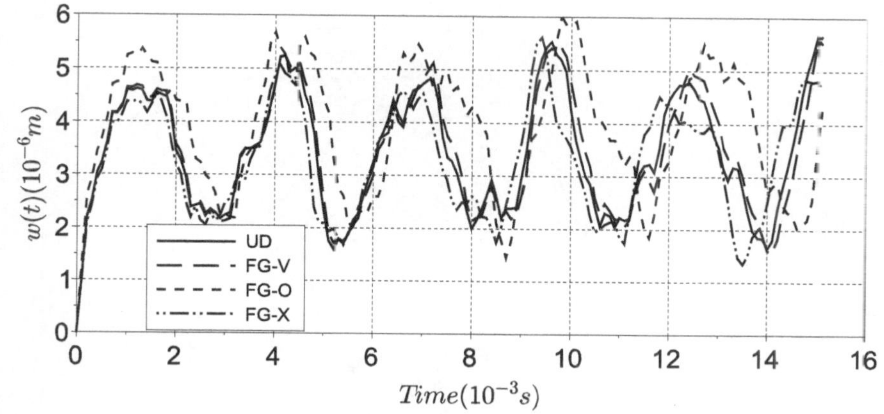

Figure 4.36 Transient center deflection of clamped FG-CNTRC spherical cap for different profiles of CNTs and with radius-to-thickness ratio (R/H = 20) and volume fraction ($V^*_{CNT} = 0.11$).

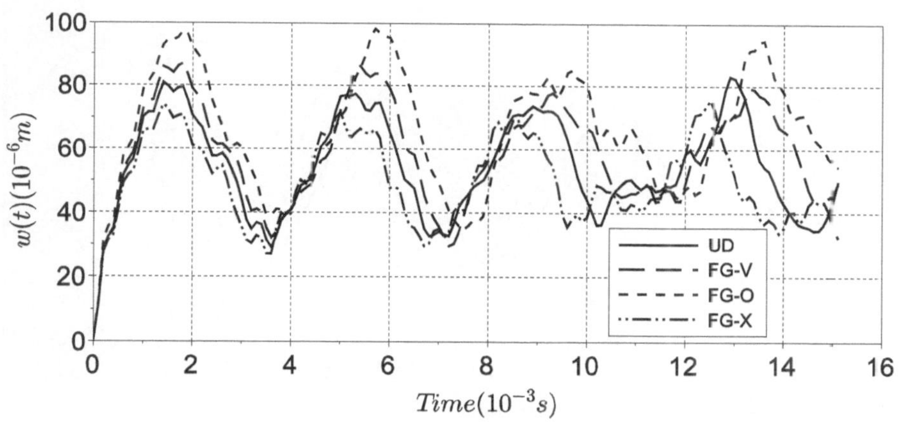

Figure 4.37 Transient center deflection of clamped FG-CNTRC spherical cap for different profiles of CNTs and with radius-to-thickness ratio (R/H = 100) and volume fraction ($V^*_{CNT} = 0.11$).

FG-X-CNTRC ring plate with a CNT volume fraction ($V^*_{CNT} = 0.17$). It can be observed that, more the thickness parameter increases, more the amplitude of vibration decreases. Hence, when the FG-CNTRC ring plate behaves as thick plate, it becomes more sensitive to the variation of the thickness parameter. The dynamic response of the FG-CNTRC ring plate under a harmonic loading of amplitude $q_0 = 4N$ and eigen-frequency excitation equal to $\omega = 0.18\ rad/s$ is equally investigated. The vertical central deflection versus time curves for various CNTs profiles of FG-CNTRC ring plate are illustrated in Fig. 4.45. As expected, the FG-X form has the minimum values of amplitude while the FG-O form has the maximum ones. In fact, when the CNT reinforcements are distributed near the top and bottom surfaces, the load capacity of the FG-CNTRC ring plate is the highest due to the high value of the stiffness of the structure. Such results can constitute a reference for dynamic analysis of FG-CNTRC structures for future investigations.

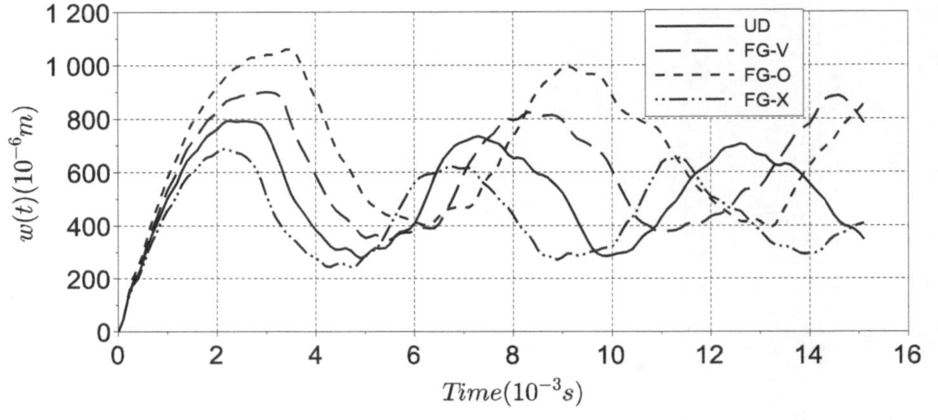

Figure 4.38 Transient center deflection of clamped FG-CNTRC spherical cap for different profiles of CNTs and with radius-to-thickness ratio (R/H = 300) and volume fraction ($V^*_{CNT} = 0.11$).

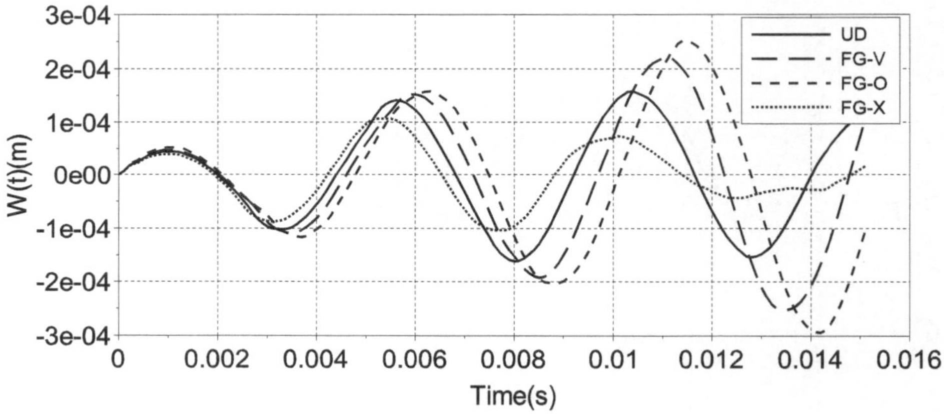

Figure 4.39 Vertical central deflection versus time curves for various CNTs profiles of FG-CNTRC spherical cap under a harmonic loading and a CNT volume fraction ($V^*_{CNT} = 0.17$).

4.6 CONCLUSION

In this chapter, various analyses of FG-CNTRC plates and shells are presented based on the double directors finite shell element (DDST) model which was developed in Chapter 1. The carbon nanotubes (CNTs) can either uniformly or gradually distributed in the thickness direction of the shell leading hence to four forms, namely UD, FG-V, FG-O and FG-X-CNTRC structures. The effective material properties of these kinds of materials are estimated using a modified rule of mixtures (MROM) with some efficiency parameters to take into account the scale dependency of CNTs. Through several investigations made on this new thematic, it is found that the CNTs reinforcements can enhance drastically the mechanical behavior of composite structures since the Young's modulus of these CNTs is over than 1TPa and their tensile strength exceeds that of steel by over one-order of magnitude which make them an alternative of reinforcements in composite structures. These revolutionary FG-CNTRC materials which are considered as the first advanced material of the twenty-first

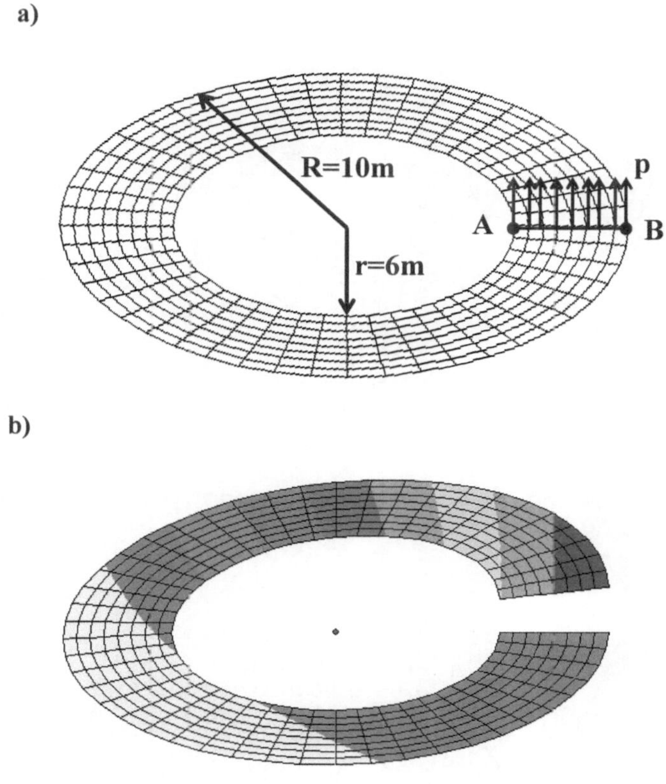

Figure 4.40 Annular ring plate. (a) geometry definition and (b) static deformed shape.

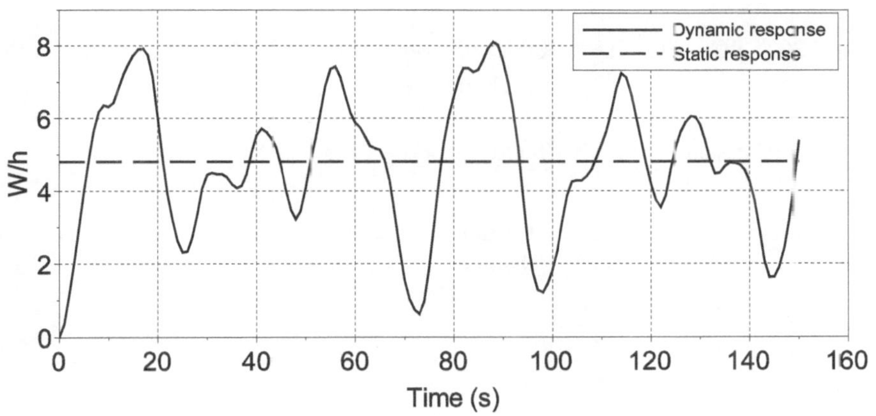

Figure 4.41 Static and dynamic responses of the isotropic annular ring plate under uniform load at its free edge.

century have been replaced day by day by the conventional materials via their very high strength and performance. The presented analysis includes static bending, stresses, free vibration and dynamic responses of FG-CNTRC structures are established in linear domain, in order to outline the potential

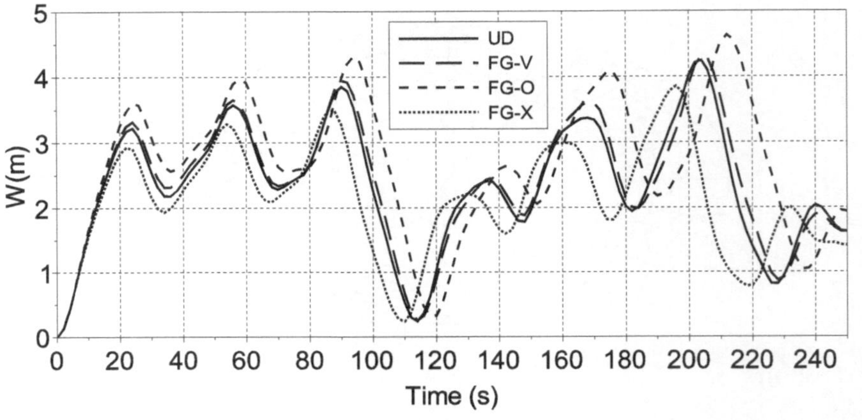

Figure 4.42 Temporal responses of FG-CNTRC ring plate for various profiles of CNTs distributions and with a CNT volume fraction ($V^*_{CNT} = 0.11$).

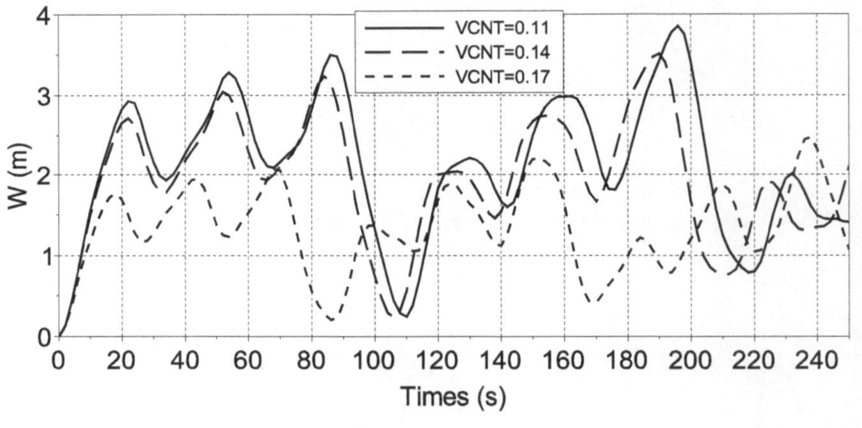

Figure 4.43 Temporal responses of FG-X-CNTRC ring plate for various CNTs volume fractions.

of the developed finite element models and solutions strategies, non-linear analysis of FG-CNTRC plates and shells are presented in the next chapter.

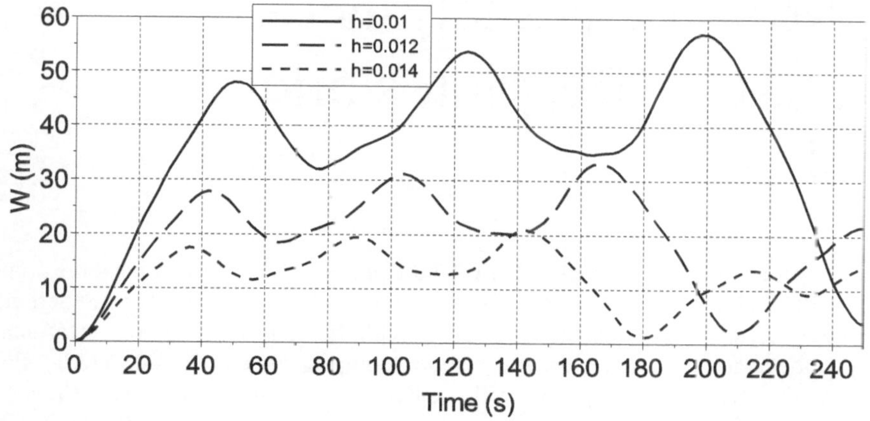

Figure 4.44 Effect of thickness parameter on temporal responses of FG-X-CNTRC ring plate under a CNT volume fraction $V^*_{CNT} = 0.17$.

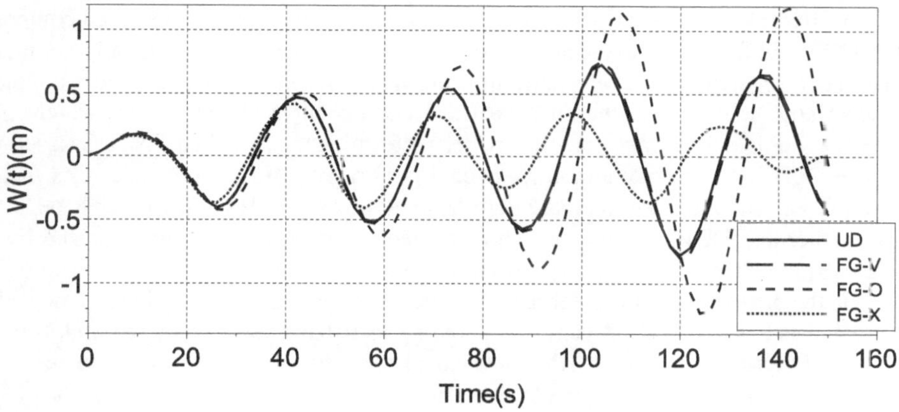

Figure 4.45 Vertical central deflection versus time curves for various CNTs profiles of FG-CNTRC ring plate under a harmonic loading and a CNT volume fraction ($V^*_{CNT} = 0.17$).

5 Non-linear analysis of FG-CNTRC plates and shells

This chapter deals with non-linear analysis of FG-CNTRC plates and shells using the three proposed kinematic models: Kirchhoff-Love model, the FSDT model and the (DD-HSDT) finite shell model. For the first model, we address the extension of the Kirchhoff shell model to non-linear analysis of functionally graded carbon nanotubes-reinforced thin composite shells (FG-CNTRCs). For the second model, the FSDT theory is employed which takes into account the transverse shear deformations and incorporates geometrical non-linearity type. For the third theory, the non-linear double directors shell model with membrane enhancement is utilized. This model induces a high-order variation of the displacement field along the thickness direction and imposes a zero transverse shear stresses condition in the top and bottom surfaces. Uniform (UD) and three graded distributions, known as FG-V, FG-O and FG-X, are considered to show the effects of carbon nanotubes (CNTs) profiles and their volume fractions on non-linear deflections of various shapes of shell structures. To illustrate these effects, a set of numerical examples including beams, plates, cylindrical panels, hyperboloid, cylindrical and hemispherical shells are presented.

5.1 INTRODUCTION

The modeling of non-linear bending behavior of plates and shells has attracted, over last years, the attention of many researchers and this concern has been twice with the discovery of advanced materials like functionally graded materials (FGMs) and carbon nanotubes which reinforced composite structures (CNTRCs). In fact, the implementation of carbon nanotubes (CNTs) into polymer matrix composites has been predicted to have a large impact on many areas of science and technology due to their superlative mechanical, thermal, electrical and chemical characteristics which make them the ultimate candidate for composite reinforcements forming the so-called CNTRC structures. Furthermore, these CNTRCs provide a multifunctional performance for various industries such as aerospace, aircraft, automobile, electrical and civil engineering[87, 192].On the other hand, functionally graded materials (FGMs) which are inhomogeneous composites are characterized by the functionally gradation of their mechanical properties along the thickness direction and can be applied to CNTs via the powder metallurgy fabrication procedure[191] and forming hence a new class of materials, known as FG-CNTRCs. As a result, a lot of papers have been developed, in last years, to study theoretically and experimentally the mechanical behavior of such advanced materials. A good review of the latest investigations made on various aspects of mechanical behavior of FG-CNTRC beams, plates and shells can be found in Liew et al.[145].

At the same time, the availability of high performance computing resources have pushed computer modeling of progress in order to provide an accurate analysis toward structural components like beams, plates and shells. The finite element technique has been widely used by many researchers as an efficient tool in the prediction of mechanical behavior of various shapes of structures. As background for understanding the modeling process, a brief review of the frequently used models for the approximation of the kinematics of shell structures is provided here. The classical shell theory (CST), which is based on the Kirchhoff-Love assumptions, constitutes a suitable choice notably for thin structures. However, this theory neglects the transverse shear deformations which is an essential parameter in modeling of plates and shells. To remedy this drawback, another theory, known as the

DOI: 10.1201/9781003483786-5

first order shear deformation theory (FSDT) is proposed by Reissner and Mindlin wherein the effect of transverse shear deformations is introduced. This theory gives more accurate results when shells become thicker and it is simple to implement it in most finite element procedures. Nevertheless, this theory face some difficulties related generally to the introduction of shear correction factors which is numerically prohibitive or suffers to the shear-locking problem in the very thin shell limit. All the limitations of CST and FSDT can be overcome by higher-order shear deformation shell theory (HSDT) which takes into account higher-order terms in the displacement fields. Even this theory provides accurate results, it can generate further computational time which is undesirable for such applications. Hence, a bridge between the accuracy and a reasonable computational time should be assured by an efficient model. In the following, the relevant works concerning non-linear deflections analysis of FG-CNTRC structures are briefly presented.

From the pioneer work of [223] concerning non-linear bending analysis of FG-CNTRC rectangular plates under transverse uniform or sinusoidal load in thermal environments, many authors have been inspired to conduct non-linear large deflections analysis of FG-CNTRCs using various finite element models and numerical solutions. In relation with this subject, Shen and his co-authors presented various works on large deflections analysis of composite laminated and FGM plates using either the HSDT theory [236, 226, 233, 227] or a semi-analytical-numerical method [300, 301]. With a six-parameter shell theory, Ansari et al. studied the non-linear behavior of shells [10]. Although, accurate results are provided, the present analysis is restricted only to the isotropic case and doesn't perform advanced materials such as FGMs or FG-CNTRCs. As well as, Zhang and his co-workers have published various papers concerning non-linear analysis of FG-CNTRCs with different shape forms [136, 315, 323, 318]. A detailed parametric study is conducted to trace the effects of CNTs distributions, their volume fractions and several geometrical parameters on non-linear deflection responses of FG-CNTRC structures. Furthermore, Mehar and his co-authors presented various analyses of non-linear bending and vibrational behavior of FG-CNTRC shells under thermal loading in the framework of the higher-order kinematics and one can be referred to these publications [161, 162, 163, 164], among others. Recently, Zghal et al. [307, 308, 310, 309, 83] have conducted various investigations on FG-CNTRC structures using a high-order model within the double directors finite element shell framework which induces a high-order distribution of the displacement field and imposes a zero condition of the transverse shear deformation on top and bottom surfaces. Both linear and non-linear aspects have been highlighted via several numerical examples of plates and shells. The obtained results have shown the effects of CNTs profiles, their volume fractions and various geometrical parameters on mechanical behavior of such structures. Moreover, Frikha et al. [80, 82, 212, 271] have presented non-linear analysis of FGM and FG-CNTRC shells using either a refined high order theory and a modified FSDT theory. As well as, Reinoso and Blazquez performed non-linear analysis of FGMs and FG-CNTRCs using a fully integrated first-order solid shell finite element [209]. Locking problem is treated by combining the enhanced assumed strain (EAS) and the assumed natural strain (ANS) methods. Even the proposed solid shell finite element allows the prediction of non-linear behavior of such structures with good accuracy, it requires, however, further computational time in the solution procedure. So, to analyze non-linear deflections analysis with low computational costs, an accurate finite element model should be provided. Further investigations related to non-linear bending, vibrational and buckling analysis of FGM structures can be found in [67, 70, 69, 171].

5.2 NON-LINEAR PROBLEM RESOLUTION AND ENHANCED ASSUMED MEMBRANE STRAINS

To deal with non-linear behavior of FG-CNTRC plates and shells, three kinematic models are used to cover a wide range of non-linear analysis from thin-moderately thick to thick shell structures as developed in Chapter 1. For the first analysis, the governing equations are established using a discrete form of the non-linear Kirchhoff model with two possibilities of discretization. Indeed, either quadrilateral or triangular finite elements are implemented and may be applied to various forms of shells. For the second range of analysis, the governing equations are elaborated within the FSDT framework taking into account the transverse shear deformations, large displacements and finite rotations and without need of any shear correction factors. For the latter analysis, a refined non-linear model based on a double directors finite element shell is used. This model takes into account the transverse shear deformations with cubic distribution of the displacement field along the thickness direction and it imposes a zero condition of the shear transverse stresses in the top and bottom surfaces enabling hence to describe the membrane, bending and shear contributions of the strains with large deformations and rotations. Furthermore, the membrane contribution of strains is enhanced to reach notably the membrane behavior for in-plane bending dominated cases. This constitutes an advantage for the presented formulation especially that the contribution of membrane apart or coupled with bending and shear parts has an essential role to describe the kinematics of the shell structure.

5.2.1 ENHANCED ASSUMED MEMBRANE STRAINS

The compatible in-plane strain is enhanced in order to improve the membrane behavior of the double directors shell element especially for in-plane bending dominated cases. As a result, the strain field can be written as:

$$e = e^c + \widetilde{e}, \quad \widetilde{e} = \widetilde{B}\alpha \tag{5.1}$$

where the enhanced membrane matrix $\widetilde{B}$ can be expressed as:

$$\widetilde{B}_{3\times4} = \frac{\sqrt{A_0}}{\sqrt{A}} T_0^{-T} e_\xi \tag{5.2}$$

A_0 and T_0 denote the surface and the transformation matrix in the mid-plane of the shell ($\xi = \eta = 0$) respectively. Their expressions are given by:

$$\sqrt{A_0} = (\sqrt{A})_{\xi=\eta=0}, \quad T_0 = T(J_0), \quad J_0 = (J)_{\xi=\eta=0} \tag{5.3}$$

$$e_\xi = \begin{bmatrix} \xi & 0 & 0 & 0 \\ 0 & \eta & 0 & 0 \\ 0 & 0 & \xi & \eta \end{bmatrix}. \tag{5.4}$$

Using the adopted notation, we obtain:

$$a = \int_A \widetilde{B}^T H_{11} \widetilde{B} dA, \quad h = \int_A \widetilde{B}^T N dA, \quad b = \int_A \widetilde{B}^T_{4\times3} \begin{bmatrix} H_{11} & H_{12} & H_{13} & 0 \end{bmatrix}_{3\times11} B_{11\times28} dA \tag{5.5}$$

After local condensation of the parameter α and using the above defined parameters *a*, *b* and *h*, the tangent matrix and the element residual can be written as:

$$K = K_M + K_G - b^T a^{-1} b, \quad r = b^T a^{-1} h + r_{ext} - r_{int} \tag{5.6}$$

5.2.2 RESOLUTION PROCEDURE

Newton solution procedure is used to obtain non-linear responses of FG-CNTRC shell structures. Hence, to outline its steps, a brief flowchart of the Newton's solution method is provided in Fig. 5.1.

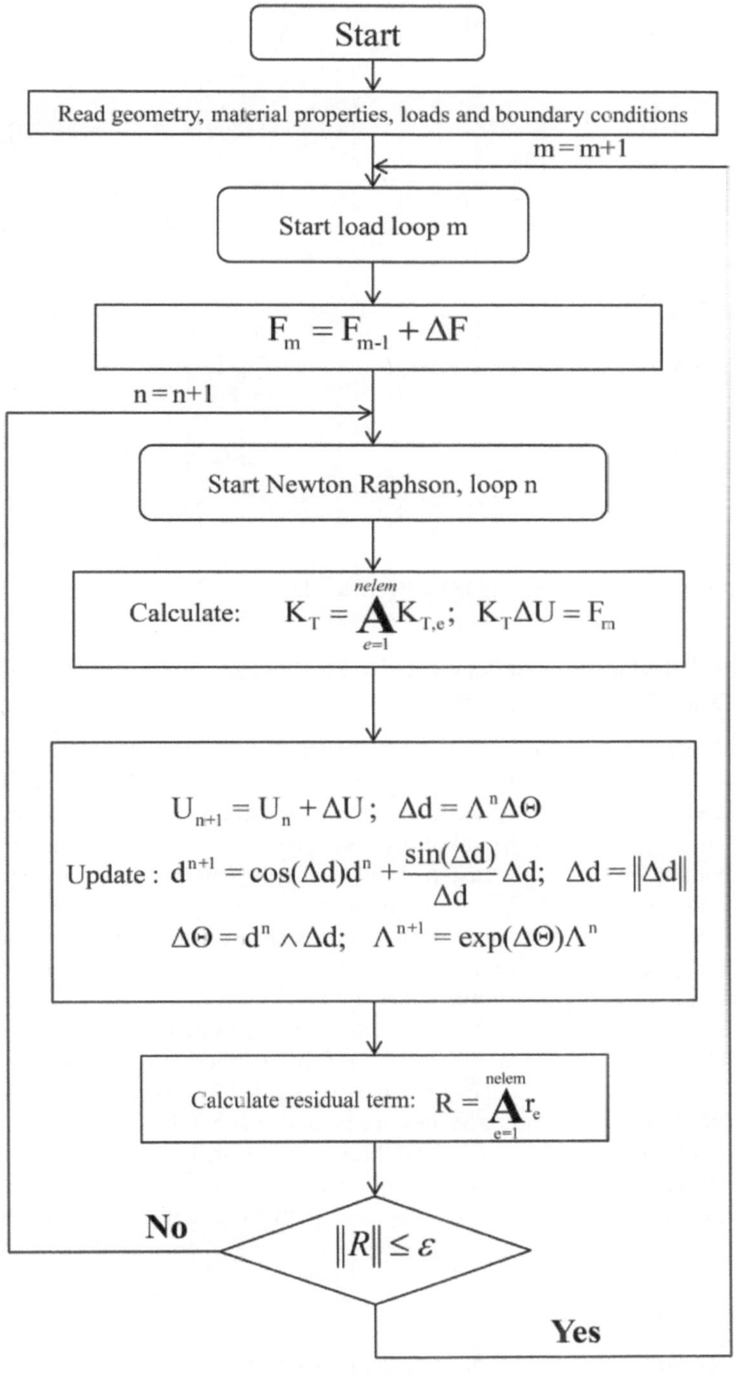

Figure 5.1 The flowchart of Newton solution procedure for non-linear analysis of FG-CNTRC structures.

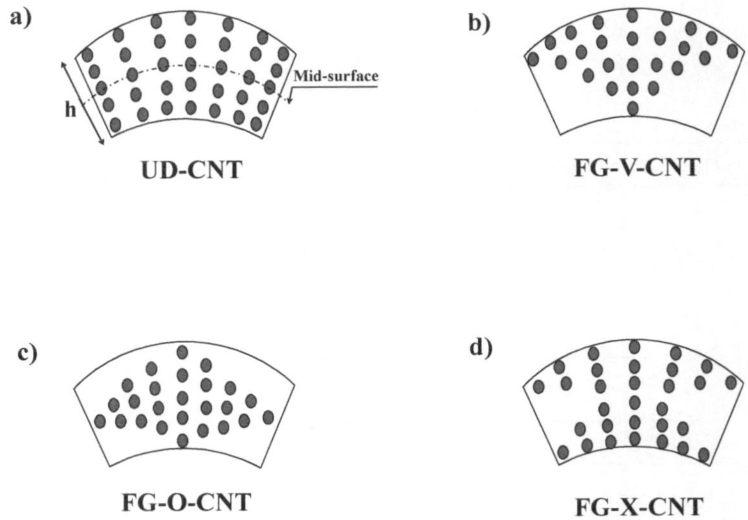

Figure 5.2 Various profiles cross sections of FG-CNTRC shells.

5.3 NON-LINEAR BENDING ANALYSIS OF FG-CNTRC STRUCTURES USING THE EXTENDED KIRCHHOFF MODEL

Based on the proposed non-linear discrete Kirchhoff shell formulation, numerical results are provided in this section in order to show the applicability and the performance of this formulation applied to FG-CNTRC structures. Since there are no existing studies on non-linear analysis of thin FG-CNTRC shell structures with the Kirchhoff model, convergence studies are first carried out for each studied geometry in the isotropic case to validate the present non-linear model. Then, a set of non-linear static results of FG-CNTRC plates, cylindrical and hyperboloidal shells are presented to show the applicability and the versatility of the non-linear discrete Kirchhoff model. In the following numerical analysis, the effective material properties of FG-CNTRC thin shells are taken as mentioned by Shen [223] where Poly(m-phenylenevinylene)-co-[(2,5-dioctoxy-p-phenylene) vinylene], known as (PmPV), is chosen for the matrix and (10,10) armchair single-walled for carbon nanotubes (SWCNTs). Their corresponding values, at room temperature (T = 300K), are listed in Tab. 4.2, Chapter 4. As well as, the efficiency parameters η_i (i = 1,2,3) are evaluated as indicated by Han and Elliot [95]. Their corresponding values are mentioned in Tab. 4.3, Chapter 4. We further assume that $G_{12} = G_{13} = G_{23}$. Uniform (UD) and three graded distributions, known as FG-V, FG-O and FG-X, are considered to show the effects of carbon nanotubes (CNTs) profiles and their volume fractions on non-linear deflections of various shapes of shell structures. Their profiles are as given in Fig. 5.2.

5.3.1 CONVERGENCE STUDIES

Convergences studies are first carried out on large deflection behavior of isotropic plates, orthotropic hyperboloidal shells and isotropic open-ended cylindrical shells to show the validity of the present formulation.

5.3.1.1 Non-linear static behavior of isotropic plates

The first test constitutes of a simply supported square isotropic plate ($\nu = 0.316$) subjected to uniform transverse load. This popular benchmark test is examined by Gorji [88], Shen [219] and recently by Lei et al. [136]. The results are performed in terms of non-dimensional parameters which are defined as follows: $W = W_0/h$, $q = \frac{q_0 a^4}{E_m h^4}$ where q_0 is the uniform transverse applied load, a and h are the length and the total thickness of the studied shell structure with $h = 2mm$. The geometry of the plate is shown in Fig. 5.3. Moreover, the whole of the plate is meshed using 15×16 of SQAD4 and STRI3 finite elements. It should be noted that the abbreviations SQAD4 and STRI3 are referred to the elaborated non-linear discrete Kirchhoff element using four-nodes and three-nodes respectively. In fact, to validate the proposed non-linear model and its aptitude to describe the non-linear behavior of shell structures with four or three-nodes elements, Fig. 5.4 depicts the non-dimensional load-deflection curves obtained using both SQAD4 and STRI3 elements and the one derived from Lei el al. solution [136]. As can be seen, the non-linear responses obtained by the present SQAD4 and STRI3 non-linear Kirchhoff model agree well with the solution of Lei et al. [136]. Hence, the present non-linear model improves its generality and applicability for both three or four-nodes shells.

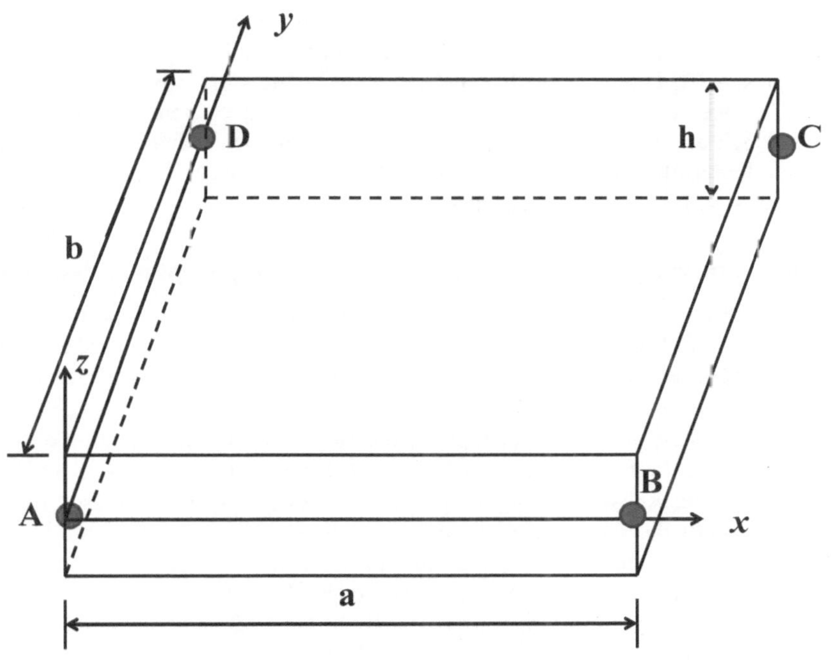

Figure 5.3 The geometry of the plate.

5.3.1.2 Non-linear static behavior of orthotropic hyperboloidal shells

The second example consists of an orthotropic hyperboloidal shell which is composed with three layers placed as follows: $[0°/90°/0°]$. The geometrical and material properties of the hyperboloidal shell are given as follows: $R_1 = 15m$, $R_2 = 7.5m$, $H = 20m$, $E_1 = 40 \times 10^6 kN/m^2$, $E_2 = 10^6 kN/m^2$, $G_{12} = G_{23} = G_{13} = 0.6 \times 10^6 kN/m^2$, $h = 0.04m$, $\nu_{12} = 0.25$, $F = 5kN$. The shell structure is subjected to two pairs of opposite loads as shown in Fig. 5.5. Due to the symmetry only one-eighth of the shell is modeled as depicted in Fig. 5.5(a) using four-node shell elements (SQAD4). Moreover,

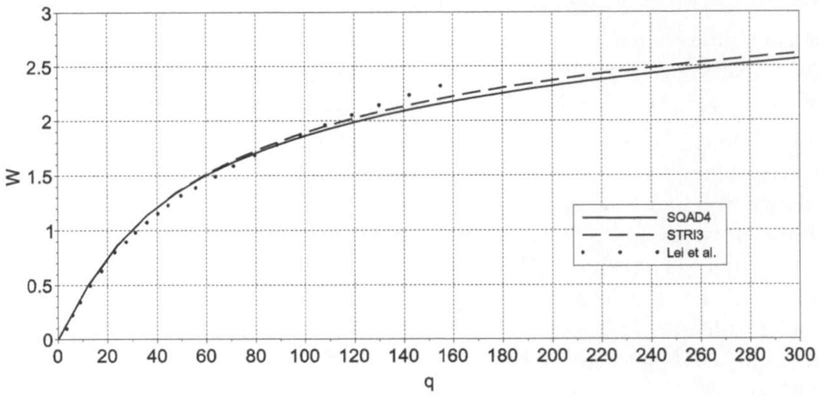

Figure 5.4 Convergence property of simply supported square isotropic plate subjected to uniform transverse load

the undeformed and deformed shapes corresponding to the whole of the shell are illustrated in Figs. 5.5(b) and 5.5(c), respectively. The obtained results illustrated in Fig. 5.6 for the displacements of the load points A, B, C and D are in good agreement with those given by Basar et al. [19] and the little difference between the results for high loads can be explained by the effect of shear deformations which are neglected in the present Kirchhoff model. Otherwise, the present non-linear Kirchhoff model has the capacity to reproduce the non-linear behavior of such arbitrary composite shell geometries with strong nonlinearities.

5.3.1.3 Non-linear static behavior of pullout of an isotropic open-ended cylindrical shells

The third example consists of a pullout of an isotropic open-ended cylindrical shell structure which is subjected to a pair of diametrically opposite concentrated tensile loads, as shown in Fig. 5.7. This popular benchmark test has been extensively used by many researchers to validate geometrically non-linear shell formulations and one can be referred to these publications [209, 37, 256] among others. The geometrical and mechanical properties of the cylindrical shell in the isotropic case are taken as indicated in [209] where: $L = 10.35m$ is the axial length, $R = 4.953m$ is the radius and $h = 0.094m$ is the thickness of the shell. The young modulus and the Poisson ratio are $10.5MPa$ and 0.3125, respectively. The value of each external applied concentrated load is of $40kN$. Due to the symmetry of the geometry, only one-eighth of the cylinder is modeled using 20×33 of SQAD4 developed finite elements. Figure 5.8 shows the obtained displacement curves compared to those given by Reinoso et al. [209] in three different load points A, B and C. A good agreement between the results leads to the validation the proposed non-linear model and draw hence its ability to predict the non-linear behavior of such shell structures with good accuracy.

5.3.2 NON-LINEAR STATIC ANALYSIS OF FG-CNTRC SHELL STRUCTURES

After the validation of the current non-linear formulation via the presented convergence studies, the non-linear static behavior is now performed for FG-CNTRC shell structures in order to show the effects of introduction of CNT reinforcements to such shell structures. For that uniform (UD) and three graded distributions, which are FG-V, FG-O and FG-X, are considered. Their mechanical properties are as indicated in Tabs. 4.2 and 4.3. The same geometry properties presented in section 5.3.1.1 are provided here with the introduction of CNT reinforcements.

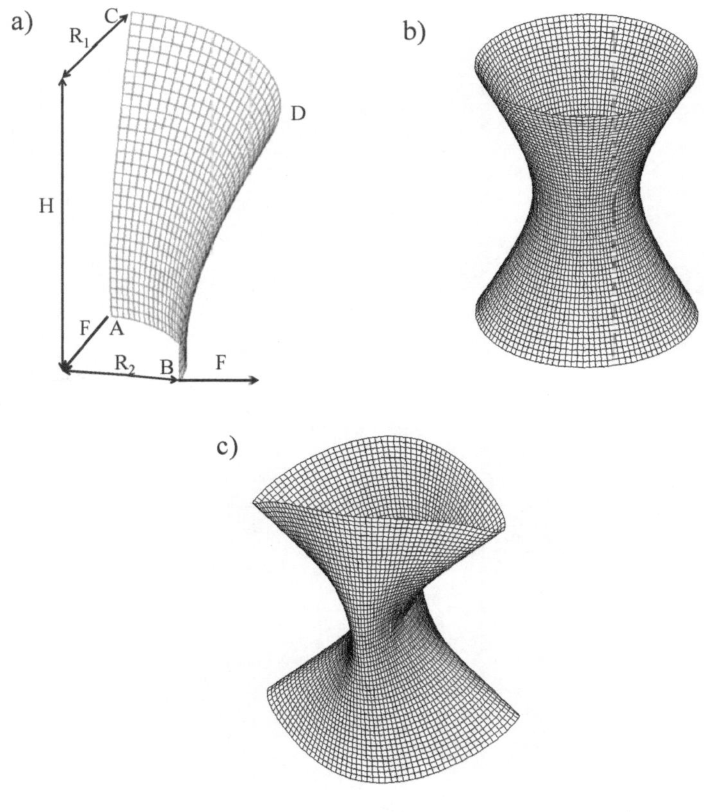

Figure 5.5 Hyperboloidal shell. (a) geometry definition, (b) undeformed shape, (c) deformed shape.

5.3.2.1 Large deflections analysis of FG-CNTRC plates

In this example, a FG-CNTRC simply supported palate is considered. Its geometrical and mechanical properties are as indicated in Fig. 5.3, Tabs. 4.2 and 4.3, respectively. The results are presented in terms of non-dimensional parameters as mentioned earlier in section 5.3.1.1. A detailed parametric study is carried out to investigate the non-linear behavior of various forms of FG-CNTRC plates under uniform transverse load.

Figure 5.9 shows the effect of CNT distributions on non-dimensional deflections of various forms of simply supported (SSSS) square FG-CNTRC plate with width-to-thickness ratio ($b/h = 10$) and volume fraction ($V^*_{CNT} = 0.11$). It can be observed that the non-dimensional central deflection of FG-V and FG-O distributions are larger than UD distribution while the FG-X plate has smaller deflection with respect to three other cases. It is due to the change of the stiffness of the FG-CNTRC plate with the form of CNT distribution in the isotropic matrix. Therefore, more the distribution is close to the top and bottom surfaces, more the stiffness of the whole of the structure is higher. On the other hand, the effects of plate aspect ratios (a/b) on non-linear central deflections for various forms of simply supported plates under uniform transverse load, width-to-thickness ratio ($b/h = 20$) and volume fraction ($V^*_{CNT} = 0.17$). Figures 5.10(a), 5.10(b), 5.10(c) and 5.10(d) illustrate the deflection-load curves of UD, FG-V, FG-O and FG-X CNTRC plates, respectively. For each distribution, it can be seen that for plate aspect ratios less or equal to 1 ($a/b = 0.5, 0.75, 1$), the deflection is weakly non-linear compared to other ratios ($a/b = 1.5, 2$). It fact, for plate aspect ratios ($a/b = 1.5, 2$), the non-linear effect is more pronounced. This fact is due to the stiffness which decreases with the increases of the plate aspect ratio ($a/b = 0.5, 2$) and the structure becomes more larger and then deforms

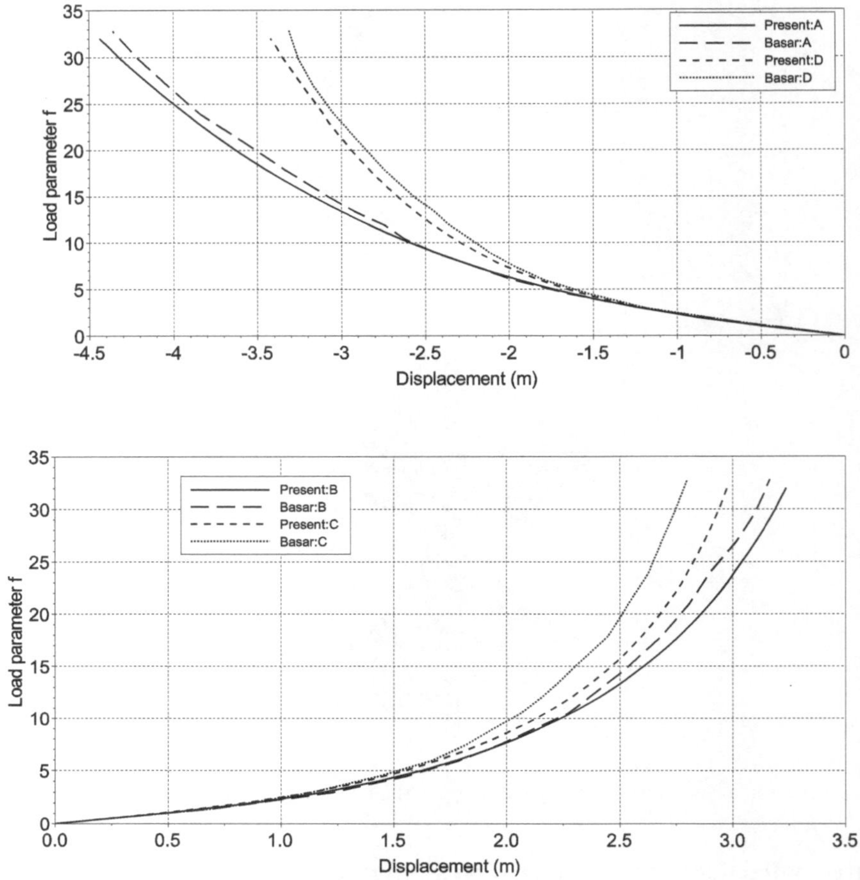

Figure 5.6 Convergence property of orthotropic hyperboloidal composite shells subjected to two pairs of opposite loads "AD" and "BC."

easily. In addition, the effects of boundary conditions on non-linear behavior for various forms of square FG-CNTRC plates is also examined. Four types of boundary conditions are considered which are: four sides (AB), (BC), (CD) and (DA) are fully clamped (CCCC), four sides are fully simply supported (SSSS), two sides (AB) and (DC) are simply supported and the others (BC) and (AD) are clamped or free (SCSC), (SFSF). Figures 5.11(a), 5.11(b), 5.11(c) and 5.11(d) show that the non-linear central deflection for each distribution of CNTRC plate with (CCCC) boundary condition is less than those with (SSSS), (SCSC) and (SFSF) boundary conditions. In fact, the fully clamped boundary condition imposes more constraints than simply supported or free boundary conditions and thus it has higher stiffness compared to three other boundary conditions which can explain the observed results.

5.3.2.2 Large deflections analysis of FG-CNTRC hyperboloidal shells

Large deflections analysis is performed here for FG-CNTRC hyperboloidal shells. The same geometry described in section 5.3.1.2 is considered with the material properties are as indicated at the

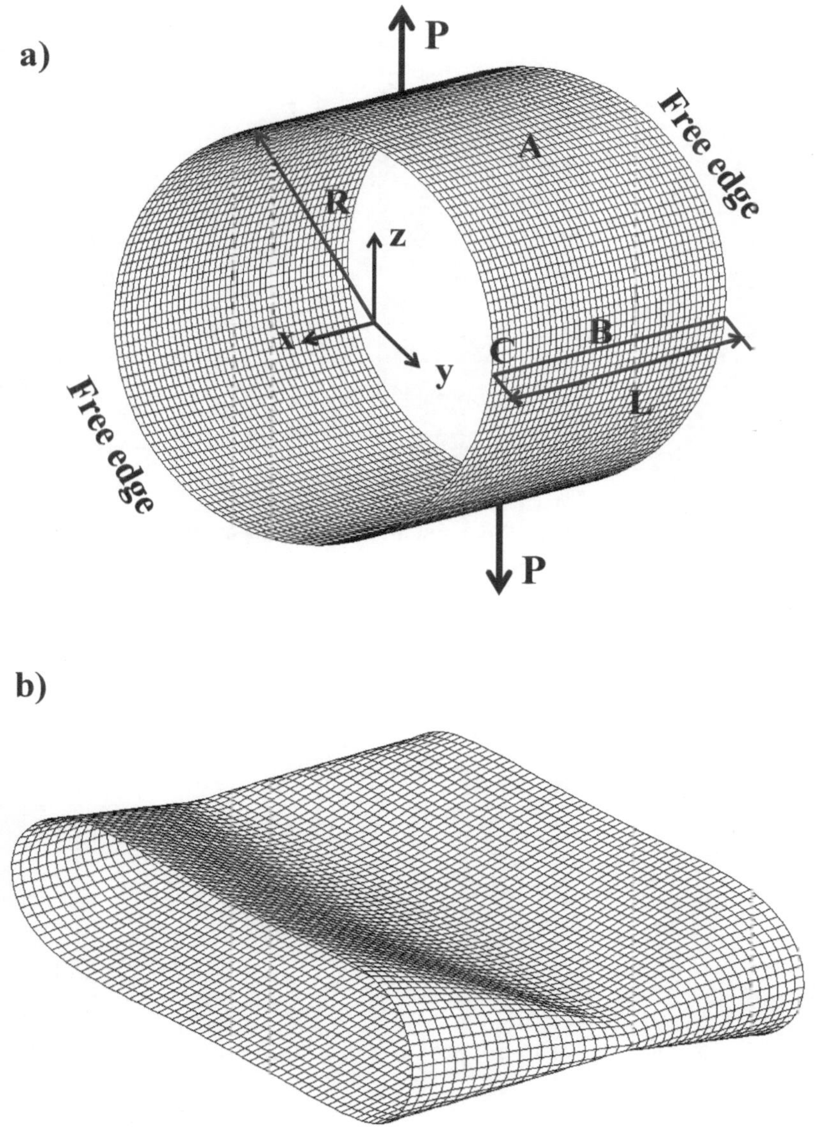

Figure 5.7 The geometry of the open-ended cylindrical shell. (a) Undeformed and (b) deformed shapes.

beginning of the section 5. In fact, the CNTs are uniaxially aligned in axial direction and functionally graded in the thickness direction of the hyperboloidal shell. Fig. 5.12 shows that the non-linear deflections of (UD) and (FG-V) are located between the FG-X and FG-O deflection curves and this is for all characteristics load points A, B, C and D which reflects the effect of profile distribution of CNTs on non-linear behavior of such hyperboloidal shells. Hence, for CNTs distributed near the top and bottom surfaces, the stiffness of the structure is higher than for CNTs distributed near mid-surface which make the designers able to adjust the desired stiffness by choosing the profile of CNT distribution.

The effects of volume fractions V_{CNT}^{*} and the shell thickness h are examined now for FG-X form because, as explained previously, the FG-X is the most efficient one. For that, Fig. 5.13 illustrates the effect of volume fractions of CNTs on non-linear behavior of FG-X-CNTRC hyperboloidal shell

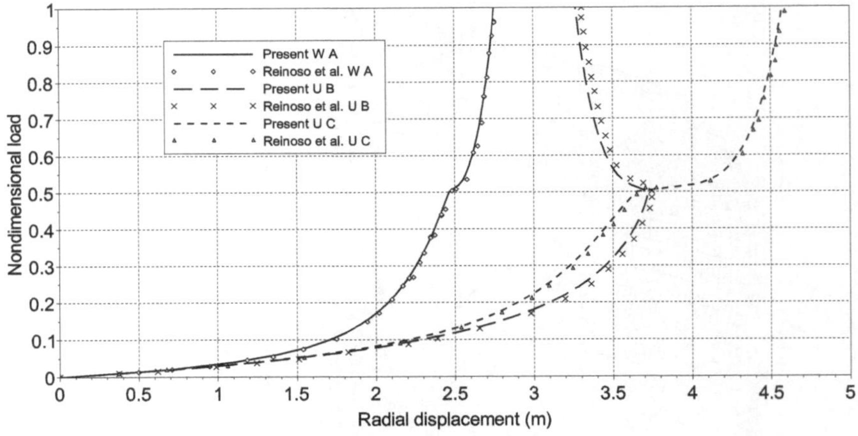

Figure 5.8 Convergence property of isotropic open-ended cylindrical shell subjected to a pair of diametrically opposite concentrated tensile loads.

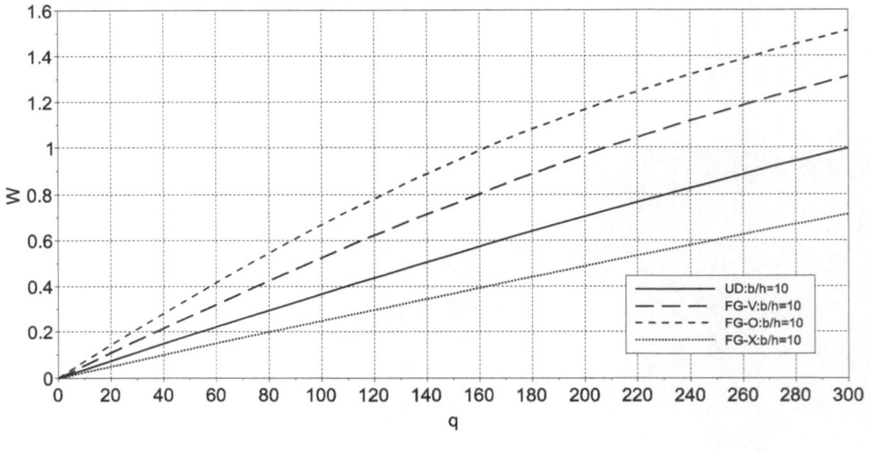

Figure 5.9 Non-dimensional deflection versus load responses for various forms of simply supported square FG-CNTRC plate under uniform transverse load and with width-to-thickness ratio $(b/h = 10)$ and volume fraction $V^*_{CNT} = 0.11$.

at load points A, B, C and D, respectively. As can be seen, for a fixed force value, the non-linear displacement of the shell increases as the volume fraction of CNTs increases and this is for the load points A and D. However, for the load points B and C, the non-linear displacement decreases with the increases of volume fraction of CNTs. This can be explained by the CNT reinforcements which influences the stiffness of the shell and induces hence the observed opposite non-linear behavior of FG-CNTRC hyperboloidal shell. Figure 5.14 shows the effect of variation of the shell thickness parameter on non-linear displacement of FG-X-CNTRC hyperboloidal shell at load points A, B, C and D and under a volume fraction of CNTs $V^*_{CNT} = 0.11$. It can be observed that the non-linear displacement of the hyperboloidal shell increases as the thickness variable increases in load points A and D while this effect is inversed in load points B and C. This implies the significant effect of

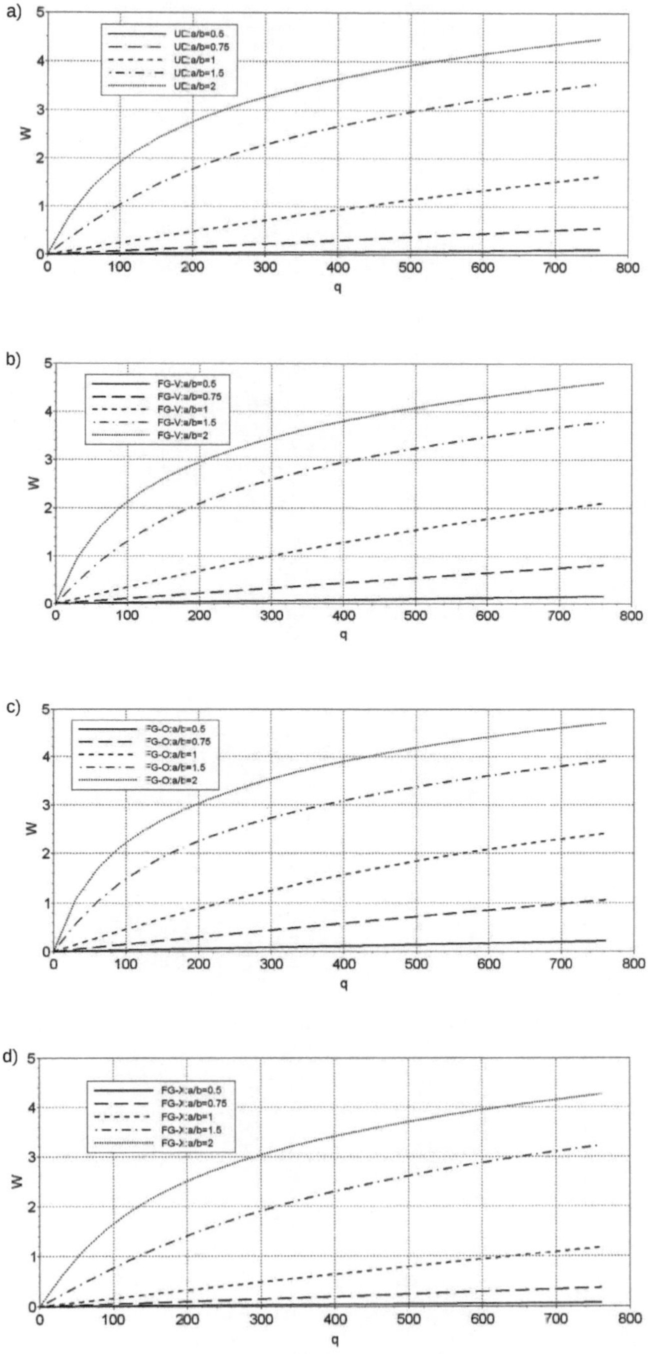

Figure 5.10 Effects of plate aspect ratios on non-linear central deflections for simply supported (a): UD, (b): FG-V, (c) FG-O and (d): FG-X CNTRC plates under uniform transverse load, width-to-thickness ratio $(b/h = 20)$ and volume fraction $(V^*_{CNT} = 0.17)$.

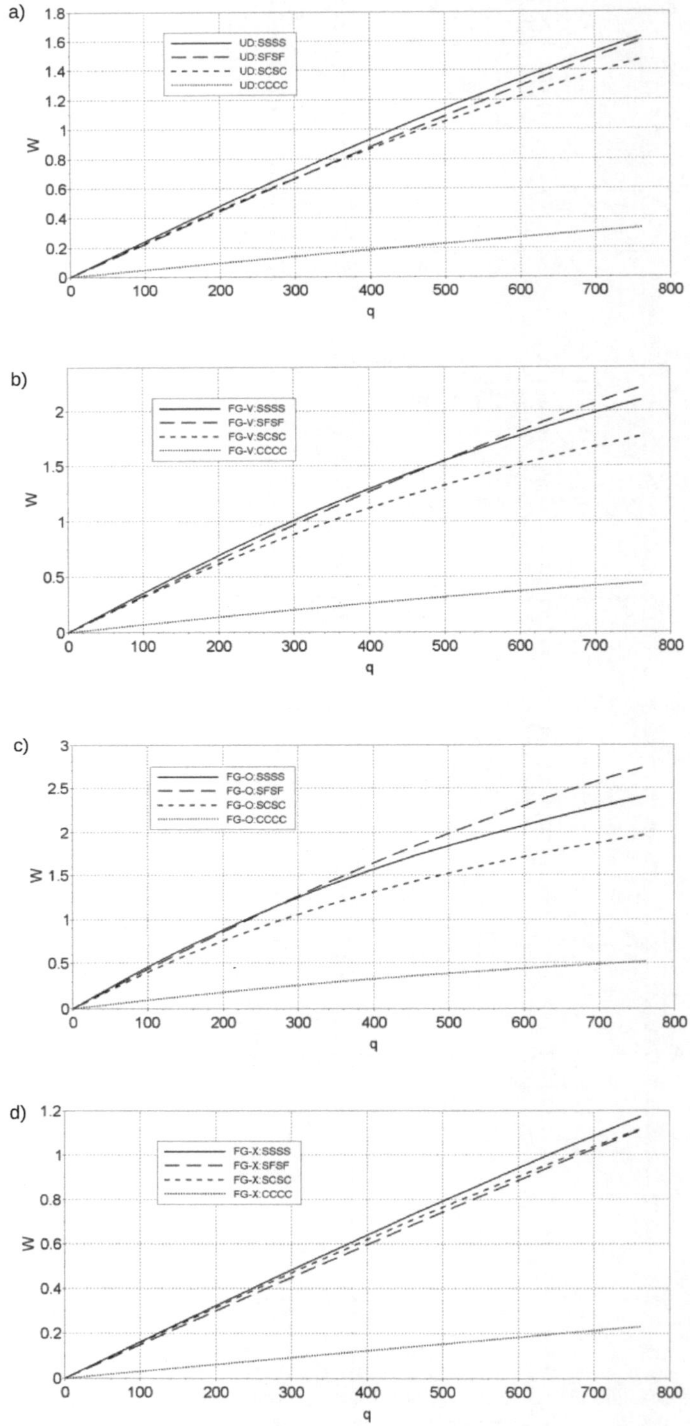

Figure 5.11 Effects of boundary conditions on non-linear central deflections for (a): UD, (b): FG-V, (c) FG-O and (d): FG-X square CNTRC plates under uniform transverse load, width-to-thickness ratio ($b/h = 20$) and volume fraction ($V_{CNT}^{*} = 0.17$).

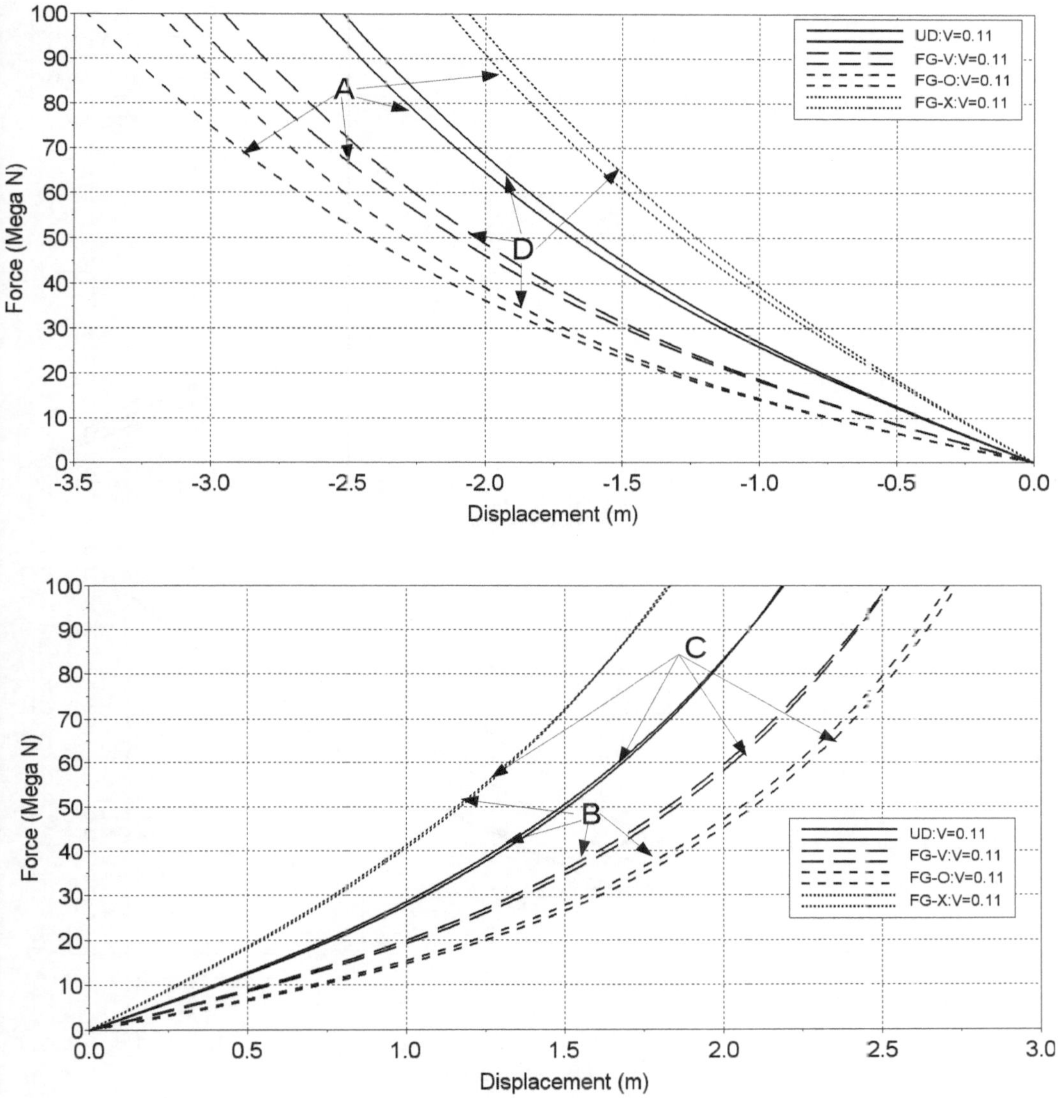

Figure 5.12 Non-linear load-displacement responses for various forms of CNTs of FG-CNTRC hyperboloidal shell at load points A, B, C and D under a volume fraction of CNT $V^*_{CNT} = 0.11$.

the shell thickness parameter on non-linear behavior of the FG-X-CNTRC shell which becomes sensitive to this geometrical parameter.

5.3.2.3 Large deflections analysis of pullout of FG-CNTRC open-ended cylindrical shells

Now, non-linear deflections' analysis are provided for FG-CNTRC open-ended cylindrical shells where the geometrical parameters are the same as indicated in section 5.1.3. For material properties, the PMMA is selected for matrix and three sets of CNT volume fractions are considered: $V^*_{CNT} = 0.12, 0.17, 0.28$ for the purpose of result comparisons. The PMMA matrix has the same characteristics of PmPV matrix as mentioned in Tab. 5.1 except the Young modulus which its value

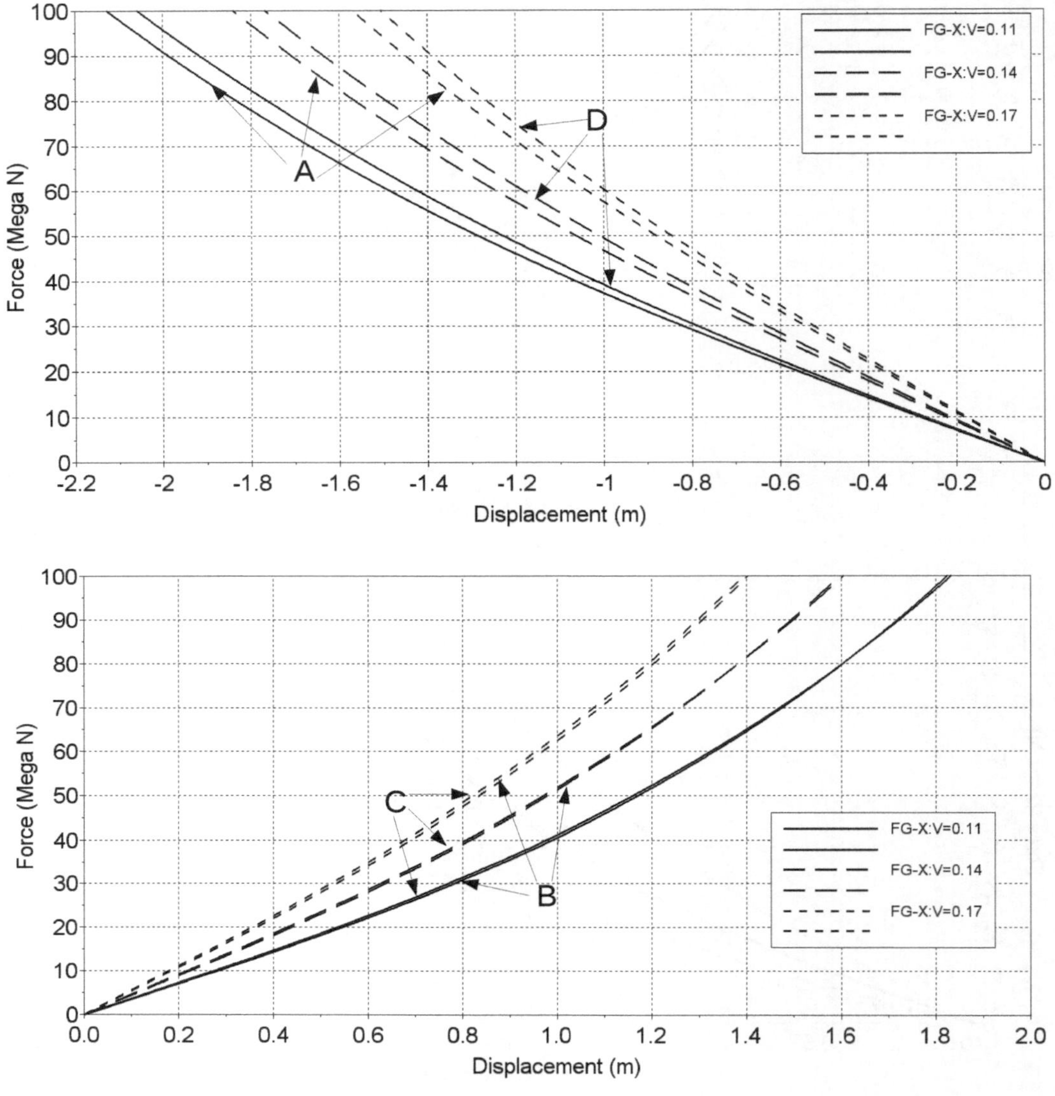

Figure 5.13 Non-linear load-displacement responses of FG-X-CNTRC hyperboloidal shell at load points A, B, C and D for various volume fractions.

equal to $E_m = 2.5GPa$. The values of efficiency parameters corresponding to the CNT volume fractions are as given in [209]. Furthermore, the CNTs are of type (10,10) SWCNTs and their material properties are as listed in Tab. 5.1. As well as, the CNTs are uniaxially aligned in the axial direction and functionally graded in the thickness direction. Figure 5.15 shows the non-linear load-displacement curves for different profiles of CNTs of the cylindrical shell depicted at load points A and B, respectively and under a volume fraction of CNTs $V_{CNT}^* = 0.12$. According to these representations, it can be seen the significant effect of CNTs distributions on non-linear behavior of the open-ended cylindrical shell. In fact, the FG-X distribution has the lowest value while the FG-O distribution has the highest one which indicates that the form of CNTs affects significantly the response of the structure. For that, the desired structural behavior can be patterned using a desired

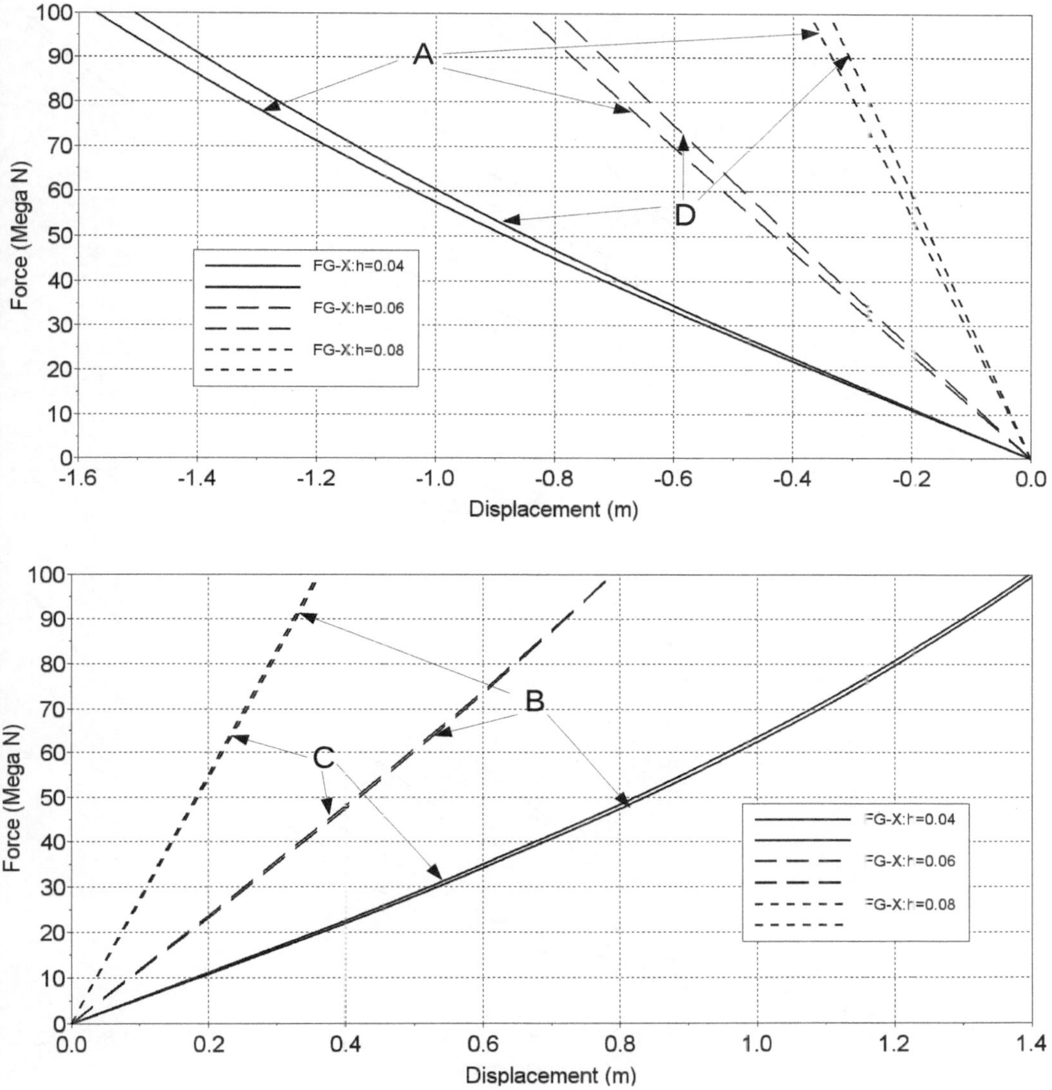

Figure 5.14 Effect of the shell thickness parameter on non-linear load-displacement curves of FG-X-CNTRC hyperboloidal shell at load points A, B, C and D and under a volume fraction of CNTs $V_{CNT}^{*} = 0.11$.

form of CNTs and thus an adjustment of the stiffness of the structure can be achieved. In addition, a correlation between the obtained results and those given in [209] can be observed which implies the effectiveness of the proposed non-linear model.

5.4 NON-LINEAR BENDING ANALYSIS OF FG-CNTRC STRUCTURES USING THE FSDT MODEL

This section is devoted to present the applicability of FSDT theory for non-linear bending analysis of FG-CNTRC plates and curved panels. For that, a comparison study is first carried out in the isotropic

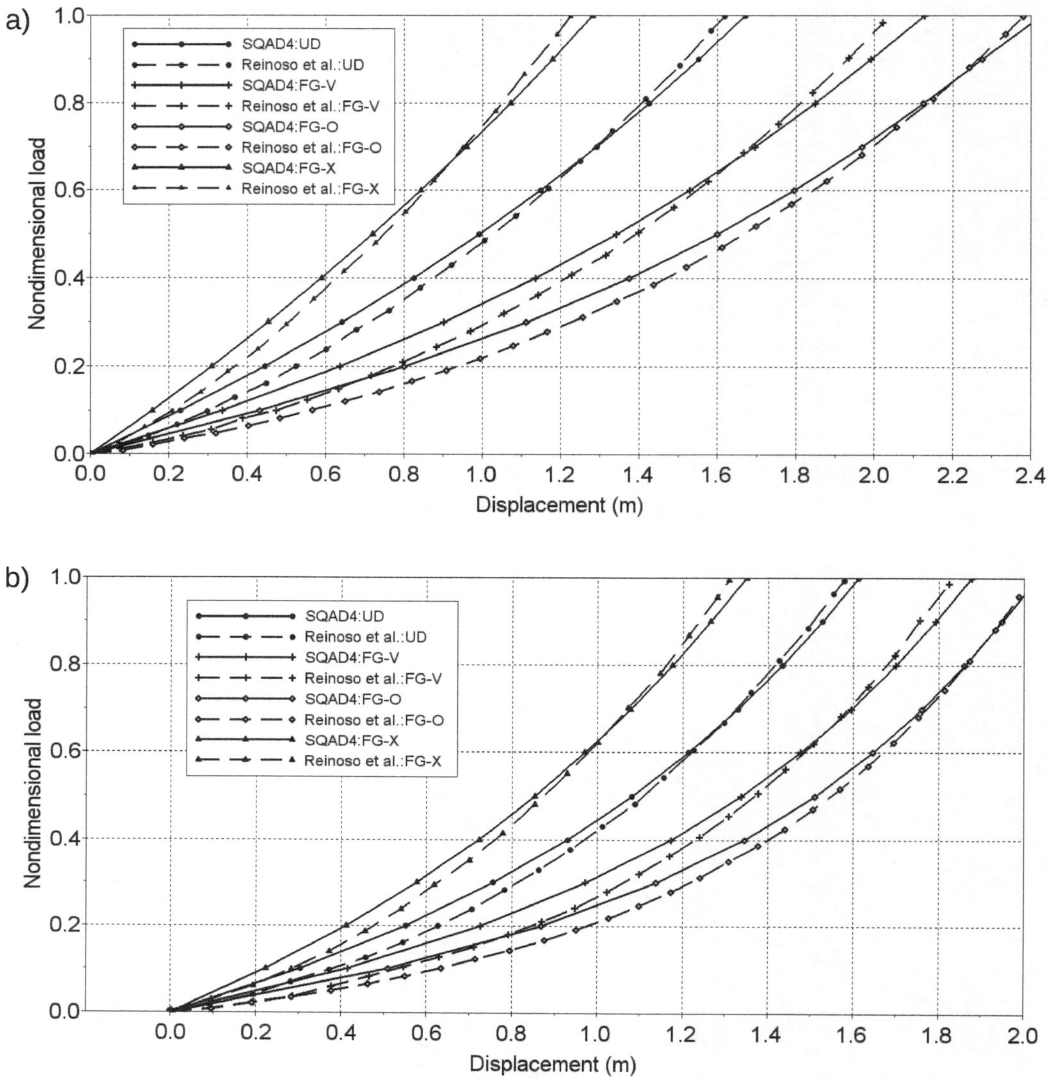

Figure 5.15 Non-linear load-displacement curves for different profiles of CNTs of an open-ended cylindrical shell at load points (a) B and (b) A under a volume fraction of CNTs $V_{CNT}^{*} = 0.12$.

case in order to verify the capacity and the ability of the present formulation in the prediction of non-linear deflection responses of such structures. Then, numerical applications are performed for composite structures with FG-CNT reinforcements leading hence to draw the usefulness and the versatility of the proposed finite element method. For FG-CNTRCs simulations, PmPV (Poly(m-phenylenevinylene)-co-[(2,5-dioctoxy-p-phenylene) vinylene]) is considered for the matrix and the (10,10) armchair single-walled carbon nanotubes (SWCNTs) for the reinforcements. Their corresponding material properties, at room temperature (T = 300 K), are reported in Tab. 4.2. Moreover, the values of CNT efficiency parameters η_i (i = 1,2,3) are listed in Tab. 4.3.

5.4.1 VERIFICATION AND COMPARISON STUDY

To verify the capacity and the aptitude of the present formulation in the prediction of non-linear bending behavior of shell structures, several comparisons are carried out. The first numerical test consists of a simply supported square plate subjected to uniform transverse load with a length side a and a total thickness $h = 2mm$ as illustrated in Fig. 5.16. Its material properties are taken as indicated in [136, 88, 219] with ($\nu = 0.316$)). For comparison purpose, the results are performed in terms of non-dimensional parameters: $W = W_0/h$ and $q = \frac{q_0 a^4}{E h^4}$ where E is the isotropic Young's modulus and q_0 is the uniform transverse applied load. The whole of plate is meshed using 16×16 of S4 finite elements where the abbreviation S4 refers to the developed non-linear FSDT element. The non-dimensional load-deflection curves are depicted in Fig. 5.17. The comparison of the present curve to the one derived by Lei et al. [136] reveals a good agreement between the results which leads to verification of the aptitude of the present formulation in the prediction of non-linear bending behavior of isotropic square plates.

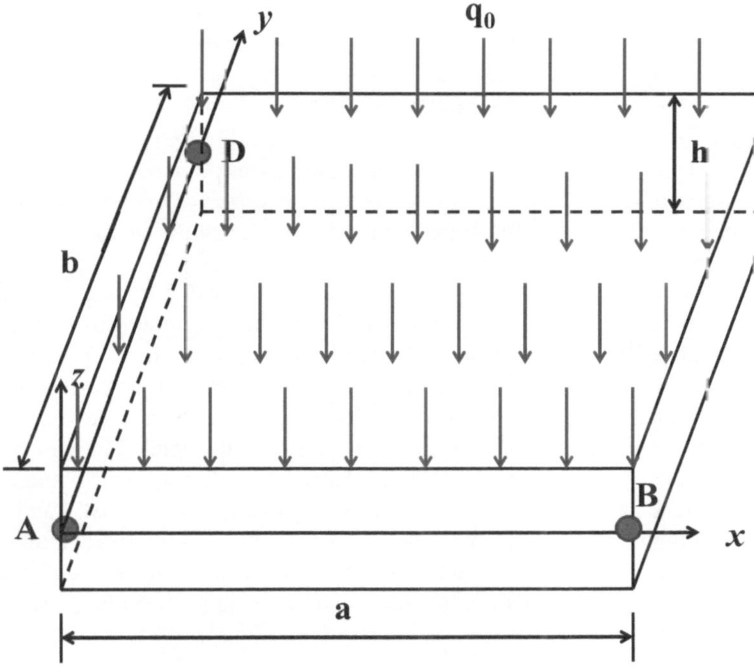

Figure 5.16 The geometry of the plate.

The second numerical test consists of an annular ring plate with a slit cut along its radial direction corresponding to the line A-B as depicted in Fig. 5.18 (a). The geometrical parameters of the ring are defined as follows: the internal radius $r = 6m$, the external radius $R = 10m$ and the shell thickness $h = 0.03m$. Its material properties corresponding of the isotropic material law are: the Young's modulus $E = 2.1 \times 10^8 kN/m^2$ and the Poisson's ratio $\nu = 0.3$, respectively. The ring plate is loaded at its free edge while the other edge is fully clamped. The maximum applied load at the free edge is fixed to $p = 0.1 f kN/m$ with an applied load factor $f = 60$ to obtain the deformed geometry as shown in Fig. 5.18 (b). The used mesh involves 8×48 of S4 finite elements. It is worth to mention

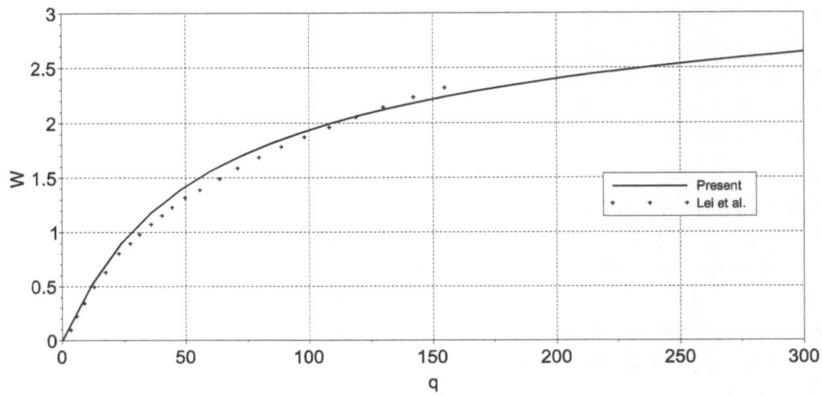

Figure 5.17 Non-dimensional load-deflection curves of a simply supported square isotropic plate subjected to uniform transverse load.

that this numerical test is considered as a good example for testing finite element shell formulations, see [209, 40, 37]. The non-linear load-deflection responses at points A, B and C versus the load factor f are depicted in Fig. 5.19. As can be seen, an excellent agreement between the present results and those reported in [40] is illustrated which proves the ability of the present formulation in the restitution of non-linear behavior of annular ring plates.

The third example considers an isotropic cylindrical panel with a concentrated transverse load at the center point of the panel as depicted in Fig. 5.20. This example illustrates a shell problem and serves to test the ability of the proposed finite element to accurately reproduce the non-linear bending behavior of curved panels. The geometrical and mechanical properties of the cylindrical panel are taken as indicated in [38] where: $L = 6in$ is the axial length, $R = 2.5in$ is the radius, $\theta_0 = 45^o$ is the span angle and $h = 0.01in$ is the thickness of the shell. The young's modulus and the Poisson's ratio are $E = 10^7 psi$ and $\nu = 0.3$, respectively. The applied boundary condition is defined as follows: (AB) and (DC) edges are simply supported (SS) and (AD) and (BC) edges are free. Due to the symmetry of the problem, only one-quarter of the cylindrical panel is modeled with 10×12 mesh. Fig. 5.21 shows a satisfactory agreement in comparison with the reference solution provided by [38]. Thereby, the present model has the aptitude to perform large deflections analysis of shell-type geometry.

5.4.2 NON-LINEAR DEFLECTIONS RESPONSES OF FG-CNTRC STRUCTURES

To investigate non-linear deflections analysis of FG-CNTRC structures, parametric studies are performed in this section. Indeed, the effects of CNTs profiles, their volume fractions and some geometrical parameters including aspect ratio, thickness parameter and boundary conditions on the non-linear bending behavior of FG-CNTRC plates and curved panels are discussed in detail through the presented numerical results. For this purpose, uniform (UD) and three graded CNT distributions, known as FG-V, FG-O and FG-X are considered as depicted in Fig. 5.2. Each configuration reflects how the CNTs are distributed through the thickness of the shell structure. In fact, for the two symmetric distributions with regard to the mid-plane surface, one can observe that the top $(z = +h/2)$ and the bottom $(z = -h/2)$ surfaces are CNT-rich for FG-X form and they are free of CNTs for FG-O form. However, for FG-V form, a maximum distribution of CNTs is displayed on the top surface $(z = +h/2)$ while its bottom surface $(z = -h/2)$ is free of CNTs. Note that for all FG-CNTRC

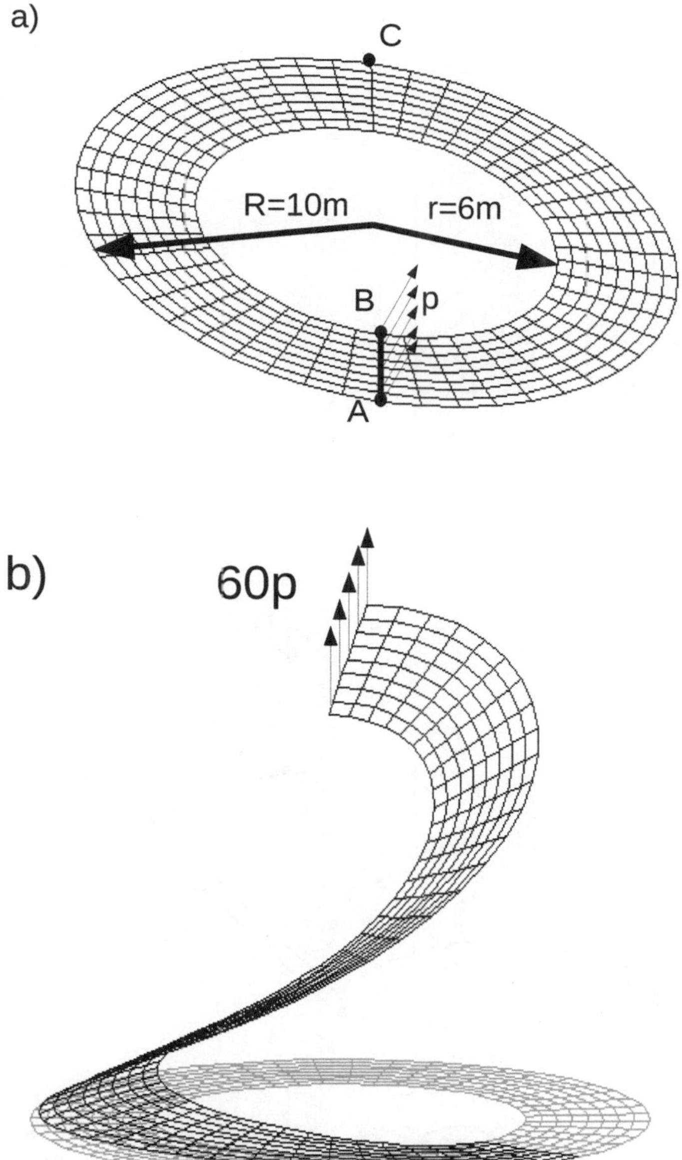

Figure 5.18 Annular ring plate. (a) geometry definition and (b) deformed shape.

presented structures, the geometric properties replicated those previously introduced in section 5.4.1, whereas the material properties of CNTs and isotropic matrix are resumed in Tabs. 4.2 and 4.3, respectively.

5.4.2.1 Non-linear responses of FG-CNTRC plates

Now, the applicability of the current formulation and its corresponding finite element implementation for non-linear analysis of plates with FG-CNT reinforcements is examined. For that a simply

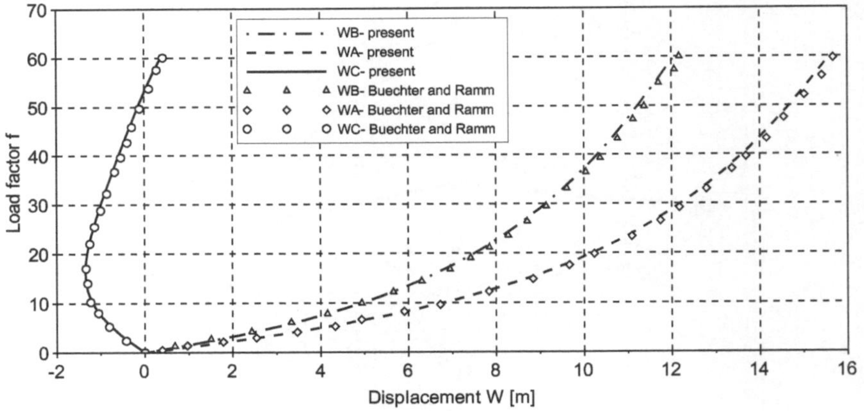

Figure 5.19 Load-deflection curves of the ring plate at load points A, B and C.

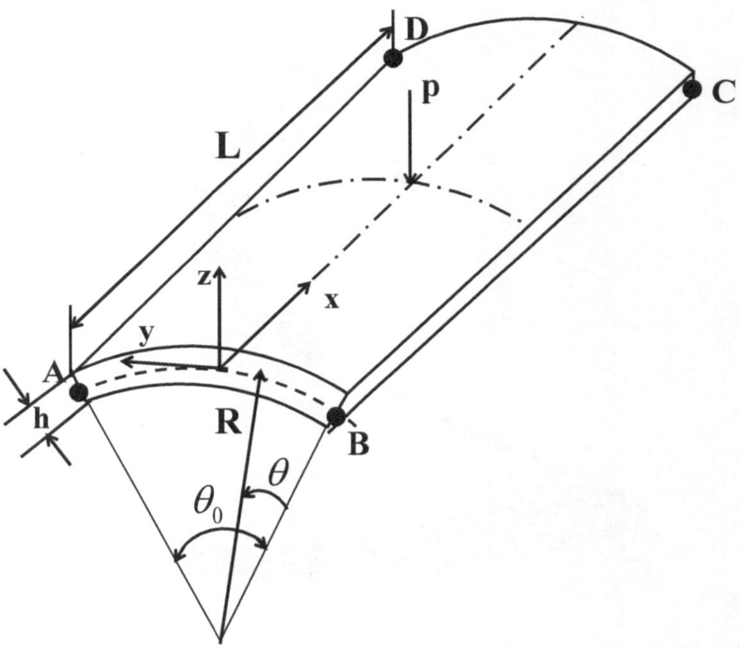

Figure 5.20 Cylindrical panel under a concentrated load.

supported FG-CNTRC plate under uniform distributed load is performed. At first, and for the verification purpose for FG-CNTRC plates, the non-dimensional central deflection responses of a simply supported (SSSS) square FG-CNTRC plate ($b/h = 50$, $V_{CNT}^* = 0.11$) subjected to uniform transverse load and for different forms of (CNTs) are compared in Fig. 5.22 to those of Lei et al. [136].

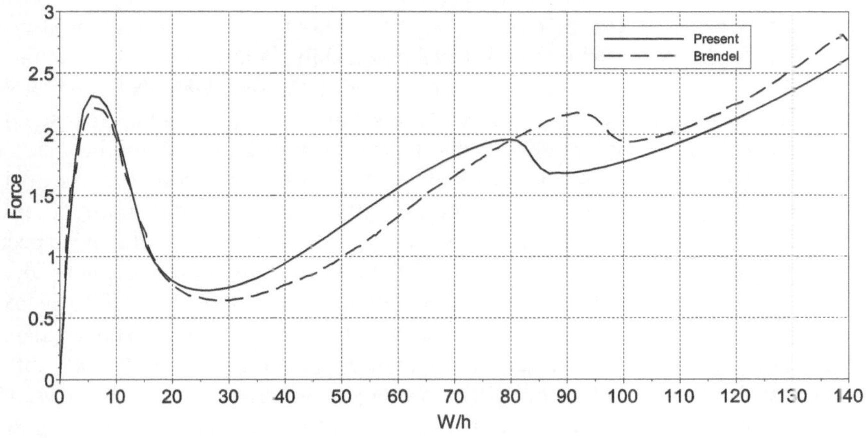

Figure 5.21 Load-deflection curves of isotropic cylindrical panel under a concentrated load.

As can be seen, the present curves agree with those of Lei et al. [136] in the range [0–100] of the non-dimensional applied load, where small deflections ($W \approx 0.5$) are performed. Beyond this, a difference between the results is observed due to the different strategies used in the two studies. In fact, in [136], the authors use the FSDT theory with Von-Karman kinematic assumptions which include only small deformations and moderate rotations, whereas, in our paper the formulation is implemented with the introduction of large displacements and finite rotations. Hence, and for a fixed load point, the present finite element model has the aptitude of the prediction of large deflections which can reach high values ($W \approx 7$) while in [136] the deflections are small and do not exceed ($W \approx 1.5$). Otherwise, the present results follow the shape of the curves derived from the literature [136]. Therefore, the present model constitutes a more general and convenient finite element model in predicting large deflections of FG-CNTRC plates.

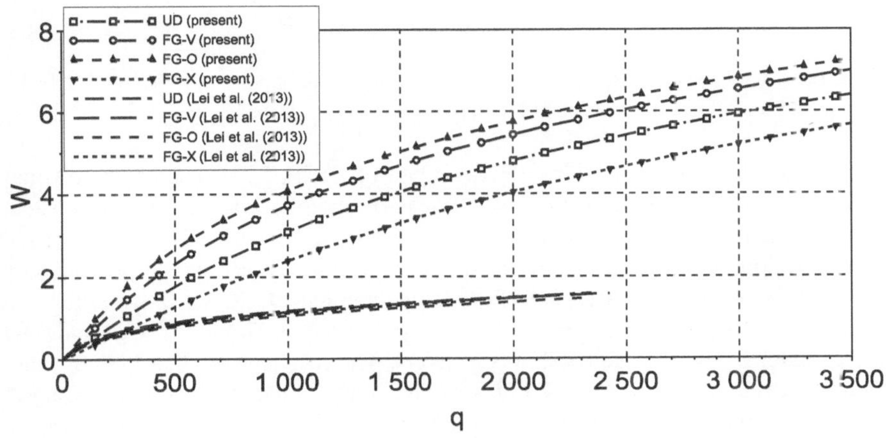

Figure 5.22 Non-dimensional central deflection for various forms of simply supported square plate subjected to uniform transverse load with ($b/h = 50$, $V^*_{CNT} = 0.11$).

Figure 5.23 displays the effects of various profiles of CNTs on large deflections responses of a simply supported (SSSS) square FG-CNTRC plate ($b/h = 10$, $V^*_{CNT} = 0.11$) subjected to uniform transverse load. In this graph, it can be observed that the FG-O-CNT configuration is predicted to experience the largest deflection while FG-X-CNT case exhibits the stiffest response among the different FG-CNT distributions. As well as, the other two configurations (UD-CNT and FG-V-CNT) which are located between these extreme cases induce a smaller central deflection for the UD form compared to the FG-V form. Hence, the CNTs distributions close to the top and bottom surfaces provide a more effective reinforcement, in terms of increasing the structural stiffness of the FG-CNTRC plates. So, designer can adjust the desired stiffness by choosing the appropriate form of CNTs. In addition, Figs. 5.24(a), 5.24(b), 5.24(c) and 5.24(d) draw the effects of plate aspect ratios (a/b) on non-linear central deflections of various distributions of simply supported plates ($b/h = 20$, $V^*_{CNT} = 0.17$) subjected to uniform transverse load. With regard to square FG-CNTRC plate aspect ratio ($a/b = 1$), two sets of aspect ratio values less than 1 ($a/b = 0.5, 0.75$) and greater than 1 ($a/b = 1.5, 2$) are considered. For the former, the value of deflection is not significantly influenced by the aspect ratio while for the latter, this effect is more pronounced. In fact, for a high aspect ratio ($a/b \succ 1$) the FG-CNTRC plate is more likely to experience the non-linear bending than that for lower values of aspect ratio ($a/b \prec 1$). Thus, it can be explained by the decrease in the flexural rigidity of the FG-CNTRC plate with the increase of the aspect ratio.

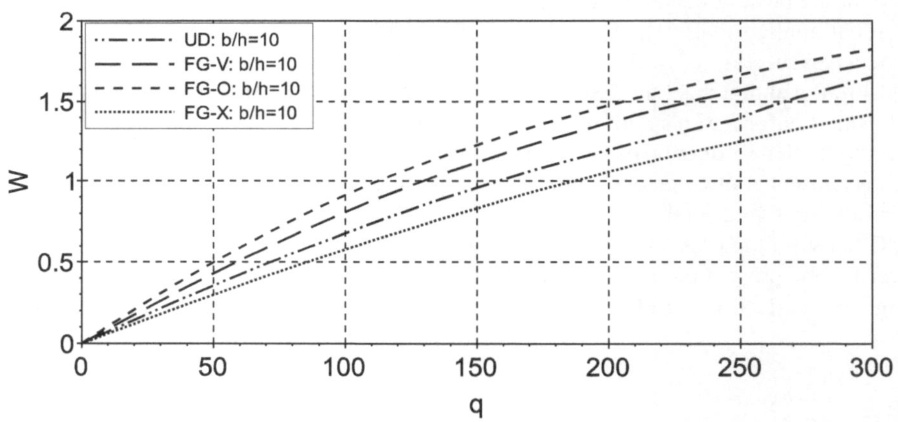

Figure 5.23 Effects of various profiles of CNTs on central large deflections responses of simply supported square plate subjected to uniform transverse load with ($b/h = 10$, $V^*_{CNT} = 0.11$).

Further conclusions can be deduced from Figs. 5.25(a), 5.25(b), 5.25(c) and 5.25(d) which illustrate the effects of boundary conditions on non-linear behavior of various forms of square FG-CNTRC plates. A sequence of boundary conditions is utilized where (C), (S) and (F) refer to clamped, simply supported or free edge, respectively. As a result, the following combinations are examined: four edges (AB), (BC), (CD) and (DA) fully clamped (CCCC) or fully simply supported (SSSS), a pair of opposite side (AB) and (DC) simply supported and left ones (BC) and (AD) fully clamped or free, namely: SCSC and SFSF, respectively. The results show that the non-linear central deflection for a fully clamped (CCCC) FG-CNTRC plate is the smallest one among the four cases, while the one related to simply supported (SSSS) case is the largest one. The reason is that the constraint of clamped boundary condition is stronger than the simply supported or free boundary condition. Hence, the type of applied boundary condition has a significant effect on non-linear bending behavior of FG-CNTRC plates.

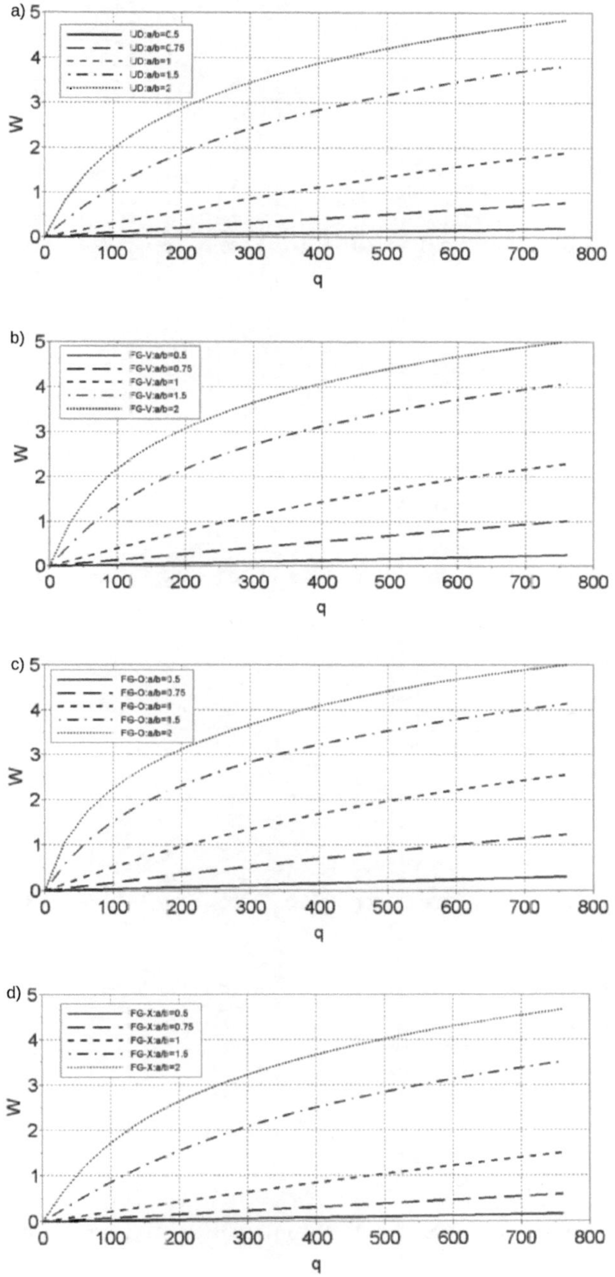

Figure 5.24 Central large deflections responses of simply supported (a): UD, (b): FG-V, (c) FG-O and (d): FG-X-CNTRC plates subjected to uniform transverse load and under different aspect ratios, with ($b/h = 20$, $V^*_{CNT} = 0.17$).

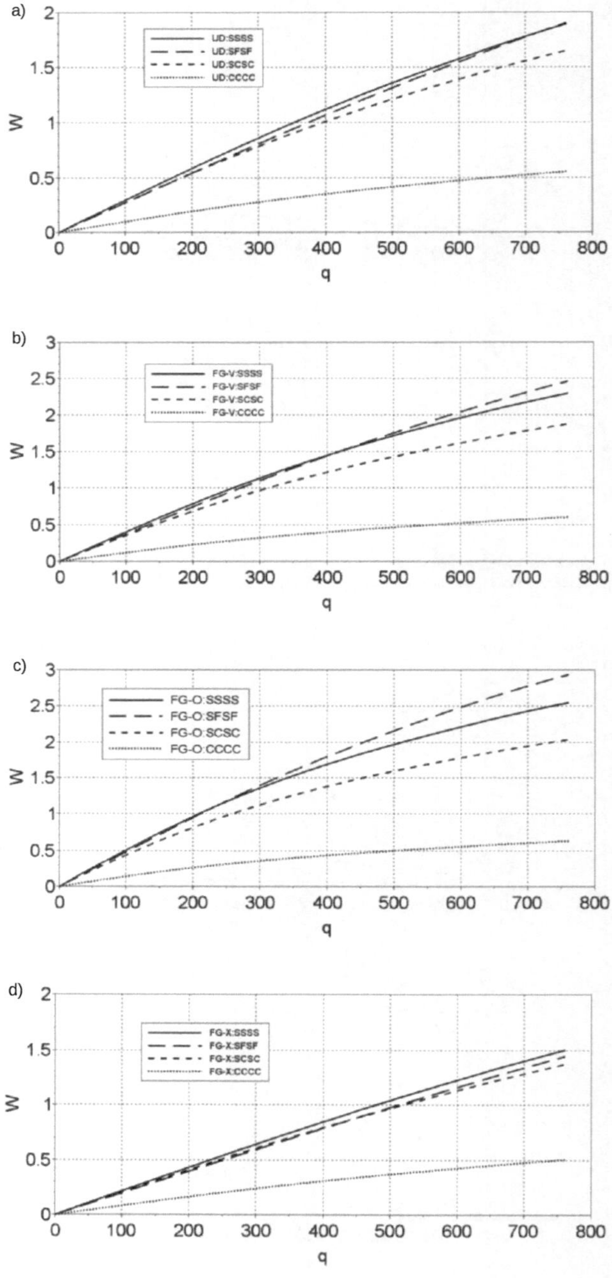

Figure 5.25 Central large deflections responses of (a): UD, (b): FG-V, (c) FG-O and (d): FG-X-CNTRC square plates subjected to uniform transverse load and under various boundary conditions, with ($b/h = 20$, $V^*_{CNT} = 0.17$).

5.4.2.2 Non-linear responses of FG-CNTRC annular ring plates

With the same strategy, non-linear deflections analysis is also performed for FG-CNTRC annular ring plates. According to the existing bibliography, large deflections analysis of FG-CNTRC annular ring plates are rather scarce, so we provide here the new first results concerning this geometry-type. Firstly, the effects of different CNTs profiles on non-linear bending behavior of FG-CNTRC annular ring plates are examined. Indeed, Fig. 5.26 depicts the load versus the vertical displacement curves at points A, B and C for various CNTs distributions. It can be seen that the values of non-linear deflections of functionally graded distributions are less than uniformly distribution and this is for all load points. Consequently, FG distributions can be chosen to optimize the stiffness of the structure via the considered form of CNTs .

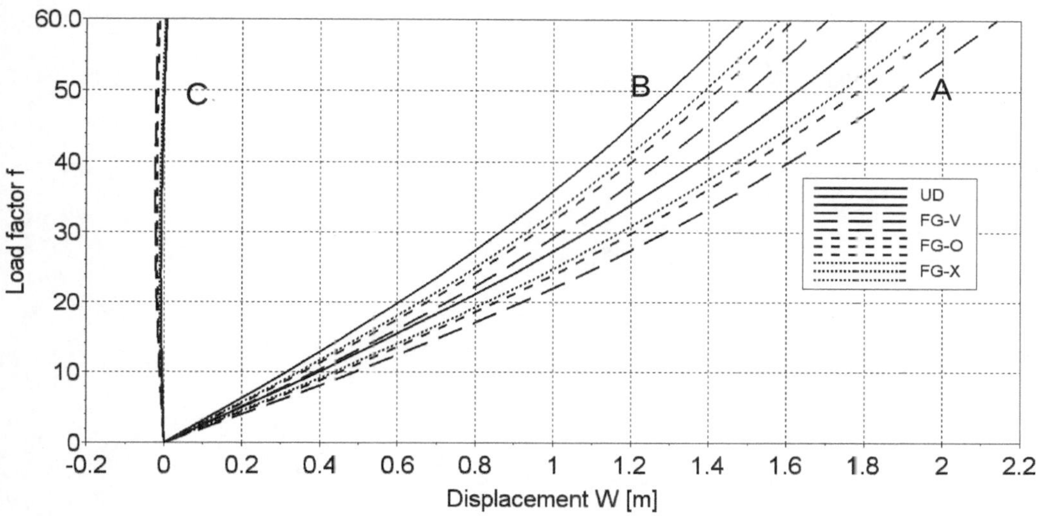

Figure 5.26 Non-linear load-displacement responses of FG-CNTRC annular ring plates at load points A, B and C for various CNTs forms ($V_{CNT}^* = 0.11$).

Furthermore, Fig. 5.27 shows the effects of variation of CNTs volume fractions ($V_{CNT}^* = 0.11, 0.14, 0.17$) on non-linear responses of UD, FG-V, FG-O and FG-X-CNTRC annular ring plates, respectively at the studied load points. It can be observed that, for a fixed load factor value f, the value of non-linear displacement decreases as the CNTs volume fraction increases and reaches its maximum with $V_{CNT}^* = 0.17$ which implies that more the ring plate is reinforced by carbon nanotubes, more the structure becomes stiffer. A combination of large CNTs volume fraction and appropriate distribution of CNTs can optimize properly their bending resistance. Moreover, Fig. 5.28 illustrates the effect of plate thickness parameter h on non-linear behavior of various forms of FG-CNTRC ring plates at load points A, B and C under a CNT volume fraction $V_{CNT}^* = 0.11$. The variation of the plate thickness geometrical parameter from $h = 0.01$ to $h = 0.03$ induces a decrease in the non-linear response of the FG-CNTRC annular ring plate and this is for all CNTs distributions. Thus, as the thickness parameter increases, as the ring plate becomes sensitive to the variation of this parameter.

5.4.2.3 Non-linear responses of FG-CNTRC cylindrical panels

Finally, the non-linear bending behavior of FG-CNTRC cylindrical panels is investigated with a uniaxially alignment of CNTs in the axial direction and functionally gradation in the thickness

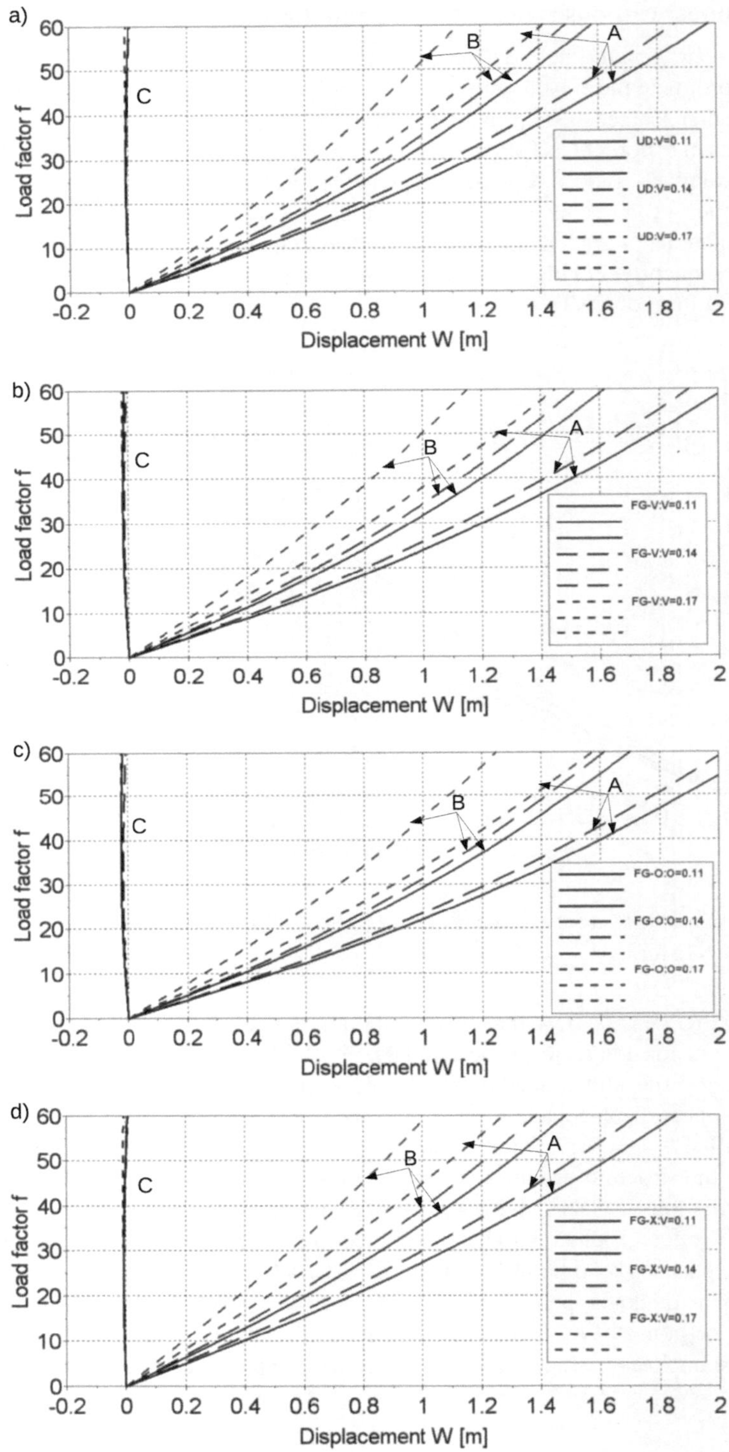

Figure 5.27 Effects of different CNTs volume fractions on large deflections responses of (a) UD, (b) FG-V, (c) FG-O and (d) FG-X CNTRC annular ring plates at load points A, B and C.

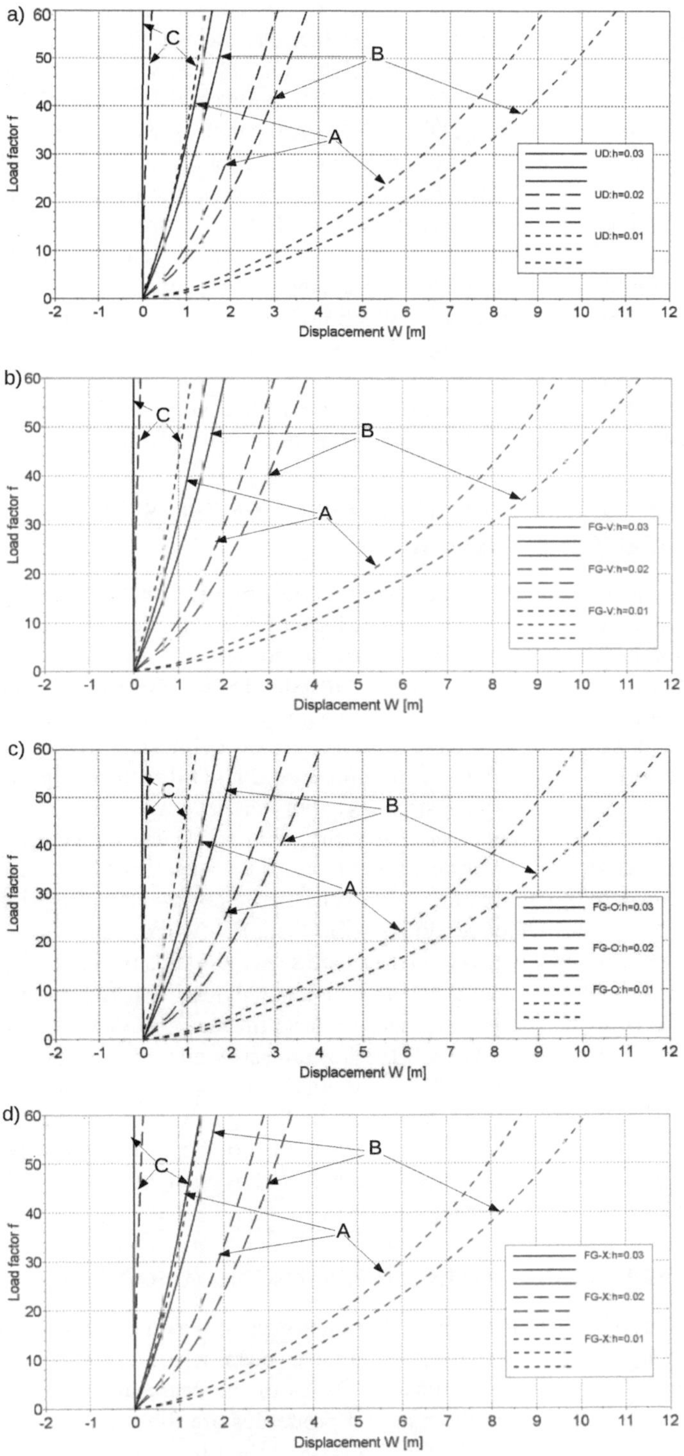

Figure 5.28 Effects of the thickness geometrical parameter on large deflections responses of (a) UD, (b) FG-V, (c) FG-O and (d) FG-X CNTRC annular ring plates at load points A, B and C with ($V^*_{CNT} = 0.11$).

direction. It should be mentioned that the present results constitute the first new results for FG-CNTRC panels with a concentrated point load. In fact, the most of papers in the literature treated the case of FG-CNTRC panels with uniform distributed load [315, 310]. A parametric study is also carried out here to outline the effects of CNTs volume fractions as well as the thickness parameter on non-linear deflections analysis of FG-CNTRC cylindrical panels. Figs. 5.29(a), 5.29(b), 5.29(c) and 5.29(d) display the effect of variation of CNTs volume fraction, from $V_{CNT}^{*} = 0.11$ to $V_{CNT}^{*} = 0.17$ on non-linear responses of FG-CNTRC panels. As can be seen, the increase of the CNTs volume fraction induces an increase in the stiffness of the FG-NTRC structure which manifest by the rise of the amplitude of the non-linear deflection responses. Thus, the mechanical behavior of composites can be considerably improved by the addition of an appropriate percentage of CNTs. Figs. 5.30(a), 5.30(b), 5.30(c) and 5.30(d) depict the load-deflections curves of UD, FG-V, FG-O and FG-X forms, respectively. It can be remarked that as the thickness decreases as the deflection of the FG-CNTRC cylindrical panel increases. In fact, for a low value of the thickness, the FG-CNTRC cylindrical panel behaves as thin structure and hence bends easily. Consequently, the variation of the thickness parameter has a significant effect on non-linear behavior of FG-CNTRC shell structures.

5.5 NON-LINEAR BENDING ANALYSIS OF FG-CNTRC STRUCTURES USING THE DOUBLE DIRECTORS SHELL MODEL

In this section, numerical results are presented for geometrically non-linear deflections analysis of FG-CNTRC shells using the double directors shell (DD-HSDT) model. Firstly, the performance of the enhanced membrane formulation is demonstrated through a bending test of a cantilever beam subjected to in-plane force, then its applicability is extended to FG-CNTRC where the effects of CNTs distributions and their volume fractions are examined for a cantilever beam under distributed end load. Secondly, three examples of FG-CNTRC plates, cylindrical panels and hemispherical shells are presented to show the ability of the proposed double directors non-linear model in the prediction of non-linear behavior of such structures. For that, a convergence studies are first carried out for isotropic cases of each studied shell structure, then non-linear deflections results are illustrated for FG-CNTRC cases in order to verify in one hand the accuracy of the present model and to draw in other hand the usefulness and the effectiveness of the present formulation. Uniform (UD) and three graded distributions, which are FG-V, FG-O and FG-X, are considered. According to examples under consideration, two types of matrices are selected in this paper which are Poly (methyl methacrylate), referred as PMMA and Poly(m-phenylenevinylene)-co-[(2,5-dioctoxy-p-phenylene) vinylene], referred as (PmPV). These two matrices have the same mass density and Poisson ratio, only the Young modulus is different. For CNTs, they are assumed to be of type of armchair (10,10) single-walled carbon nanotubes (SWCNTs) . The material properties of both matrices and CNTs are given at room temperature (300K) as indicated in Tab. 5.1. In addition, the efficiency parameters η_i (i = 1,2,3) are evaluated as indicated by Han and Elliot [95]. Their corresponding values are listed in Tab. 4.3. According to the investigations reported by Shen et al. [223, 235], the following assumption is used in this paper $G_{12} = G_{13} = G_{23}$.

5.5.1 NON-LINEAR IN-PLANE ENHANCED BENDING BEHAVIOR OF A CANTILEVER BEAM

The performance of the enhanced membrane formulation is outlined here through the assessment of the non-linear in-plane bending behavior of a cantilever beam which is subjected to in-plane force as shown in Fig. 5.31. The geometrical characteristics of the beam include: length $L = 1$, width $b = 0.1$ and thickness $h = 0.1$ while the material properties are those of an isotropic constitutive law and they are given as: $E = 1 \times 10^7$, $\nu = 0.3$. The applied in-plane force (F) is equal to 1000. The beam is meshed using 10×2 SHO4 finite elements where the abbreviation SHO4 denotes the present double directors shell element. Figure 5.32 depicts the Force-displacement response at the free end of the beam which demonstrates the aptitude of the SHO4 enhanced element to predict

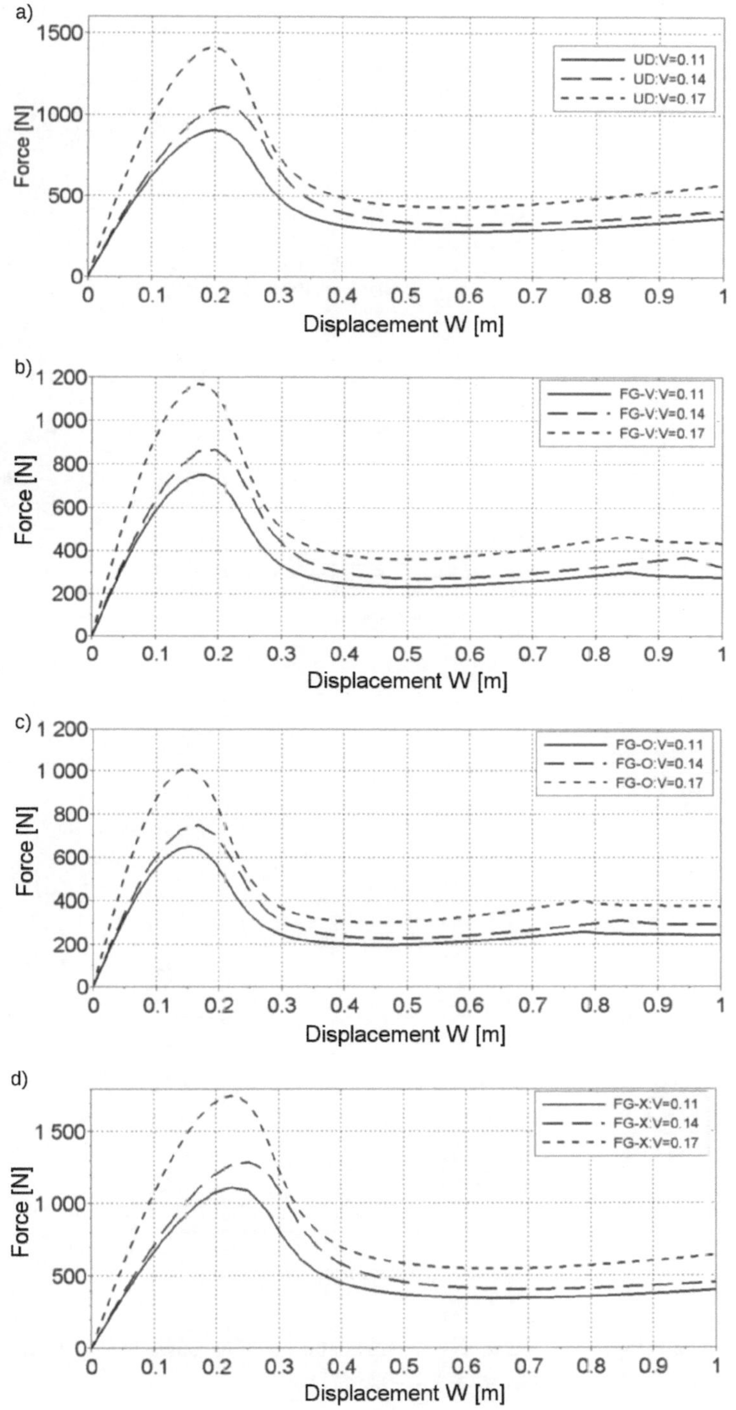

Figure 5.29 Effects of different CNTs volume fractions on large deflections responses cf (a) UD, (b) FG-V, (c) FG-O and (d) FG-X CNTRC cylindrical panels.

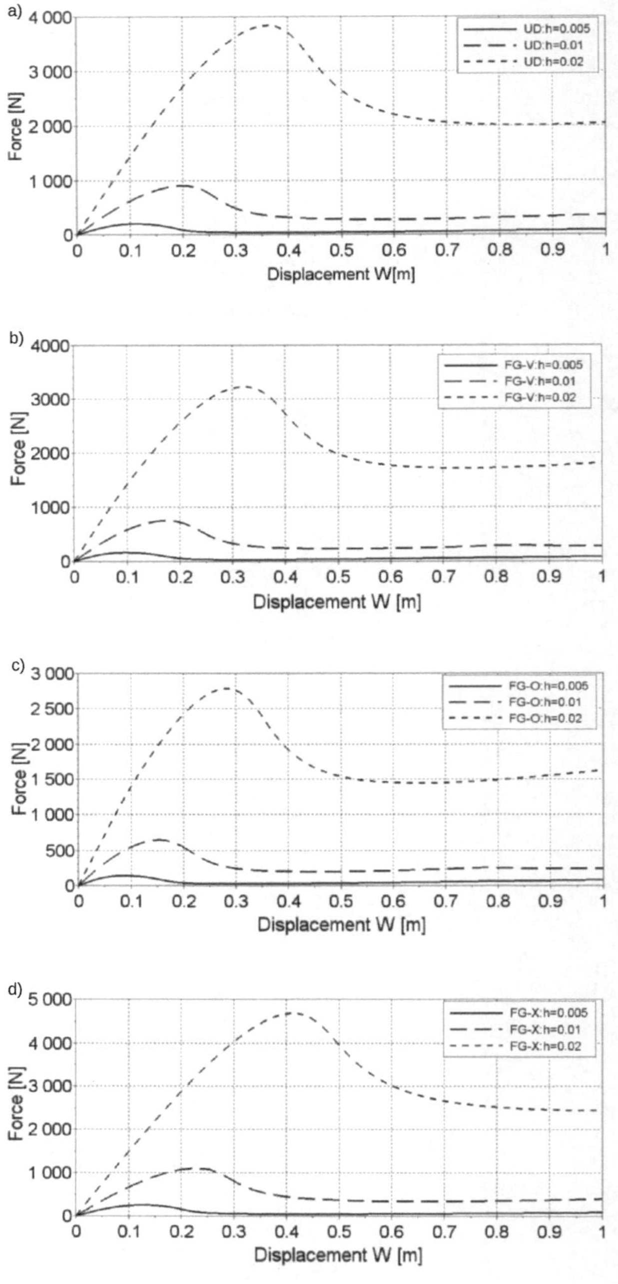

Figure 5.30 Effects of the thickness parameter on large deflections responses of (a) UD, (b) FG-V, (c) FG-O and (d) FG-X CNTRC cylindrical panels, with ($V^*_{CNT} = 0.11$).

Table 5.1
Material properties of SWCNTs , PmPV and PMMA matrices at room temperature (300K) [223, 235].

(10,10) SWCNTs	PmPV matrix	PMMA matrix
$E_{11}^{CNT} = 5.6466TPa$	$E_m = 2.1GPa$	$E_m = 2.5GPa$
$E_{22}^{CNT} = 7.0800TPa$	$\rho^m = 1150Kg/m^3$	-
$G_{12}^{CNT} = 1.9445TPa$	$\nu_m = 0.34$	-
$\rho^{CNT} = 1400Kg/m^3$		
$\nu_{12}^{CNT} = 0.175$		

the non-linear bending behavior with high accuracy over the SHO4 element without enhancement. Therefore, for in-plane bending case, the enrichment of the membrane part has a significant effect in the improvement of non-linear deflection responses.

5.5.2 NON-LINEAR ANALYSIS OF FG-CNTRC CANTILEVER BEAM UNDER DISTRIBUTED END LOAD

The previous analysis is extended here to FG-CNTRC cantilever beam while the applied load considered is now distributed at the end of the beam. The longitudinal direction of the beam is taken as the reference material orientation due to the anisotropic character of FG-CNTRC constitutive material laws. For the purpose of validation of the present formulation for FG-CNTRC materials, the geometrical properties are taken as indicated by Reinoso et al. [209] where their values are: length $L = 10m$, width $b = 1m$ and thickness $h = 0.1m$. In this example, the (10,10) SWCNT are chosen as reinforcements for a PMMA matrix. Their characteristics are as listed in Tab. 5.1. Uniform (UD) and three graded distributions of CNTs, FG-V, FG-O and FG-X, are considered here with three volume fractions($V_{CNT}^* = 0.12, 0.17, 0.28$). Their associated efficiency parameters are given in Tab. 4.19. In this case, we further assume that, $G_{12} = G_{13}$ and $G_{23} = 1.2G_{12}$. Figure 5.33 illustrates the obtained load-displacement responses compared to those given by Reinoso et al. [209] for a cantilever beam subjected to distributed end forces with different profiles of CNTs and under a volume fraction $V_{CNT}^* = 0.12$. A good agreement between the results leads to validation of the proposed SHO4 finite element. Besides, it can be that the FG-O-CNTRC beam has the highest value of deflection whereas the FG-X-CNTRC beam has the lowest one. This can be explained by the form of distribution of reinforcements which can affect the stiffness of the FG-CNTRC beams. In addition, the effect of variation of volume fractions of CNTs on non-linear deflections responses of FG-O-CNTRC and FG-V-CNTRC beams is examined. Figures 5.33(a) and 5.33(b) show that non-linear deflections derived from the proposed model converge to those of Reinoso et al. [209] which reflects the ability of the current SHO4 element to draw the non-linear behavior of FG-CNTRC beams with good accuracy. Figure 5.34. demonstrate that the present displacement-load curves of FG-O-CNTRC and FG-V-CNTRC beams obtained from the developed SHO4 finite element are well correlated with those of reference [209], for different values of CNTs volumes fractions. As a consequence, the correctness and reliability of the developed SHO4 finite element is well highlighted.

5.5.3 NON-LINEAR BENDING ANALYSIS OF FG-CNTRC PLATES

5.5.3.1 Non-linear analysis of isotropic plates

Now, geometrically non-linear bending analysis is performed for isotropic plates to verify the ability of the present double directors shell model to predict the non-linear behavior of plates. For that, a simply supported square plate subjected to uniform transverse load is considered ($\nu = 0.316$) as shown in Fig. 5.36. The obtained non-linear adimensional load-deflection curves are compared in

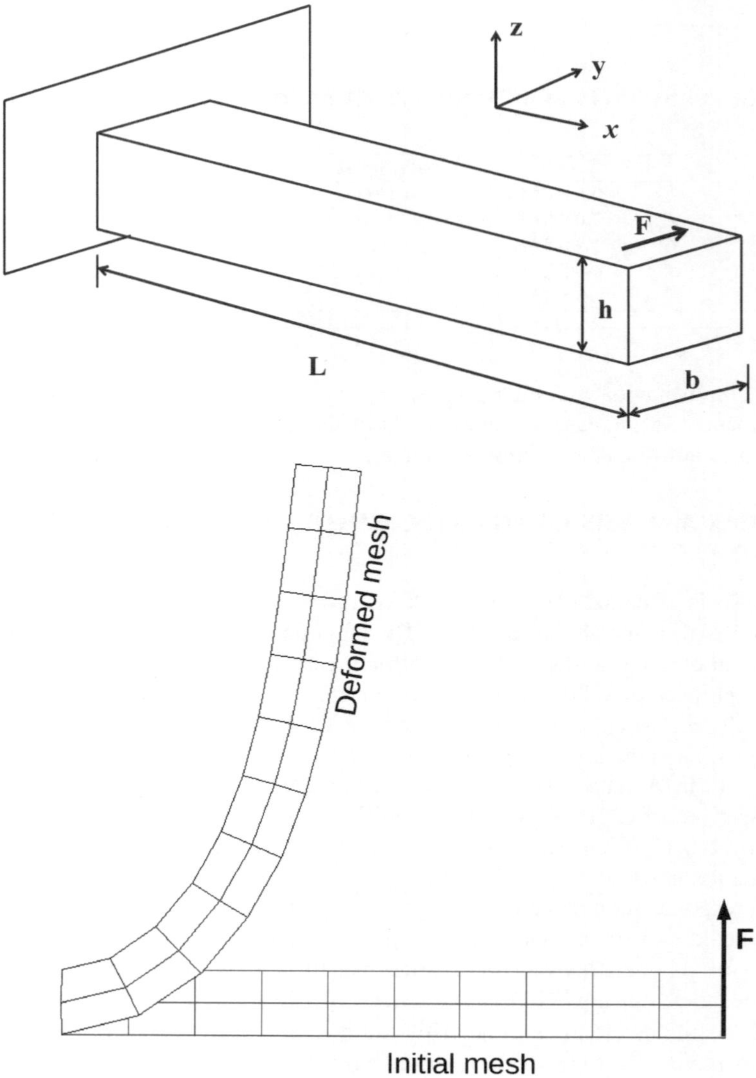

Figure 5.31 Cantilever beam subjected to in-plane force with its two initial and real deformed configurations.

Fig. 5.35 to the solutions of Lei et al. [136] which are based on first-order shear deformations (FSDT) theory and Von Karaman strains. The used adimensional parameters are: $W = W_0/h$, $q = \frac{q_0 a^4}{E h^4}$ where a, h are the length and the thickness of the isotopic plate, respectively. q_0 is the transverse applied load and E is the Young modulus of the isotropic square plate as indicated in [136, 88, 219]. It can be observed from Fig. 5.35 that the result obtained by the present double directors shell model agrees well with the solution of Lei et al. [136] which leads to validate the present model of the present model in the prediction of non-linear behavior of isotropic plates.

5.5.3.2 Non-linear analysis of FG-CNTRC plates

Secondly, the non-linear bending analysis is presented for FG-CNTRC plates. In fact, PmPV is considered as the matrix and the reinforcements are of type (10,10) armchair SWCNTs where their characteristics are as given in Tab. 5.1. Four types of CNTs distributions are considered : UD, FG-V,

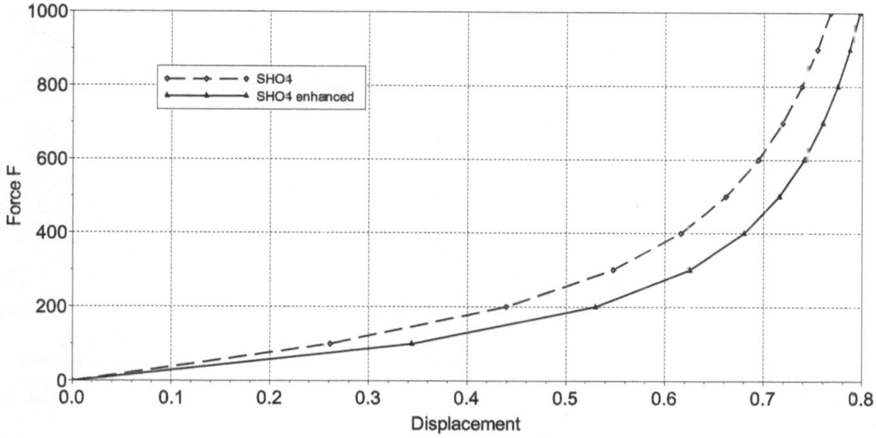

Figure 5.32 Force-Displacement responses of cantilever beam.

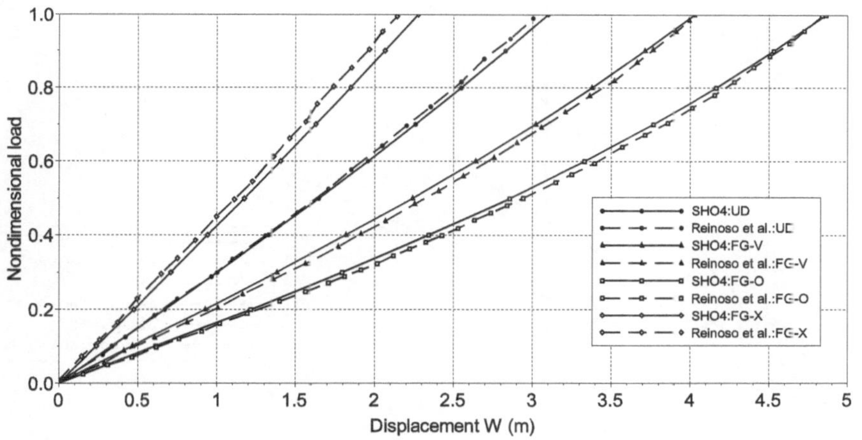

Figure 5.33 Load-Displacement responses of cantilever beam subjected to distributed end forces for different profiles of CNTs and under a volume fraction $V_{CNT}^* = 0.12$.

FG-O and FG-X with three volume fractions ($V_{CNT}^* = 0.11, 0.14, 0.17$). The plate is characterized by its length side a, width b and thickness h as depicted in Fig. 5.36. Non-dimensional parameters including $W = W_0/h$, $q = \frac{q_0 a^4}{E_m h^4}$ are defined to describe the geometrical non-linear results. The used mesh involves 16×16 SHO4 finite elements. The abbreviation SHO4 referred to the developed double directors finite element shell model. The effects of some parameters are presented through various numerical examples. For that, the effect of variation of CNTs profiles on non-linear responses of simply supported square CNTRC plates under transverse uniform load with plate width-to-thickness ratio $b/h = 10$ and $V_{CNT}^* = 0.11$ is first investigated. In fact, Fig. 5.37 shows that FG-O distribution has the highest value of non-linear central deflection while FG-X has the lowest one. The non-linear responses of FG-V and UD distributions are between FG-O and FG-X responses. Hence, when the CNT reinforcements are distributed near the top and the bottom surfaces of the

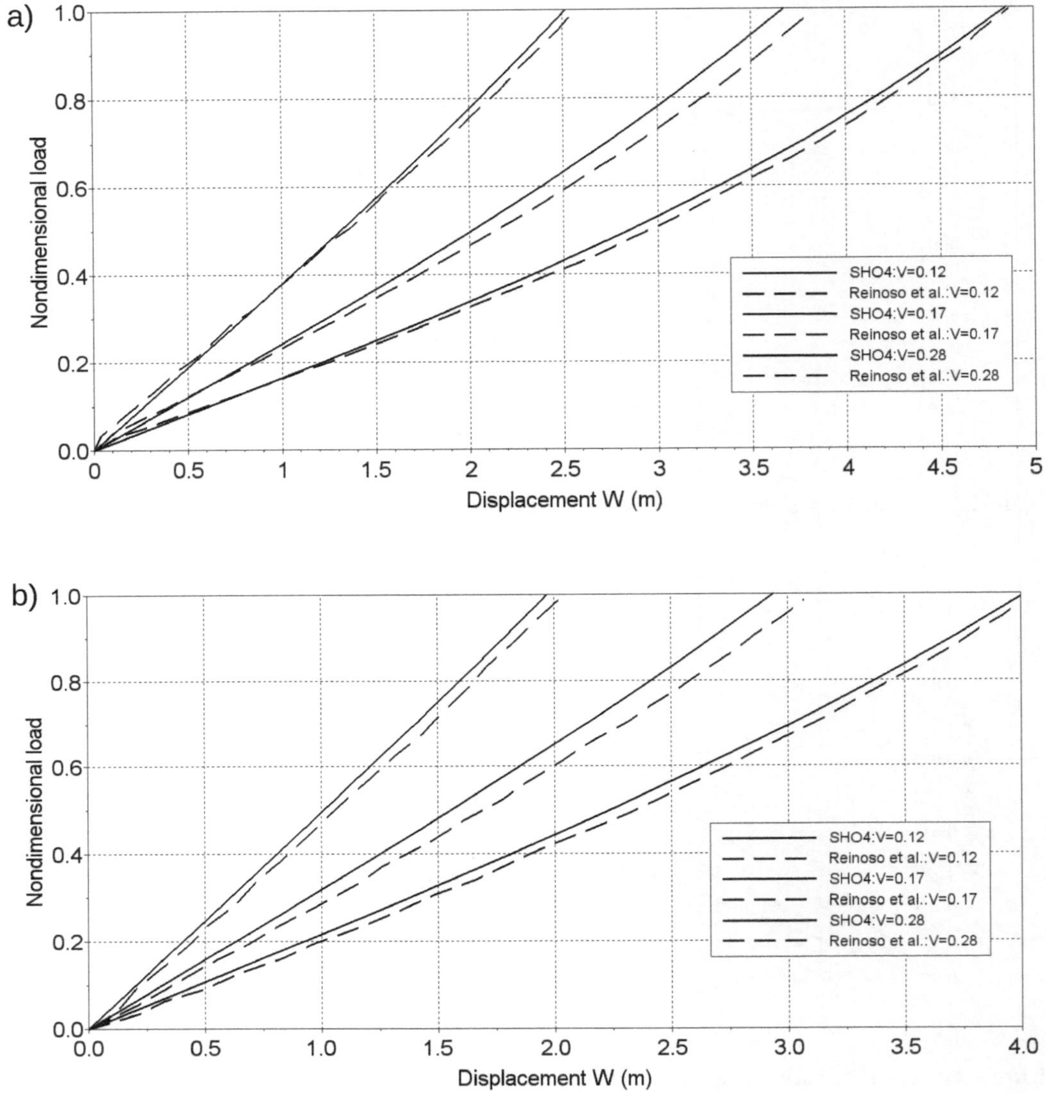

Figure 5.34 Load-Displacement responses of (a): FG-O-CNTRC and (b): FG-V-CNTRC cantilever beams subjected to distributed end forces for various volume fractions of CNTs .

structure, the plate has the highest flexural stiffness. So to adjust the stiffness of the shell structure, one can select the appropriate form of CNT distribution. As well as, Fig. 5.38 shows the linear and non-linear load-deflection curves of simply supported FG-X-CNTRC square plate under uniform transverse load and with width-to-thickness ratio ($b/h = 100$) for various CNTs volume fractions. It is clear that, for each CNT volume fraction, the load-deflection curve is linear for linear case while, it is non-linear for non-linear case. In fact, the linear response is tangent at zero point to its corresponding non-linear curve which indicates clearly the effect of introduction of geometrical non-linearity into the model. Moreover, when the CNT volume fraction increases, the deflection

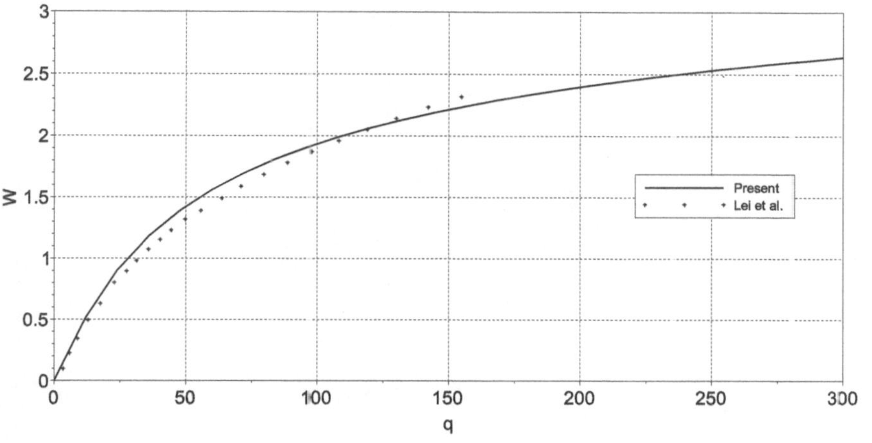

Figure 5.35 Convergence property of simply supported isotropic square plate subjected to uniformly transverse load.

decreases and this is observed for both linear and non-linear cases which implies that more the plate is reinforced by CNTs, more it becomes stiffer. On the other hand, the effect of plate aspect ratio a/b on non-linear response of simply supported UD and the three FG-CNTRC plates under a uniform transverse load is examined. The volume fraction of CNTs is taken as $V_{CNT}^{*} = 0.17$, with width-to-thickness ratio $b/h = 20$ and thickness $h = 2mm$. It can be observed from Figs. 5.39(a), 5.39(b), 5.39(c) and 5.39(d) that for a plate aspect ratio less or equal to 1 ($a/b = 0.5, 0.75, 1$), the central deflection varies weakly non-linearly with the load. Beyond this value, the non-linear effect is clearly pronounced and this is for all CNTs distributions. Furthermore, the value of central deflection increases as the plate aspect ratio rises from 0.5 to 2. In fact, for rectangular ($a/b = 0.5, 0.75$) or square ($a/b = 1$) plates, the non-linear behavior does not be reached while for larger plates ($a/b \succ 1$), this

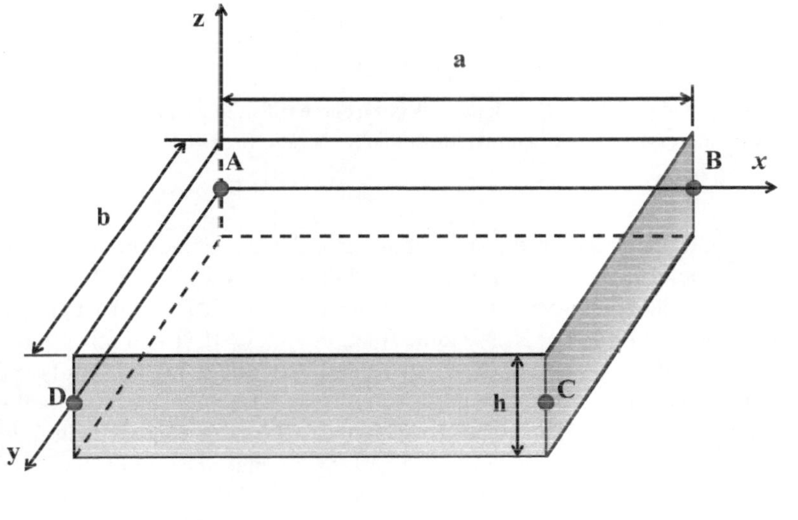

Figure 5.36 Geometry of the plate.

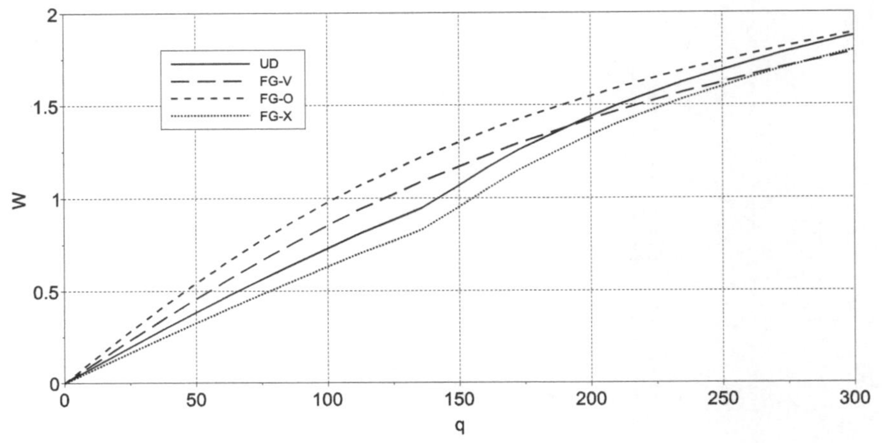

Figure 5.37 Non-linear response for various types of simply supported square CNTRC plates with width-to-thickness ratio ($b/h = 10$) and volume fraction $V^*_{CNT} = 0.11$.

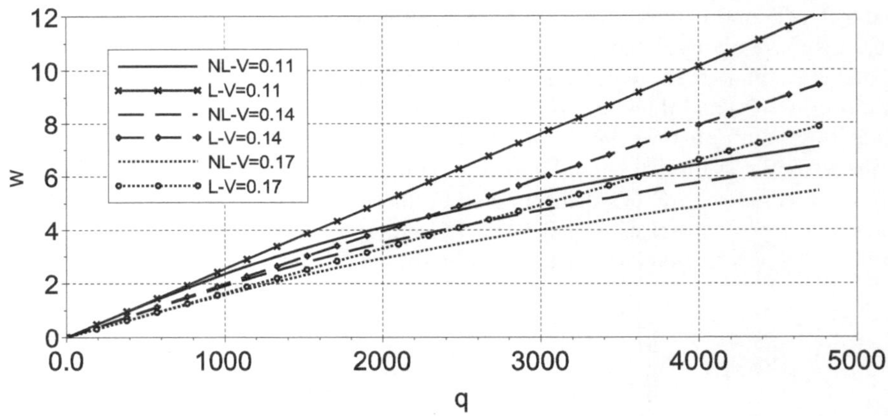

Figure 5.38 Linear (L) and non-linear (NL) load-deflection curves of simply supported FG-X-CNTRC square plate under uniform transverse load and with width-to-thickness ratio ($b/h = 100$) for various CNTs volume fractions.

effect starts clearly and this is due to the shear transverse deformations which become significant for larger plates that deform easily. Figures 5.40(a), 5.40(b), 5.40(c) and 5.40(d) depict the geometrically non-linear deflection curves of various distributions of CNTs under uniform transverse load, width-to-thickness ratio ($b/h = 20$), volume fraction $V^*_{CNT} = 0.11$ and for different boundary conditions: all sides are fully clamped (CCCC), two sides (AB) and (CD) are simply supported and the others sides (BC) and (AD) are clamped (SCSC) or two sides (AB) and (CD) are simply supported and the others are free (SFSF). As can be seen, the value of non-linear central deflection for (CCCC) FG-CNTRC plate is notably lower than SSSS, SCSC and SFSF FG-CNTRC plates for all types of CNTs distributions. Indeed, the clamped boundary condition induces stronger constraints compared to simply supported or free boundary conditions and thus the structure becomes more stiffer. It should be mentioned that the obtained results by the proposed formulation are visually

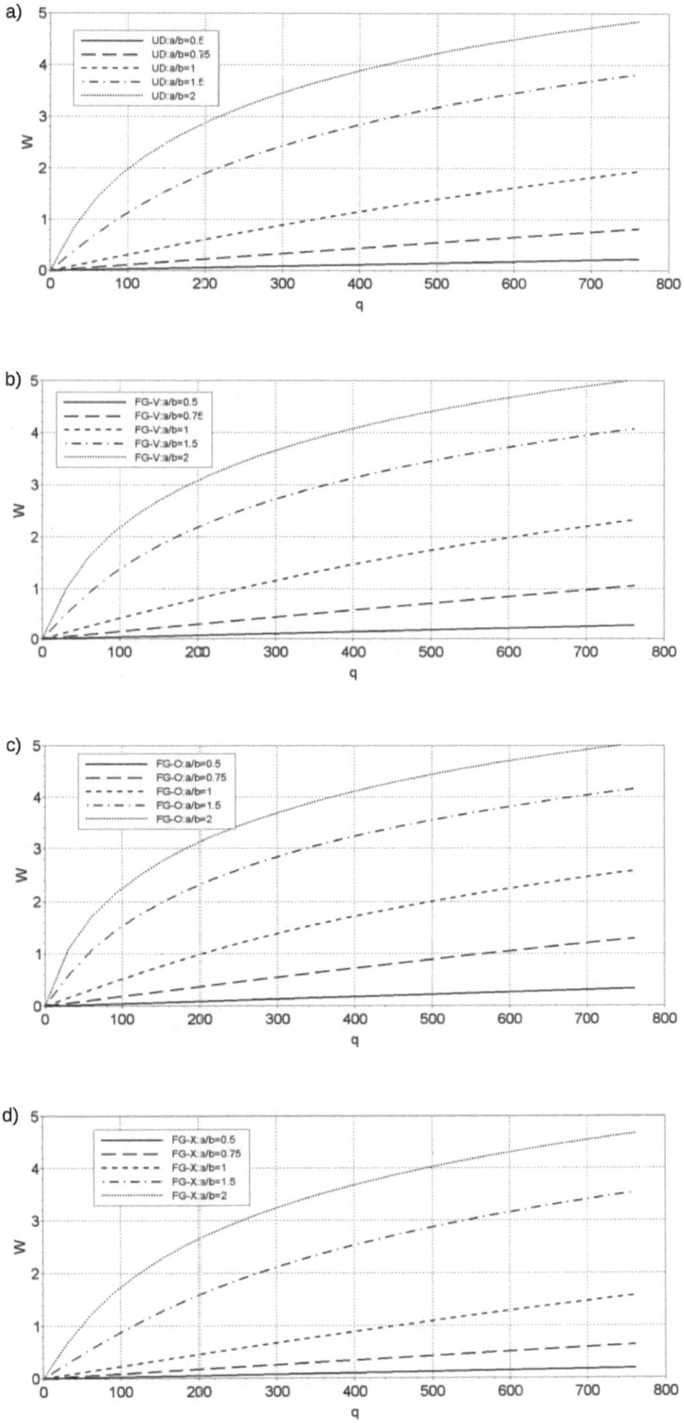

Figure 5.39 Effects of plate aspect ratios on geometrical non-linear central deflections for simply supported (a): UD, (b): FG-V, (c) FG-O and (d): FG-X CNTRC plates under uniform transverse load, volume fraction ($V^*_{CNT} = 0.17$) and width-to-thickness ratio ($b/h = 20$).

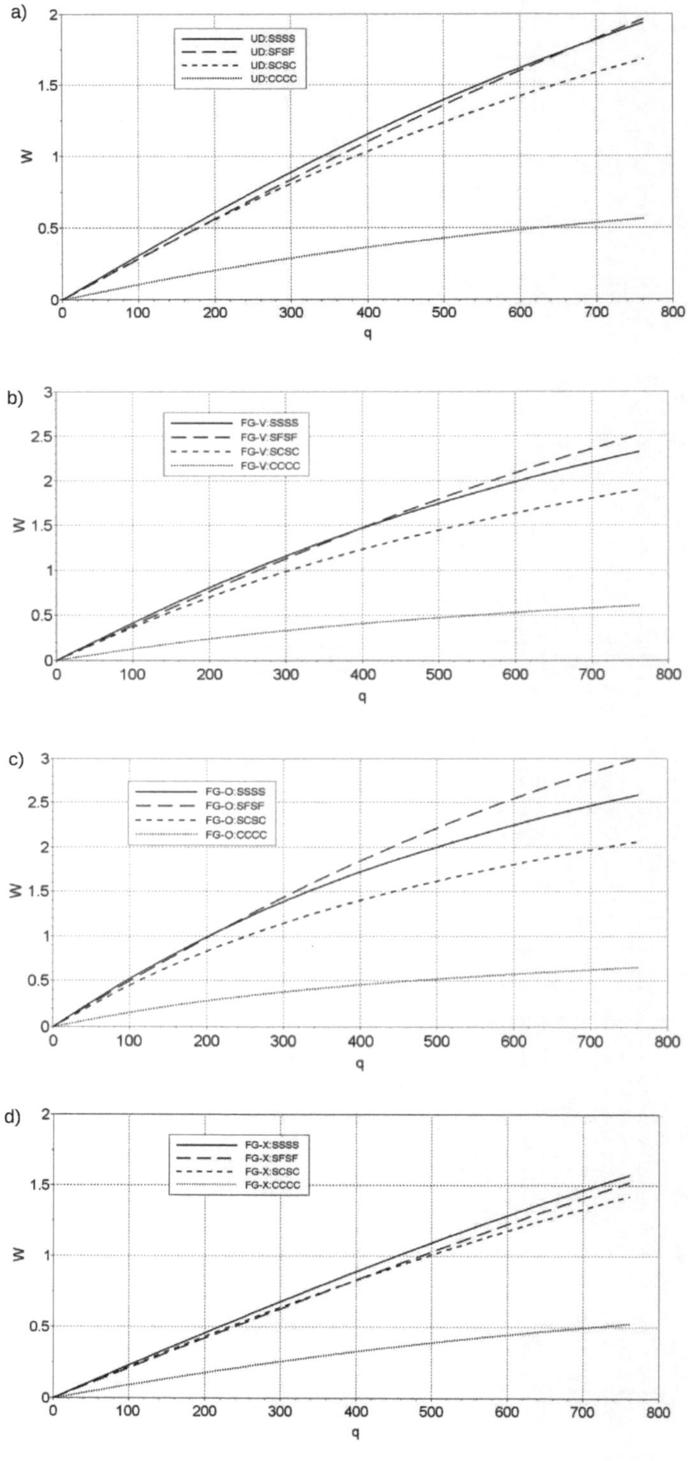

Figure 5.40 Effects of boundary conditions on geometrical non-linear central deflection for various CNTs distributions (a): UD, (b): FG-V, (c): FG-O and (d): FG-X square CNTRC plates under uniform transverse load, width-to-thickness ratio ($b/h = 20$), volume fraction $V_{CNT}^{*} = 0.17$.

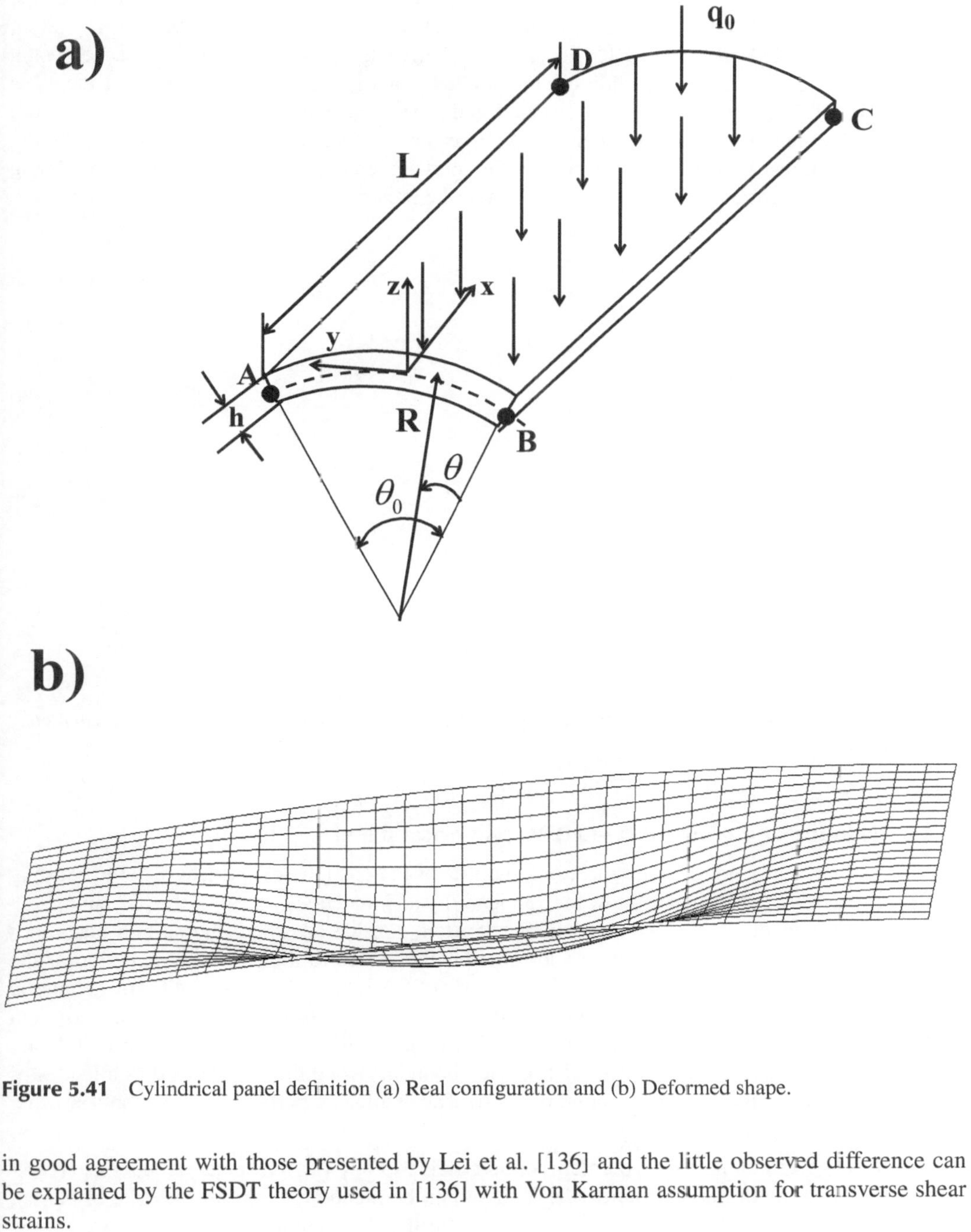

Figure 5.41 Cylindrical panel definition (a) Real configuration and (b) Deformed shape.

in good agreement with those presented by Lei et al. [136] and the little observed difference can be explained by the FSDT theory used in [136] with Von Karman assumption for transverse shear strains.

5.5.4 NON-LINEAR BENDING ANALYSIS OF CYLINDRICAL PANELS

5.5.4.1 Non-linear analysis of isotropic cylindrical panel

A comparison study for an isotropic cylindrical panel is first carried out in order to show the validity and the accuracy of the present formulation in the prediction of the non-linear behavior of such structures. The geometry of the studied cylindrical panel and its corresponding deformed shape are shown in Fig. 5.41 and its material properties are given as [315, 193]: $\theta_0 = 0.1rad$, $L = 20in$, $R = 100in$, $h = 0.125in$, $E = 4.5 \times 10^5 psi$ and $\nu = 0.3$. Furthermore, this panel is meshed using 32×20 SHO4 finite elements. In addition, the studied panel is fully clamped such that the sides (AB), (BC), (CD) and (DA) are all constrained and it is subjected to a uniform radial pressure distribution. The obtained non-linear deflection curves are presented in Fig. 5.42. As can be seen, the present non-linear SHO4 deflection curve is in good agreement with those presented by Zhang et al. [315] and Palazotto et al. [193] which validates the present double directors shell model.

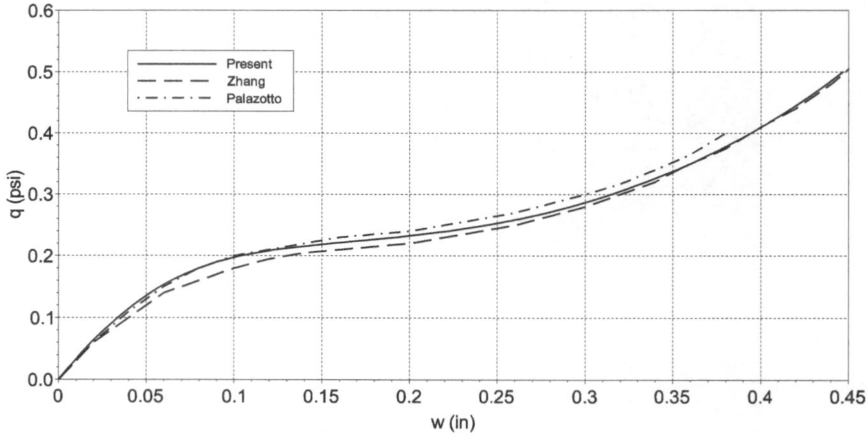

Figure 5.42 Convergence property for fully clamped isotropic cylindrical panel under a uniform radial pressure distribution.

5.5.4.2 Non-linear analysis of FG-CNTRC cylindrical panel

Now, the geometrically non-linear deflections analysis of FG-CNTRC cylindrical panel by varying some parameters are performed to show the applicability and the effectiveness of the present model. In fact, the same geometry as indicated in Fig. 5.41 is considered where the carbon nanotubes are uniaxially aligned in axial direction and functionally graded in the thickness direction of the panel. Uniform (UD) and three profiles of distributions of CNTs are also considered. The geometric characteristics of the FG-CNTRC cylindrical panel are: $\theta_0 = 0.1rad$, $h = 0.002m$, $R = 1m$ and $L = 0.1m$. The mechanical properties of FG-CNTRC cylindrical panel are as indicated in Tab. 4.2 and Tab. 4.19 with the PMMA matrix. The applied boundary condition is clamped in all sides of the panel with uniformly distributed radial pressure. The results are performed using the following adimensional parameters: $W = W_0/h$, $q = \frac{q_0 L^4}{E_m h^4}$.

The effects of variation of the volume fraction of CNTs V^*_{CNT} on non-linear deflections are examined for each type of distribution of CNTs. It should be mentioned that for the sake of clarity, the volume fraction of CNTs V^*_{CNT} is denoted by V in the illustrations. Figure 5.43 presents the non-linear central deflections with load for UD, FG-V, FG-O and FG-X distributions respectively by varying the volume fractions of CNTs. As can be remarked, the value of the non-linear central deflection decreases when the value of volume fraction of CNTs increases and this is for all types of CNTs distributions. In fact, the increase of CNT volume faction induces a high stiffness of the whole of the structure and thus a decrease in the non-linear central deflection. As well as, Fig. 5.44 demonstrate the effect of variation of the panel radius R on non-linear central deflection for different

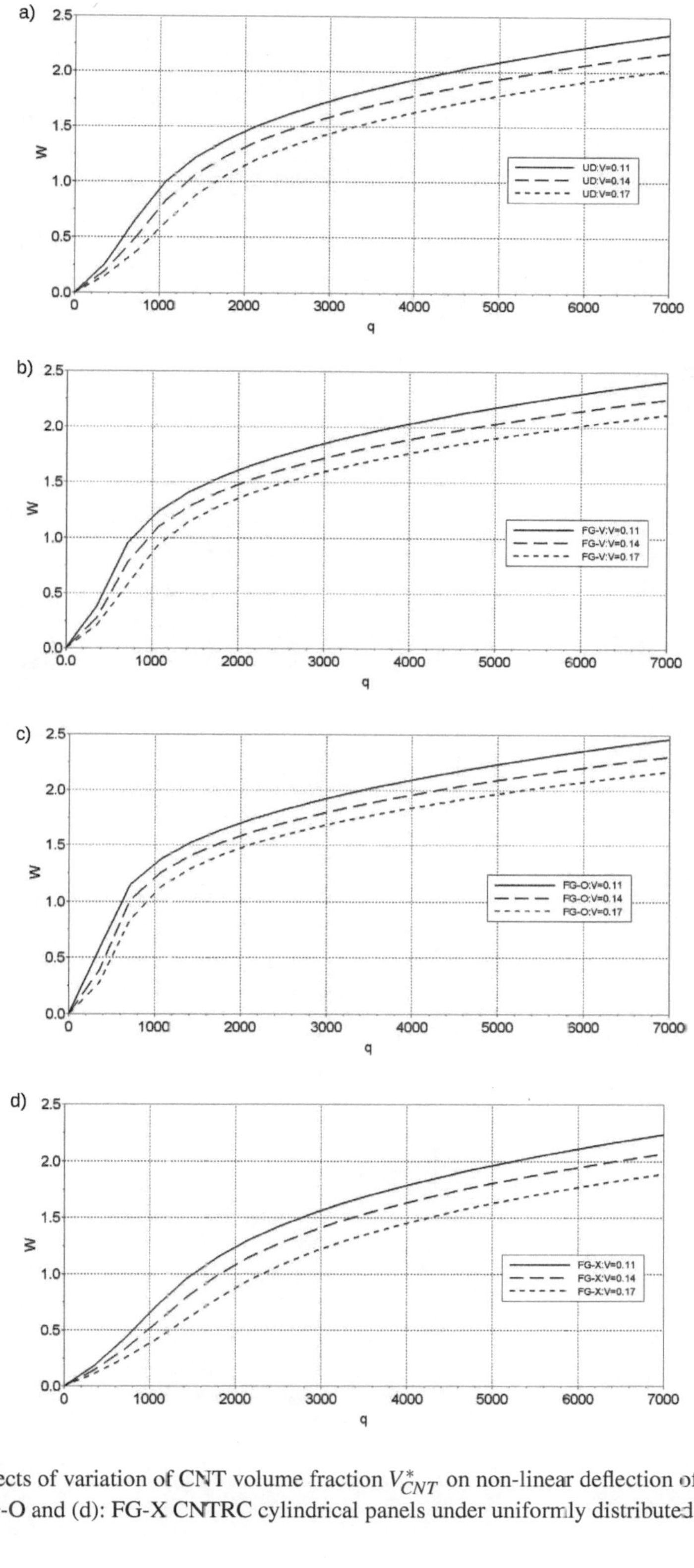

Figure 5.43 Effects of variation of CNT volume fraction V_{CNT}^{*} on non-linear deflection of clamped (a): UD, (b): FG-V, (c): FG-O and (d): FG-X CNTRC cylindrical panels under uniformly distributed radial pressure.

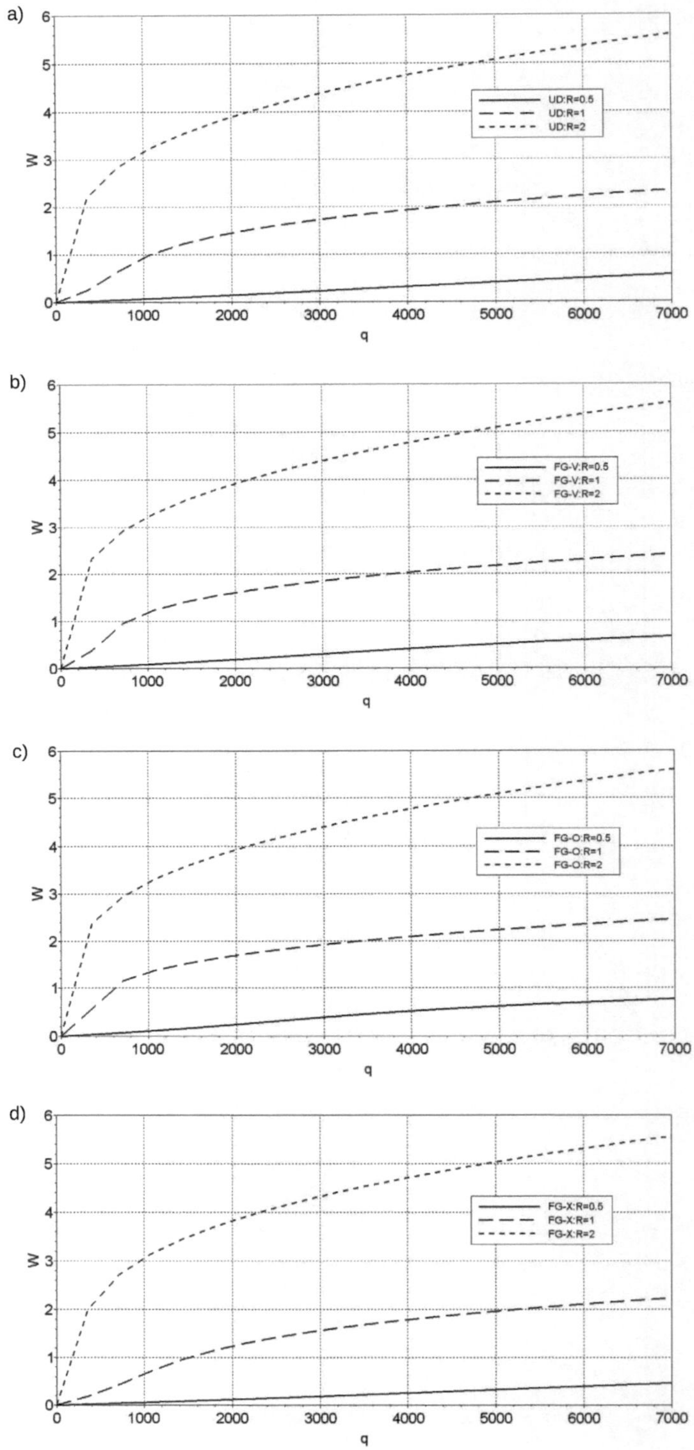

Figure 5.44 Effect of variation of panel radius on non-linear deflection of clamped (a): UD, (b): FG-V, (c): FG-O and (d): FG-X CNTRC cylindrical panels under uniformly distributed radial pressure and volume fraction $V_{CNT}^{*} = 0.11$.

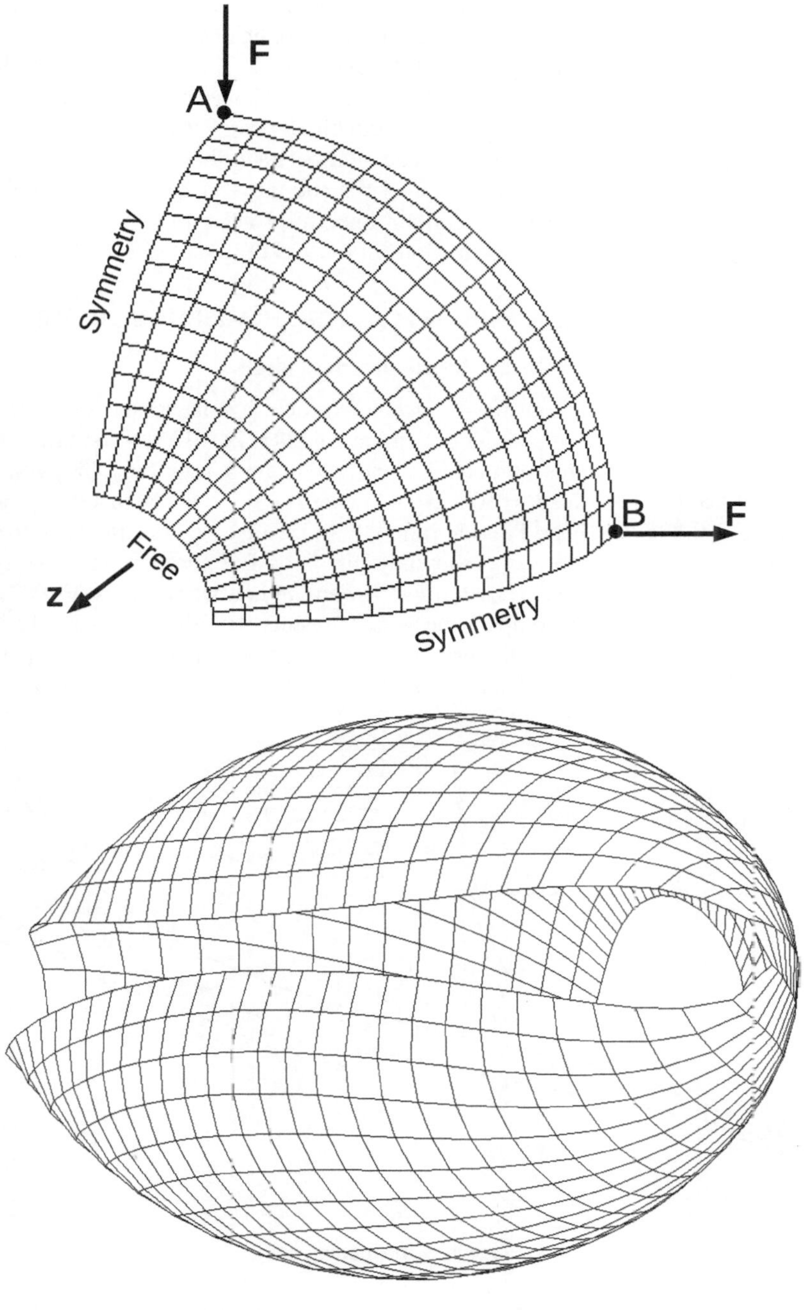

Figure 5.45 A pinched hemispherical shell with the deformed geometry obtained under a load $F = 360$.

profiles of CNTs distributions of fully clamped cylindrical panels under uniformly distributed radial pressure. As can be seen, for all CNTs distributions, and for a radius $R = 0.5$, the central deflection varies linearly with the load pressure while the non-linear effect is slightly reached with a radius equal to $R = 1$ or $R = 2$. Furthermore, when the value of the radius increases, the value of non-linear deflection increases quickly. Moreover, the increases of the load manifest rapidly by an increase in the non-linear central deflection with the variation of the radius of the FG-CNTRC cylindrical panel. Consequently, as the radius of the FG-CNTRC cylindrical panel increases, the whole of the structure deforms easily which induces high values in the non-linear central deflections. As well as, the non-linear deflections obtained by the present double directors shell model are similar to the results presented in [315]. The little difference generated between the results is due to the FSDT theory which is restricted in its application to thin shells and the Von Karman assumptions which are limited for small to moderately geometrical non-linearity. Hence, the proposed refined model constitutes an alternative to analyze large deflections of FG-CNTRC shells.

5.5.5 NON-LINEAR BENDING ANALYSIS OF FG-CNTRC HEMISPHERICAL SHELL

5.5.5.1 Non-linear analysis of isotropic hemispherical shell

A convergence study is also carried out for a pinched isotropic hemispherical shell. This popular benchmark test for shell problems which is also examined in [243, 214, 125] constitutes an excellent example to show the ability of the present model in handling the finite rotations. The geometry of the spherical shell is defined by the radius $R = 10$ and the thickness $h = 0.04$. The material properties are Young's modulus $E = 6.825 \times 10^7$ and Poisson ratio $\nu = 0.3$. Due to the symmetry of the problem, only the quadrant of the shell is modeled as shown in Fig. 5.45. Furthermore, the shell is disposed with an 18° hole at the top and it is subjected to two inward and outward forces (F) 90° apart. The total load $F = 450$ is considered to compare the present results with those of Simo [243], Saleeb [214] and Kim [125]. The validation of the present formulation with this example is performed in [79] and the obtained results are in good agreement with those reported in the literature.

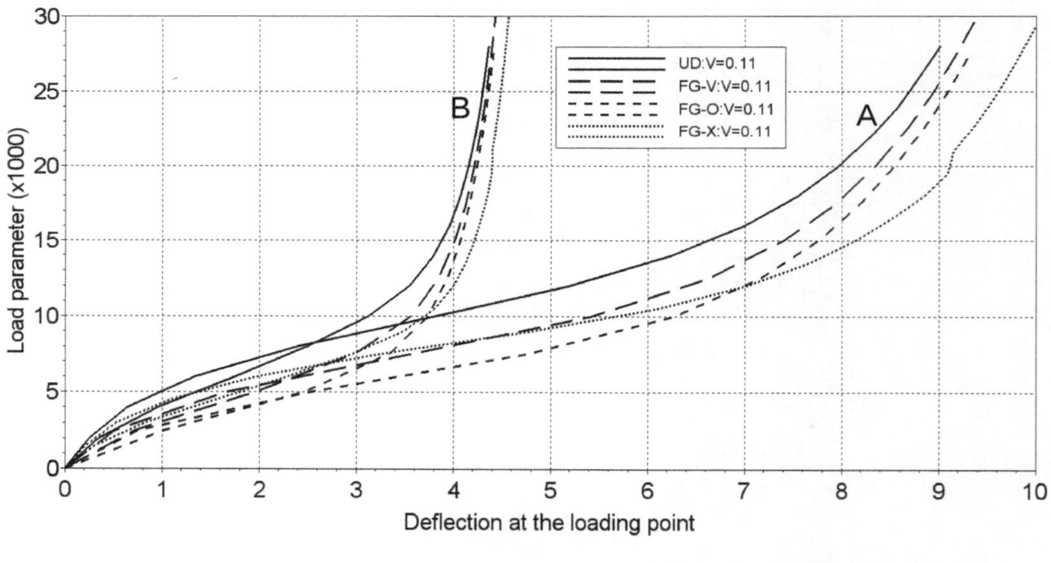

Figure 5.46 Non-linear load-deflection responses for various forms of FG-CNTRC hemispherical shell under a volume fraction of CNTs $V^*_{CNT} = 0.11$.

5.5.5.2 Non-linear analysis of FG-CNTRC hemispherical shell

Now, the CNTs are introduced to the hemispherical shell forming a FG-CNTRC shell and its configuration is as shown in Fig. 5.45 and the material properties are the same as mentioned for FG-CNTRC plates. Since no results are available in the literature for non-linear analysis of FG-CNTRC

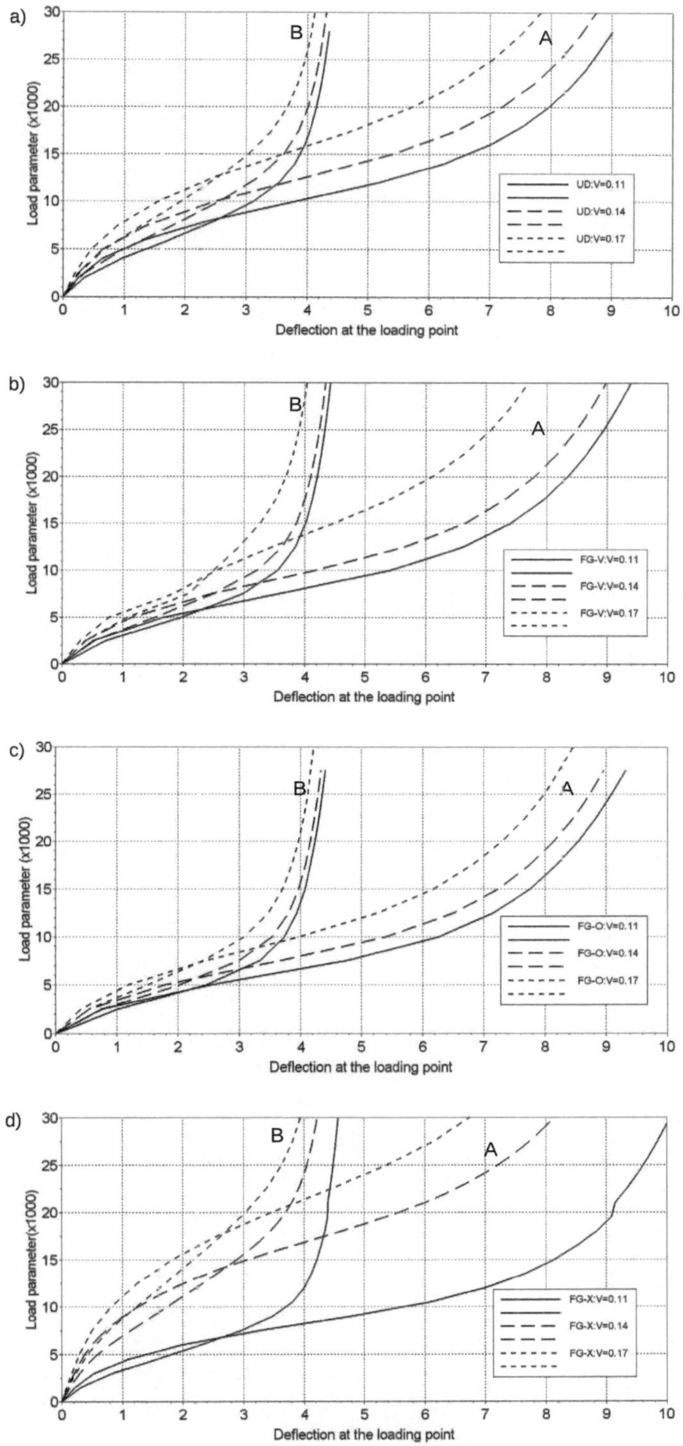

Figure 5.47 Effects of volume fractions of CNTs on non-linear central deflections for (a): UD, (b): FG-V, (c) FG-O and (d): FG-X hemispherical shells.

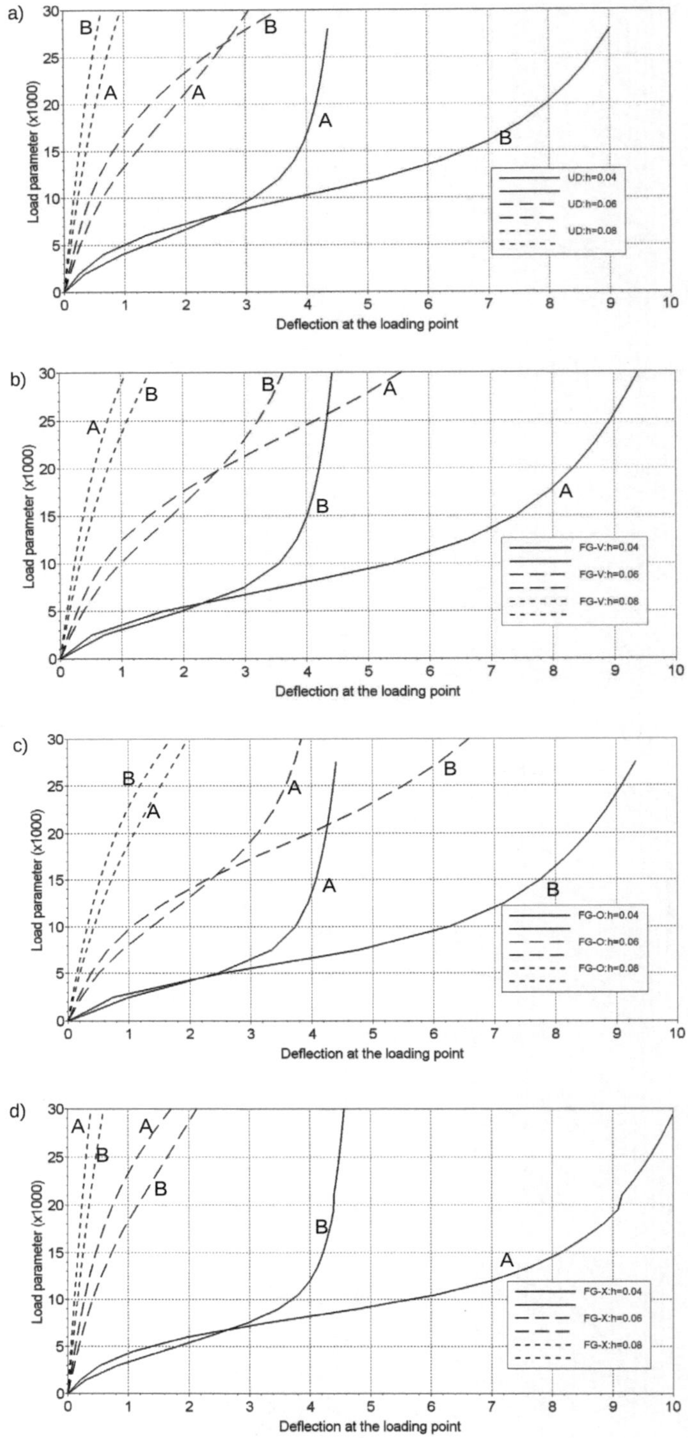

Figure 5.48 Effects of the thickness parameter on non-linear central deflections for (a): UD, (b): FG-V, (c) FG-O and (d): FG-X hemispherical shells under a volume fraction of (CNT) $V^*_{CNT} = 0.11$.

spherical shells, and after the validation of the proposed non-linear double directors model in the isotropic case, we provide through this numerical study the very first and useful information on the investigation of the effects of some geometrical parameters on the non-linear behavior of FG-CNTRC spherical shells. Figures 5.46 shows the effects of CNT profiles on non-linear deflections of FG-CNTRC hemispherical shell under a volume fraction $V^*_{CNT} = 0.11$. It can be seen that for both loading points A and B, the value of non-linear deflection of FG-X form is less than UD, FG-V and FG-O distributions. This leads to conclude that the CNT distributions close to top and bottom surfaces are more efficient in increasing the stiffness of the hemispherical shell than those close to mid-surface or uniformly distributed. Hence, the desired stiffness can be adjusted by regulating the distributions of CNTs . Figures 5.47(a), 5.47(b), 5.47(c) and 5.47(d) show the effects of variation of CNT volume fractions on non-linear deflections of UD, FG-V, FG-O and FG-X spherical shells. It can be remarked that, as the volume fraction of CNTs increases, the non-linear deflection decreases and this is for the two loading points (A) and (B). Thus, an important percentage of CNT ($V^*_{CNT} = 0.17$) induces a high stiffness of the spherical shell which implies a decrease in the central deflection. Consequently, the CNTs constitute a good solution of composite reinforcements in the enhancement of the mechanical behavior of spherical shells.

As well as, the effect of variation of shell thickness is also examined for various profiles of CNT distributions. Indeed, Figs. 5.48(a), 5.48(b), 5.48(c) and 5.48(d) demonstrate that the increases of the shell thickness parameter induces a decrease in the non-linear deflection of FG-CNTRC hemispherical shells and this is for various forms of CNT distributions. Hence, for a thicker FG-CNTRC spherical shell, the stiffness of the structure is higher and thereby the whole of the structure does not deform easily which can explain the decreases in the deflection of the structure.

5.6 CONCLUSION

Non-linear deflections analysis of FG-CNTRC plate and shell structures is presented using a discrete form of Kirchhoff finite element model, the FSDT theory and the double directors finite shell model, respectively. For the first modeling analysis, the approximation of the displacement field is carried out using four-node and three-node finite elements. The Kirchhoff finite element model enables to bring up two levels of analysis: computational and application sides. On the computational side, the present model constitutes an alternative to analyze large displacements and rotations of thin FG-CNTRC shell structures without locking problems and with low computational effort. On the application side, the non-linear Kirchhoff finite element model bridges the numerical framework to advanced applications of multifunctional composites structures using FG-CNTRCs . For the second modeling analysis, a non-linear formulation based on the FSDT theory is proposed to model the large deflections responses of FG-CNTRC structures. The proposed model incorporates the transverse shear deformations and it is free from shear-locking problem via the utilization of the assumed natural transverse shear strain (ANS) method. Geometrical non-linearity type is also introduced into the formulation to accurately describe large displacements and finite rotations. Finally, a non-linear double directors shell model is used to investigate large deflections analysis of FG-CNTRC plates, cylindrical panels and hemispherical shells. The (DDST) model implies a high-order distribution of the displacement field through the thickness direction and satisfies a zero condition of the transverse shear stresses on top and bottom surfaces. In these investigations, the CNTs are assumed to be graded in the thickness direction of the shell structure and the effective material properties are determined using the modified rule of mixtures which introduces some efficiency parameters to take into account the scale dependance of the CNTs. The enhancement of the membrane strains is highlighted via in-plane bending test of a cantilever beam. As well as, out-of plane distributed load for a FG-CNTRC cantilever beam is also examined. Convergence studies are carried out for studied plates and shells leading to verification of the ability of the present model in the prediction of non-linear behavior of such structures. Then, parametric studies are performed for various shapes of FG-CNTRC structures in order to show the effects of CNTs profiles, their volume fractions, some geometrical parameters and boundary conditions on non-linear behavior of FG-CNTRC shells. The obtained results outline the significant effect of CNT reinforcements in the enhancement of the mechanical behavior of composite shell structures. For geometrical non-linearity, the enhancement of the membrane part of the strain field can improve significantly in-plane dominated cases but it remains small for FG-CNTRC which represents a high flexural rigidity. Further this can open a future work with the introduction of the membrane enhancement for geometrical non-linearity using an update procedure of the present model.

6 Buckling and post-buckling of FG-CNTRC plates and shells

This chapter aims to investigate buckling and post-buckling behavior of functionally graded carbon nanotubes-reinforced composite (FG-CNTRC) plates and shells subjected to in-plane compressive loads. The governing equations are established using a double directors finite element shell model which induces a high-order distribution of the displacement field and takes into account the effect of transverse shear deformations. In resolution procedure, the critical buckling load is determined via the resolution of the buckling eigenvalue problem and the post-buckling responses are obtained using the Newton-Raphson with arc length control solution strategy. The extended rule of mixture with some efficiency parameters is used to estimate the mechanical properties of FG-CNTRC shell structures. UD and four types of CNTs distributions are considered which are: FG-V, FG-A, FG-O and FG-X. Convergence and comparison studies are first carried out to validate the accuracy of the present model. Then, the results are extended FG-CNTRC shell structures to perform the effectiveness and usefulness of the present model in the prediction of the critical buckling behavior of such structures.

6.1 INTRODUCTION

To ensure high performance in the next generation of aerospace, naval, civil, nuclear and vehicle structures, it is necessary to develop reliable, high-fidelity analysis tools for predicting the buckling and post-buckling of plate and shell structures. Recently, and with the discovery of new materials such as functionally graded (FG) materials [129] and carbon-nanotubes reinforced composites (FG-CNTRC) [104] which have superior mechanical, thermal and chemical properties, many studies have been conducted to analyze as well as to predict the buckling and post-buckling responses of such structures. However, the most of published papers focus on linear post-buckling responses and they are subjected to in-plane compression load, thermal or thermo-mechanical loadings. Further attention must be paid to mechanical loadings-type especially for composite plates and curved panels which are often subjected to various types of mechanical loads. Moreover, to accurately modeling the structural responses of composite structures, an efficient shell model should be provided in contrast to the usual existing models in the literature which are based generally to the classical shell theory (CST), the first-order shear deformation theory (FSDT) or higher-order shear deformation theory (HSDT) [297, 205, 292, 42, 50, 92, 317]. Unfortunately, these models are implemented within the Von-Karman kinematic assumptions which are limited to small deformations and moderate rotations and cannot hence perform large deformations and rotations. On the other hand, highly accurate predictions of buckling and post-buckling responses of nanocomposites reinforced by carbon-nanotubes require rigorous experimental and numerical validations. However, validation with experimental results is rather scarce and not yet fully mature for these advanced materials due to high costs required in both elaboration and experimental devices. For instance, for compression-loaded panels, the present post-buckling results are compared to experimental ones from the literature [261] which can bring up the high accuracy of the proposed high-order shear deformable model. In the following, a brief review on buckling and post-buckling works for FG-CNTRC structures is presented.

DOI: 10.1201/9781003483786-6

Lei et al. [135, 138, 324, 325], Jafari-Mehrabadi [106], Malekzadeh et al. [196] and Kiani [121, 122] have presented various analyses on buckling behavior of FG-CNTRC plates using the FSDT theory. Parametric studies are also carried out leading to perform the effect of CNT profiles, their volume fractions and various geometrical parameters on critical buckling of such structures. On the other hand, Ansari et al. have employed an efficient numerical method within the framework of variational formulation to study the buckling and vibration of axially-compressed FG-CNTRC conical shells [11, 12]. Moreover, Mehri et al. [165] have developed a semi-analytical solution on vibration and buckling analysis of FG-CNTRC truncated conical shell under an external pressure and axial compression simultaneously. The governing equations are based on the trigonometric expansion and the harmonic differential quadrature (HDQ) method. Besides, Jam et al. [107] have used the Donnell shell theory assumptions consistent with the FSDT theory for buckling analysis of pressurized FG-CNTRC conical shells. The obtained results have shown that CNTs volume fraction and CNTs distribution law are important factors on the buckling mode and buckling load of the FG-CNTRC conical shells. Najafizadeh and his co-authors have presented buckling and vibration analysis of FG-CNTRC beams under axial load [187]. The stability and motion equations are based on the two-dimensional elasticity theory and Hamilton principle. The generalized differential quadrature method (GDQM) is used to discretize the equations of stability and motion and to implement the boundary conditions. Using the HSDT theory, Sobhaniaragh et al. [252] have presented mechanical buckling behavior of the continuously CNT-reinforced shells stiffened by stringer and rings. A detailed parametric study is conducted to highlight the impacts of various geometrical and material parameters involved on the critical mechanical buckling of the SCS aggregated by graded CNTs. Recently, Garcia-Macias et al. have performed the linear buckling analysis of FG-CNTRC cylindrical curved panels under compressive and shear loading [85]. The effective material properties are estimated using the Eshelby-Mori-Tanaka approach or the extended rule of mixtures and the results have demonstrated that the change of fiber orientation, CNTs distribution, panel aspect ratio, loading condition and temperature have noticeable effects on the buckling strength and buckling modes of FG-CNTRC curved panels. Equally, Garcia-Macias et al. have studied the post-buckling behavior of axially compressed FG-CNTRC cylindrical panels with the Eshelby-Mori-Tanaka approach [86]. A detailed parametric analysis including the effects of filler content, fiber orientation, distribution along the thickness, and misalignment of fibers are also examined. Shen and his co-authors have conducted various works on postbuckling behavior of FG-CNTRC plates and shells using the HSDT theory and one can be referred to these publications [224, 235, 234, 321, 322, 320] among others.

6.2 BUCKLING AND POST-BUCKLING PROBLEMS

6.2.1 BUCKLING PROBLEM RESOLUTION

The governing equations are established here using the double directors finite shell element model as developed previously in Chapter 1. We recall, that the buckling eigenvalue problem can be written as:

$$(K_M + \lambda_{cr} K_G) U_n = 0. \tag{6.1}$$

where K_M and K_G represent the linear stiffness matrix, and the geometric stiffness matrix, respectively. For the linear stiffness matrix K_M, it is constructed from the material tangent operator as follows:

$$K_M = \int_A B^T H_T B dA. \tag{6.2}$$

with H_T and B are the elastic modulus and the displacement-strain vector respectively, whose expressions are as given in Eq. (1.104) and Eq. (1.113), respectively. For the geometrical stiffness matrix, K_G, it can be expressed for a couple of nodes (I,J) as follows:

$$K_{G_{IJ}} = \left\{ \begin{array}{ccc} UUM_{IJ}I + UUF_{2IJ} & (UBF_{1IJ} + UBC_{1IJ})I & UBF_{2IJ} \\ (BUF_{1IJ} + BUC_{1IJ})I & (BBF_{1IJ} + BBC_{1IJ})I & 0 \\ BUF_{2IJ} & 0 & BBF_{2IJ} \end{array} \right\}, \tag{6.3}$$

where UUM, $(UBF_1, BBF_1, UUF_2, UBF_2, BBF_2)$ and (UBC_1, BBC_1) represent the membrane, the bending and shear parts of the strain field, respectively and their corresponding expressions are given as indicated in Tab. 1.6.

In Eq. (6.1), λ_{cr} is the buckling load parameter and denotes the proportional increase in the load needed to reach neutral equilibrium. By applying to the structure a reference level of loading, N_{ref}, and carrying out a standard linear static analysis, the membrane stresses in the elements will be obtained. Hence, the buckling load can be written as:

$$N_{cr} = \lambda_{cr} N_{ref}. \tag{6.4}$$

6.2.2 POST-BUCKLING EQUILIBRIUM PATH RESOLUTION

The post-buckling equilibrium path for FG and nanocomposite shell structure is obtained via the resolution of the following global non-linear system equation:

$$K_T \Delta U = F_n. \tag{6.5}$$

where K_T is the global tangent stiffness matrix of the structure and F_n represents the external load vector. The global stiffness of the shell structure which regroups the material and geometrical contributions of the tangent operator can be expressed as follows:

$$K_T = K_M + K_G. \tag{6.6}$$

The matrices K_M and K_G are as defined previously. Newton-Raphson with arc length control solution strategy is used to resolve the equilibrium path responses as indicated in Fig. 3.34, Chapter 3.

6.3 BUCKLING OF FG-CNTRC PLATES AND SHELLS

6.3.1 MECHANICAL BUCKLING BEHAVIOR OF FG-CNTRC PLATES

Here, a functionally graded carbon-nanotubes reinforced composite (FG-CNTRC) plate is considered. This plate is composed by an isotropic matrix of Poly(m-phenylenevinylene)-co-[(2,5-dioctoxy-p-phenylene) vinylene], referred as (PmPV) and a single-walled carbon-nanotubes (SWCNTs) of type armchair (10,10). The material properties of both matrix and SWCNTs are given in Tab. 4.2 at room temperature ($300K$). Uniform (UD) and three graded distributions of CNTs are considered: FG-V, FG-O and FG-X in order to examine the effect of their profiles as well as their volume fraction V_{CNT}^* on mechanical buckling behavior of FG-CNTRC plates. For that, three ratios of CNTs volume fractions are used ($V_{CNT}^* = 0.11, 0.14, 0.17$) and their corresponding efficiency parameters are listed in Tab. 4.3. The geometry of the plate with uniform distribution of the CNTs is depicted in Fig. 6.1.

First, we provide the effect of CNTs profile distributions on non-dimensional critical buckling load $\overline{N}_{cr}^{CNT} = N_{cr}a^2/E_m h^3$ of a simply supported (SSSS) FG-CNTRC square plate ($a = b$) subjected to different compressive loadings: (a) uniaxial compression load in x direction, (b) uniaxial compression load in y direction and (c) biaxial compression load. As well as, various volume fractions of CNTs are considered which range from 0.11 to 0.17. For the sake of comparison, the thickness-to-length ratio $h/a = 0.01$ and the two types of loading cases (a) and (c) are considered. Tab. 6.1 and

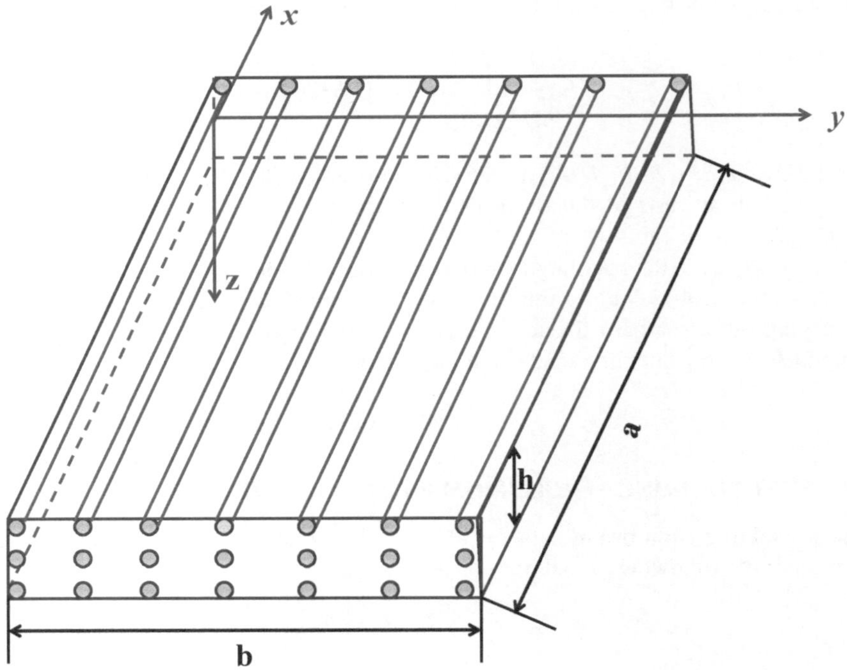

Figure 6.1 FG-CNTRC plate with uniform distribution of CNTs.

Tab. 6.3 shows the obtained values of non-dimensional critical buckling load $\overline{N}_{cr}^{CNT}$ of the square FG-CNTRC plate under the two loading types (a) and (c) for various profiles and volume fractions of CNTs which are compared to those of Zhang et al. [324]. In fact, Zhang et al. [324] investigate the buckling behavior of thick FG-CNTRC skew plate using an element free-approach and the comparison is carried out here for a skew angle equal to 0° in order to verify the accuracy and efficiency of the present double directors finite shell model in the prediction of critical buckling behavior of FG-CNTRC square plates. Secondly, we provide the critical buckling load results for the simply supported FG-CNTRC square plate subjected to in-plane loading in the y direction under various forms of CNTs distributions: UD, FG-V, FG-O and FG-X and different volume fractions ratios ($V_{CNT}^* = 0.11, 0.14, 0.17$) in order to examine the effect of loading type on the buckling behavior of such nanostructures (see Tab. 6.2). From Fig. 6.2, it can be observed that the variation of the CNT volume fraction V_{CNT}^* from 0.11 to 0.17 induces an increase in the critical buckling parameter and this can be explained by the increase of the stiffness of the structure when it is more reinforced by carbon nanotubes. Besides, it can be seen that FG-X-CNTRC plate has the highest value of the critical buckling parameter while FG-O-CNTRC plate has the lowest one. This observation is similar to those for loading types (a) and (c) illustrated in Fig. 6.2, which affirms that the reinforcements distributed near the top and bottom surfaces are more efficient than those distributed near the mid-plane surface due to their increasing of the stiffness of the FG-CNTRC plate. Furthermore, when comparing the influence of loading type on critical buckling load, it can be concluded that the effect of uniaxial compression load in x direction is more prominent than those in y direction or in both x and y directions as shown in Fig. 6.2. In fact, the FG-CNTRC plate is more stiffer in x direction because the carbon-nanotubes are uniaxially aligned in this direction which implies a high stiffness of the FG-CNTRC plate in x direction compared to others directions.

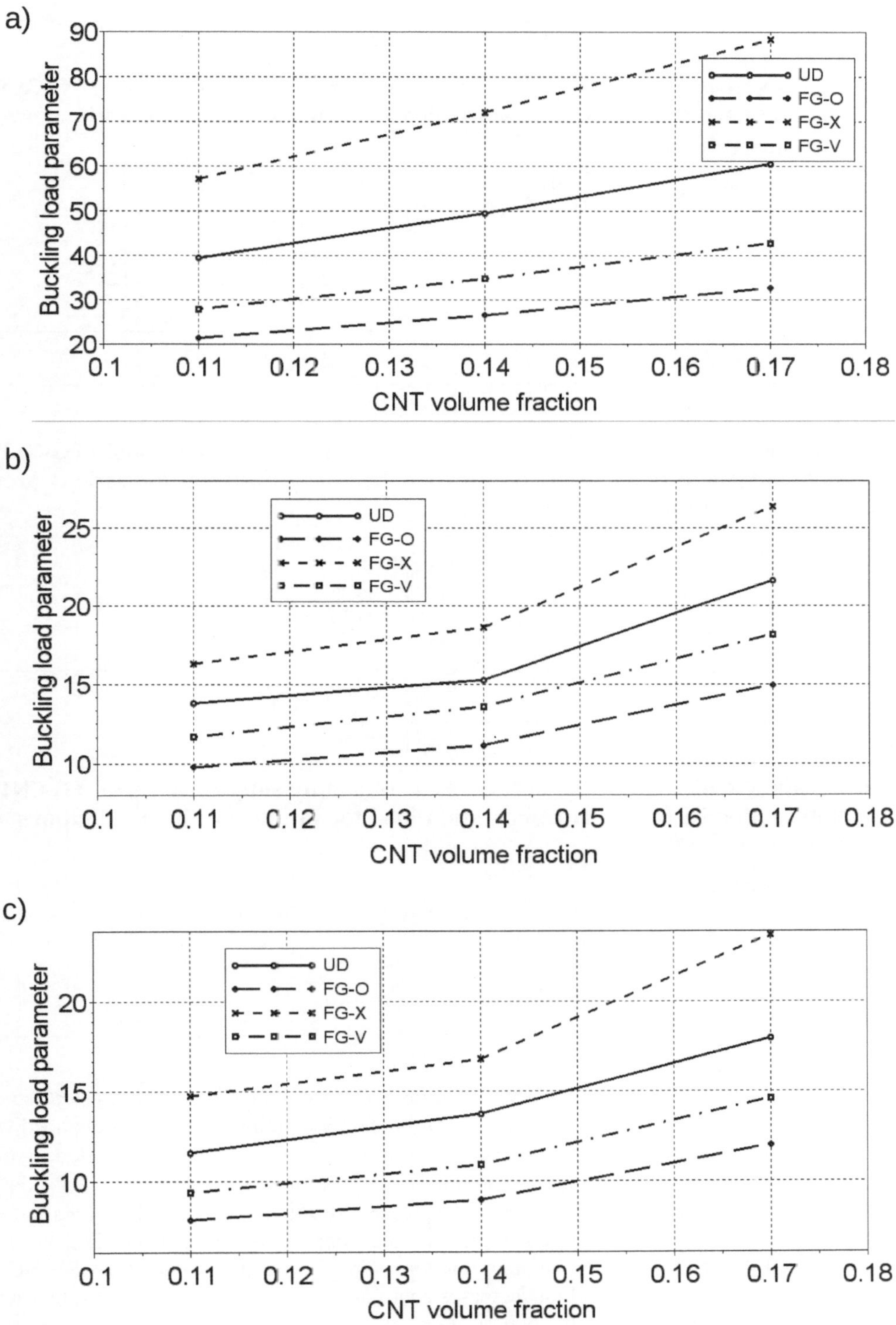

Figure 6.2 Effect of CNT volume fractions on buckling load parameter of simply supported FG-CNTRC square plate under (a) uniaxial in x direction, (b) uniaxial in y direction and (c) biaxial compression loads.

Table 6.1
Critical buckling load factor $\overline{N}_{cr}^{CNT} = N_{cr}a^2/E_mh^3$ of simply supported (SSSS) FG-CNTRC square plate subjected to uniaxial compression load in x direction for various profile distributions and volume fractions of CNTs.

CNT profile	V_{CNT}^*	0.11		0.14		0.17	
		Present	Ref.[324]	Present	Ref.[324]	Present	Ref.[324]
UD		39.45	37.85	49.52	49.14	60.62	57.54
FG-V		27.98	-	34.80	-	42.79	-
FG-O		21.50	21.32	26.60	26.39	32.76	31.22
FG-X		57.23	56.71	72.17	71.76	88.28	83.73

Table 6.2
Critical buckling load factor $\overline{N}_{cr}^{CNT} = N_{cr}a^2/E_mh^3$ of simply supported (SSSS) FG-CNTRC square plate subjected to uniaxial compression load in y direction for various profile distributions and volume fractions of CNTs.

CNT profile	V_{CNT}^*	0.11	0.14	0.17
UD		13.80	15.29	21.65
FG-V		11.69	13.58	18.18
FG-O		9.79	11.16	14.96
FG-X		16.34	18.61	26.38

Table 6.3
Critical buckling load factor $\overline{N}_{cr}^{CNT} = N_{cr}a^2/E_mh^3$ of simply supported (SSSS) FG-CNTRC square plate subjected to biaxial compression load for various profile distributions and volume fractions of CNTs.

CNT profile	V_{CNT}^*	0.11		0.14		0.17	
		Present	Ref.[324]	Present	Ref.[324]	Present	Ref.[324]
UD		11.57	11.45	13.72	13.37	17.95	17.82
FG-V		9.37	-	10.89	-	14.58	-
FG-O		7.84	7.73	8.94	8.81	11.98	12.09
FG-X		14.73	14.34	16.78	16.37	23.79	23.79

On the other hand, the effect of boundary conditions is also examined for the FG-CNTRC square plate with $a/b = 1$ and $h/a = 0.01$. In fact, a set of boundary conditions including clamped "C", simply supported "S" and free "F" edges are tested and the obtained results are reported in Tabs. 6.4 and 6.5. The relative error between the present results and those of Zhang et al. [324] do not exceed 5% which guarantee the effectiveness of the proposed formulation. As well as, Tabs. 6.4, 6.5 and 6.6 reveal that when the boundary condition changes from the fully clamped (CCCC) to fully simply supported (SSSS) by varying the nature of constraint in one support edge (CSCS), (SCSC) or (CCCF), the value of critical buckling decreases continuously until it reaches lower values when an edge becomes free. Hence, more the plate is constrained, more the flexural rigidity of the plate becomes higher which leads to high buckling load.

The variation of buckling load parameter versus the length-to-thickness ratio (a/h) for the simply supported FG-CNTRC square plate under three studied compression loads and CNT volume

Table 6.4

Effect of boundary conditions (CCCC, CSCS) on critical buckling load factor $\overline{N}_{cr}^{CNT} = N_{cr}a^2/E_mh^3$ for different types of FG-CNTRC square plates under uniaxial compression load in x direction and CNT volume fraction ratio ($V_{CNT}^* = 0.11$).

CNT-profile	BC	CCCC		CSCS	
		Present	Ref.[324]	Present	Ref.[324]
UD		148.08	142.94	145.44	142.94
FG-V		104.67	-	101.787	-
FG-O		79.78	77.08	77.06	77.08
FG-X		213.48	206.29	210.88	206.29

Table 6.5

Effect of boundary conditions (SCSC, CCCF) on critical buckling load factor $\overline{N}_{cr}^{CNT} = N_{cr}a^2/E_mh^3$ for different types of FG-CNTRC square plates under uniaxial compression load in x direction and CNT volume fraction ratio ($V_{CNT}^* = 0.11$).

CNT-profile	BC	SCSC		CCCF	
		Present	Ref.[324]	Present	Ref.[324]
UD		43.10	42.77	144.19	137.61
FG-V		-	31.78	-	100.32
FG-O		25.12	24.84	75.55	72.59
FG-X		60.97	60.62	209.85	199.94

Table 6.6

Effect of boundary conditions (SSSS, SSSF, SFSF) on critical buckling load factor $\overline{N}_{cr}^{CNT} = N_{cr}a^2/E_mh^3$ for different types of FG-CNTRC square plates under uniaxial compression load in x direction and CNT volume fraction ratio ($V_{CNT}^* = 0.11$).

CNT-profile	BC	SSSS		SSSF		SFSF	
		Present	Ref.[324]	Present	Ref.[324]	Present	Ref.[324]
UD		39.45	39.11	37.12	38.04	36.71	37.39
FG-V		27.98	-	25.55	-	25.14	-
FG-O		21.50	21.35	19.25	20.24	18.86	19.65
FG-X		57.23	56.74	54.79	55.63	54.36	54.92

fraction ($V_{CNT}^* = 0.11$) for various CNT profile distributions is shown in Figs. 6 3 (a), 6.2 (b) and 6.2 (c), respectively. The results in Fig. 6.3 indicate that as the length-to-thickness ratio of the plate increases, the buckling load parameters also increase. Hence, when the length-to-thickness ratio increases and reaches 50, the FG-CNTRC structure behaves as thin plate and the effect of CNT profile distributions becomes prominent whereas when the FG-CNTRC structure behaves as thick plate (for $a/h = 10$), this effect becomes weaker. This observation reflects the influence of CNT reinforcements in the enhancement of the stiffness of the structure through a thin FG-CNTRC plate more than thick FG-CNTRC plate and this is for all types of loadings.

Effects of loading types on buckling load parameters of various types of simply supported FG-CNRTC square plates by varying the length-to-thickness ratio and under CNT volume fraction ($V_{CNT}^* = 0.11$) are shown in Fig. 6.4. It can be observed that, the compressive load in x direction

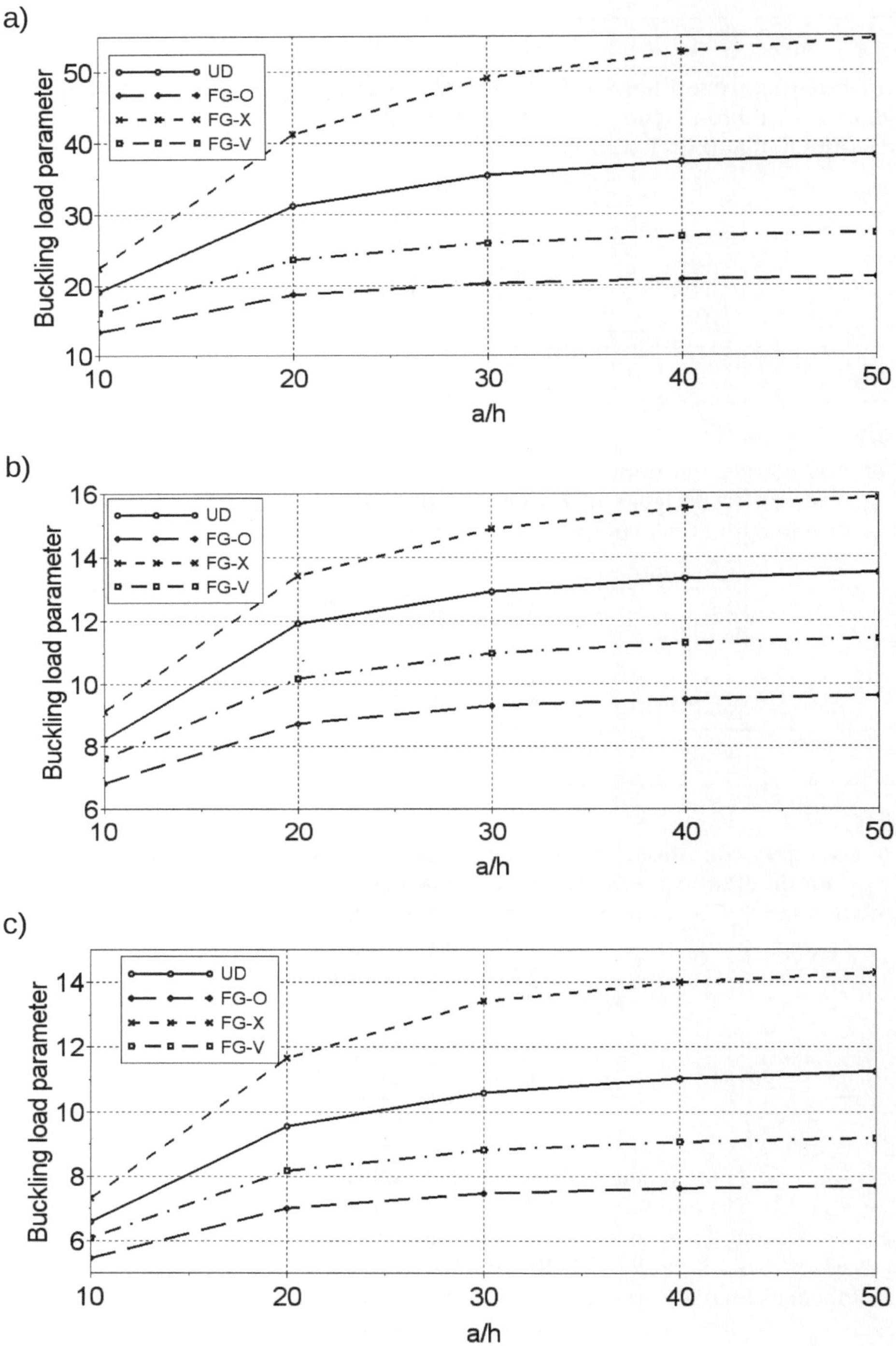

Figure 6.3 Buckling load parameter versus the length-to-thickness ratio (a/h) of simply supported FG-CNTRC square plate under (a) uniaxial in x direction, (b) uniaxial in y direction and (c) biaxial compression loads and a CNT volume fraction $V_{CNT}^{*} = 0.11$.

has the most influence on critical buckling load parameter for various types of CNT distributions compared to others types of loadings. As well as, the effect of loadings is more pronounced for FG-X-CNTRC plate compared to others CNT distributions which manifest by highest values of buckling load parameter for both low ($a/h = 10$) and high ($a/h = 50$) values of length-to-thickness ratios. This affirms that the CNT distributions close to top and bottom surfaces are more efficient in the enhancement of the stiffness of the structure than those near the mid-surface of the plate. In addition, in x direction, where the CNT reinforcements are aligned, the FG-CNTRC plate is more stiffer which induces hence the high values of buckling load parameter observed in this direction.

6.3.2 MECHANICAL BUCKLING BEHAVIOR OF FG-CNTRC PANELS

Extending the previous analysis to FG-CNTRC cylindrical panels, we provide here a new results concerning mechanical buckling behavior of FG-CNTRC cylindrical panels. In fact, according to the existing bibliography, the only reference in mechanical buckling analysis for FG-CNTRC cylindrical panels is provided by Gracia-Macias et al. [85]. However, in this study, the used model is based on the FSDT theory which requires the introduction of the transverse shear correction factors. Therefore, to perform such analysis with high level of accuracy, the proposed high-order model based on double directors finite shell element is applied here. The geometrical characteristics of the cylindrical panel are the same as mentioned for the FGM case in section 3.3.3.2, Chapter 3. The CNTs are uniaxially aligned in the axial direction and functionally graded in the thickness direction of the panel. UD and four types of CNTs are considered: FG-V, FG-A, FG-O and FG-X. Their mechanical properties, as well as, their corresponding efficiency parameters are as reported earlier in Table 5.1 and Table 4.19, respectively. Figure 6.5 shows the variation of the non-dimensional critical buckling load $\overline{N}_{cr}^{CNT} = N_{cr}L^2/E_m h^3$ versus the CNT volume fraction of the different profiles of the simply supported (SSSS) cylindrical panel subjected to uniaxial compression in x direction, uniaxial compression in y direction and biaxial compression loads, respectively. It is clear that, the FG-X form induces the highest value of critical buckling load while the FG-O form has the lowest one and this is for all loading types. As well as, the buckling load intensity increases as the CNT volume fraction rises which implies an increase in the stiffness of the panel. Hence, it is concluded that reinforcements distributed close to the top and bottom surfaces induce higher stiffness values. Figures 6.6(a), 6.6(b), 6.6(c), 6.6(d) and 6.6(e) demonstrate the effects of boundary conditions on critical buckling load of UD, FG-V, FG-A, FG-O and FG-X-CNTRC cylindrical panels subjected to uniaxial compression load and with a CNT volume fraction ($V_{CNT}^* = 0.11$). Different combinations of boundary conditions are examined where (C), (S) and (F) refer to clamped, simply supported and free edge, respectively. For all CNT distributions, it can be remarked that fully clamped (CCCC) or clamped in three edges and free in one edge (CCCF) FG-CNTRC cylindrical panels induces higher values of critical buckling load compared to other boundary conditions. In fact, more an edge is clamped, more constraints are imposed which induces high values of the stiffness of the FG-CNTRC cylindrical panel. On the other hand, the effect of boundary conditions is more significant for a length-to-thickness ratio ($L/h \succ 50$) which implies that more the FG-CNTRC cylindrical panel becomes larger, more it becomes sensitive to the change of boundary conditions.

Figure 6.7 shows that the critical buckling load decreases as the radius-to-thickness ratio increases for all CNT distributions of the simply supported cylindrical panel under different CNT volume fractions and subjected to uniaxial compression load along its curved edges which reflects that the variation of this geometrical parameter affects the mechanical buckling behavior of the FG-CNTRC cylindrical panel. Furthermore, the effect of variation of the span angle is also examined in Fig. 6.8. Indeed, it can be seen that for small values of span angle less than 0.2, a maximum value of the critical buckling load is reached. Beyond this value, the critical buckling load decreases and becomes then constant. This can be explained by the effect of curvature which becomes insignificant for high values of span angle and the cylindrical panel hence becomes flat.

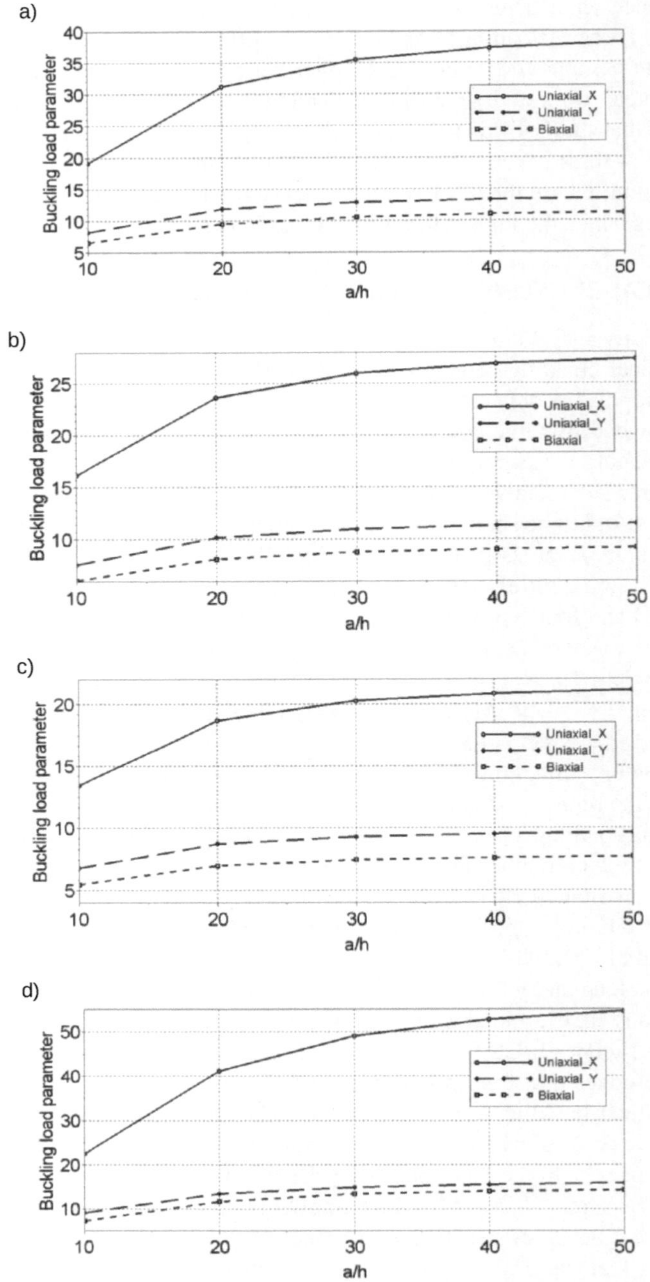

Figure 6.4 Buckling load parameter versus the length-to-thickness ratio (a/h) of simply supported (a) UD, (b) FG-V, (c) FG-O and (d) FG-X CNTRC square plates under uniaxial compression load in x direction and CNT volume fraction ($V^*_{CNT} = 0.11$).

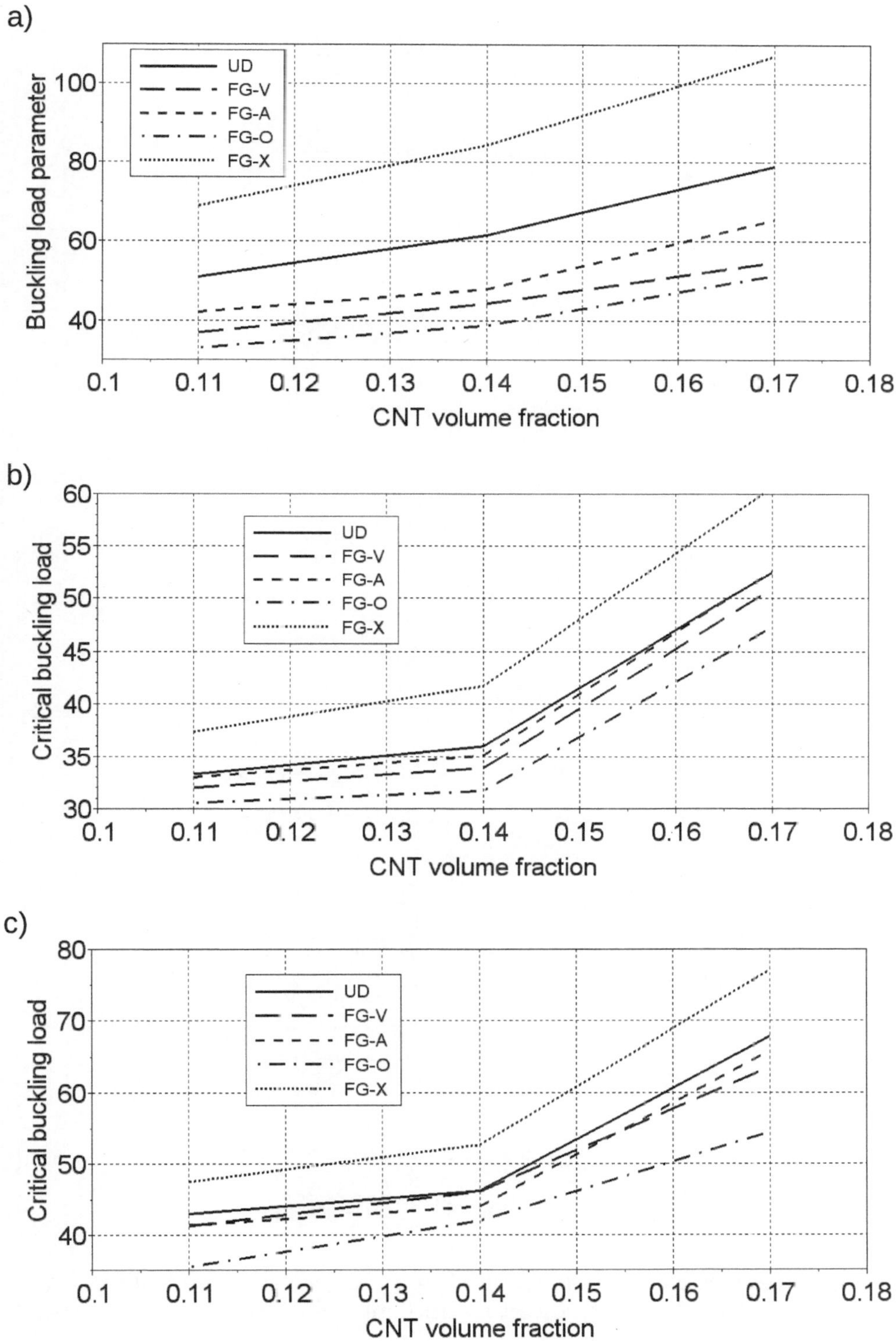

Figure 6.5 Effect of CNT volume fractions on critical buckling load parameter of the simply supported (SSSS) FG-CNTRC cylindrical panel under (a) uniaxial in x direction, (b) uniaxial in y direction and (c) biaxial compression loads.

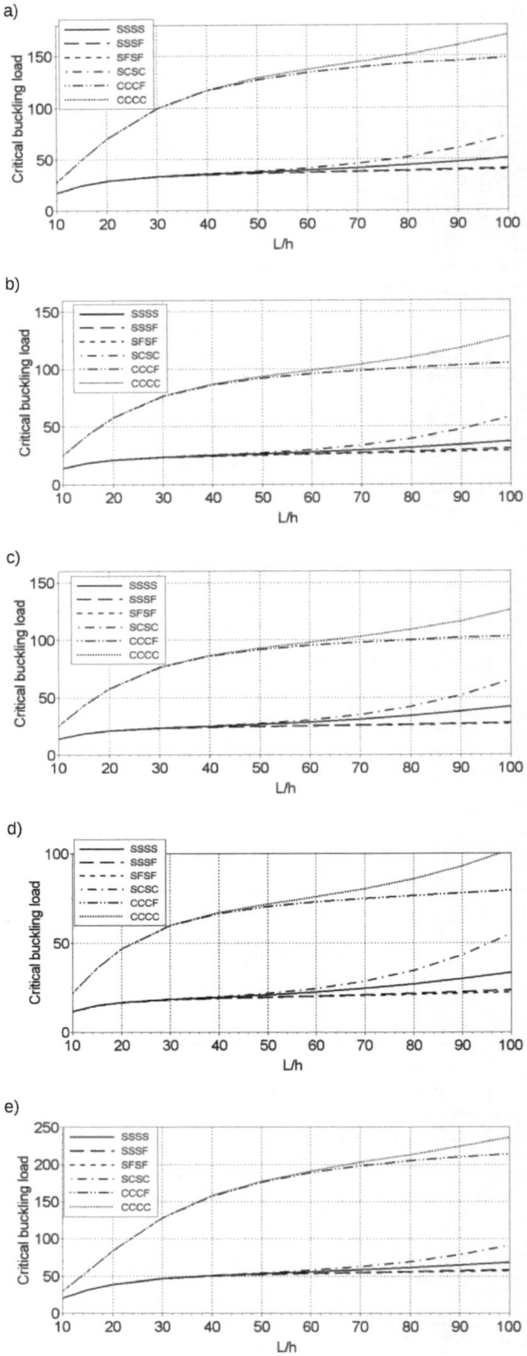

Figure 6.6 Effects of boundary conditions on critical buckling load parameter of (a) UD, (b) FG-V, (c) FG-A, (d) FG-O and (e) FG-X CNTRC cylindrical panels under uniaxial compression load along its curved edges and with CNT volume fraction ($V^*_{CNT} = 0.11$).

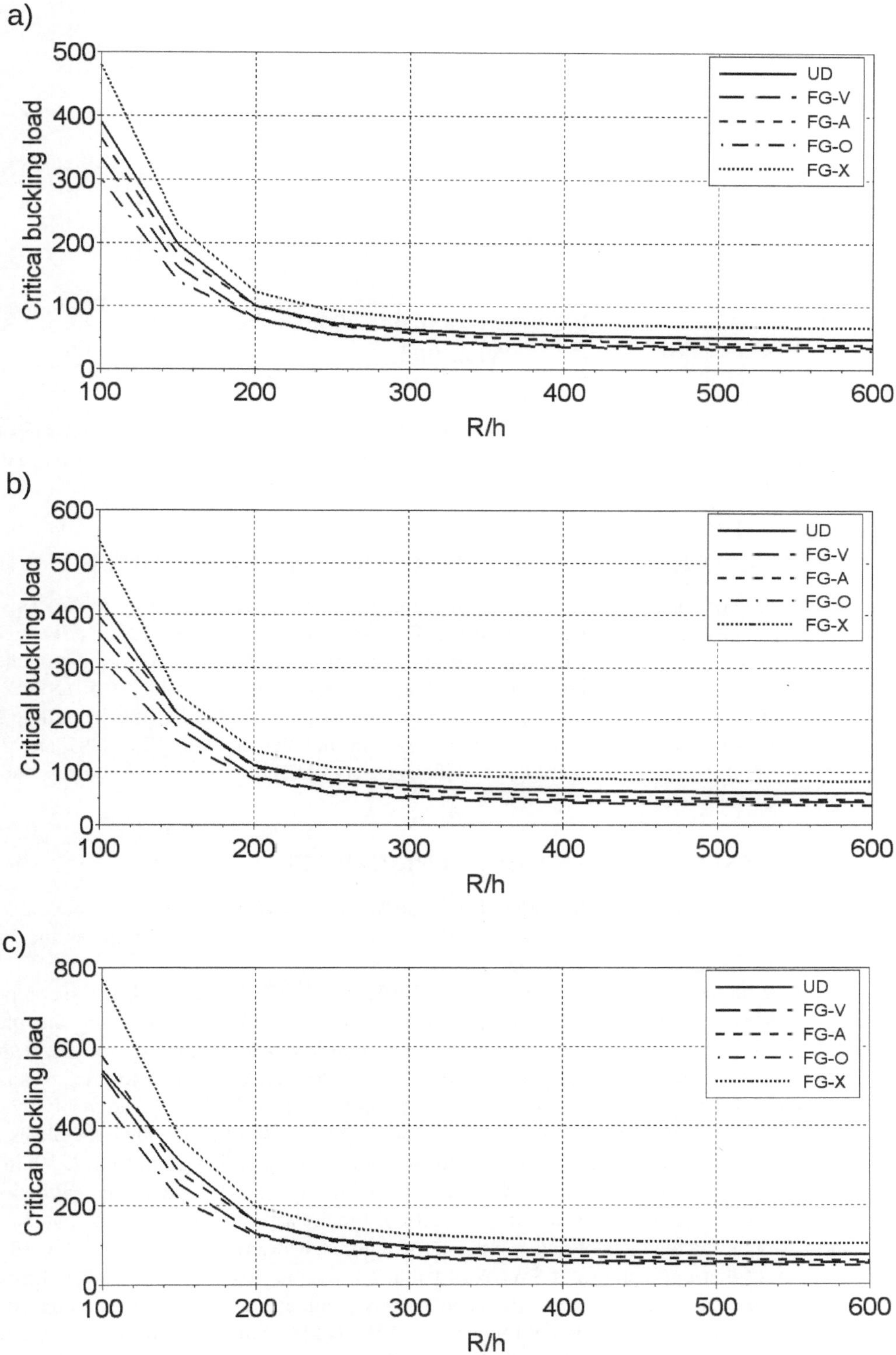

Figure 6.7 Variation of the critical buckling load parameter versus the radius-to-thickness ratio R/h of the simply supported (SSSS) FG-CNTRC cylindrical panel under (a) 0.11, (b) 0.14 and (c) 0.17 CNTs volume fractions and subjected to uniaxial compression load along its curved edges.

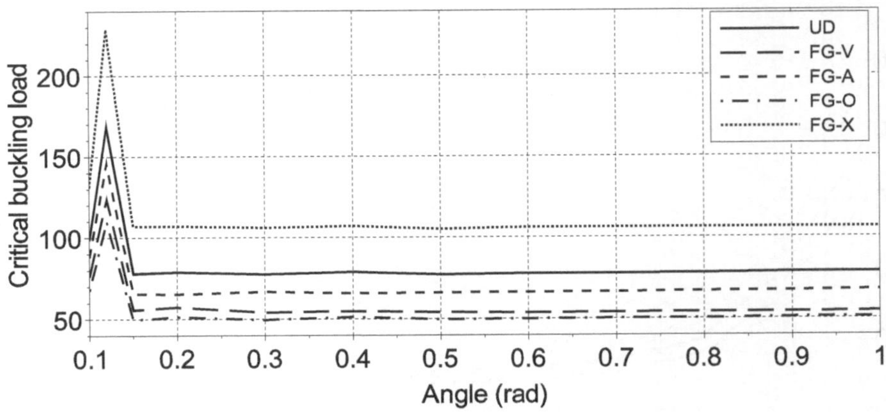

Figure 6.8 Variation of the critical buckling load parameter versus the span angle of the simply supported (SSSS) FG-CNTRC cylindrical panel under uniaxial compression load along its curved edges and with a CNT volume fraction ($V_{CNT}^{*} = 0.17$).

6.4 POST-BUCKLING RESPONSES OF FG-CNTRC PLATES AND SHELLS

In this section, the post-buckling responses of nanocomposite plates reinforced by SWCNTs subjected to different mechanical loads are provided. For that, four types of SWCNTs (UD, FG-V, FG-O and FG-X), with aligned distribution along the x direction and functionally graduation along the z direction, are considered as shown in Fig. 5.2. The material properties of the (10,10) SWCNTs and the PMMA (Poly (methyl methacrylate)) isotropic matrix are reported in Tab. 5.1. Three sets of volume fractions of SWCNTs are taken into account in the analysis with 12%, 17% and 28% percentages for reinforcements. The SWCNTs efficiency coefficients η_i (i = 1,2,3) corresponding to these rates are listed in Tab. 4.19.

6.4.1 POST-BUCKLING RESPONSES OF FG-CNTRC PLATES

As indicated earlier, the post-buckling equilibrium paths of FG-CNTRC plates are performed in the function of non-dimensional quantities where the non-dimensional load factor has the same expression given in section 3.4.1.2, Chapter 3.

Figures 6.9, 6.10 and 6.11 illustrate the effects of different SWCNTs volume fractions on post-buckling responses of simply supported square (UD, FG-V, FG-O and FG-X) nanocomposite plates subjected to uniaxial in-x direction, uniaxial in-y direction and biaxial compression loads, respectively. The nanocomposite plate is assumed to be square ($a/b = 1$) with length-to-thickness ratio ($a/h = 100$) and all its edges are simply supported as reported in [319]. In order to verify the capacity of the present model in the prediction of post-buckling behavior of nanocomposite plates, the present results are compared to those available in the literature given by Zhang and Liew [319] in the case of uniaxial in x direction compression load as depicted in Fig. 6.9. A good agreement between the results is revealed which outlines again the ability of the present model to furnish accurate solutions for post-buckling analysis of nanocomposite structures. Moreover, from Figs. 6.9, 6.10 and 6.11, it is apparent that the increase of SWCNTs volume fractions (from 0.12 to 0.28) induces an increase in the post-buckling strength of the nanocomposite plates due to the enhancement of the stiffness of the structure by the addition of high rates of SWCNTs and this is for all types of mechanical loadings. For the sake of clarity, it should be noted that in all illustrations, the symbol V denotes the SWCNTs volume fraction (V_{CNT}^{*}). Figure 6.12 traces the effect of different profiles of SWCNTs

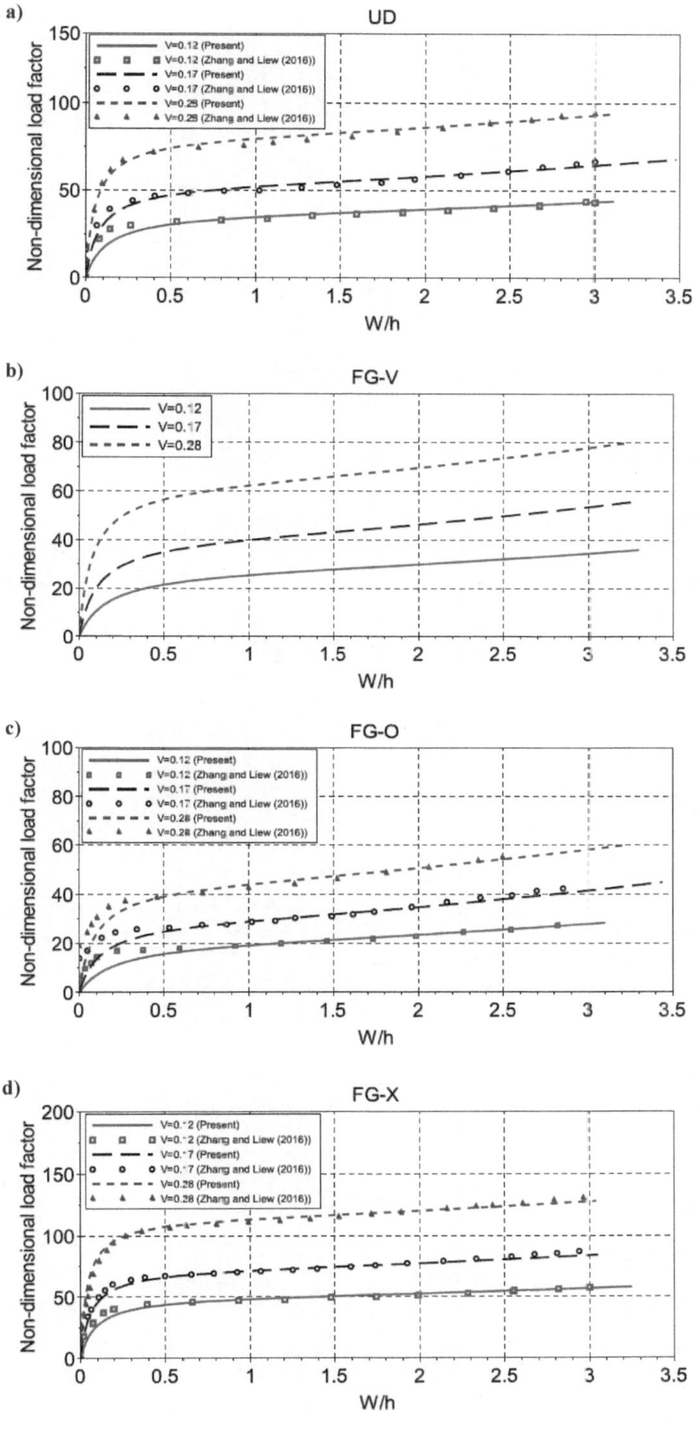

Figure 6.9 The influence of SWCNTs volume fractions on post-buckling responses of (a) UD, (b) FG-V, (c) FG-O and (d) FG-X nanocomposite plates subjected to uniaxial in x direction compression load (SSSS, $a/h = 100$).

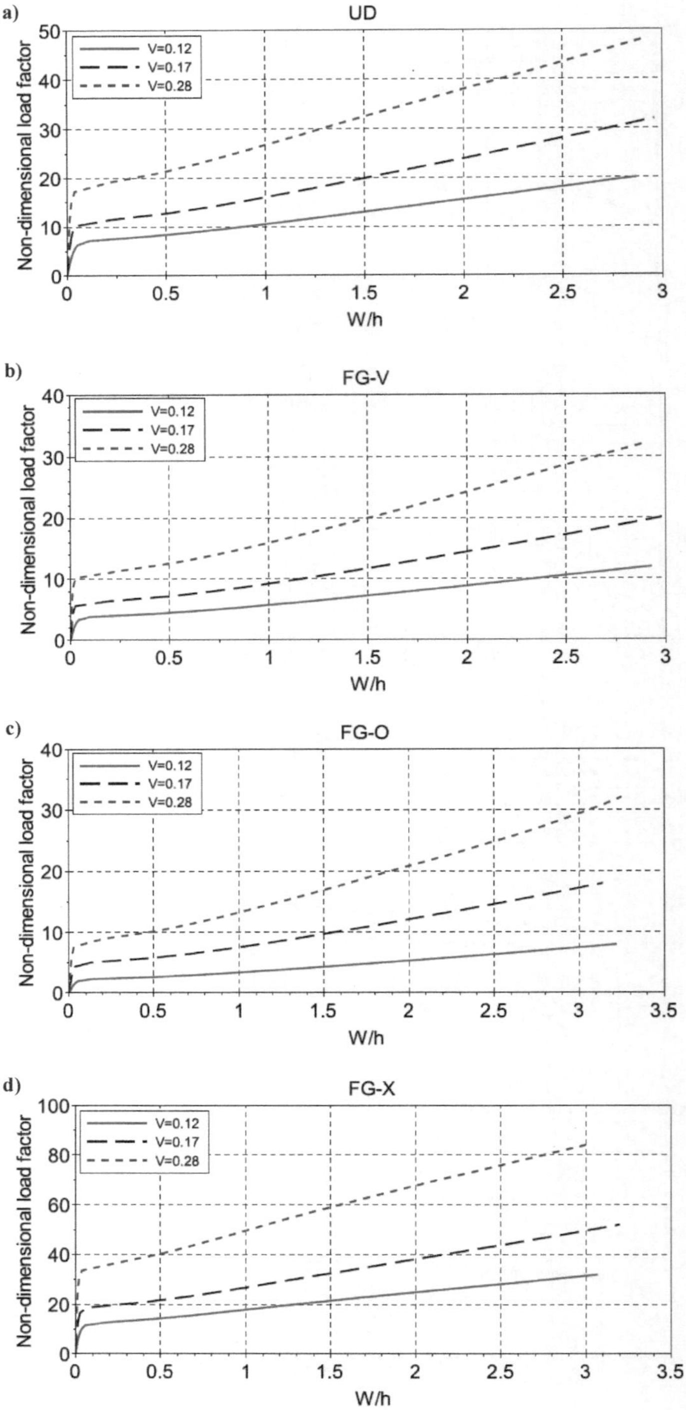

Figure 6.10 The influence of SWCNTs volume fractions on post-buckling responses of (a) UD, (b) FG-V, (c) FG-O and (d) FG-X nanocomposite plates subjected to uniaxial in y direction compression load (SSSS, $a/h = 100$).

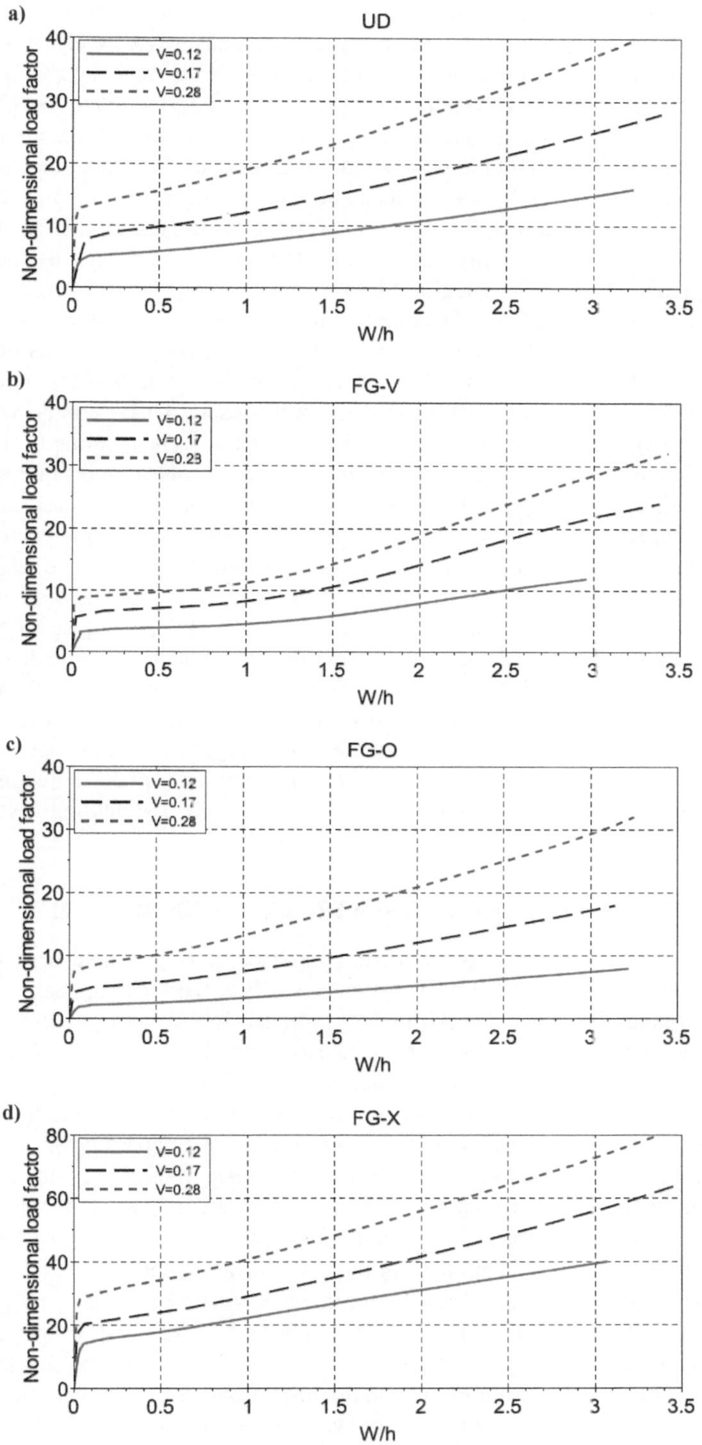

Figure 6.11 The influence of SWCNTs volume fractions on post-buckling responses of (a) UD, (b) FG-V, (c) FG-O and (d) FG-X nanocomposite plates subjected to biaxial compression load (SSSS, $a/h = 100$).

on post-buckling responses of simply supported (SSSS) square nanocomposite plates under uniaxial in x direction, uniaxial in y direction and biaxial compression loads, respectively. It is clear that the FG-X form yields the minimum post-buckling deflection response whereas the FG-O form induces the maximum one. The others forms (UD and FG-V) are setting intermediately between these two extreme cases. Thus, the SWCNTs distributions close to top and bottom surfaces are more efficient in the enhancement of the stiffness and the post-buckling strength of the nanocomposite plates than those distributed near the mid-plane. Moreover, it can be seen that the post-buckling strength of FG-CNTRC plate subjected to uniaxial in x direction compression load is higher than the one subjected to uniaxial in y direction or biaxial compression loads. This can be explained by the fact that the SWCNTs reinforcements which are aligned in the axial direction of the structure contribute in the rise of the flexural rigidity of the FG-CNTRC plate which becomes more stiffer. Therefore, more the FG-CNTRC plate is reinforced by SWCNTs, more its post-buckling strength is improved.

Further remarks can be deduced from Figs. 6.13(a), 6.13(b) and 6.13(c) where the influence of applied boundary conditions on post-buckling responses of FG-X nanocomposite plate ($a/h = 100, V_{CNT}^* = 0.28$) subjected to different mechanical loadings is examined. The FG-X form is considered since, according to the previous results, this type of distribution induces the stiffest mechanical post-buckling equilibrium path. Four types of applied boundary conditions with different combinations of simply supported (S) and clamped (C) edges are considered (CCCC, SSSS, CCSS, CSSS). It can be seen that nanocomposite plates with simply supported edges exhibits more deflection than the ones with clamped edges. Indeed, more the nanocomposite plate is constrained, more the post-buckling strength of the plate is enhanced. Figure 6.14 depicts the post-buckling equilibrium path of a simply supported (SSSS) FG-X nanocomposite plate ($V_{CNT}^* = 0.28$) with different plate aspect ratios (b/a) and under different mechanical loads. For all applied mechanical loadings, the post-buckling responses of nanocomposite plates decrease as the plate aspect ratio increases. Indeed, for high values of aspect ratio, the plate becomes more larger and the intensity of the applied load becomes more pronounced which enhances the nanocomposite plate deflection. Therefore, the variation of the plate aspect ratio has a significant effect on post-buckling behavior of such structures.

6.4.2 POST-BUCKLING RESPONSES OF FG-CNTRC CYLINDRICAL PANELS

Now, the post-buckling behavior of nanocomposite cylindrical panels is investigated. The geometry of the studied panel is defined as shown in Fig. 3.45. The cylindrical panel is reinforced by single-walled carbon-nanotubes (SWCNTs) which are aligned in the axial direction and functionally graded in the thickness direction. Four CNTs distributions (UD, FG-V, FG-O and FG-X) are considered as mentioned in section 5.1.3. Their material properties, as well as the isotropic PMMA matrix, are as indicated in Tab. 5.1. Furthermore, three rates of SWCNTs volume fractions are used in the analysis (12%, 17% and 28%) and their corresponding efficiency parameters are listed in Tab. 4.19. The external axial compression loads are also assumed to be imposed only on the curved edges (AB) and (CD), respectively. The post-buckling responses are evaluated in terms of non-dimensional parameters which are central deflection $\overline{W} = \frac{W}{h}$ and load parameter $\overline{N} = NL^2/E_m h^3$. The effects of CNTs distributions on post-buckling responses of nanocomposite cylindrical panels (CCSS) under uniaxial compression load and or different CNTs volume fractions are reported in Fig. 6.15.

The geometrical characteristics of the panel are taken as follows [144]: $L = 0.2m$, $\theta_0 = 0.2rad$, $R = 1m$ and $h = 0.004m$. From Figure 6.15, it can be seen that the FG-X nanocomposite panel has the highest post-buckling strength while the FG-O has the lowest one. The post-buckling responses of UD and FG-V nanocomposite panels are located between these two extremes. This affirms that the nanocomposite cylindrical panels with top and bottom surfaces enriched by SWCNTs have a high flexural rigidity than those distributed near the mid-surface which reflects the significant effect of SWCNTs reinforcements in the enhancement of the post-buckling behavior of such curved panels. Moreover, Fig. 6.16 shows the effects of CNTs volume fractions on post-buckling responses of

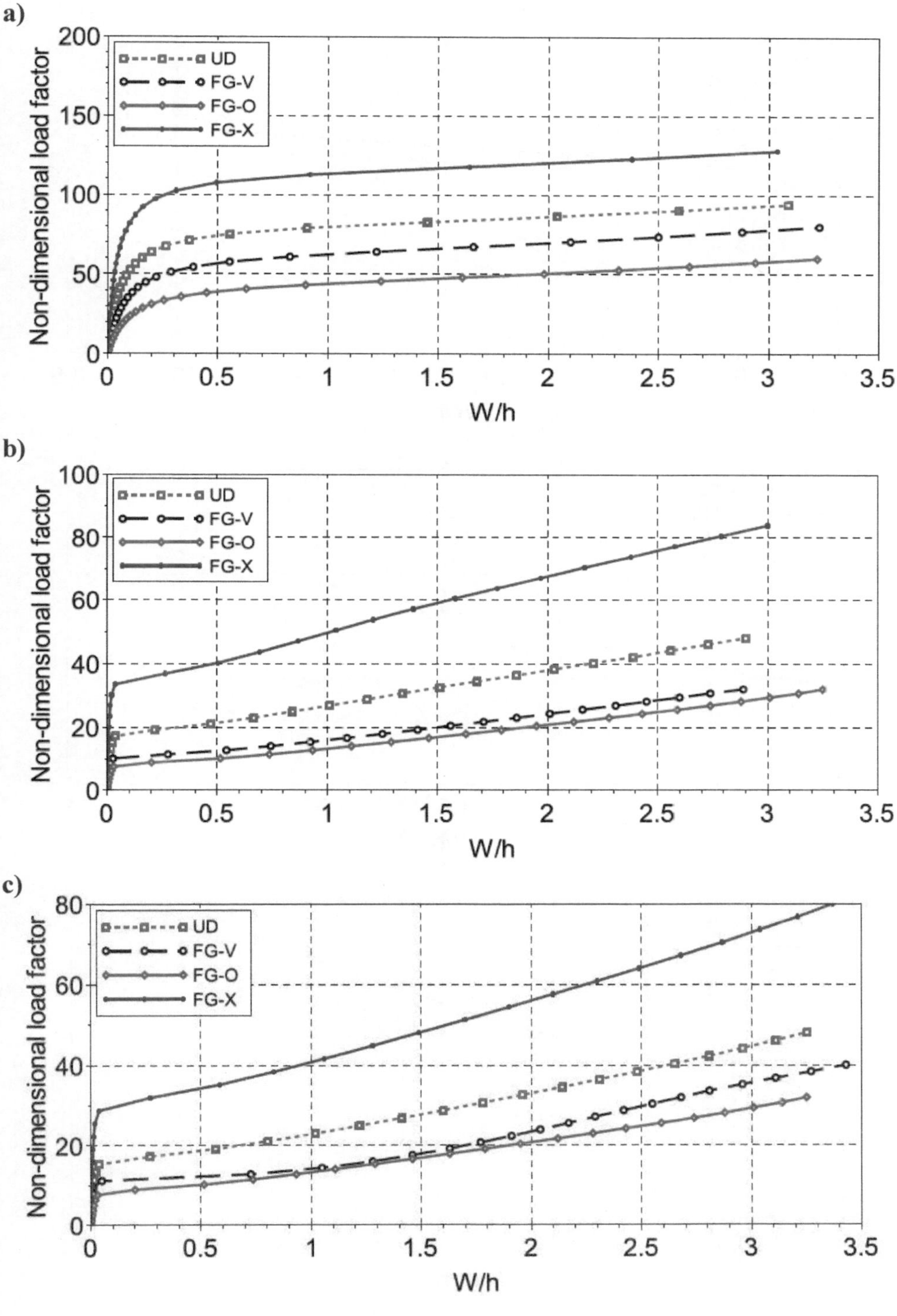

Figure 6.12 Post-buckling responses of various profiles of nanocomposite plates subjected to (a) uniaxial in x direction, (b) uniaxial in y direction and (c) biaxial compression loads with a SWCNT volume fraction ($V_{CNT}^* = 0.28$) and (SSSS, $a/h = 100$).

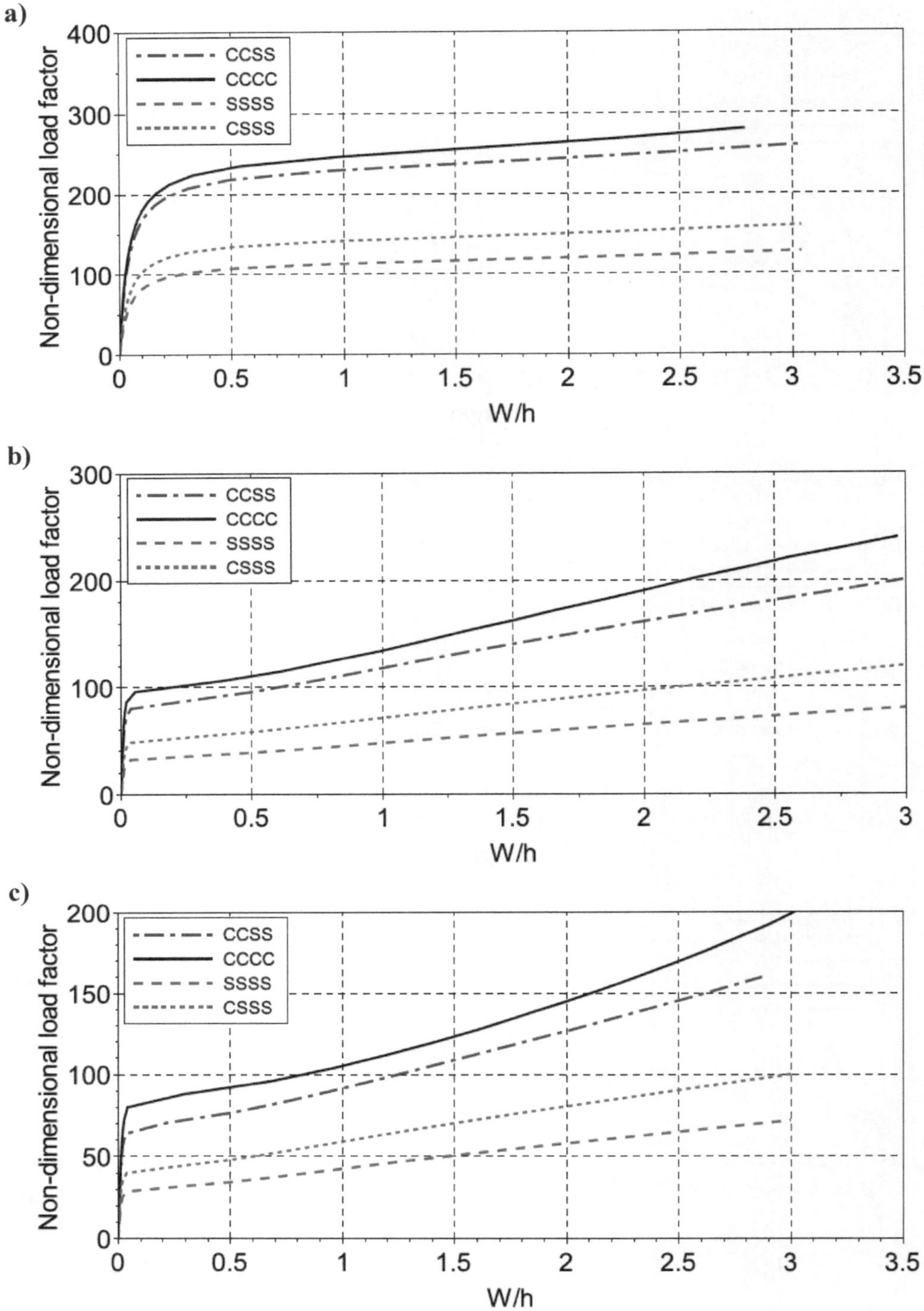

Figure 6.13 The influence of applied boundary conditions on post-buckling responses of FG-X nanocomposite plate subjected to (a) uniaxial in x direction, (b) uniaxial in y direction and (c) biaxial compression loads with a SWCNT volume fraction ($V_{CNT}^{*} = 0.28$) and ($a/h = 100$).

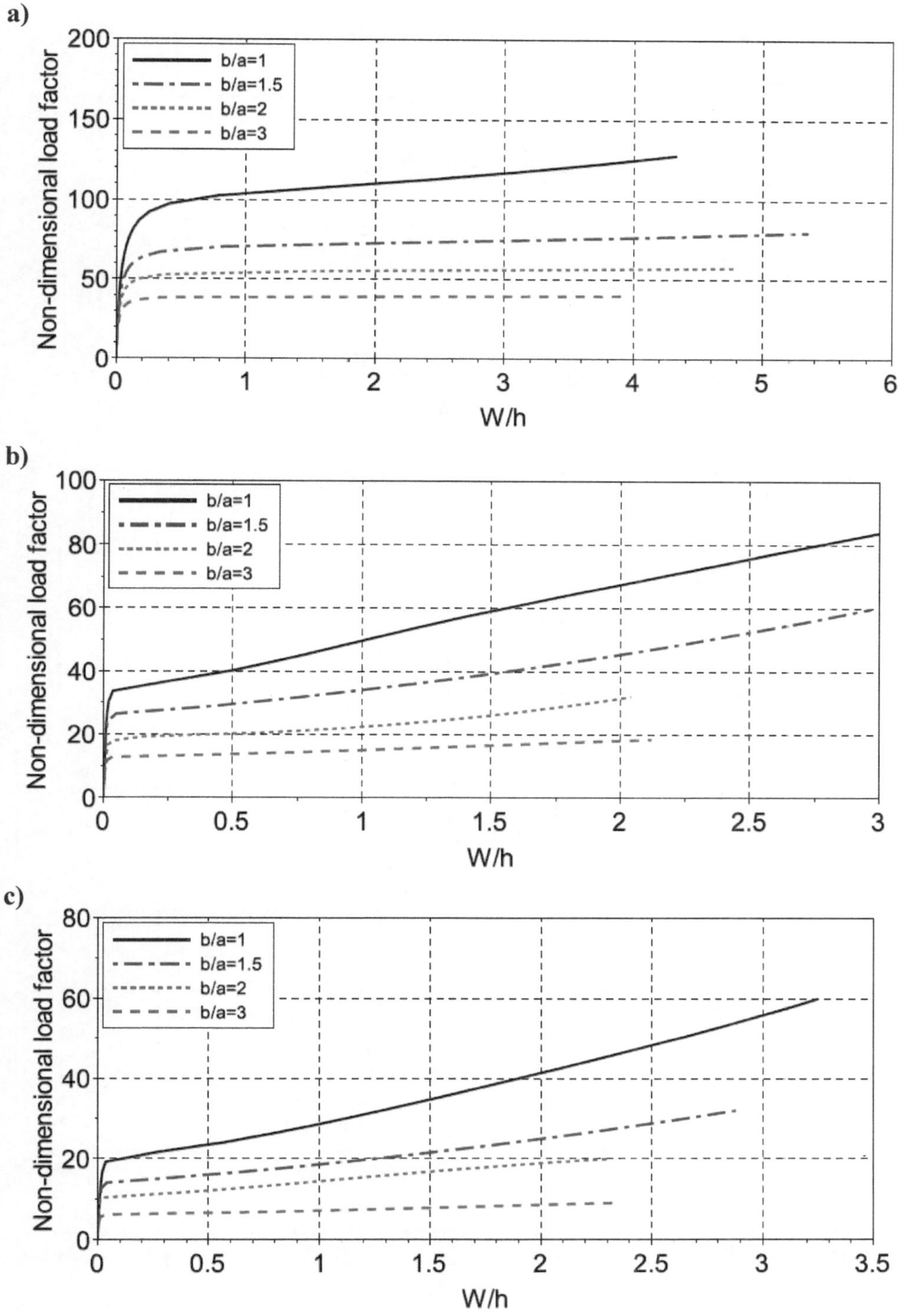

Figure 6.14 Effects of plate aspect ratios (b/a) on post-buckling responses of FG-X nanocomposite plate subjected to (a) uniaxial in x direction, (b) uniaxial in y direction and (c) biaxial compression loads with a SWCNT volume fraction ($V_{CNT}^{*} = 0.28$) and ($a/h = 100$).

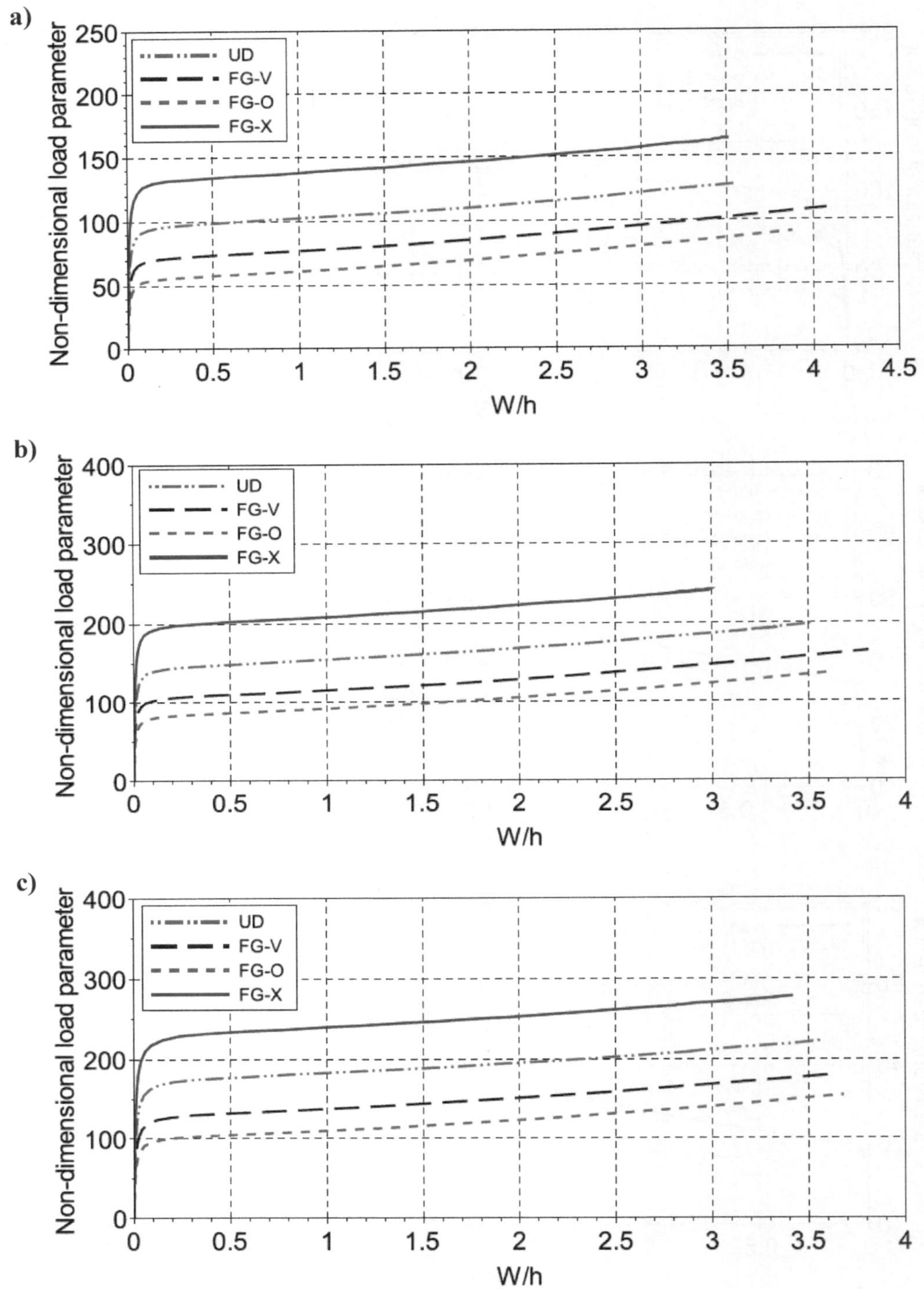

Figure 6.15 Effects of CNTs distributions on post-buckling responses of the nanocomposite cylindrical panels (CCSS) under uniaxial compression load for different CNTs volume fractions: (a) $V^*_{CNT} = 0.12$, (b) $V^*_{CNT} = 0.17$ and (c) $V^*_{CNT} = 0.28$.

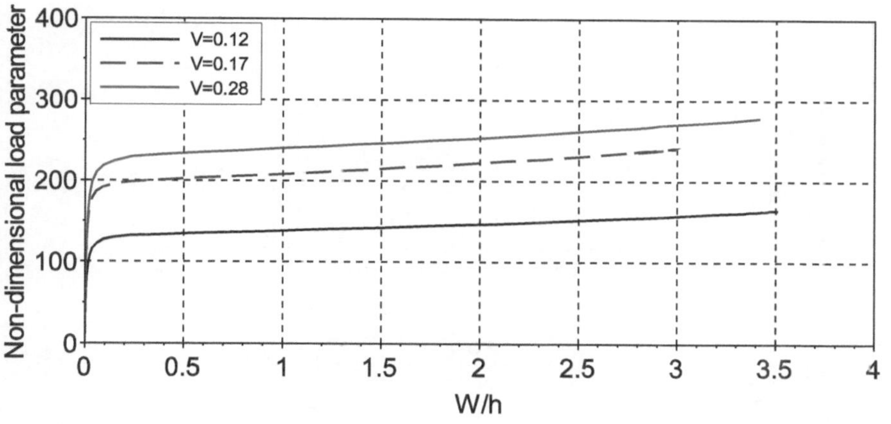

Figure 6.16 Effects of CNTs volume fractions on post-buckling responses of the FG-X nanocomposite cylindrical panel (CCSS) under uniaxial compression load.

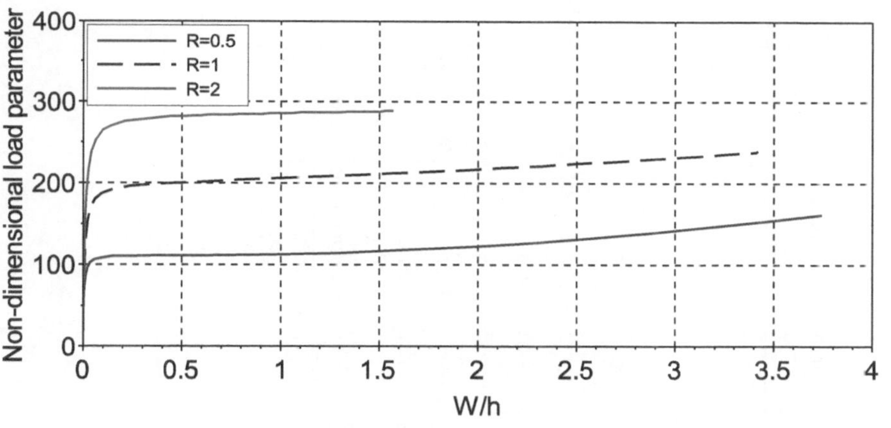

Figure 6.17 Effects of radius parameter on post-buckling responses of the FG-X nanocomposite cylindrical panel (CCSS, $V^*_{CNT} = 0.28$) under uniaxial compression load.

FG-X nanocomposite panels (CCSS) for which the increase of CNTs volume fraction (from $V^*_{CNT} = 0.12$ to $V^*_{CNT} = 0.28$) induces an increase in the post-buckling strength of the nanocomposite curved panel. This is to be expected since the augmentation of CNTs volume fraction yields to an increase of the stiffness of the panel which becomes more stiffer. The variation of the radius parameter is also examined where the post-buckling responses of the FG-X nanocomposite cylindrical panel (CCSS, $V^*_{CNT} = 0.28$) for three values of the radius parameter ($R = 0.5$, $R = 1$ and $R = 2$) are illustrated in Fig. 6.17. It can be observed that the decrease in the radius parameter induces a decrease in the post-buckling strength of the panel. Indeed, with a low value of the radius parameter ($R = 0.5$), the panel becomes more curved and it is stimulated to buckle easily since the applied compression load is more intensive on the curved edges. Subsequently, the effects of three types of applied boundary conditions (CCSS, SSSS, CCCC) on post-buckling responses of the nanocomposite panel for different CNTs distributions and with a CNT volume fraction ($V^*_{CNT} = 0.12$) are illustrated in Fig. 6.18. As can be seen, the fully clamped (CCCC) or the clamped in two sides and simply supported

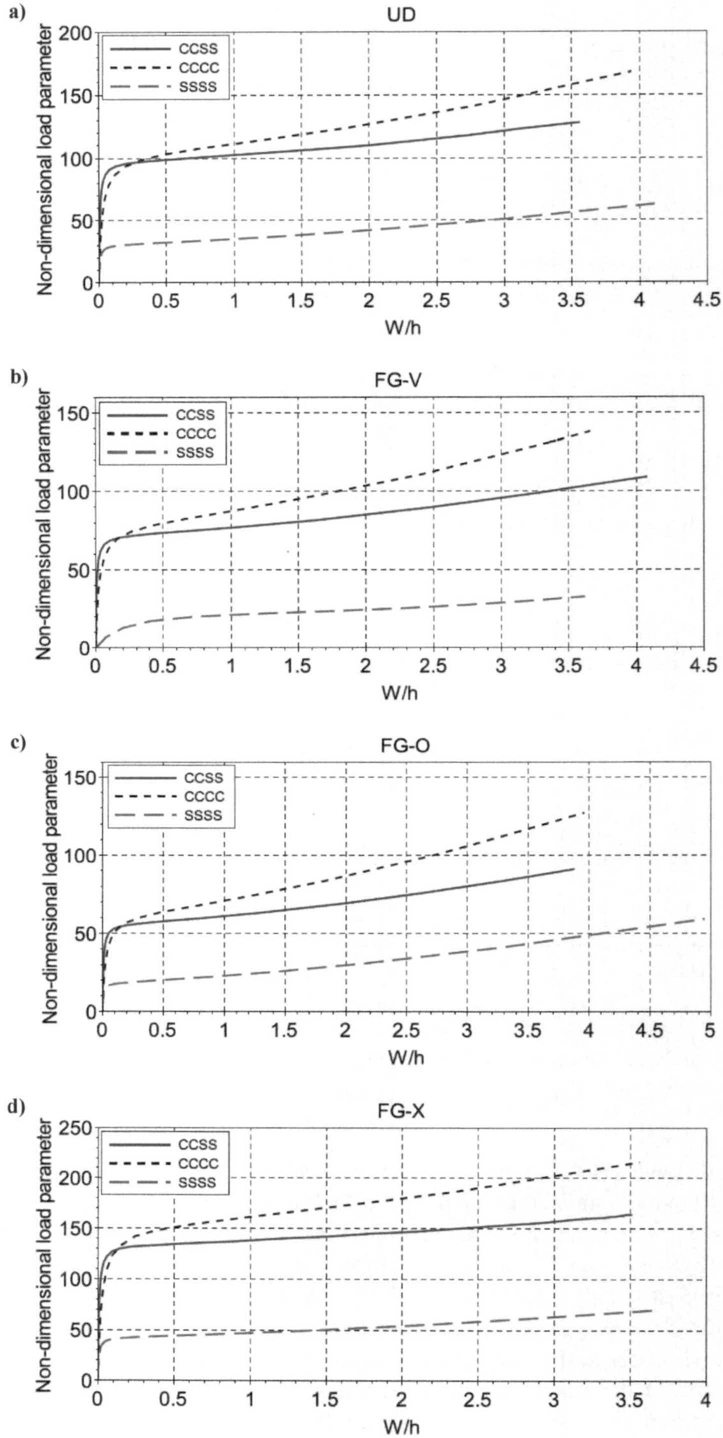

Figure 6.18 Effects of applied boundary conditions on post-buckling responses of the (a) UD, (b) FG-V, (c) FG-O and (c) FG-X nanocomposite cylindrical panels under uniaxial compression load and with a CNT volume fraction ($V^*_{CNT} = 0.12$).

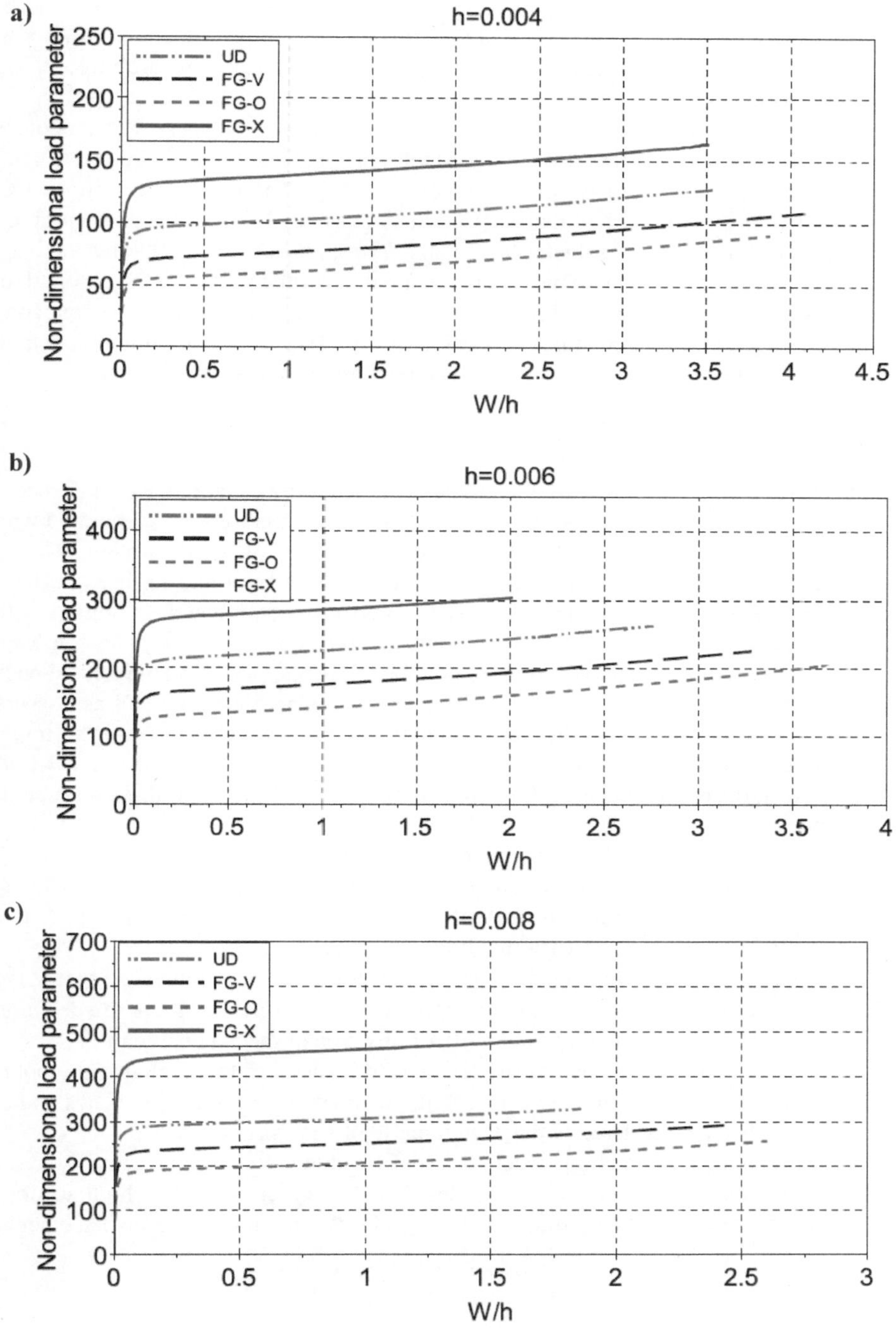

Figure 6.19 Effect of thickness parameter on post-buckling responses of the nanocomposite cylindrical panels (CCSS, $V^*_{CNT} = 0.12$) under uniaxial compression and with: (a) $h = 0.004$, (b) $h = 0.006$ and (c) $h = 0.008$.

in the other ones (CCSS) nanocomposite panels have the highest post-buckling strength, whereas, the fully simply supported (SSSS) panels present the lowest values of post-buckling responses and this is for all CNTs distributions which highlight that when more constraints are imposed on the cylindrical panel, more its flexural rigidity becomes higher.

Finally, the effect of the thickness parameter on the post-buckling behavior of nanocomposite cylindrical panels is investigated where two thicker panels are considered ($h = 0.006$ and $h = 0.008$). Figures 6.19(a), 6.19(b) and 6.19(c) illustrate the post-buckling responses of the different CNTs forms of the nanocomposite cylindrical panels (CCSS, $V^*_{CNT} = 0.12$) under uniaxial compression load and for three sets of thickness parameter ($h = 0.004$, $h = 0.006$ and $h = 0.008$), respectively. As can be remarked, the increase of the thickness parameter enhances the post-buckling strength of the nanocomposite panels where the post-buckling responses of the different forms of the nanocomposite panels reach their maximum with a thickness value ($h = 0.008$). Therefore, more the nanocomposite cylindrical panel becomes thicker, more its post-buckling response is enhanced.

6.5 CONCLUSION

Buckling and post-buckling responses of functionally graded carbon-nanotubes reinforced (FG-CNTRC) plates and curved panels are presented. The governing equations are elaborated within a double directors shell model which induces high-order distribution of the displacement field and guarantees the zero condition of the transverse shear stresses on the extreme surfaces of the shell. This finite element shell model is implemented in the sense of non-linear formulation with the introduction of large deformations and finite rotations. In resolution procedure, Newton-Raphson with arc length control solution strategy is used for post-buckling responses and the critical buckling load is determined via the resolution of the buckling eigenvalue problem. As well as, the present formulation provides accurate results with low computational costs, which constitutes an important feature of the proposed methodology. Equally, several complex equilibrium paths of FG-CNTRC panels and plates with different mechanical loadings have been first provided that could be useful for future references.

From the parametric study, the following major remarks can be drawn:

(i) The CNTs reinforcements enhance strongly the stiffness of the FG-CNTRC plates and shells which manifest by a high post-buckling strength of such structure. In particular, the FG-X form of CNTs distributions induces the highest post-buckling response.

(ii) The mechanical post-buckling responses of FG-CNTRC plates are influenced by the type of the applied mechanical load where the bi-axial compression load stimulates more the deflection of the structure due to the high load-carrying capability of this type of mechanical load.

(iii) The geometrical parameters such as aspect ratio, thickness, radius and also the boundary conditions have a significant effect on the post-buckling behavior of FG-CNTRC plates and cylindrical panels which must be considered in the design phase of such structures.

(iV) The present double directors high-order shear deformable model has the capacity to predict the mechanical post-buckling behavior of FG and FG-CNTRC structures with a high accuracy. In fact, the model incorporates the shear strains with a good description of large displacements and rotations.

7 Bending and free vibration analysis of FG porous beams

This chapter presents static bending and free vibration analysis of FG porous beams using a refined mixed finite element beam model. This model consists on the approximation of the displacements, stress and strain fields independently. Furthermore, the strain field is decomposed into two parts in order to perform low-order finite beam elements. Both linear and a quadratic interpolations are used for the discretization of the displacement, strain and stresses fields. Shear locking problems are alleviated using the assumed natural strain (ANS) interpolations. The material properties of functionally graded porous beams are estimated using a modified power law distribution with two different types of porosity namely even and uneven distributions. The potential of the proposed model is highlighted via a comparison study. Then, a parametric study is carried out to show the effects of power law index, porosity coefficient, boundary conditions and types of porosity distributions on deflections, stresses and natural frequencies of the studied FG porous beams.

7.1 INTRODUCTION

In recent years, many investigations have been conducted on mechanical behavior of functionally graded (FG) beams, plates and shells due to their interesting features such as lack of delamination and cracking phenomenons observed in conventional materials. The earliest FGMs were introduced by a Japanese scientists group in the mid-1980s as ultra-high-temperature-resistant materials for aerospace applications. Recently, these materials have found other uses in electrical devices, energy transformation and biomedical engineering. By gradually varying the volume fraction of constituent materials, the material properties of FGM structures obey a smooth and continuous change from one surface to another leading to eliminate the interface problems and generate a new class of advanced materials able to withstand dynamic and thermal loadings. However, these materials can exhibit some deficiencies such as porosity during the manufacturing process. So, for a good knowledge of porosity effect on mechanical behavior of FG structures, an overview on recent works related to this subject are presented.

According to the open literature, many researchers have paid attention to the analysis of FGM structures with porosities [288, 289, 287, 47, 211, 90, 116]. Further works on mechanical behavior of FG beams can be found in [275, 58, 97]. However, papers on bending static analysis of FG porous structures are rather scarce compared to vibrational or buckling analysis. Therefore, a further effort must be paid to this topic notably that the porosity constitutes a crucial parameter in the design of modern structures. In the following, few published papers on bending static behavior of FG structures with porosity are presented.

Atmane et al. presented bending, free vibration and buckling responses of FG porous beams resting on elastic foundations [13] via an efficient quasi-3D theory. The proposed theory incorporates both shear deformation and thickness stretching effects. The material properties of FG beams are assumed to vary across the thickness direction according to a power-law-distribution with evenly and unevenly types of porosity distributions. The effect of porosity and some geometrical parameters on mechanical behavior of FG beams are also examined. Recently, Zenkour [305] investigates the bending responses of porous functionally graded (FG) single-layered and sandwich thick rectangular plates using a quasi-3D shear deformation theory. Both the effect of shear strain and normal

DOI: 10.1201/9781003483786-7

deformation are included in the formulation. Effects of the power-law index and porosity factor are highlighted. Based on high-order shear deformation theory (HSDT), Merdaci and Belghoul studied the bending responses of FG porous thick rectangular plates [167]. The effects of various parameters, such as thickness ratio, gradient index and volume fraction of porosity on the flexural behavior of FG ceramic–metal plates are discussed. Subsequently, Kaddari et al. presented statics and free vibration analysis of functionally graded porous plates resting on elastic foundations using a new quasi-3D model [113]. The influences of the porosity parameter, power-law index, aspect and thickness ratios and the foundation parameters on bending and vibration of porous FG plates is also discussed. As well as, the effect of porosity on dynamic behavior of FG plates [3] and FG-CNTRC sandwich plates [160] is performed by Tounsi et al. where the governing equations are derived using either a quasi 3D HSDT theory or the energy principle. Equally, Bourada et al. investigated the vibrational behavior of porous FG beams using a high-order trigonometric deformation theory [35]. Non-linear static, buckling and vibration analysis of viscoelastic micro-composite beam with geometrical imperfection is performed by Alimirzaei et al. [7]. In this study, the equations of motion are established using the variational principle in combination with the modified strain gradient theory (MSGT). The results have shown that the geometrical imperfection parameter, damping coefficient and various boundary conditions have a significant effect on non-linear frequency parameter of micro-beam. For porous FG nano-beams, a nonlocal nth-order shear deformation theory is proposed by Berghouti et al. [27] to study the vibrational behavior of FG nano-beams. In the context of non-structures, and in order to take into account the small-scale effect several continuum theories have been developed, in the last decades, such as the Eringen nonlocal continuum theory, the strain gradient elasticity and the modified couple stress theory and one can be referred to these papers [118, 117, 120], among others.

As can be remarked, the number of papers treating flexural bending behavior of FG structures with porosity is so limited especially for beams where the only investigation on bending behavior of FG porous beams is conducted by Atmane et al. as shown in the presented overview. Furthermore, the most of papers related to free vibration or buckling analysis of FG porous beams use generally the Euler-Bernoulli theory, the Timoshenko theory, the first-order shear deformation theory (FSDT), quasi-3D theory and the high-order shear deformation theories (HSDT). A brief review on works conducted based on these theories is presented here. Draiche et al. presented static analysis of laminated reinforced composite plates [59] using a simple FSDT (S-FSDT) while Draoui et al. performed the static and dynamic behavior of FG-CNTRC sandwich plates [60] using the standard FSDT theory. Although these models introduce few kinematics unknowns and allow low computational time, it require generally the $C1$ continuity which can be prohibitive for such applications. More accurate results can be obtained from the HSDT theories. In fact, Sahla et al. presented free vibration analysis of laminated composite and sandwich plates [213] using a simple four-variable trigonometric (HSDT) theory. Equally, Balubaid et al. studied the free vibrational behavior of FG nano-plates based on two variables refined plate theory [17]. In an other study, Chaabane et al. performed static and free vibration analysis of FG beams using a hyperbolic (HSDT) theory [44]. Further works on static and free vibrational behavior of FG beams and plates using the HSDT theory can be found in [24, 166]. In other works, a quasi 3D shear deformation theory is employed in order to perform bending, vibrational and buckling analysis of FG beams, plates and shells [34, 33, 304, 153]. This theory allows the introduction of the transverse shear deformations and the thickness stretching parameter effects. However, each of the mentioned theories and models presents some drawbacks and shortcomings. In fact, The Euler-Bernoulli or the classical beam theory (CBT) ignores the shear deformation and provides acceptable solutions only for thin beams. One of the shortcomings of the Timoshenko or the FSDT theory is in the transverse shear strain which is constant and it doesn't satisfy the stress-free boundary conditions at the top and bottom surfaces of the structures, so this model requires a shear correction factors. As well as, the quasi-3D theories which include the effects of shear deformation and thickness stretching can be numerically prohibitive for complex geometries. Therefore, the development of a refined beam model for static analysis of FG porous beams appears attractive. Motivated by our previous works concerning static, free vibration, dynamic, buckling, post-buckling analysis of FG and functionally graded carbon-nanotubes reinforced composite (FG-CNTRC) shells [307, 308, 309, 83, 82, 212, 271], we propose in this chapter a new refined mixed beam model which is based on a high-order shear deformation theory and it accounts for parabolic distribution of the transverse shear strain through the thickness of the beam. A discrete constraint is used to get the stress free condition on the top and bottom surface and to avoid the $C1$ transverse continuous displacement. From this overview, it can be remarked that the number of works on static bending and vibration analysis of FGM beams including porosity is rather limited and the major of works use the FSDBT theory which needs generally the inclusion of shear correction factors. However, the computation of these factors is difficult for such applications and especially for composites and advanced materials. Hence, we propose in this paper a robust finite beam element based on a mixed variational formulation. The key idea of this model is to provide approximations of the displacements, stress and strain fields independently. Furthermore, the strain field is decomposed into two parts in order to perform low-order finite beam elements. Both linear and a quadratic interpolations are used for the discretization of the displacement, strain and stresses fields. Shear locking problems are alleviated using the assumed natural strain (ANS) interpolations. Numerical examples are provided in order to show the robustness of the present finite element model and parametric studies are conducted for the effects of porosity parameter, types of distributions and other geometrical parameter on natural frequencies and modes' shapes of FGM porosity dependent beams. Results demonstrate that the porosity parameter is an important factor on deflections, stresses and vibrational behavior of FGM porous beams that should be taken into account in both design and control of vibrations stages.

7.2 MATERIAL PROPERTIES OF FUNCTIONALLY GRADED POROUS BEAMS

A functionally graded (FG) porous beam of length L, width b and thickness h is considered as shown in Fig. 7.1. The beam is produced by continuously grading ceramic and metal phases through the thickness direction where the top surface ($z = +h/2$) is metal-rich and the bottom surface ($z = -h/2$) is ceramic-rich. The volume fraction of metal phase, V_m, follows a power-law distribution which can be expressed as:

$$V_m = \left(\frac{1}{2} + \frac{z}{h}\right)^p, \quad z \in [-h/2, h/2] \tag{7.1}$$

where p is the power-law index and z is the coordinate in the thickness direction.

Due to technical and infiltration problems in the fabrication process, porosities may appear as an imperfection in the FG beams leading to two types of porosity, namely even and uneven distributions as shown in Fig. 7.1. According to the investigations of Wattanasakulpong et al. [289, 287], the effective material properties of the FG beam, such as the Young's modulus E, the mass density ρ and the Poisson's ratio ν, are determined using the modified rule of mixture in which the porosity volume fraction, noted by α, affects averagely the material volume fraction of each constituent. As a result, the material property $P(z,\alpha)$ can be written for each type of porosity as follows:

$$P(z,\alpha) = \begin{cases} (P_m - P_c)V_m + P_c - \frac{\alpha}{2}(P_m + P_c), & (Even\ \ porosity) \\ (P_m - P_c)V_m + P_c - \frac{\alpha}{2}(P_m + P_c)\left(1 - \frac{2|z|}{h}\right), & (Uneven\ \ porosity) \end{cases}, \tag{7.2}$$

As mentioned earlier, α refers to the porosity coefficient which is defined as the ratio between the void volume and the total volume ($0 \leq \alpha \prec 1$). The subscripts m and c denote the metal and ceramic phases, respectively where their volumes fractions are related by:

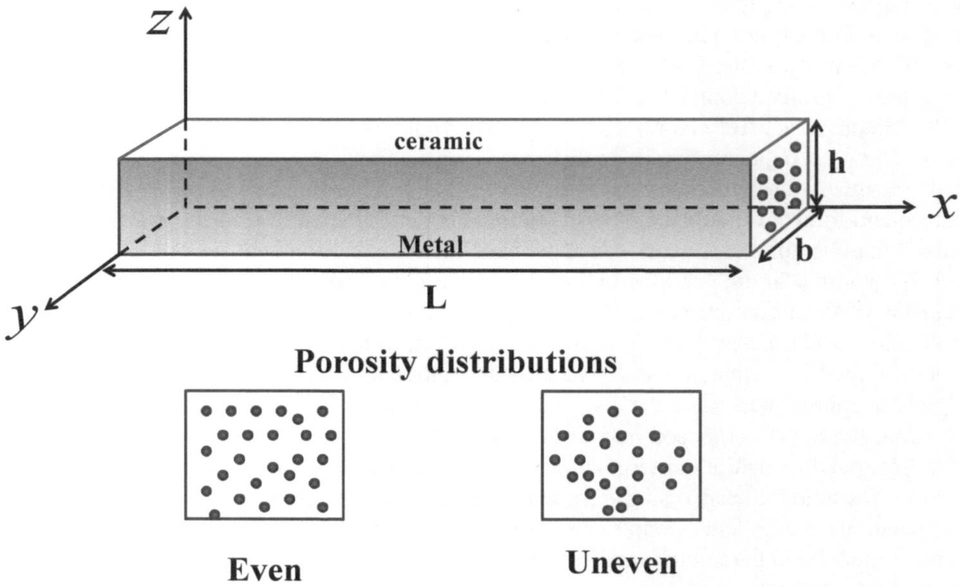

Figure 7.1 Schematic of FGM beam with evenly and unevenly porosity distributions.

$$V_m + V_c = 1. \tag{7.3}$$

From Fig. 7.1, it can be mentioned that for even-type the porosity phases are uniformly distributed along the beam cross section whereas for uneven-type, the porosity phases are spreading mostly around the middle surface of the beam cross section and vanish in the top and bottom surfaces.

7.3 KINEMATIC DESCRIPTION OF A HIGH-ORDER MIXED BEAM MODEL

In this section, the kinematics of the proposed high-order mixed beam model are briefly presented. In fact, the proposed model induces a high-order distribution of the displacement field and imposes a zero condition of the transverse shear stress at the top and bottom surfaces. In the following, the displacement and strain fields, the mixed formulation and the finite element procedure related to the proposed high-order mixed beam model are also presented.

7.3.1 DISPLACEMENT FIELD

According to the high-order shear deformation theory [98, 81], the displacement field of FG beam with rectangular cross section can be expressed as follows:

$$\begin{gathered} u(x,z) = u(x) + f_1(z)\beta(x) + f_2(z)\theta(x), \\ w(x,z) = w(x) \end{gathered} \tag{7.4}$$

where $u(x)$ and $w(x)$ represent the axial and the transverse displacements of the beam, respectively. θ defines the angle between the initial fiber and the normal to the deformed mid-plane surface while the angle β defines the tangent to the cubic fiber at the reference surface ($z = 0$) as shown in Fig. 7.2. The functions $f_1(z)$ and $f_2(z)$, which reflect a high-order distribution, are expressed as:

$$f_1(z) = z - f_2(z), \quad f_2(z) = \frac{4z^3}{3h^2}. \tag{7.5}$$

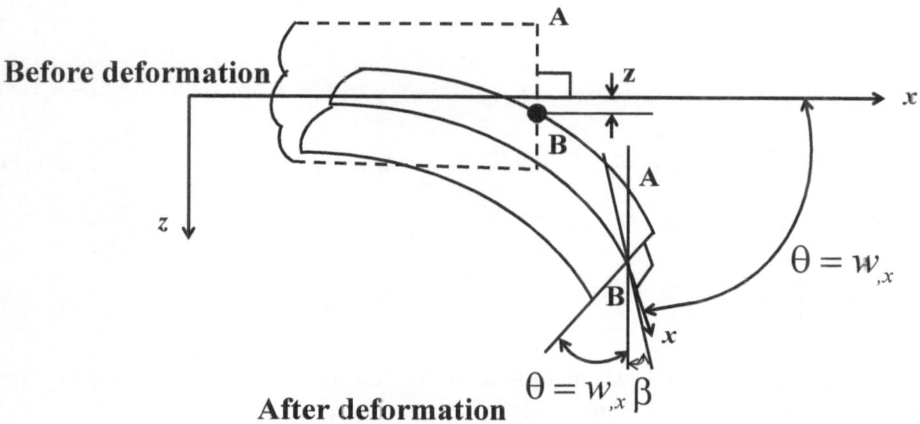

Figure 7.2 Kinematic model for the FGM beam.

7.3.2 STRAIN FIELD

The strain field, which contains the axial normal and the transverse shear strains, can be expressed as follows:

$$\begin{cases} \varepsilon_{xx} = u_{,x} + f_1(z)\beta_{,x} + f_2(z)\theta_{,x} \\ \gamma_{xz} = w_{,x} + f_1'(z)\beta + f_2'(z)\theta \end{cases}, \tag{7.6}$$

The symbols (,) and (′) denote the derivatives with respect the quantity following it and the derivatives with respect z, respectively. The transverse shear strain γ_{xz} can be rewritten as:

$$\begin{cases} \gamma_{xz} = f_1'(z)\gamma_1 + f_2'(z)\gamma_2 \\ \gamma_1 = w_{,x} + \beta \\ \gamma_2 = w_{,x} + \theta \end{cases}, \tag{7.7}$$

As well as, it should be mentioned that the transverse shear stress vanishes on the top and bottom beam surfaces which verify the following condition:

$$\begin{cases} \gamma_{xz} = f_1'(z)\gamma_1 \\ \gamma_2 = 0 \end{cases}, \tag{7.8}$$

The stress-strain relationships, with neglecting the deformation along the z axis, are given by:

$$\begin{cases} \sigma_{xx} = E(z)\varepsilon_{xx} \\ \tau_{xz} = G(z)\gamma_{xz} \end{cases}, \tag{7.9}$$

where the pairs (σ_{xx}, τ_{xz}) and ($E(z)$, $G(z)$) correspond to normal and shear stresses and Young's and shear moduli of the FG beam, respectively.

7.3.3 GOVERNING EQUATIONS

The governing equations of the FG porous beam are derived using the variational principle for which the total potential energy is defined as:

$$\pi = \frac{1}{2}\int_V (\sigma_{xx}\cdot\varepsilon_{xx} + \tau_{xz}\cdot\gamma_{xz})\,dV + \pi_{ext} = \pi_{int} + \pi_{ext} \tag{7.10}$$

where π_{ext} is the external potential energy done by external loads. The corresponding variation of the total potential energy $\delta\pi$ can be reads as:

$$\delta\pi = \int_0^L (\delta\varepsilon^T P + \delta\gamma_1 T_1)\,dx + \delta\pi_{ext} = 0, \tag{7.11}$$

where P is the stress resultants vector and T_1 is the shear stress resultant which can be given as:

$$P = \begin{Bmatrix} N \\ M_1 \\ M_2 \end{Bmatrix} = b\int_{-h/2}^{h/2} \sigma_{xx}\begin{Bmatrix} 1 \\ f_1 \\ f_2 \end{Bmatrix} dz\ ,\ T_1 = b\int_{-h/2}^{h/2} f_1^{'}\tau_{xz}. \tag{7.12}$$

where N refers to stress resultant while M_1 and M_2 refer to moments resultants. Subsequently, the variations of the strain vector $\delta\varepsilon$ and the shear part $\delta\gamma_1$ are given as:

$$\delta\varepsilon = \begin{Bmatrix} \delta u_{,x} \\ \delta\beta_{,x} \\ \delta\theta_{,x} \end{Bmatrix},\ \delta\gamma_1 = \delta w_{,x} + \delta\beta. \tag{7.13}$$

7.3.4 CONSTITUTIVE RELATIONS

The constitutive relations of the studied beam can be expressed as:

$$P = H\varepsilon,\ T_1 = H_s\gamma_1 \tag{7.14}$$

In matrix form, the linear elastic matrix H can be given by:

$$H = \begin{bmatrix} H_{11} & H_{12} & H_{13} \\ H_{12} & H_{22} & H_{23} \\ H_{13} & H_{23} & H_{33} \end{bmatrix}, \tag{7.15}$$

where:

$$(H_{11}, H_{12}, H_{13}, H_{22}, H_{23}, H_{33}) = b\int_{-h/2}^{h/2} (1, f_1, f_2, f_1^2, f_1 f_2, f_2^2)\,E(z)dz. \tag{7.16}$$

For the shear part, its corresponding elastic matrix H_s is given by:

$$H_s = b\int_{-h/2}^{h/2} \left(f_1^{'}\right)^2 G(z)dz. \tag{7.17}$$

In the case of isotropic material, H and H_s are expressed as:

$$H = \frac{E}{105}\begin{bmatrix} 105A & 0 & 0 \\ 0 & 68I & 16I \\ 0 & 16I & 5I \end{bmatrix},\ H_s = \frac{8}{15}GA \tag{7.18}$$

where A and I are the cross section properties of the beam.

7.4 FINITE ELEMENT APPROXIMATIONS

Now, a finite element approximation constructed by means of isoparametric elements is carried out using a two-nodes beam element with 4 DOFs/node. A linear interpolation is used for the nodal variables (u, w, β) while a quadratic interpolation is adopted for the variable θ. As a result, the approximations of the displacement field, in the (x, z) orthogonal coordinate system, are given by:

$$\left\{\begin{array}{ll} x = N_1 x_1 + N_2 x_2; & \theta = N_1 \theta_1 + N_2 \theta_2 + N_m \theta_m \\ u = N_1 u_1 + N_2 u_2; & \beta = N_1 \beta_1 + N_2 \beta_2 \\ w = N_1 w_1 + N_2 w_2 & \end{array}\right. , \tag{7.19}$$

where N_1, N_2 and N_m are the standard isoparametric shape functions satisfying:

$$N_1 = \frac{1-\xi}{2}, \quad N_2 = \frac{1+\xi}{2}, \quad N_m = N_1 N_2 \tag{7.20}$$

The second part of the shear strain γ_2 is imposed, in integral form, as follows:

$$\int_0^L \gamma_2 dx = \int_0^L (w_{,x} + \theta)\, dx = 0, \tag{7.21}$$

The parameter θ_m can be then deduced as:

$$\theta_m = \frac{6}{L}\left(w_1 - w_2 - \frac{L}{2}(\theta_1 + \theta_2)\right) = -6 B_{f0} U. \tag{7.22}$$

where the vectors U and B_{f0} are given by:

$$\left\{\begin{array}{l} U = [u_1 \ w_1 \ \beta_1 \ \theta_1 \ u_2 \ w_2 \ \beta_2 \ \theta_2]^T \\ B_{f0} = \left[0 \ -\frac{1}{L} \ 0 \ \frac{1}{2} \ 0 \ \frac{1}{L} \ 0 \ \frac{1}{2}\right] \end{array}\right. , \tag{7.23}$$

Substituting Eqs. (7.19) and (7.22) into Eq. (7.13), the variation of the strain vector can be rewritten as:

$$\delta\varepsilon = B\delta U = \left(B_{mf} + \frac{6\xi}{L} B_f\right)\delta U \tag{7.24}$$

The expressions of the strain-displacement matrices B_{mf} and B_f are given by:

$$B_{mf} = \begin{bmatrix} -\frac{1}{L} & 0 & 0 & 0 & \frac{1}{L} & 0 & 0 & 0 \\ 0 & 0 & -\frac{1}{L} & 0 & & 0 & 0 & \frac{1}{L} \\ 0 & 0 & 0 & -\frac{1}{L} & 0 & 0 & 0 & \frac{1}{L} \end{bmatrix}, \quad B_f = \begin{bmatrix} 0 \\ 0 \\ B_{f0} \end{bmatrix}_{3\times 8}. \tag{7.25}$$

The variation of the shear strain $\delta\gamma_1$ can be written as:

$$\delta\gamma_1 = B_s \delta U = (B_{s0} + \xi B_{s1})\delta U. \tag{7.26}$$

where the shear strain-displacement matrices B_{s0} and B_{s1} are given by:

$$B_{s0} = \left[\begin{array}{cccccccc} 0 & -\frac{1}{L} & \frac{1}{2} & 0 & 0 & \frac{1}{L} & \frac{1}{2} & 0 \end{array}\right], \quad B_{s1} = \left[\begin{array}{cccccccc} 0 & 0 & -\frac{1}{2} & 0 & 0 & 0 & \frac{1}{2} & 0 \end{array}\right]. \tag{7.27}$$

Combining Eqs. (7.11), (7.12), (7.24) and (7.26), the stiffness matrix is constructed, yielding to:

$$K = K_{mf} + K_s \tag{7.28}$$

where the membrane-bending K_{mf} and the shear K_s parts of the stiffness matrix are calculated as:

$$\left\{\begin{array}{l} K_{mf} = \frac{L}{2}\int_{-1}^{1} B^T H B d\xi = B_{mf}^T H B_{mf} L + \frac{12 H_{33}}{L} B_{f0}^T B_{f0} \\ K_s = \frac{L}{2}\int_{-1}^{1} B_s^T H B_s d\xi \end{array}\right. , \tag{7.29}$$

7.5 MIXED FORMULATION

In order to perform low-order elements with enhanced accuracy, the mixed assumed strain method is herein adopted. Accordingly, a quadratic interpolation is used for the variable β with a constant shear resultant T_1 while the others parameters are interpolated in the same manner as mentioned in the previous finite element section. Hence, the variable β is interpolated as:

$$\left\{ \begin{array}{l} \beta = N_1\beta_1 + N_2\beta_2 + N_m\beta_m; \quad N_m = N_1 N_2 \\ T_1 = T_0 - F(x); \quad F(x) = \int_0^x q(s)ds; \quad \delta T_1 = \delta T_0 \end{array} \right. , \tag{7.30}$$

Based on a mixed formulation in terms of displacements and transverse stress resultants, the variation of the total potential energy $\delta\pi$ is given by:

$$\delta\pi = \int_0^L \left(\delta\varepsilon^T P + \delta\gamma_1 T_1 + \delta T_1(\gamma_1 - T_1/H_s)\right) dx - \int_0^L \delta w q dx = 0, \tag{7.31}$$

where q is the transverse distributed load. The expressions of $\delta\varepsilon$ and $\delta\gamma_1$ are a follows:

$$\left\{ \begin{array}{l} \delta\varepsilon = B\delta U + B_m \delta\beta_m \\ \delta\gamma = B_s \delta U + N_m \delta\beta_m \end{array} \right. , \tag{7.32}$$

where $B_m = \begin{bmatrix} 0 \\ N_{m,x} \\ 0 \end{bmatrix}$.

In matrix form, the total potential energy can be written as:

$$\delta\pi = \left[\delta U^T \;\; \delta T_0 \;\; \delta\beta_m\right] \cdot \left(\begin{bmatrix} K_{mf} & K_{uT} & K_{u\beta} \\ K_{uT}^T & -L/H_s & K_{T\beta} \\ K_{u\beta}^T & K_{T\beta} & K_{\beta\beta} \end{bmatrix} \begin{bmatrix} U \\ T_0 \\ \beta_m \end{bmatrix} - \begin{bmatrix} f_u \\ f_T \\ f_\beta \end{bmatrix} \right). \tag{7.33}$$

The components of the stiffness matrix, as well as the loadings variables are given in Tab. 7.1.

Table 7.1
Expressions of the stiffness matrix components and loadings variables.

Stiffness matrix components	
$K_{uT} = \int_{-1}^{1} B_s^T \frac{L}{2} d\xi = \int_{-1}^{1} (B_{s0} + \xi B_{s1})^T \frac{L}{2} d\xi = LB_{s0}^T$,	$K_{u\beta} = \int_{-1}^{1} B^T H B_m \frac{L}{2} d\xi = -\frac{2H_{23}}{L} B_{f0}^T$
$K_{\beta\beta} = \int_{-1}^{1} B_m^T H B_m \frac{L}{2} d\xi = \frac{H_{22}}{3L}$,	$K_{T\beta} = \int_{-1}^{1} N_m \frac{L}{2} d\xi = \frac{L}{6}$
Loadings variables	
$f_u = \int_0^L N_w^T q dx + \int_0^L B_s^T F dx$, $\quad f_T = -\frac{1}{H_s} \int_0^L F(x) dx$,	$f_\beta = \int_0^L N_m(x) F(x) dx$
$N_w = \left[0 \;\; N_1 \;\; 0 \;\; 0 \;\; 0 \;\; N_2 \;\; 0 \;\; 0 \right]$	

The stress resultant T_0 and rotation β_m are eliminated at element level in two steps. As a result, the elimination of the rotation β_m leads to:

$$\beta_m = \frac{1}{K_{\beta\beta}} \left(f_\beta - K_{u\beta}^T U - K_{T\beta} T_0 \right), \tag{7.34}$$

Substituting Eq.(7.34) into Eq.(7.33), the expression of the total potential energy becomes:

$$\delta\pi = \left[\delta U^T \;\; \delta T_0\right] \cdot \left(\begin{bmatrix} K_{11} & b^T \\ b & -a \end{bmatrix} \begin{bmatrix} U \\ T_0 \end{bmatrix} - \begin{bmatrix} f_1 \\ f_2 \end{bmatrix} \right). \tag{7.35}$$

The expressions of K_{11}, b, a, f_1 and f_2 are given by:

$$\begin{cases} K_{11} = K_{mf} - \frac{3L}{H_{22}} K_{u\beta} K_{u\beta}^T, & a = \frac{L}{H_s} + \frac{L^3}{12H_{22}} \\ b^T = K_{uT} - \frac{L^2}{2H_{22}} K_{u\beta}, & b = L\left(B_{s0} + \frac{H_{23}}{H_{22}} B_{f0}\right) \\ f_1 = f_u + \frac{6H_{23}}{H_{22}} f_\beta B_{f0}^T, & f_2 = f_T - \frac{L^2}{2H_{22}} f_\beta \end{cases}, \tag{7.36}$$

The elimination of the stress resultant T_0 leads to:

$$T_0 = \frac{1}{a}(bU - f_2) \tag{7.37}$$

As a result, the stiffness matrix can be written as:

$$K = K_{11} + b^T a^{-1} b \tag{7.38}$$

Substituting Eqs.(7.36) into Eq.(7.28), the final form of the stiffness matrix reads as:

$$K = B_{mf}^T H B_{mf} L + b^T a^{-1} b + \frac{12}{L}\left(H_{33} - \frac{H_{23}^2}{H_{22}}\right) B_{f0}^T B_{f0} \tag{7.39}$$

For the distributed load vector, its expression is given by:

$$f = f_1 + \frac{f_2}{a} b^T \tag{7.40}$$

In matrix form, the distributed load vector reads as:

$$f = \begin{bmatrix} 0 \\ \int_0^L N_1 q dx - \frac{1}{L}\left(B - A + \int_0^L F dx\right) \\ \int_0^L N_1 F dx - \frac{A}{2} \\ B \\ 0 \\ \int_0^L N_2 q dx + \frac{1}{L}\left(B - A + \int_0^L F dx\right) \\ \int_0^L N_2 F dx - \frac{A}{2} \\ B \end{bmatrix} \tag{7.41}$$

The expressions of A and B are listed in Tab. 7.2.

Table 7.2
Load vector components.

Components of Eq. (7.41)				
$A = \int_0^L \frac{6N_m+\phi}{1+\phi} F dx,$	$B = (6f_\beta - A)\frac{H_{23}}{H_{22}},$	$\phi = \frac{12H_{22}}{L^2 H_s}$		
Case of uniform load				
$F = qx,$	$f_\beta = \frac{qL^2}{12},$	$\int_0^L F(x) dx,$	$A = \frac{qL^2}{2},$	$B = 0$
$f = \begin{bmatrix} 0 & \frac{qL}{2} & -\frac{qL^2}{12} & 0 & 0 & \frac{qL}{2} & \frac{qL^2}{12} & 0 \end{bmatrix}$				

Table 7.3
Non-dimensional parameters used in our numerical assessment.

Parameter	Expression
Adimensional deflection for an isotropic beam	$\overline{w_1}(L/2) = w_1(L/2)w_{1,Ref}(L/2)$
$w_{1,Ref}$	$w_{1,Ref}(L/2) = 100\frac{E_c I}{qL^4}$
Adimensional deflection for FG porous beam in Ref.(11)	$\overline{w_2}(L/2) = w2(L/2)w_{2,Ref}(L/2)$
$w_{2,Ref}$	$w_{2,Ref}(L/2) = 100\frac{E_m h^3}{qL^4}$
Adimensional deflection for FG porous beam in our study	$\overline{w}(L/2) = \frac{w(L/2)}{h}$
Dimensionless axial stress	$\overline{\sigma_{xx}} = \frac{\sigma xx}{\sigma_{xx,Ref}}$
$\sigma_{xx,Ref}$	$\sigma_{xx,Ref} = \frac{3qL^2}{4bh}$
Dimensionless shear stress	$\overline{\tau_{xz}} = \frac{\tau_{xz}}{\tau_{xz,Ref}}$
$\tau_{xz,Ref}$	$\tau_{xz,Ref} = \frac{3qL}{4bh}$

7.6 NATURAL FREQUENCIES AND MODES SHAPES

The determination of the natural frequencies of the FGM porous beam is carried out by the resolution of the following eigenvalue problem:

$$\left(\mathbf{K} - \omega^2\mathbf{M}\right)\{\Lambda\} = 0. \tag{7.42}$$

where M and K are the global mass and stiffness matrices of the FGM beam, respectively. ω is the natural frequency of vibration and Λ is its corresponding mode shape. The mass and stiffness matrices are constructed from the variations of the kinetic and potential energies after assembling the elementary matrices at global level. The stiffness matrix is as given in Eq. (7.39) while the expression of mass matrix is given by:

$$M = bL\int_{-h/2}^{h/2} \rho \begin{bmatrix} 1 & 0 & f_1 & f_2 \\ 0 & 1 & 0 & 0 \\ f_1 & 0 & f_1^2 & f_1 f_2 \\ f_2 & 0 & f_1 f_2 & f_2^2 \end{bmatrix} dz \tag{7.43}$$

7.7 STATIC BENDING AND DEFLECTION RESPONSES OF FG POROUS BEAMS

In this section, the proposed finite mixed beam model is applied for FG porous beam with two different types of distributions: Even and Uneven porosities. The study focuses on the static analysis of a simply supported (SS) FG porous beam. The FG beam is made of Aluminium/Alumina composite (Al/Al_2O_3) with the following material properties:

metal (Aluminum, Al): $E_m = 70 \times 10^9 N/m^2$; $\rho_m = 2702Kg/m^3$; $\nu = 0.3$

ceramic (Alumina, Al_2O_3): $E_c = 380 \times 10^9 N/m^2$; $\rho_c = 3960Kg/m^3$; $\nu = 0.3$

We need to specify that, in the current study, the bottom surface is ceramic-rich while the top surface is metal-rich. As well as, we specify the various non-dimensional parameters which are listed in Tab. 7.3.

Table 7.4
Non-dimensional deflection of an isotropic beam subjected to uniform pressure.

Length-to-thickness ratio (L/h)	Present	Ref.(11)	Ref.(43)	Ref.(44)
5	1.43184	1.31338	1.42024	1.42026
15	1.31358	1.30226	1.31527	1.31528
120	1.30229	1.30090	1.30229	1.30229

Table 7.5
Non-dimensional transverse deflection of FGM beam with different values of porosity coefficient and with ($p = 2$ and $L/h = 5$).

Porosity coefficient α	Present	Ref.(11)	Error (%)
0	5.20	5.35	2.80
0.1	5.82	6.22	6.43
0.2	6.63	7.38	10.16

7.7.1 VALIDATION AND CONVERGENCE STUDY

As a first step, the validity and the merit of the proposed model is checked first for an isotropic beam. In fact, the obtained non-dimensional central deflection $\overline{w_1}(L/2)$ from the present new mixed beam model is compared with others' beam theories [13, 302, 48] for three sets of length-to-thickness ratios ($L/h = 5, 15, 120$) as mentioned in Tab. 7.4. An excellent agreement is revealed between the results which outlines the high accuracy and reliability of the present model in the prediction of flexural behavior of isotropic beams.

The second assessment aims to verify the accuracy and the reliability of the present model for FG porous beams under a uniform load $q = 1 \times 10^6 N/m^2$. The values of non-dimensional central deflection $\overline{w_2}(L/2)$ for FG beam ($L/h = 5$) with three values of porosity coefficient ($\alpha = 0$, $\alpha = 0.1$ and $\alpha = 0.2$) and a power law-index ($p = 2$) are reported in Tab. 7.5 in comparison with those given by Atmane et al. [13]. It can be observed that for perfect FG beam ($\alpha = 0$), the relative error between the results does not exceed 3% whereas this error can reach 7% and 10% for imperfect FG beams ($\alpha = 0.1$, $\alpha = 0.2$) and this discrepancy can be explained by the different beam theories used in both studies. Indeed, in Ref.[13], an analytical solution based on the FSDT theory with thickness stretching parameter is used while in our paper a finite element solution based on the HSDT theory is performed. Therefore, the present model constitutes a more general theory to analyze as well as to predict the flexural behavior of FG porous beams. In the following, the present beam model is adopted to perform static analysis of FG porous beams. In a second step of this work, a convergence study of FG beam under uniformly distributed load $q = 1 \times 10^6 N/m^2$ and with length-to-thickness ratio ($L/h = 5$) and a power law-index ($p = 2$) is carried out where the normalized central deflection is plotted versus the mesh size for different sets of porosity coefficients ($\alpha = 0$, $\alpha = 0.1$ and $\alpha = 0.2$) as depicted in Fig. 7.3. It is revealed that, as the number of elements per side increases, as the normalized central deflection converges well with the reference solution with 32 elements. In addition, 16 elements are sufficient to reach accurate bending static results. Accordingly, this mesh size is adopted to perform static analysis of FG porous beams.

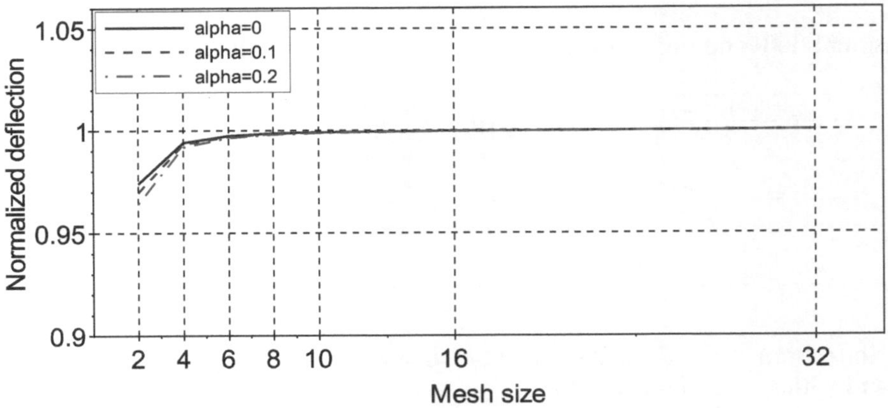

Figure 7.3 Convergence property of non-dimensional transverse deflection of FG porous beam ($L/h = 5$, $p = 2$) subjected to uniform load.

7.7.2 PARAMETRIC STUDY

First, the effect of porosity parameter on bending behavior of FG beams subjected to uniform transverse load $q = 1 \times 10^6 N/m^2$ is performed. For that, the non-dimensional central deflection $\overline{w}(L/2)$ versus the porosity coefficient of the FG beam (L/h = 5) with different power-law-index values and for even and uneven types of porosity distributions are plotted in Fig. 7.4. It can be remarked that the non-dimensional central deflection $\overline{w}(L/2)$ increases with the rise of the porosity coefficient and this effect is more pronounced with high values of porosities (from 20% to 40% of porosity rate) and this is due to the decrease in the flexural rigidity of the FG beam with high rates of porosity. As well as, the observed effect is more significant for metallic beam ($p = 0$) where the stiffness of the beam becomes lower with porosities. In addition, it can be seen that as the power-law index increases, the central non-dimensional deflection decreases and this is for even or uneven types of porosity. In fact, for a power-law-index equal to zero ($p = 0$), the beam is fully metallic with lower stiffness and when the power-law-index (p) tends to infinity, the beam is fully ceramic with high stiffness which manifest by a decrease in the central deflection.

Secondly, the effect of the slenderness ratio (L/h) on non-dimensional deflection $\overline{w}(L/2)$ of FG porous beams with three sets of porosity coefficients ($\alpha = 0$, 0.1 and 0.2) is shown in Fig. 7.5. It can be concluded that the increase in the slenderness ratio (L/h) induces an increase in the non-dimensional deflection. Equally, the non-dimensional deflection increases with the rise of the porosity coefficient. In fact, when the length-to-thickness ratio (L/h) increases, the beam behaves as thin structure and it bends hence easily. On the other hand, with high values of porosity ($\alpha = 0.2$), the elastic modulus of the FG beam becomes weaker which leads to an increase in the adimensional displacements.

Figure 7.6 illustrates the non-dimensional transverse deflection $\overline{w}(L/2)$ of perfect and imperfect FG beam with different length-to-thickness ratios and even and uneven porosity distributions while fixing $p = 1$ and $\alpha = 0.2$. To start with, the more the slenderness ratio (L/h) increases, the more the non-dimensional deflection increases and this is due to the decrease in the stiffness of the FG beam. Furthermore, moving from perfect to imperfect beam symbol of the existence of porosity, it can be observed that the imperfect beam has a higher deflection which is explained by the influence of material stiffness that decreases with porosity defects. Ultimately, for a fixed length-to-thickness ratio, the curves are organized in terms of flexural rigidity in this way: Perfect-Uneven distribution-Even distribution. The maximum of deflection is observed for even distribution since the repartition

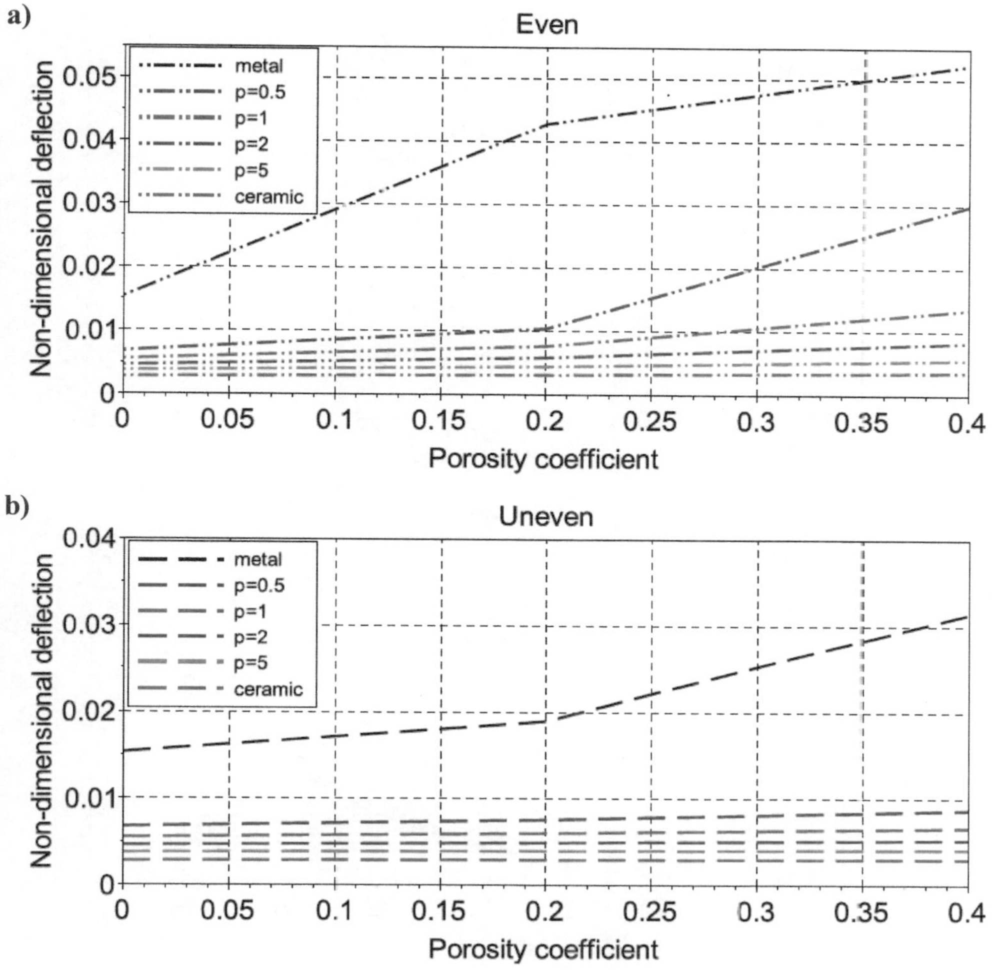

Figure 7.4 Non-dimensional central deflection versus the porosity coefficient of the FG beam ($L/h=5$) under uniform transverse load and with different power-law-index values for (a) Even and (b) Uneven distributions.

of porosity is uniform across the beam section in this case, causing hence a pronounced decrease in the stiffness of the FG beam than in the case of uneven distribution for which the porosity is randomly distributed.

In Figure 7.7, we plotted the non-dimensional transverse deflection $\overline{w}(L/2)$ of the FG beam for different boundary conditions (SS): simply supported and (CC): clamped-clamped and with even and uneven porosity distributions for a length-to-thickness ratio ($L/h=5$), power-law-index ($p=2$) and a porosity coefficient ($\alpha=0.4$). From the porosity distribution side, the non-dimensional deflection is higher in the case of even distribution than in the case of uneven distribution. In fact, the inclusion of porosity leads to a decrease in the modulus of elasticity since it represents a tiny void in the material (pores) and thus in the material stiffness resulting in a higher deflection. For even type of porosity , the distribution of pores is homogenous and uniform through the thickness of the FG beam while for uneven type of porosity , the distribution of pores is random and concentrates more in the mid-surface and free in the top and bottom surfaces. Therefore, the degree of repartition of porosity has a significant effect on the stiffness of the FG beam and consequently on the bending

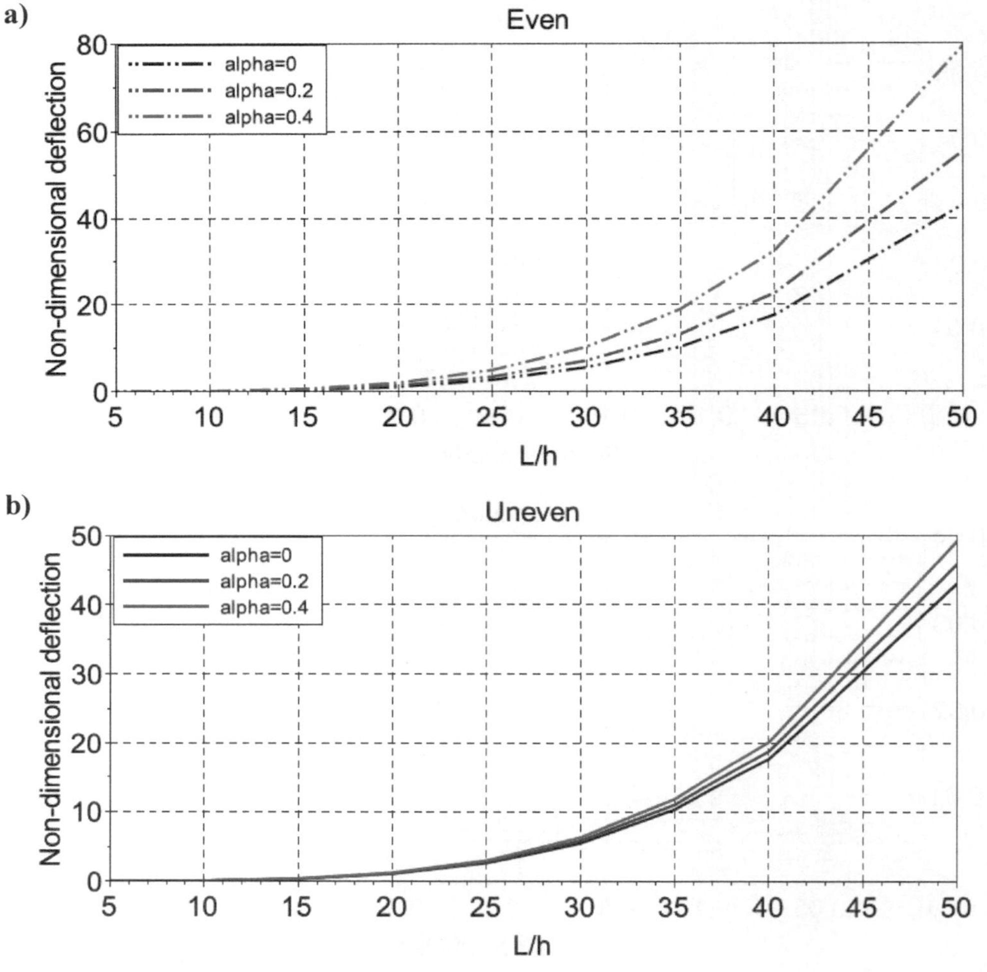

Figure 7.5 Non-dimensional central deflection versus the slenderness ratio (L/h) of FG porous beam under uniform transverse load and with different porosity coefficients for (a) Even and (b) Uneven distributions ($p = 2$).

behavior of FG beams. From the boundary conditions side, it can be seen that the non-dimensional deflection $\overline{w}(L/2)$ of simply supported (SS) beam is higher than clamped-clamped (CC) beam which is expected since the clamped boundary condition imposes more constraints leading hence to less deflections. Figure 7.8 displays the dimensionless axial stress $\overline{\sigma_{xx}}$ of the FG beam across the beam thickness for three values of porosity coefficients ($\alpha = 0$, $\alpha = 0.2$ and $\alpha = 0.4$) and with even and uneven distributions ($L/h = 5$, $p = 2$). It can be remarked that an increase in the index of porosity (α) leads to an increase in stress. This can be justified by the fact that the porosity reduces the rigidity of the FG beam. As well as, Figure 8(a) can be divided into three zones: in zone 1, $z/h \in [-h/2, -h/4]$, the axial stress increases almost in a linear way with (z/h) and decreases with the porosity coefficient, in zone 2, $z/h \in [-h/4, h/4]$, the axial stress increases in a parabolic shape with (z/h) and also increases with the porosity coefficient and in zone 3, $z/h \in [h/4, h/2]$, the axial stress decreases with (z/h) and with the porosity coefficient. So, the stresses are tensile near the mid-plane and compressive near the top and bottom surfaces which can be explained by

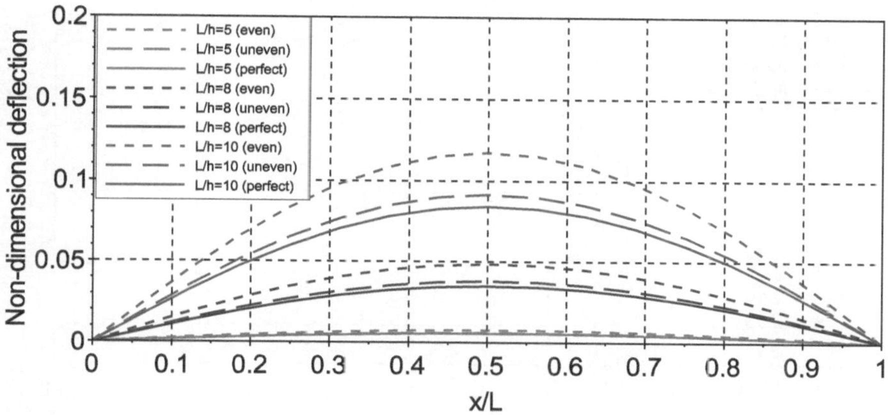

Figure 7.6 Non-dimensional transverse deflection of FG porous beam under uniform transverse load for different length-to-thickness ratios and with even and uneven porosity distributions ($p=1$. $\alpha=0.2$).

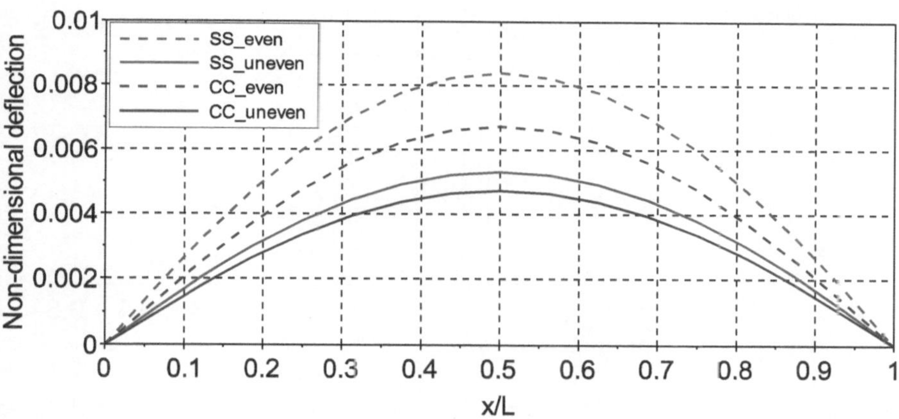

Figure 7.7 Non-dimensional deflection of the FG beam subjected to uniform transverse load for different boundary conditions and with even and uneven porosity distributions ($L/h=5$, $p=2$, $\alpha=0.4$).

the degree of repartition of porosity in even type. For uneven type, an antisymmetric distribution of axial stresses relative to the mid-plane surface is observed. In fact, the axial stresses are tensile above the mid-plane and compressive below the mid-plane. Therefore, the degree of repartition of porosity influence significantly in axial stresses of FG beams and the porosity parameter must be considered in design phase. Figure 7.9 shows the dimensionless shear stress $\overline{\tau_{xz}}$ through the transverse thickness distribution of the FG beam with different porosity coefficients ($\alpha=0$, $\alpha=0.2$ and $\alpha=0.4$) and for even and uneven distributions ($L/h=5$, $p=2$). It can be seen that the shear stresses are parabolic and vanish in top and bottom surfaces for both even and uneven distributions. Furthermore, the shear stresses increase with the increase of the porosity coefficient above the mid-plane and decrease below the mid-plane. So, the manner and the percentage of porosity distribution affects significantly the shear stresses of FG beams and this parameter should be considered in the fabrication process of FG beams.

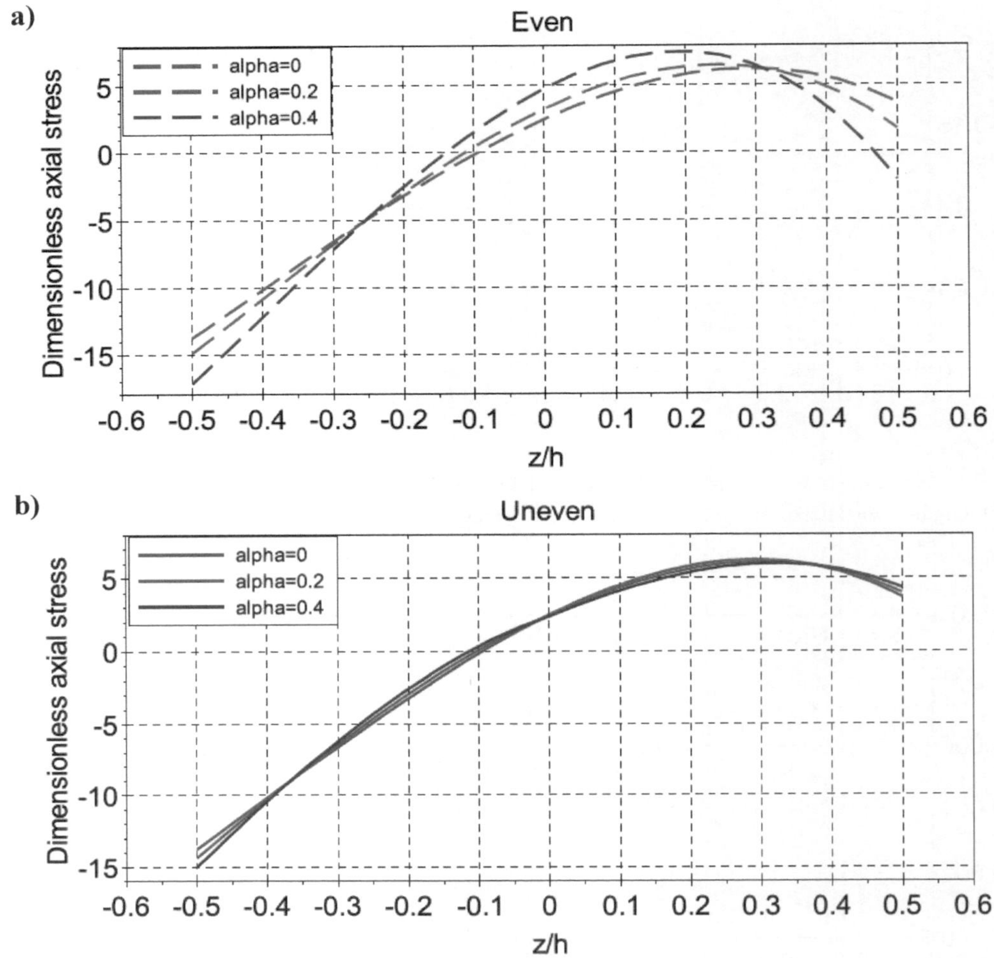

Figure 7.8 Dimensionless axial stress of the FG beam subjected to uniform transverse load and with different porosity coefficients for (a) Even and (b) Uneven distributions ($L/h = 5$, $p = 2$).

7.8 FREE VIBRATION ANALYSIS OF FG POROUS BEAMS

The present section, dedicated to the presentation of results and their preliminary discussion, contains in first a verification study, after which a set of case studies considering different porosity distributions models are presented and analyzed concerning their influence on natural frequencies and modes shapes of the FGM porous beams.

7.8.1 VERIFICATION STUDIES

7.8.1.1 Comparison studies

The first verification study concerns the determination of the natural frequencies of perfect FGM beams without porosities. To this end, a simply supported (SS) FGM beam ($L/h = 5$, $p = 1$) made of Aluminum (Al) and Alumina (Al_2O_3) phases is considered. The material properties of each phase are taken as: $E_{Al} = 70 \times 10^9 N/m^2$; $\rho_{Al} = 2702 Kg/m^3$; $E_{Al_2O_3} = 380 \times 10^9 N/m^2$; $\rho_{Al_2O_3} = 3960 Kg/m^3$. In this study, the Poisson's ratio is assumed to be constant $\nu = 0.3$. Table 7.6 provides the first

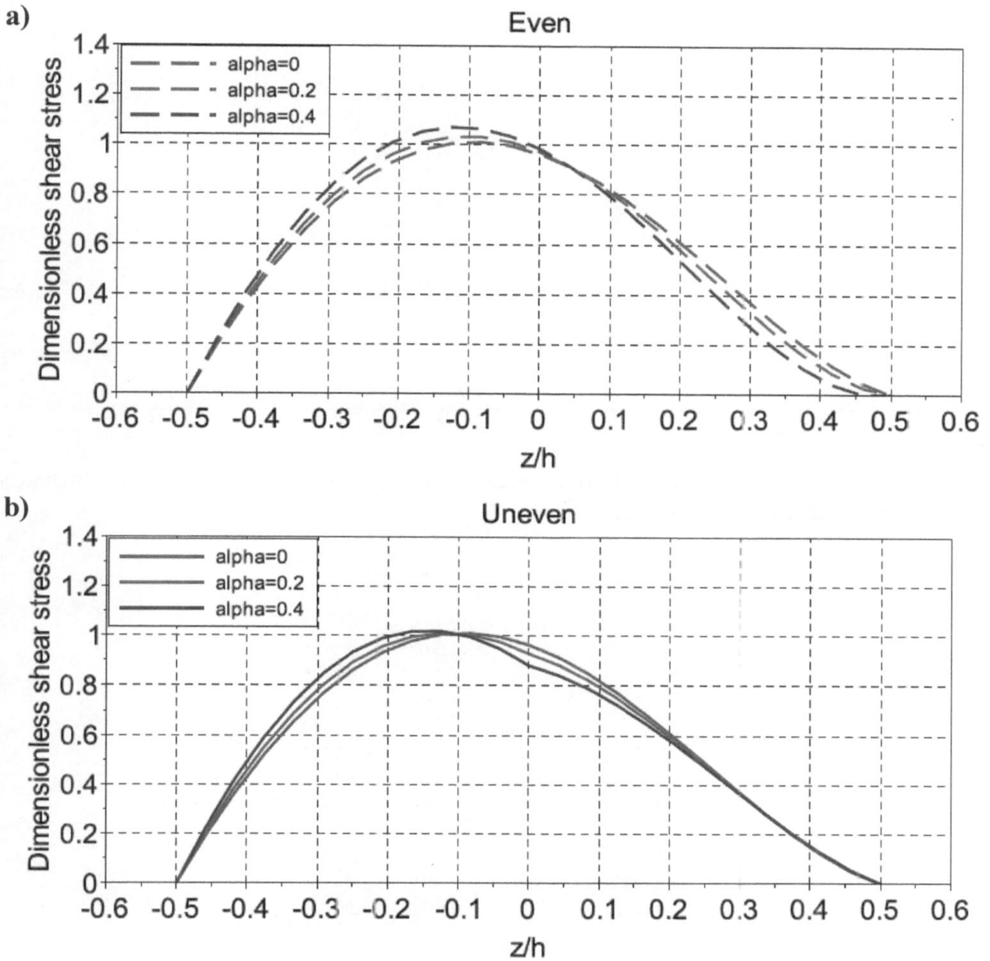

Figure 7.9 Dimensionless shear stress of the FG beam subjected to uniform transverse load and with different porosity coefficients for (a) Even and (b) Uneven distributions ($L/h = 5$, $p = 2$).

four natural frequencies of perfect FGM beam (SS, $L/h = 5, p = 1$) where an excellent agreement with the reference results is shown. Continuously, a comparative study is also conducted for perfect FGM beams ($L/h = 20$) for different gradient indexes (p) and boundary conditions. Three sets of boundary conditions are considered, namely simply supported (SS), clamped-free (CF) and clamped-clamped (CC). The results are evaluated in terms of adimensional frequency parameter $\overline{\omega_1} = \frac{\omega L^2}{h}\sqrt{\frac{\rho_m}{E_m}}$ as presented in Tabs. 7.7 and 7.8. The small relative differences observed make it possible to conclude that there is a very good agreement between the present model and the reference results, although different theories are applied in ([245]) like the CBT, FSDBT and HSDBT theories referred to Th1, Th2 and Th3, respectively. Therefore, the present mixed finite beam element model is efficient and reliable for vibrations analysis of perfect FGM beams.

The second assessment concerns porous FGM beams where the adimensional natural frequency parameter $\overline{\omega_2} = \omega L\sqrt{I_{00}/A_{110}}$ is computed for different gradient indexes, length-to-thickness (L/h) ratios and types of porosity distributions while fixing the porosity coefficient ($\alpha = 0.2$) and boundary condition to clamped-clamped (CC). Moreover, the computation of the dimensionless natural

Table 7.6
First four natural frequencies (rad/s) of a perfect FGM beams (SS, $L/h = 5, p = 1$).

Theories	Mode			
	1	2	3	4
Current	6425.1	21670	40053	60099
Ref.[143]	6457.93	21603.18	40145.42	59779.01
%error	0.5080	0.3101	0.2302	0.5352
Ref.[245]	6443.78	21493.99	39909.87	59509.80
%error	0.2907	0.8122	0.3573	0.9803

Table 7.7
Natural frequencies $\overline{\omega_1}$ of perfect FGM beams ($L/h = 20$) for various gradient indexes ($p = 0, 0.5, 1$) and boundary conditions.

BC	Theories	gradient index (p)		
		0	0.5	1
SS	Current	5.46466	4.64664	4.20734
	Th1.[245]	5.47773	4.66458	4.21634
	Th2.[245]	5.46032	4.65137	4.20505
	Th3.[245]	5.46030	4.65159	4.20503
CC	Current	12.2329	10.4199	9.44150
	Th1.[245]	12.4142	10.5713	9.55538
	Th2.[245]	12.2235	10.4263	9.43135
	Th3.[245]	12.2238	10.4287	9.43158
CF	Current	1.94944	1.65748	1.50094
	Th1.[245]	1.95248	1.66265	1.50293
	Th2.[245]	1.94957	1.66044	1.50104
	Th3.[245]	1.94954	1.66049	1.50106

frequency $\overline{\omega_2}$ is carried out by setting: $I_{00} = \rho_m h$ and $A_{110} = E_m h$. The obtained results are reported in Tab. 7.9. For the two even and uneven types of porosity distributions, the present model is found valid and accurate. In fact, the discrepancy between the present results and those in Ref. [287] does not surpass 1% which draws the robustness and effectiveness of the present mixed finite beam element in the analysis of vibrations of FGM porous beams.

7.8.1.2 Convergence rate

The consistency of the formulation is also verified via convergence analysis where the normalized frequency parameter relative to the reference solution versus the mesh size for the three first modes of vibrations of perfect FGM beams (CC, $L/h = 20$) is depicted in Fig. 7.10. It is shown that the present mixed finite beam elements has a good convergence rate with mesh refinement for different values of gradient indexes ($p = 0, 1, 2$) and 32 elements per side are enough to obtain accurate vibration results. As a result, this mesh size is used to investigate the vibrations of FGM porous beams in the rest of numerical examples.

Table 7.8

Natural frequencies $\overline{\omega_1}$ of perfect FGM beams ($L/h = 20$) for various gradient indexes ($p = 5, 10, \infty$) and boundary conditions.

BC	Theories	Gradient index (p)		
		5	10	Metal
SS	Present	3.64976	3.54107	2.83944
	Th1.[245]	3.66283	3.55465	2.84618
	Th2.[245]	3.65088	3.54156	2.83713
	Th3.[245]	3.6485	3.53896	2.83716
CC	Present	8.15345	7.89333	6.35600
	Th1.[245]	8.30064	8.0556	6.45034
	Th2.[245]	8.16985	7.91275	6.35123
	Th3.[245]	8.1446	7.88576	6.35139
CF	Present	1.30329	1.26455	1.01291
	Th1.[245]	1.30574	1.26713	1.01449
	Th2.[245]	1.30375	1.26495	1.01297
	Th3.[245]	1.30332	1.26453	1.01301

Table 7.9

Natural frequencies $\overline{\omega_2}$ for evenly and unevenly porous FGM beams ($\alpha = 0.2$, CC) with different gradient indexes and length-to-thickness ratios.

L/h	p	Even		Uneven	
		Current	Ref.[287]	Current	Ref.[287]
5	0.2	1.9306	1.9205	1.9258	1.9238
	0.5	1.7573	1.7402	1.7785	1.7695
	1	1.5347	1.5210	1.6006	1.5955
	2	1.2826	1.2815	1.4130	1.4254
	5	1.0574	1.0933	1.2286	1.2942
10	0.2	1.1046	1.1092	1.1111	1.1166
	0.5	0.9972	0.9956	1.0093	1.0202
	1	0.8631	0.8606	0.9107	0.9142
	2	0.7196	0.7193	0.8147	0.8174
	5	0.6156	0.6229	0.7599	0.7597
20	0.2	0.5759	0.5797	0.5812	0.5847
	0.5	0.5184	0.5186	0.5266	0.5329
	1	0.4473	0.4465	0.4745	0.4765
	2	0.3727	0.3722	0.4262	0.4262
	5	0.3234	0.3241	0.4037	0.3996

7.8.2 PARAMETRIC ANALYSIS

In this section, a set of parametric studies is performed in order to show the influence of several parameters on free vibration behavior of FGM porous beams. Numerical illustrations are provided for each case study with the emphasis on the performed parameter.

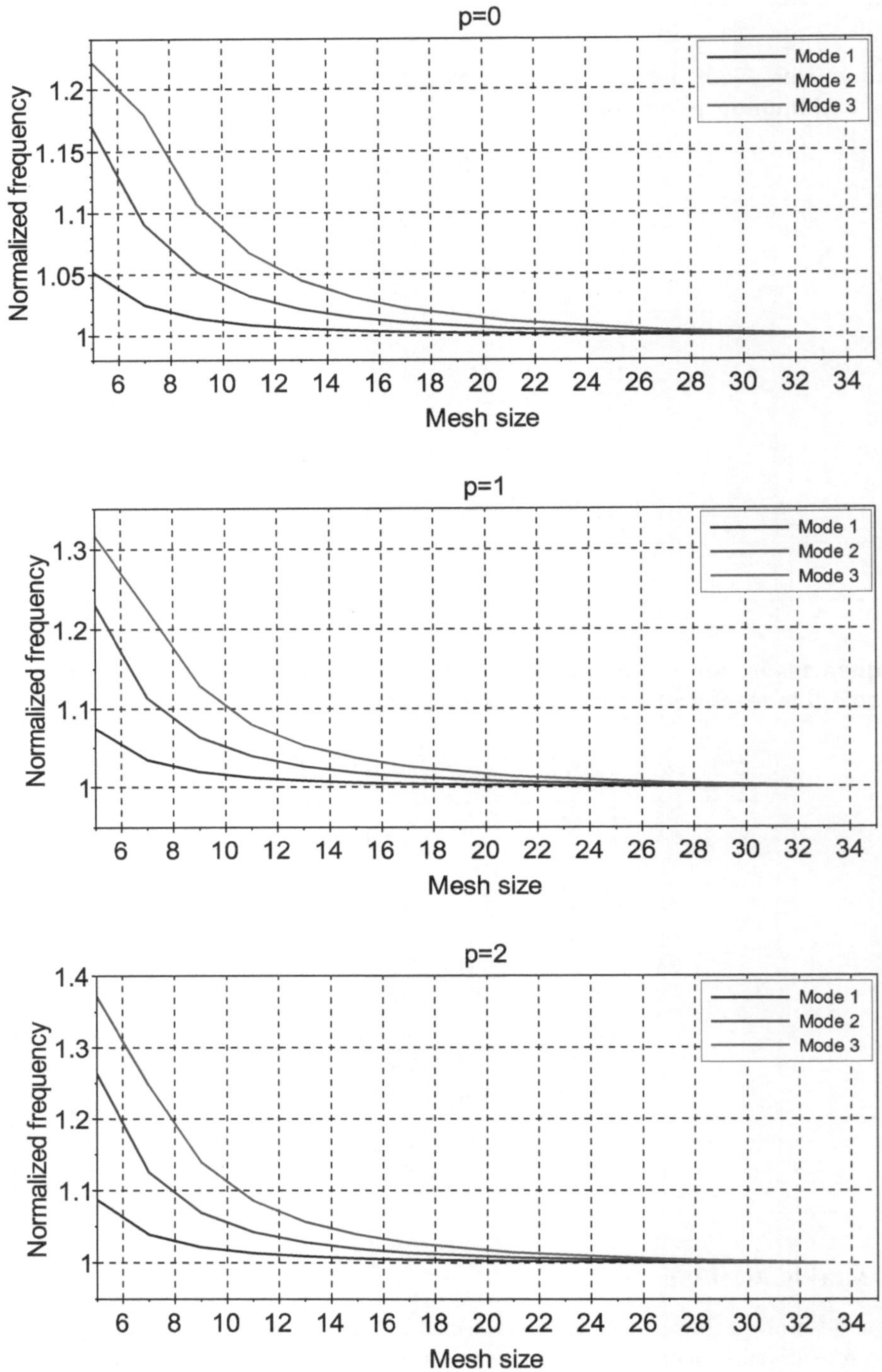

Figure 7.10 Convergence rate of frequency parameter of FGM beams (CC, $L/h = 20$) for different gradient indexes (p).

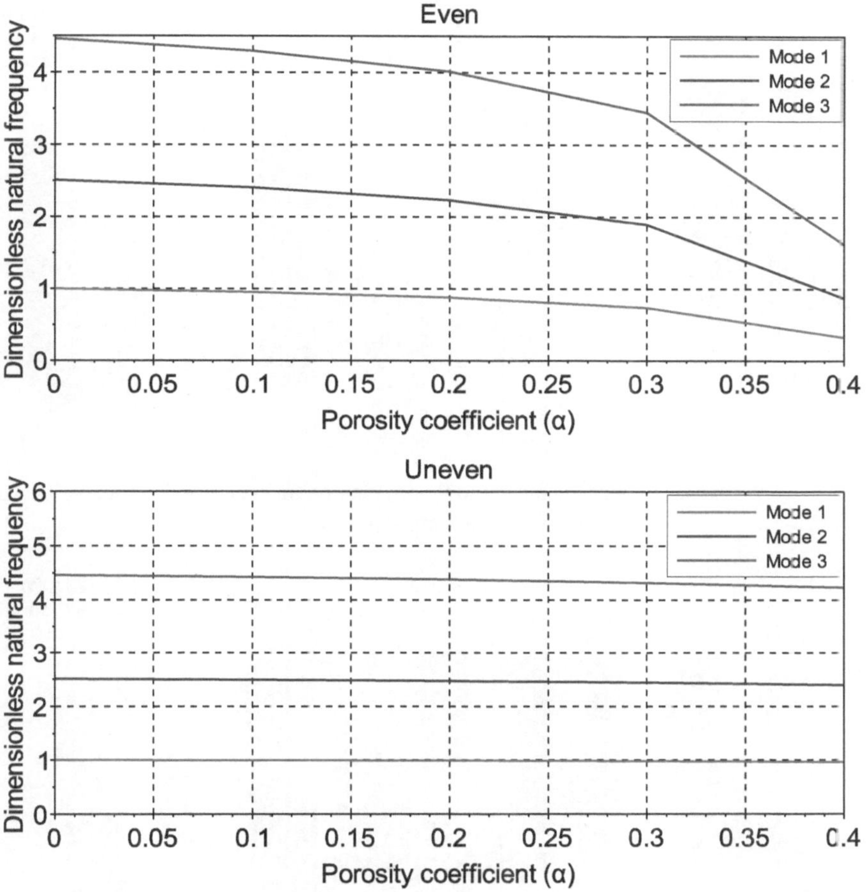

Figure 7.11 Natural frequencies of FGM porous beams (CC, $p = 2$, $L/h = 8$) versus the porosity coefficient (α) for even and uneven types of porosity distributions.

7.8.2.1 Influence of porosity coefficient and types of distributions

The influence of porosity coefficient and the types of distributions is first investigated through Fig. 7.11 which draws the natural frequency parameter $\overline{\omega_2} = \omega L\sqrt{I_{00}/A_{110}}$ versus the porosity coefficient (α) for the first three modes of vibrations of evenly and unevenly FGM porous beams (CC, $p = 2$, $L/h = 8$). As can be seen, the increase of the porosity coefficient induces a reduction in the natural frequencies for evenly FGM porous beams, in contrast to unevenly type for which this variation is quasi-constant. This phenomenon is observed for high values of porosity parameter ($\alpha \geq 0.25$). So, for a prescribed value of gradient index, and with increasing porosity coefficient, ratio of mass to stiffness also changes. So, the values of dimensionless natural frequency strongly depend on which part of the ratio is more reduced. Continuously, the impact of porosity distributions can be highlighted via the histogram of repartition of three first modes of vibration for evenly and unevenly FGM porous beams as shown in Fig. 7.12. It is demonstrated that unevenly FGM beams possess high natural frequencies compared to evenly FGM beams and this is more marked with the increase of the porosity coefficient. In fact, when the porosity coefficient reaches a certain up value ($\alpha = 0.3$), the gap between the frequencies values becomes more prominent. Therefore, the type of porosity distribution has a marked effect on vibrational behavior of FGM porous beams where beams with concentrated pores are more resistant to vibrations than beam with uniform distributed

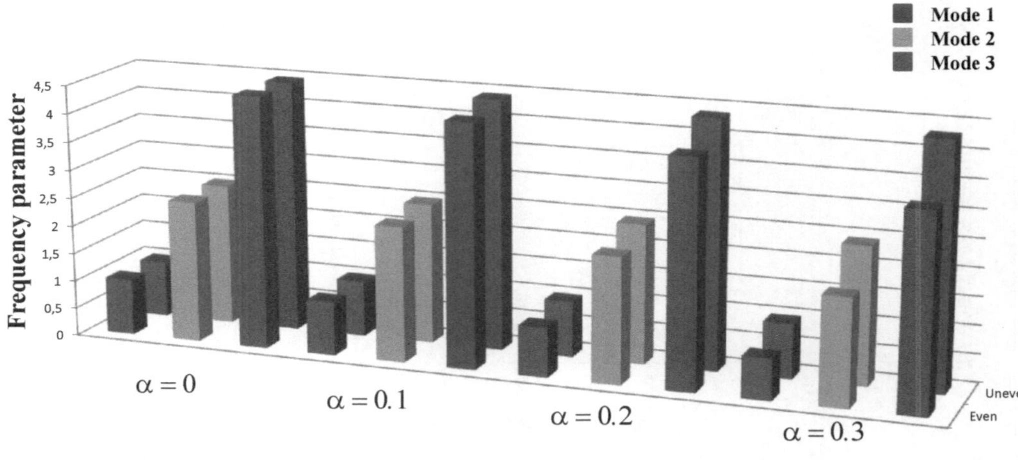

Figure 7.12 Histogram of repartition of three first modes of vibration for evenly and unevenly FGM porosity dependent beams $(CC, p = 2, L/h = 8)$ under different porosity coefficients.

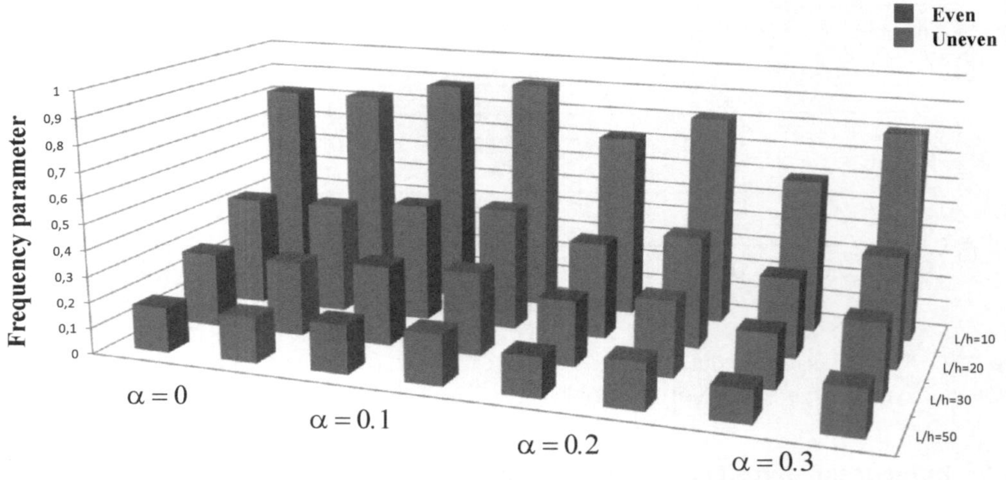

Figure 7.13 Histogram of repartition of natural frequencies for evenly and unevenly FGM porosity dependent beams $(CC, p = 1)$ under different slenderness ratios and porosity coefficients.

pores. So, beams with unevenly pores are more stiffer than beams with evenly pores due to their relative high structural stiffness. Another emphasis on the impact of porosity distributions and coefficients on natural frequencies of FGM beams is highlighted in the histogram given in Fig. 7.13. In this analysis, four configurations of FGM beams are considered: beams with slenderness ratio $(L/h = 10, 20, 30, 50)$. It is shown that as the slenderness ratio (L/h) rises, as the natural frequency of the beam is reduced. Furthermore, by comparing the frequency parameter associated to evenly FGM beams with those of unevenly ones, it can be remarked that the former induces less natural frequencies and this is due to the reduction of the flexural rigidity of the beam caused by the uniform distribution of the porosity. Meanwhile, for the latter, the FGM beam has enough resistance with concentrated pores in the mid-surface. On the other hand, it can be seen that the difference in frequency parameter between Even and Uneven forms increases while the porosity coefficient

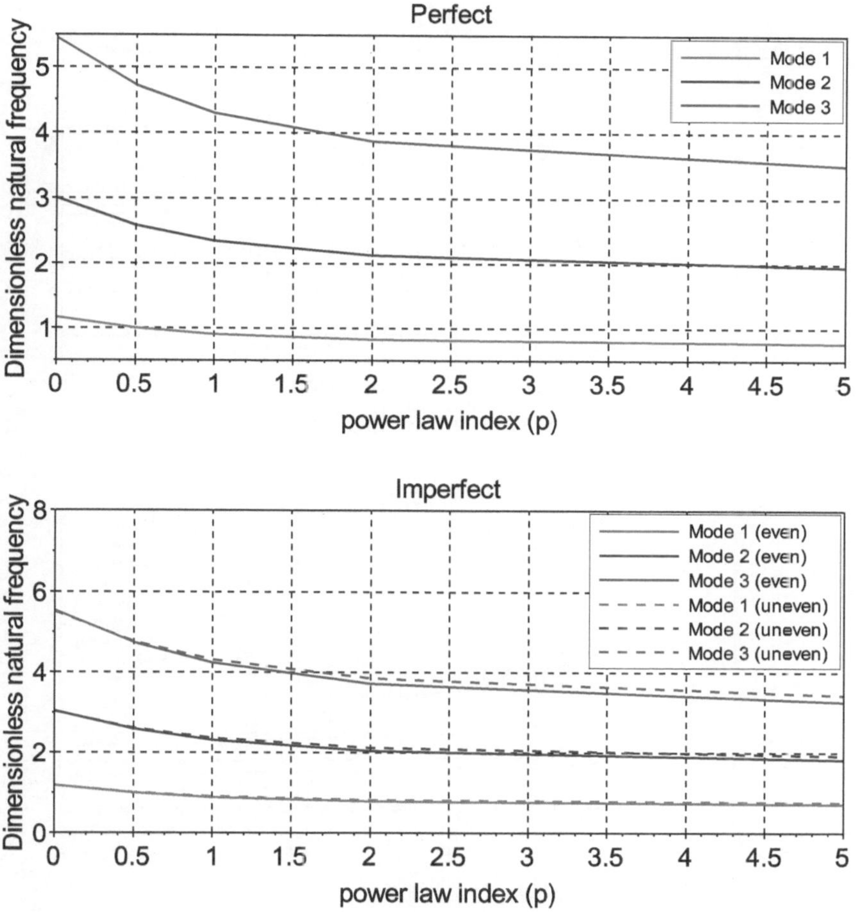

Figure 7.14 Variation of the natural frequencies in the function of the gradient index (p) for the three first modes of vibrations of perfect and imperfect FGM beams (CC, $L/h = 10$, $\alpha = 0.1$).

increases and the natural frequency reaches its maximum for a fixed value of porosity coefficient ($\alpha = 0.1$). This phenomenon can be explained by the change of the ratio between the mass and stiffness where the decreasing of the structural stiffness of the beam overshadows the decreasing inertia at this point, which leads to a downward trend for natural frequency after this value of the porosity coefficient ($\alpha = 0.1$). Therefore, for a prescribed value of gradient index, and with increasing porosity coefficient, ratio of mass to stiffness also changes. So, the values of frequency parameter strongly depend on which part of the ratio is more reduced. In conclusion, the manner of porosity distribution through the thickness direction of the FGM beams has a significant effect on vibrations of such structures and this parameter should be considered in the control phase of vibrations as a key design factor.

7.8.2.2 Influence of gradient index

The influence of the gradient index (p) on natural frequencies of FGM beams (CC, $L/h = 10$) is presented in Fig. 7.14. The imperfection parameter of the FGM beam is considered as $\alpha = 0.1$.

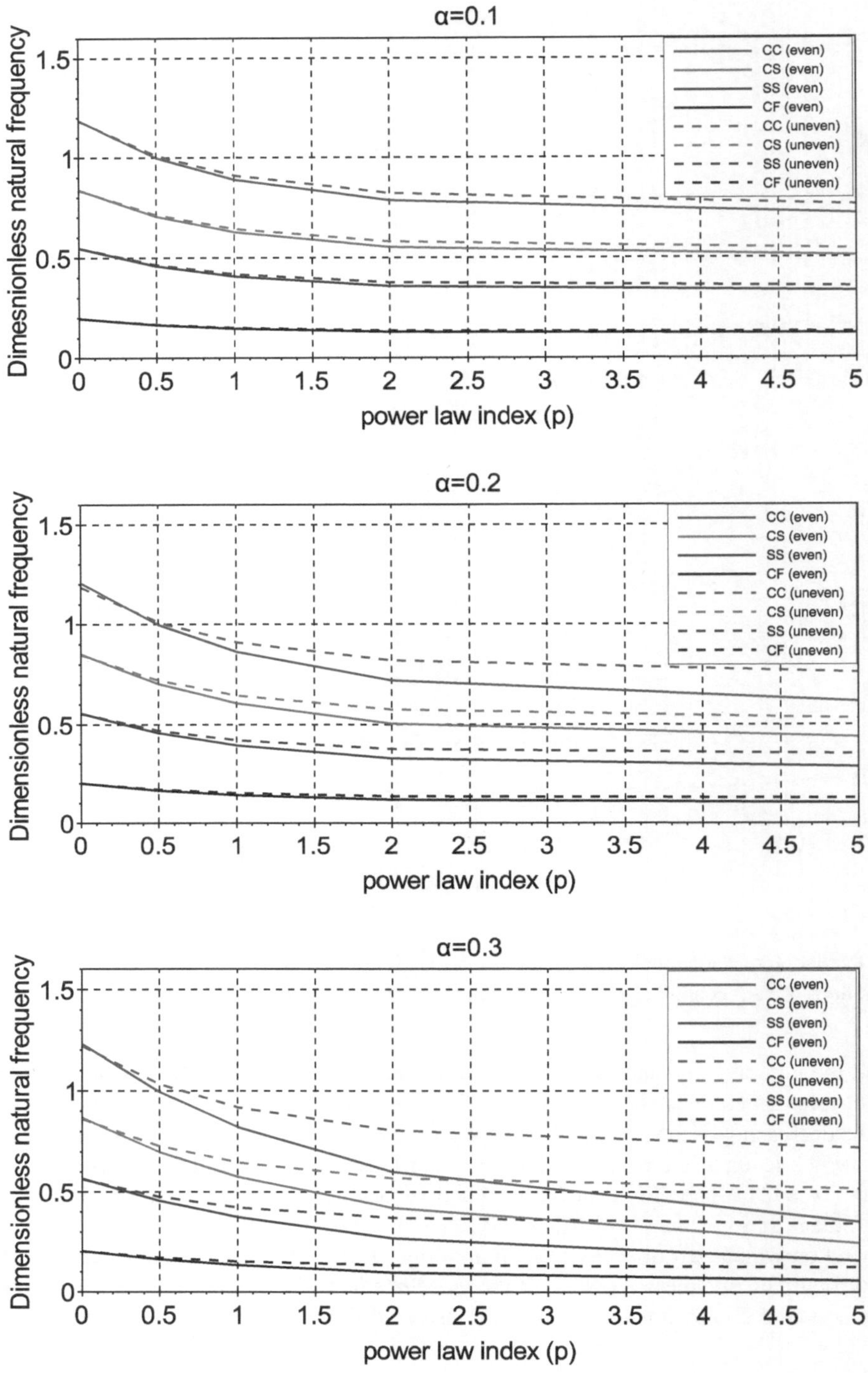

Figure 7.15 Natural frequencies of different constrained FGM porosity dependent beams ($L/h = 10$) under different porosity coefficients ($\alpha = 0.1, 0.2, 0.3$) and with evenly and unevenly porosity distributions.

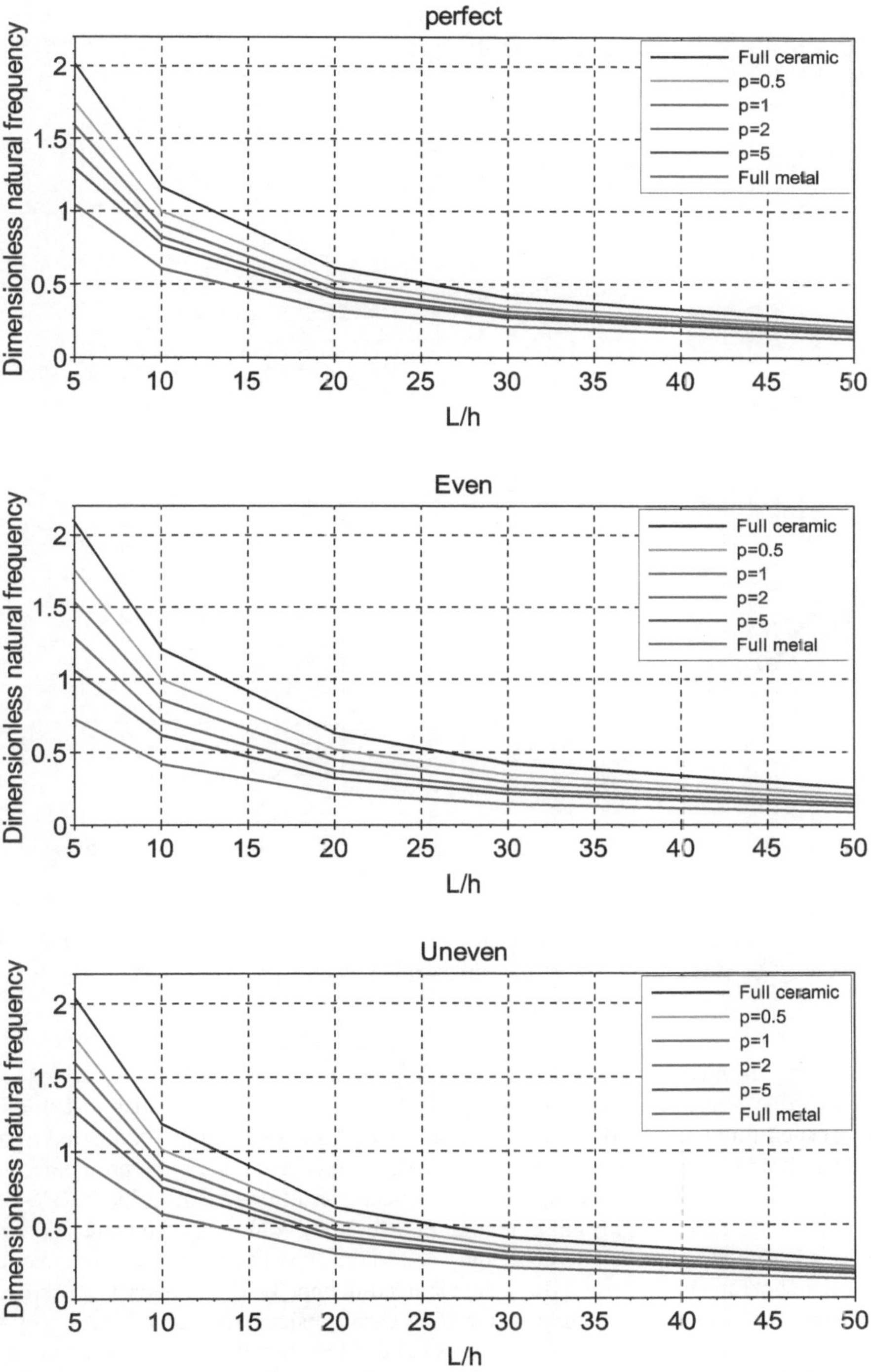

Figure 7.16 Natural frequencies versus the slenderness ratio (L/h) of FGM beams (CC, $\alpha = 0.2$) for different gradient indexes.

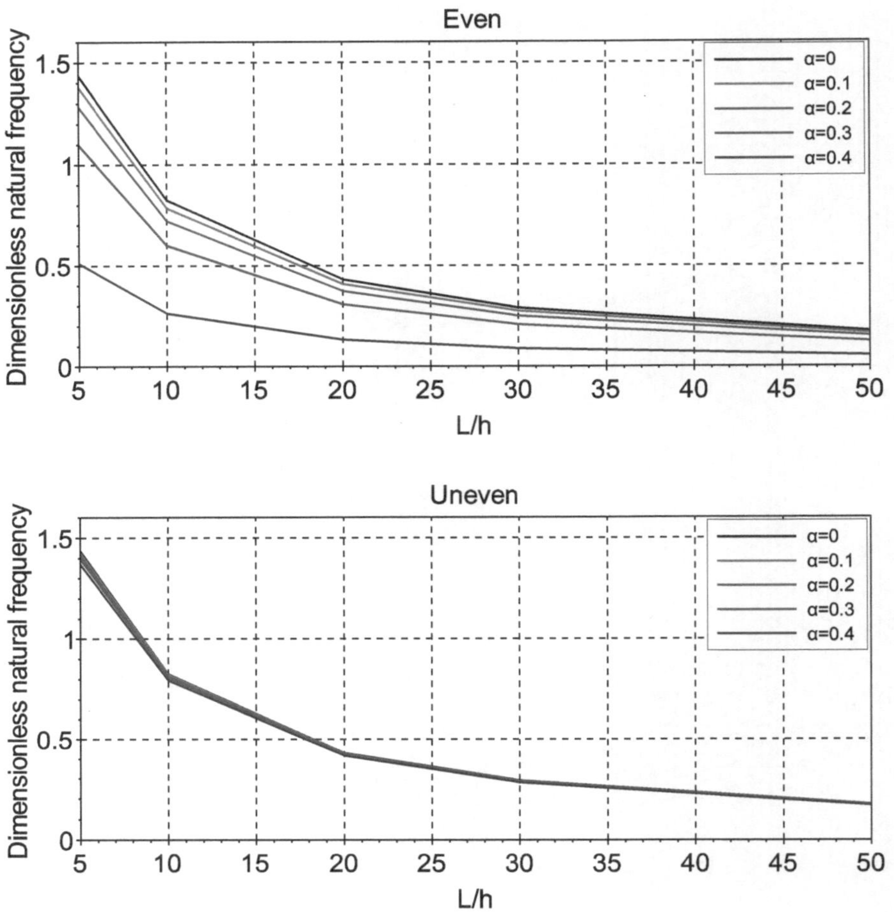

Figure 7.17 Variation of the natural frequencies in the function of the slenderness ratio (L/h) of FGM porosity dependent beams (CC, $p = 2$) for different porosity coefficients and with evenly and unevenly porosity distributions.

Besides, results are presented for both even and uneven porosity distributions in the case of imperfect beams. As can be remarked, for each case (perfect or imperfect), high values of natural frequencies are detected for ceramic portions which possess high flexural rigidity compared to metal ones. In addition, the difference between even and uneven curves becomes more pronounced with the rise of the gradient index. To outline more on this effect, for different constrained FGM beams (SS, CS, CC, CF), the obtained results are plotted in Fig. 7.15. By comparing the values of the natural frequencies associated to four applied boundary conditions, it could be seen that the maximum of natural frequency parameter is achieved for the clamped-clamped (CC) boundary condition and the minimum ones are associated to clamped-free (CF) case. Besides, the natural frequency from uneven to even drops drastically as the gradient index (p) and the porosity coefficient increase. This phenomenon is related to the decrease of the structural stiffness of the FGM beam caused by two essential factors. The former lied with the rise of the gradient index for which the FGM beam is more reinforced by metal portions and the latter with the high percentage of porosity included in the structure which amplifies the weakness of the stiffness of the beam and induces hence low natural frequencies.

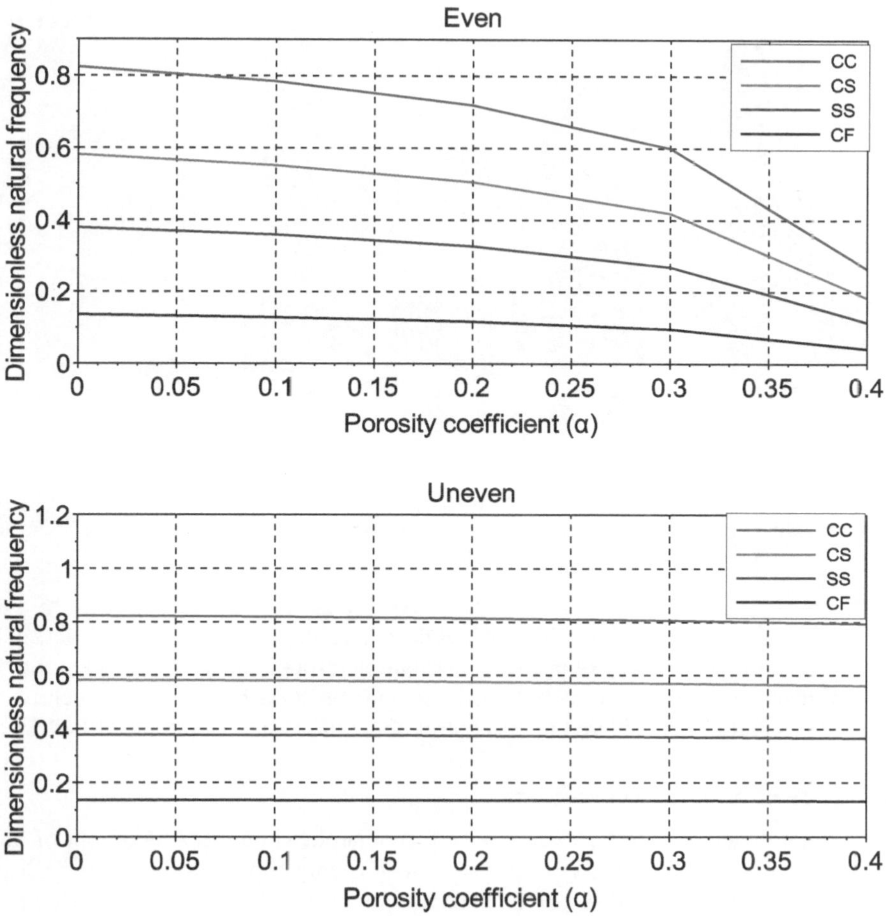

Figure 7.18 Variation of the natural frequency parameter of FGM porosity dependent beam ($L/h = 10$, $p = 2$) for different boundary conditions and with evenly and unevenly porosity distributions.

7.8.2.3 Influence of beam slenderness ratio

In this parametric study, the influence of beam slenderness ratio on natural frequencies of the FGM beams is examined. Figure 7.16 shows the variations of the adimensional frequency parameter $\overline{\omega_2}$ versus the slenderness ratio (L/h) for different values of the gradient index (p) of both perfect and porous FGM beams (CC, $\alpha = 0.2$). As indicated in the previous section, the increase of the gradient index (p) from zero to infinity leads to a reduction in the natural frequencies of the FGM beams due to highest stiffness of ceramic portions compared to metallic ones. Furthermore, the rise of the slenderness ratio induces a drastic decrease of the natural frequencies of the FGM beams in the range $L/h \in [5 - 25]$. Beyond this value $(L/h \prec 25)$, the frequencies tend to be constant. In fact, with increasing the beam slenderness ratio, beams become slender with low flexural rigidity and perform hence small vibration frequencies. Continuously, Figure 7.17 highlights the effect of slenderness ratio on vibrational behavior of porous clamped (CC) FGM beams when varying the porosity coefficient and fixing the gradient index $(p = 2)$. As can be seen, the difference between the curves is clearly shown with the increase of the porosity coefficient in the case of even porosity

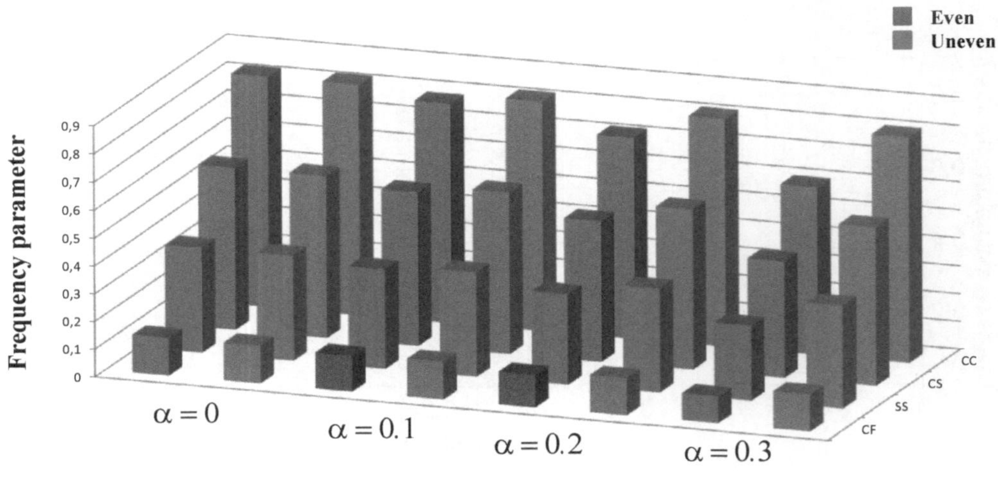

Figure 7.19 Histogram of repartition of natural frequencies for evenly and unevenly FGM porosity dependent beams ($L/h = 10$, $p = 2$) under different boundary conditions.

distribution, in contrast to uneven porosity distribution where the curves are more sticked. This observation affirms again that the way of porosity distribution through the thickness direction of the beam has a significant influence on the vibrational behavior of FGM beams. Consequently, these material parameters should be considered in the control phase of passive vibrations of FGM beams.

7.8.2.4 Influence of boundary conditions

Continuously with provided parametric studies, the effect of applied boundary conditions on natural frequencies of porous FGM beams ($L/h = 10, p = 2$) with evenly and unevenly porosity distributions is presented in Fig. 7.18. It can be seen that beams with clamped-clamped edges (CC) present high natural frequencies compared to clamped-free (CF) ones. The other ones associated to simply-supported (SS) or clamped-simply supported (CS) are located between these extreme cases. This can be explained by the fact that fully constrained edges are stronger than simply supported or free ones and the beams perform hence high natural frequencies. Furthermore, the natural frequencies drop drastically with high values of porosity coefficient where the stiffness of the FGM beams is reduced by the added percentage of porosity. This effect is more prominent for even type of porosity than uneven one. As explained previously, the type of porosity distribution has a significant effect on natural frequencies of FGM porous beams. An overall presentation of the effect of boundary condition on frequency parameter of FGM porous even and uneven beams is shown in Fig. 7.19. From this graph, three points can be concluded: fully clamped beams possess the highest natural frequency parameter; Natural frequencies associated to uneven FGM beams are relatively high to even ones; The increase of the porosity coefficient induces a reduction in the natural frequencies for evenly FGM beams more than unevenly ones. So, the type of applied boundary condition, the percentage of porosity included in the structure and the way of distribution are influential parameters on vibrations of FGM beams which must be considered in both design and control phases. Finally, some modes of vibration related even and uneven porous FGM beams are presented in Figs. 7.20, 7.21, 7.22 and 7.23, respectively.

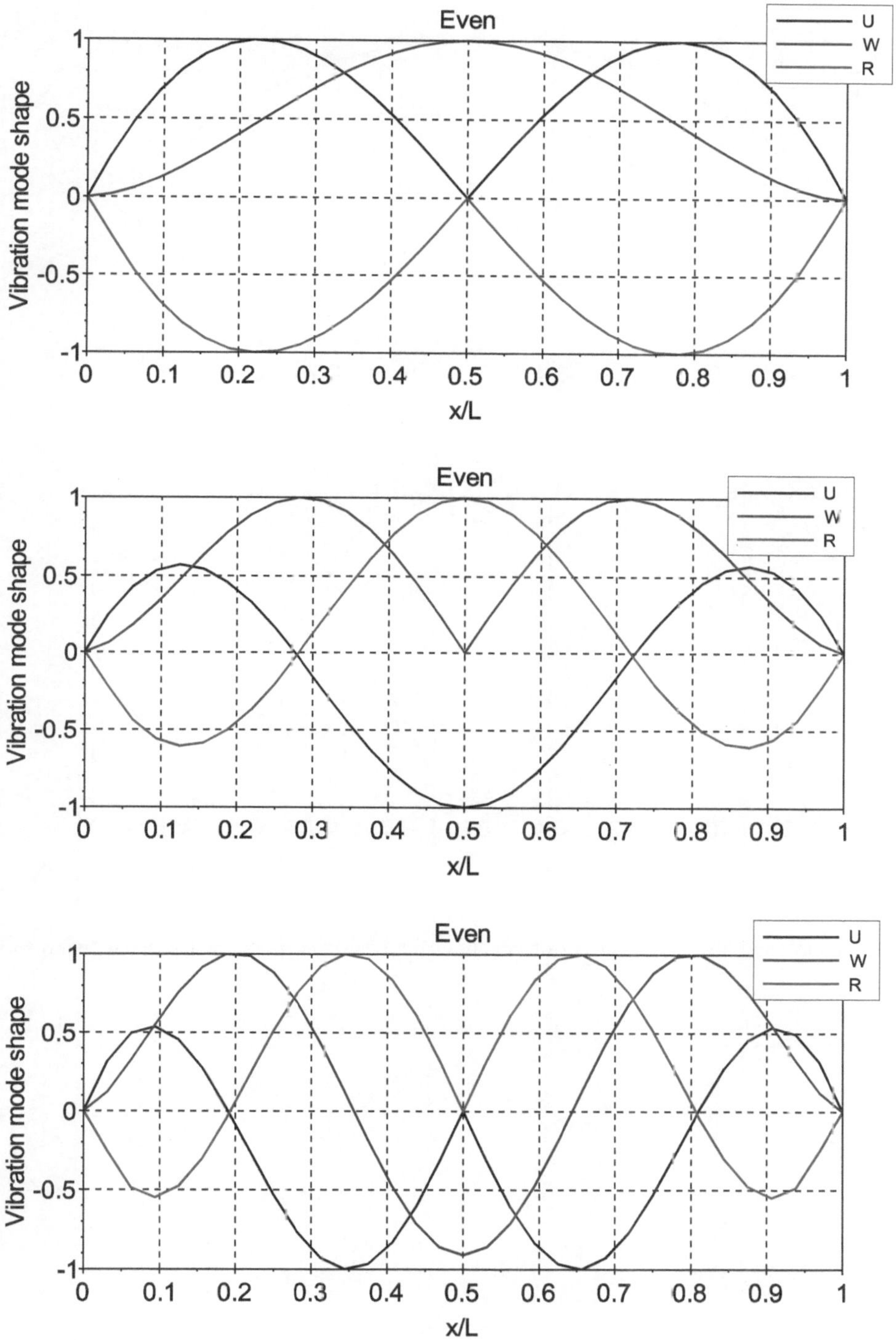

Figure 7.20 First, second and third modes shapes of vibrations for even FGM porosity dependent beams ($p = 2$, $\alpha = 0.1$, $L/h = 10$).

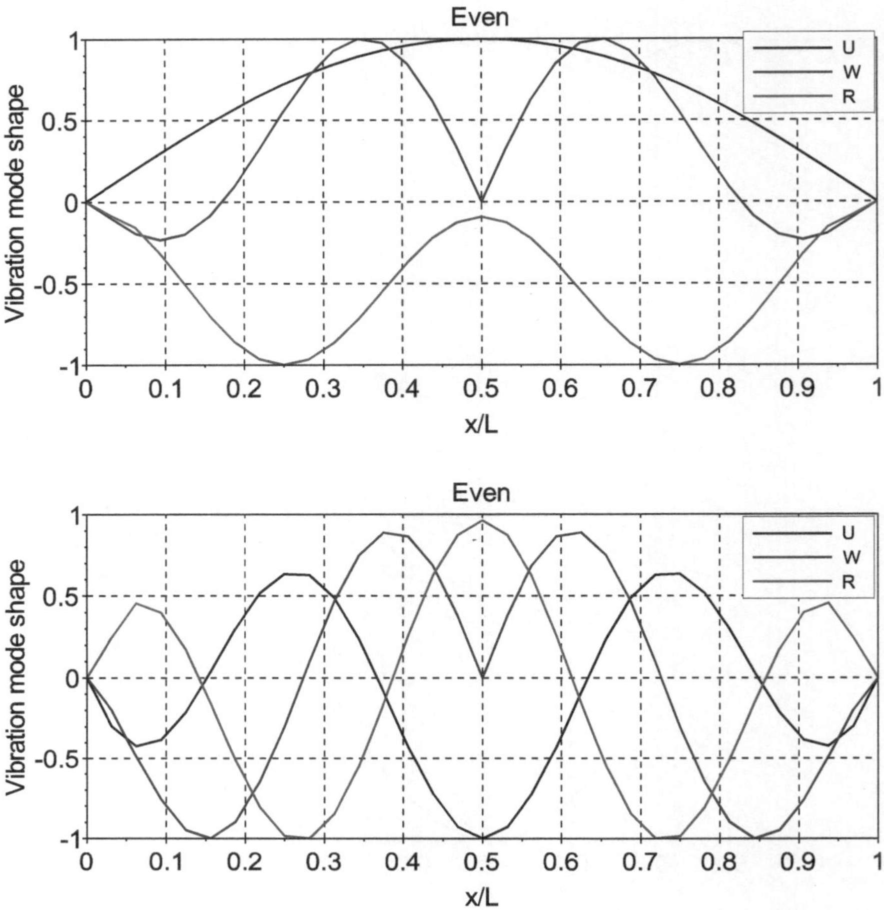

Figure 7.21 Fourth and fifth modes shapes of vibrations for even FGM porosity dependent beams ($p = 2$, $\alpha = 0.1, L/h = 10$).

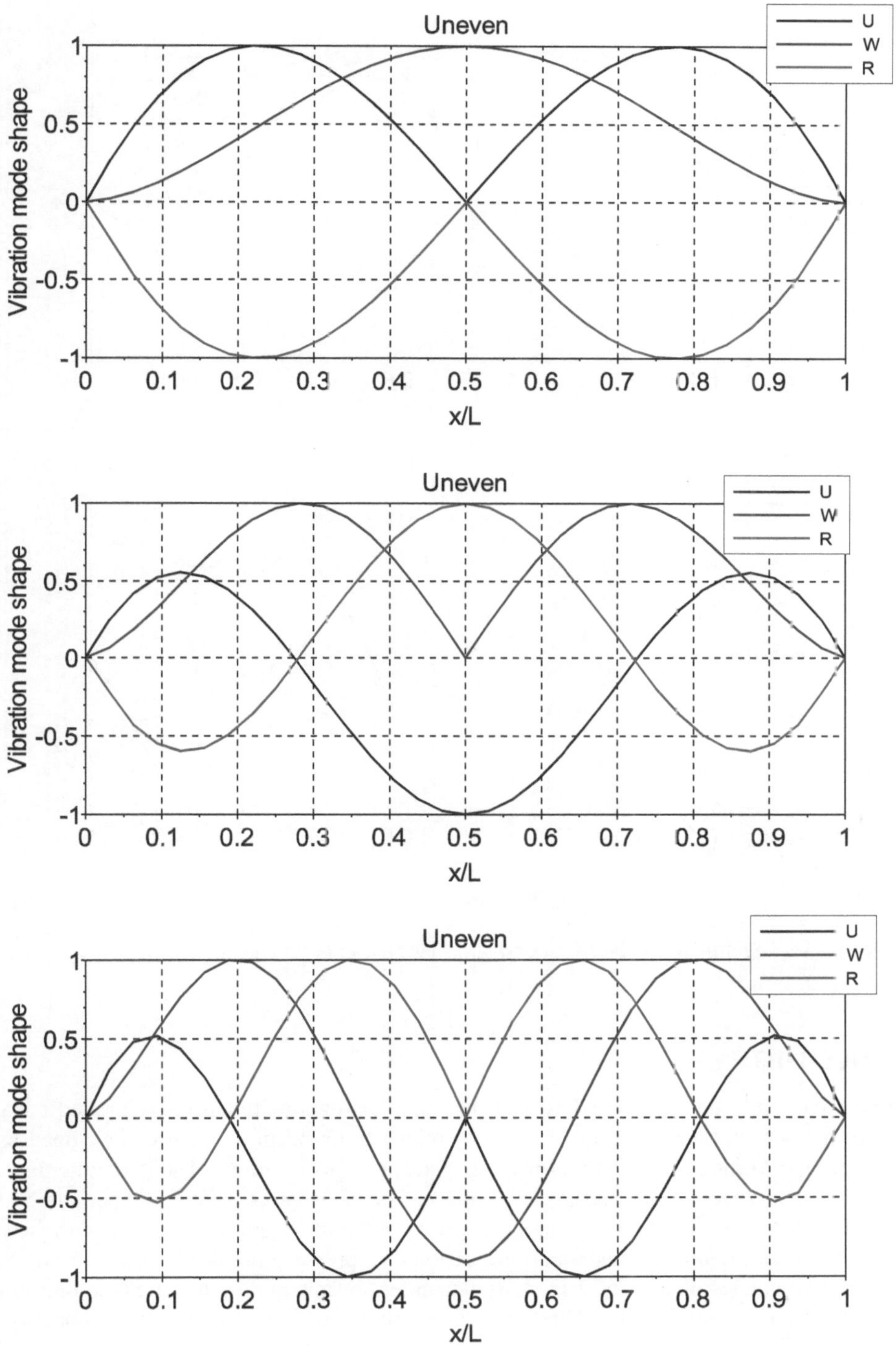

Figure 7.22 First three modes shapes of vibrations for uneven FGM porosity dependent beams ($p = 2$, $\alpha = 0.1$, $L/h = 10$).

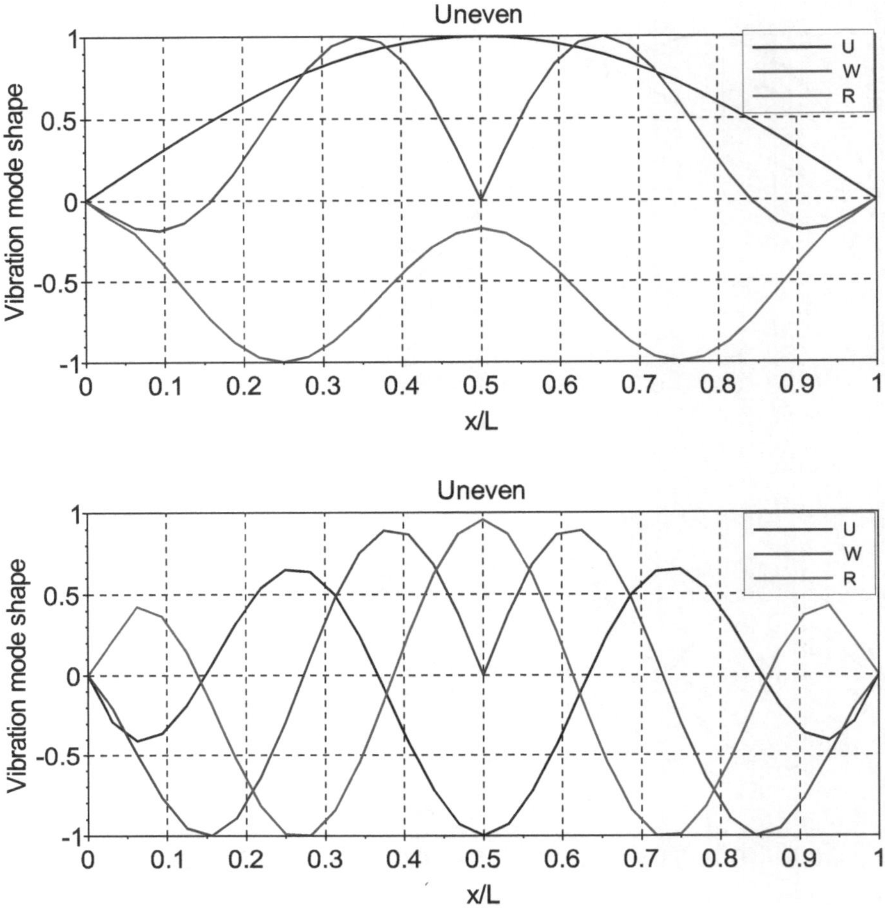

Figure 7.23 Fourth and fifth modes shapes of vibrations for uneven FGM porosity dependent beams ($p = 2$, $\alpha = 0.1$, $L/h = 10$).

7.9 CONCLUSION

A comprehensive study on static bending and natural vibrations of functionally graded porosity dependent beams is investigated in this chapter. A robust finite beam element is developed using a mixed variational formulation and high-order displacement distributions. The parabolic distribution of the transverse shear strains and the annulation of the shear stresses at the upper and bottom surfaces of the beam are verified. The development steps of the formulation beginning with the variational principle until the finite element approximations are presented in detail. The determination of the natural frequencies of the FGM porosity dependent beams and their corresponding modes shapes is carried out by solving the vibrational eigenvalue equation. The impact of porosity parameter, types of distributions and geometrical parameters are detailed via a parametric study. It is concluded that deflections and natural frequencies for unevenly FGM beams are higher than evenly ones. Low values of slenderness ratio induce high values of natural frequencies. The increase of the gradient index leads to reduced natural frequencies due to added metal portions compared to ceramic ones. The raise in the porosity coefficient reduces further the natural frequencies. Hence, for a selected slenderness ratio, gradient index, boundary condition, percentage of porosity included in the structure and type of distribution, vibrations may be controlled passively. Further investigations of FG porous plates and shells will be presented in the next chapter to highlight more the influence of the porosity parameter and its distributions on mechanical responses of such structural components.

8 Mechanical analysis of FG porous structures

This chapter deals with numerical analysis on vibrational, dynamic and mechanical and thermal buckling behaviors of functionally graded porous plates and shells using the improved (FSDT) shell model, which is already developed in Chapter 1. Unlike standard (FSDT) theory, the (I-FSDT) model presents an improvement of the FSDT via the introduction of a quadratic shear function allowing the exact representation of the parabolic distribution of the shear strains and avoiding the utilization of shear correction factors which are computationally costs. The finite element method is used for the discretization of the equations of motion and the assumed natural transverse shear strain (ANS) method is employed to avoid shear locking problems. The effect of porosity parameter and its distributions, namely even and uneven forms, are investigated through different numerical illustrative examples including free vibration, dynamic and buckling responses of FG porous structures.

8.1 INTRODUCTION

Recently and continuously with the functionally graded concept, an important category of lightweight porous materials has received worldwide interests due to their excellent multi-functionality offered by low specific weight, reduced thermal and electrical conductivity and efficient capacity of energy dissipation. The combination between functionally graded (FGM) concept and porous materials has led to a novel FGM in which FG porous materials are characterized by the graded distribution of internal pores in the micro-structure. As a result, FGM structures can be divided into two categories: perfect and imperfect FGMs. The structure is said to be "imperfect" when porosities appears as defects in the micro-structural section. Therefore, the investigation of their mechanical characteristics is of great interest for engineering design and manufacture. Wattanasakulpong et al. [288, 289, 287] studied the vibrational behavior of FGM porous beams using either the classical beam theory (CBT) within the Love-Kirchoff kinematic assumptions or the Timoshenko beam theory with the Chebyshev collocation method. In another work, Chen et al. [46, 47, 127] investigated the mechanical behavior of FGM porous beams using the Timoshenko beam theory. In numerical resolution, the Ritz method is utilized to solve the equations of motion and to determine the natural frequencies and the transient dynamic deflections of FGM-porous beams. Based on quasi-3D theory, Atmane et al. [13] presented static and vibration analysis of FGM porous beams, taking into account the transverse shear deformations and the thickness stretching effects. Continuously, free vibration and buckling analysis of FGM porous plates are performed by Rezaei et al. [211] and Trinh et al. [274] using either analytical or semi-analytical approaches with the aid of some numerical methods like Monte Carlo simulation and the Levy-type solution. Using a state space approach, Demirhan and Taskin [56] studied the vibrational behavior of FGM porous plates with the emphasis on the effect of porosity parameter on natural frequencies of such structures. In other works, and based on high-order shear deformation plate theories (HSDTs), many researchers have studied and analyzed bending, vibration, buckling and post-buckling behavior of FGM porous plates [172, 195, 167] and shells [5]. Based on the HSDT theory, Adhikari et al. [4] studied the buckling behavior of porous FGM sandwich plates under various types of nonuniform loads. The impact of nonuniform type of load, porosity coefficient and others various parameters on critical buckling load of FGM porous plates is presented and discussed. Furthermore, buckling analysis of FGM porous plates are performed by Trinh et al. [274] using a semi-analytical approach

DOI: 10.1201/9781003483786-8

with the aid of Monte Carlo simulation. In another work, Singh and Harsha studied analytically the vibrations and buckling responses of sigmoid (S-FGM) plates. The Galerkin Vlasov's method is employed to obtain governing differential equations. The effects of porosity coefficient, type of porosity distributions and boundary conditions are also examined and presented. On the other hand, Magnucki and his co-authors presented various analyses of rectangular [152] and circular plates [31] with varying mechanical properties. In these studies, non-linear hypothesis of the deformations are introduced and either analytical and finite element solutions are provided. As well as, bending, dynamic and buckling analysis of rectangular plates are performed in the sense of non-linear theory of deformation of the straight line normal to the neutral surface [55, 151, 150]. This improvement of plate theories covers the family of structures from the sandwich ones with symmetrically variable mechanical properties of the cores to the homogeneous structures. These works are based on HSDTs or Quasi-3D elasticity theories which provide generally accurate solutions but they include complicated and time-consuming calculations which are not efficient for complex geometries. In contrast, the first-order shear deformation theory (FSDT) is usual a simple and efficient theory that able to provide accurate results with low computational time. Another point consists of locking phenomena which occurs generally when the formulation considers transverse shear deformations or incompressibility in 3D constitutive material law. Within the context of linear Mindlin-Reissner plate theory, several interpolation schemes have been proposed to ameliorate the well-known shear locking effect which typically arises as the thickness of the plate goes to zero. So, to avoid the shear locking phenomenon that dominates the performance of displacement finite element formulations in the thin shell limit, several methods have been developed such as the assumed natural transverse shear strain (ANS) and the enhanced assumed strain (EAS) methods. For further details about these methods, one can be referred to these references [149, 22, 242, 119] among others. From the above literature, it can be remarked that there is a lack on the investigations of the buckling behavior of FGM porous plates and in particular shells. So, the present work has been done on the above limitations to bring the completeness of analysis of buckling responses of FGM porous structures. Furthermore, the most of studies use frequently the standard FSDT or the HSDT theories which present with aforementioned some limitations and drawbacks. So, as an effective compromise between accurate results and fast computational time, we provide in this paper a simple and efficient finite element model based on the FSDT theory with an improvement of the shear part. In fact, the present model introduces a quadratic shear function that able to describe accurately the realistic distribution of the transverse shear strains and to alleviate the shear locking problem the assumed natural transverse shear strain (ANS) method is employed. This model follows in its kinematic the standard (FSDT) theory but does not require the utilization of shear correction factors. As well as, this model is simple to implement it in the most finite element codes. So, we provide here new vibrational, dynamic and buckling results for FGM porous plates and shells undergoing different types of applied loadings where the material properties of these structures are characterized by a smooth and continuous variations of its constituents in the thickness direction. The pores can be uniformly or randomly distributed across the section of the plate/shell leading hence to two forms, namely even and uneven porosity distributions. The influence of porosity parameter, patterns of distributions, gradient index, boundary conditions, geometrical design parameters and different types of edge compression loads is investigated. Results show that the critical buckling load is significantly influenced by the porosity parameter, its distributions and the type of applied load. So, the percentage of porosity, the manner of its spread and the type of edge compression load play an important role in the determination of the load carrying capability of the FGM structures. Thereby, these parameters should be considered in the design and control phases of such structural components.

8.2 MATERIAL PROPERTIES OF FG POROUS STRUCTURES

We consider a functionally graded structure in which porosity is spread following two types of distributions, namely even and uneven patterns as illustrated in Fig. 8.1. The upper surface of the

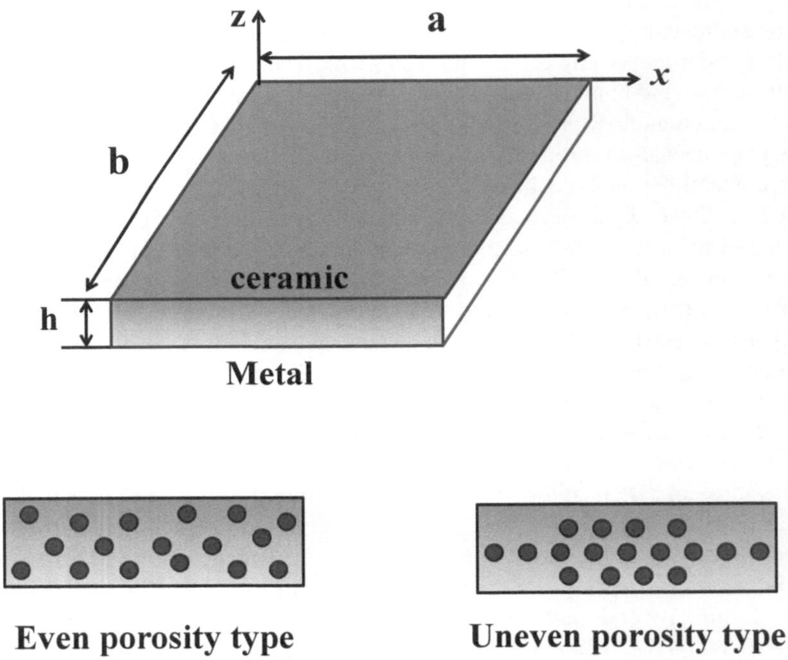

Figure 8.1 Schematic of FGM structure with evenly and unevenly porosity distributions.

structure is ceramic-rich while the bottom surface is metal-rich. To deal with porosity concept, the material property P of the FGM structure can be expressed as [289, 287]:

$$P(z,\alpha) = \begin{cases} (P_c - P_m)V_c + P_m - \frac{\alpha}{2}(E_c + E_m), & (Even\ \ porosity) \\ (P_c - P_m)V_c + P_m - \frac{\alpha}{2}(E_c + E_m)\left(1 - \frac{2|z|}{h}\right), & (Uneven\ \ porosity) \end{cases}, \tag{8.1}$$

where P represents Young's modulus and density of the structure, and the subscripts m and c denote metal and ceramic layers, respectively. Poisson's ratio is stated constant; V_c and α designate the volume fraction of ceramic layer and the porosity volume fraction (or also porosity coefficient), respectively. The volume fraction V_c can obey a power-law distribution (P-FGM) or sigmoid function (S-FGM) which can be written for each form as follows:

P-FGM

$$V_c = \left(\frac{1}{2} + \frac{z}{h}\right)^p, \quad z \in [-h/2, h/2] \tag{8.2}$$

where p is the gradient index which vary from zero to infinity. By setting $p = 0$, a fully ceramic structure is obtained, whereas $p \approx \infty$ a fully metallic structure is considered.

S-FGM

The volume fraction of sigmoid function (S-FGM) is defined by:

$$\begin{aligned} V_f^1(z) &= 1 - \tfrac{1}{2}\left(\tfrac{h/2 - z}{h/2}\right)^p, \quad 0 \leq z \leq \tfrac{h}{2} \\ V_f^2(z) &= 1 - \tfrac{1}{2}\left(\tfrac{h/2 + z}{h/2}\right)^p, \quad \tfrac{-h}{2} \leq z \leq 0 \end{aligned} \tag{8.3}$$

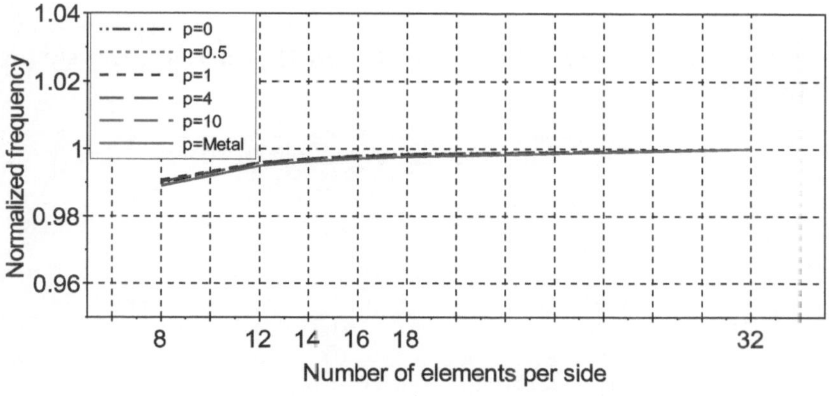

Figure 8.2 Convergence illustration of perfect (Al/Al_2O_3) FGM square plates ($a/h = 5$, SSSS) with different gradient indexes.

8.3 VIBRATION CHARACTERISTICS OF FG POROUS PLATES AND SHELLS

In this section, vibration characteristics of FG porous plates and shells are performed using the (I-FSDT) finite shell model developed in Chapter 1.

8.3.1 FREQUENCY ANALYSIS OF FG POROUS PLATES

Different numerical illustrations are presented, where the FG materials are formed by Aluminum (Al) and Alumina (Al_2O_3) constituents, otherwise noted:

metal (Aluminum, Al): $E_m = 70 \times 10^9 N/m^2$; $\rho_m = 2702 Kg/m^3$; $\nu = 0.3$

ceramic (Alumina, Al_2O_3): $E_c = 380 \times 10^9 N/m^2$; $\rho_c = 3800 Kg/m^3$; $\nu = 0.3$

8.3.2 RECTANGULAR PLATES

8.3.2.1 Perfect FGM plates

A check out of the present model for perfect (Al/Al_2O_3) FGM plates is carried out. A convergence illustration is provided in Fig. 8.2 for a simply supported (SSSS) FGM square plate ($a/h = 5$), where a monotonic convergence trend is observed for different gradient indexes. Next, the non-dimensional natural frequency parameter $\overline{\omega} = \omega h\sqrt{\rho_c/E_c}$ is computed for three sets of length-to-thickness ratios ($a/h = 5, 10, 20$) and different gradient indexes. The obtained results are reported in Tabs. 8.1 and 8.2 and they are compared with the standard FSDT model, [211], analytical solutions [96], element-free kp Ritz method [291], HSDT theory [159] and the refined plate theories (RPT) [26, 56]. A good agreement is revealed among the results which affirms the effectiveness and correctness of the present model.

8.3.2.2 FGM plates with pores imperfections

The second check out is for FGM plates with graded pores in the thickness direction as illustrated in Fig. 8.3. The computation of non-dimensional natural frequencies $\omega^* = \omega h\sqrt{\rho_m/E_m}$ of (Al/Al_2O_3) rectangular plates for even and uneven pores distributions with varying gradient indexes (p), length-to-thickness ratios (a/h) and boundary conditions are reported in Tabs. 8.3 and 8.4. Compared with a simple FSDT theory [211] and a HSDT theory [56] for SSSS and SCSC boundary conditions, the

Table 8.1

Natural frequencies versus the gradient index (p=0, 0.5, 1) of perfect FGM simply supported (SSSS) square plates.

a/h	*Theories*	Gradient index (p)		
		0	0.5	1
5	Present(Improved FSDT)	0.2110	0.1846	0.1712
	FSDT.[211]	0.2127	0.1816	0.1640
	FSDT-analytical.[96]	0.2112	0.1806	0.1650
	FSDT-based element-free kp Ritz.[291]	0.2055	0.1757	0.1587
	HSDT.[159]	0.2121	0.1819	0.1640
	RPT.[26]	0.2112	0.1806	0.1628
	RPT.[56]	0.2218	0.1887	0.1699
10	Present(Improved FSDT)	0.05766	0.05017	0.04666
	FSDT.[211]	0.05780	0.04910	0.04420
	FSDT-analytical.[96]	0.05769	0.04920	0.04454
	FSDT-based element-free kp Ritz.[291]	0.05673	0.04818	0.04346
	HSDT.[159]	0.05777	0.04917	0.04427
	RPT.[26]	0.05769	0.04900	0.04417
	RPT.[56]	0.05850	0.04960	0.04470
20	Present(Improved FSDT)	0.01479	0.01284	0.01195
	FSDT.[211]	0.01480	0.01250	
	FSDT-analytical.[96]	0.01480	0.01281	0.01150
	FSDT-based element-free kp Ritz.[291]	0.01464	0.01241	0.01118
	HSDT.[159]	-	-	-
	RPT.[26]	0.01480	0.01254	0.01130
	RPT.[56]	0.01490	0.01260	0.01130

present results are in good agreement with those reported in the literature and the some discrepancies can be explained by the different theories and strategies solutions. It should be noted that for SFSF and SCSF boundary conditions Tabs. 8.5 and 8.6, only the reference [211] is available. In general, the present model draws a good accuracy for free vibration analysis of FGM porous plates. To further improve the robustness and the efficiency of the present model, the CPU time relative to the following theories (CPT; FSDT (standard); FSDT improved); HSDT) is presented in Fig. 8.4. It is seen that the improved model spent less computational time than the HSDT and averagely the same as the standard FSDT ($2.37s$ in comparison of $2.34s$ for standard model) and a quite more than the CPT. So, this method can bring up the improvement of the shear part of the transverse strains without additional computational time compared to the standard FSDT which requires the calculations of the shear correction factors and it constitutes an advantageous choice compared with the HSDT theory which requires more computational time. So, within the compromise accurate results/low computational time, this model is an alternative to solve free vibration responses of FGM porous structures.

8.3.2.3 Parametric study

In this part, we start with the influence of porosity coefficient, porosity distributions, gradient index (p) on natural frequencies of FGM porous plates. The non-dimensional frequency parameter $\omega^* = \omega h\sqrt{\rho_m/E_m}$ versus the porosity coefficient (e) for FGM square plates ($a/h = 20$) with different gradient indexes (p) and boundary conditions is illustrated in Fig. 8.5. In these figures, with the increase of the gradient index (p), the curves related to uneven porosity distribution tend to move away and lie above the ones corresponding to even porosity distribution, gradually.

Table 8.2

Natural frequencies versus the gradient index (p=4, 10, ∞) of perfect FGM simply supported (SSSS) square plates.

a/h	*Theories*	Gradient index (p)		
		4	10	∞
5	Present(Improved FSDT)	0.1470	0.1337	0.1074
	FSDT.[211]	0.1405	-	-
	FSDT-analytical.[96]	0.1371	0.1304	0.1075
	FSDT-based element-free kp Ritz.[291]	0.1356	0.1284	-
	HSDT.[159]	0.1383	0.1306	0.1077
	RPT.[26]	0.1375	0.1300	0.1075
	RPT.[56]	0.1451	-	-
10	Present(Improved FSDT)	0.04121	0.03786	0.02934
	FSDT.[211]	0.03830	-	-
	FSDT-analytical.[96]	0.03825	0.03527	0.02936
	FSDT-based element-free kp Ritz.[291]	0.03757	0.03591	-
	HSDT.[159]	0.03811	0.03642	0.02933
	RPT.[26]	0.03804	0.03635	0.02936
	RPT.[56]	0.03870	-	-
20	Present(Improved FSDT)	0.01066	0.00982	0.00752
	FSDT.[211]	0.00980	-	-
	FSDT-analytical.[96]	0.01013	0.00963	-
	FSDT-based element-free kp Ritz.[291]	0.00970	0.00931	-
	HSDT.[159]	-	-	-
	RPT.[26]	0.00980	0.00940	-
	RPT.[56]	0.00980	-	-

Additionally, the natural frequency of the FG porous plate rises, as the gradient index (p) is small and it reaches its maximum with $(p=0)$ where the plate is fully ceramic with high rigidity. Nevertheless, the natural frequency drops as (p) is increased to a prescribed value for FGM plates with evenly distributed pores. This matter can eventually be observed for uneven distribution as power law index gets larger. The phenomenon stems from the fact that the effect of decreasing structural stiffness overshadows the decreasing inertia at this point, which leads to a downward trend for natural frequency. Therefore, for a prescribed value of gradient index, and with increasing porosity coefficient, the ratio of mass to stiffness may also change. So, the values of dimensionless natural frequency strongly depend on which part of the ratio is more reduced. As a result, since the values of the mass to stiffness ratio is very sensitive to the changes of gradient index (p), it is observed that for $(p=1)$, the fluctuations of natural frequencies are more prominent than for low values of gradient indexes $(p=0,0.1,0.5)$ and an increase in natural frequency can be detected for a predefined porosity coefficient.

Next, the effects of porosity coefficient, different distributions and gradient index (p) on high modes of vibration are investigated. Figs. 8.6(a) and 8.6(b) depict the variation of the non-dimensional frequency parameter versus the gradient index for the first five modes of vibrations for a simply supported (SSSS) square FGM porous plates $(a/h=10)$ with evenly and unevenly porosity distributions. It can be seen that the increase of porosity coefficient affects significantly the high modes of vibration more than the first ones which manifest by a pronounced decrease in the natural frequency parameter. To highlight more on this effect, the variation of the frequency parameter versus the gradient index for the fourth and the five high modes of vibration of the FG porous plate $(a/h=10)$ with two sets of porosity coefficients $(e=0.1,0.2)$ and for even and uneven types of porosity distributions is depicted in Fig. 8.7. It is revealed that, the natural frequency decreases

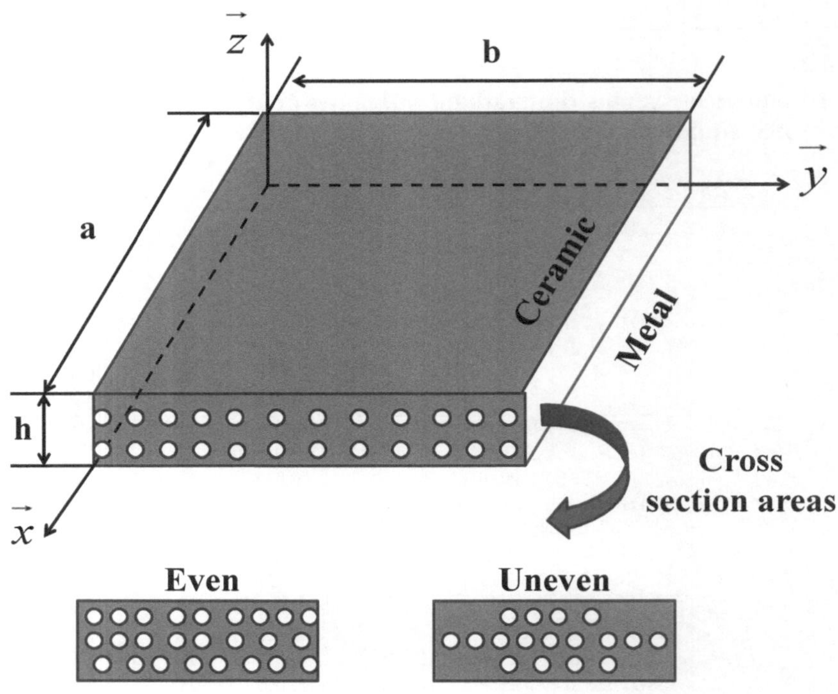

Figure 8.3 Geometry of imperfect FGM porous plate with different porosity distributions.

as the gradient index (p) increases and this effect is more prominent with even porosity distribution than uneven porosity type. In fact, with the rise of the gradient index, the plate behaves as metallic structure with low rigidity and with even porosity type, the pores are uniformly distributed through the whole of the plate whereas for uneven type of porosity the pores are concentrated in the mid-plane surface and vanish in the extreme surfaces. As a result, the elasticity modulus of the FG porous plate with even distribution is lower than the uneven one. The reason for the variation of the natural frequencies with respect to the gradient index is the effect of bending-stretching coupling coefficients (H_m, H_{mb}, H_b, H_s) which depends on porosity volume fraction and types of distributions. As well as, it is shown that the difference between the frequencies of even and uneven FG porous plates step up as the gradient index and porosity coefficient rise and higher frequencies are observed for uneven type when the porosity is enhanced in the mid-surface where stiffer and denser material is assumed, in contrast with even type of porosity which makes the plate softer and lighter. Consequently, the type of porosity distribution has a significant effect on natural frequencies and vibration modes of FG porous plates which must be considered in the design phase. Now, the effects of boundary conditions on natural frequency parameter of square FG porous plates $(a/h = 20$, $\mathrm{p} = 0.5)$ are investigated where the non-dimensional frequency parameter versus the porosity coefficient (e) for square FG porous plates with evenly and unevenly porosity distributions is depicted in Fig. 8.8. Four types of boundary conditions are considered: all edges are simply supported (SSSS), two edges are simply supported and the others are clamped (SCSC), two edges are simply supported and the others are free (SFSF) and a combination of simply supported, clamped and free (SCSF). As observed, the natural frequencies of SCSC FG porous plates are the highest ones whereas those of SFSF FG porous plates are the lowest ones. The curves corresponding to SSSS and SCSF FG porous plates are located between these two extreme cases. In fact, when the structure contains more constraints, its rigidity is high and manifest hence by high values of natural frequencies. It is

Table 8.3

Natural frequencies of imperfect FGM porous plates with different porosity distributions, length-to-thickness ratios, power indexes (p = 0,0.1) and boundary conditions (SSSS, SCSC).

a/h				p = 0		p = 0.1	
	BC	*e*	Ref.	Even	Uneven	Even	Uneven
20	SSSS	0	Present	0.02908	0.02908	0.02804	0.02804
			FSDT.[211]	0.02912	0.02912	0.02802	0.02802
			HSDT.[56]	0.02922	0.02922	0.02811	0.02811
		0.2	Present	0.02949	0.02984	0.02831	0.02876
			FSDT.[211]	0.03003	0.02999	0.02875	0.02884
			HSDT.[56]	0.03014	0.03013	0.02885	0.02891
		0.4	Present	0.03042	0.03077	0.02904	0.02964
			FSDT.[211]	0.03137	0.03101	0.02985	0.02980
			HSDT.[56]	0.03148	0.03129	0.02995	0.02993
	SCSC	0	Present	0.04213	0.04213	0.04059	0.04059
			FSDT.[211]	0.04240	0.04240	0.04081	0.04081
			HSDT.[56]	0.04264	0.04264	0.04104	0.04104
		0.2	Present	0.04277	0.04322	0.04101	0.04162
			FSDT.[211]	0.04373	0.04366	0.04188	0.04199
			HSDT.[56]	0.04398	0.04395	0.04211	0.04222
		0.4	Present	0.04414	0.04454	0.04208	0.04287
			FSDT.[211]	0.04568	0.04513	0.04348	0.04338
			HSDT.[56]	0.04594	0.04564	0.04372	0.04370
10	SSSS	0	Present	0.11338	0.11338	0.10940	0.10940
			FSDT.[211]	0.11369	0.11369	0.10945	0.10945
			HSDT.[56]	0.11514	0.11514	0.11082	0.11082
		0.2	Present	0.11515	0.11628	0.11062	0.11214
			FSDT.[211]	0.11726	0.11699	0.11233	0.112569
			HSDT.[56]	0.11875	0.11862	0.11373	0.11388
		0.4	Present	0.11891	0.11979	0.11364	0.11548
			FSDT.[211]	0.12249	0.12085	0.11662	0.11621
			HSDT.[56]	0.12405	0.12314	0.11807	0.11787
	SCSC	0	Present	0.15873	0.15873	0.15322	0.15322
			FSDT.[211]	0.16236	0.16236	0.15640	0.15640
			HSDT.[56]	0.16577	0.16577	0.15963	0.15963
		0.2	Present	0.16174	0.16269	0.15541	0.15693
			FSDT.[211]	0.16745	0.16695	0.16054	0.16072
			HSDT.[56]	0.17097	0.17063	0.16386	0.16163
		0.4	Present	0.16751	0.16743	0.16004	0.16141
			FSDT.[211]	0.17491	0.17230	0.16670	0.16579
			HSDT.[56]	0.17860	0.17710	0.17016	0.16963
5	SSSS	0	Present	0.41510	0.41510	0.40139	0.40139
			FSDT.[211]	0.41821	0.41821	0.40322	0.40322
			HSDT.[56]	0.43624	0.43624	0.42035	0.42035
		0.2	Present	0.42349	0.42500	0.40783	0.41073
			FSDT.[211]	0.43133	0.42933	0.41396	0.41357
			HSDT.[56]	0.44993	0.44885	0.43150	0.43141
		0.4	Present	0.43910	0.43684	0.42074	0.42194
			FSDT.[211]	0.45055	0.44227	0.42994	0.42539
			HSDT.[56]	0.46998	0.46522	0.44808	0.44589
	SCSC	0	Present	0.53210	0.53210	0.51588	0.51588
			FSDT.[211]	0.56436	0.56436	0.54505	0.54505
			HSDT.[56]	0.59187	0.59187	0.57943	0.57943
		0.2	Present	0.54705	0.54426	0.52821	0.52718
			FSDT.[211]	0.58207	0.57833	0.55973	0.55815
			HSDT.[56]	0.61044	0.61790	0.59528	0.57955
		0.4	Present	0.57107	0.55844	0.54868	0.54042
			FSDT.[211]	0.60801	0.59457	0.58158	0.57344
			HSDT.[56]	0.63764	0.63839	0.61875	0.61473

Table 8.4

Natural frequencies of imperfect FGM porous plates with different porosity distributions, length-to-thickness ratios, power indexes (p = 0.5,1) and boundary conditions (SSSS, SCSC).

a/h				p = 0.5		p = 1	
	BC	*e*	Ref.	Even	Uneven	Even	Uneven
20	SSSS	0	Present	0.02526	0.02526	0.02350	0.02350
			FSDT.[211]	0.02467	0.02467	0.02224	0.02224
			HSDT.[56]	0.02475	0.02475	0.02230	0.02230
		0.2	Present	0.02499	0.02582	0.02265	0.02390
			FSDT.[211]	0.02456	0.02524	0.02104	0.02247
			HSDT.[56]	0.02463	0.02542	0.02110	0.02271
		0.4	Present	0.02487	0.02653	0.02107	0.02440
			FSDT.[211]	0.02423	0.02590	0.01817	0.02269
			HSDT.[56]	0.02429	0.02606	0.01821	0.02288
	SCSC	0	Present	0.03593	0.03593	0.03257	0.03257
			FSDT.[211]	0.03595	0.03595	0.03241	0.03241
			HSDT.[56]	0.03614	0.03614	0.03256	0.03256
		0.2	Present	0.03528	0.03661	0.03058	0.03283
			FSDT.[211]	0.03580	0.03677	0.03069	0.03275
			HSDT.[56]	0.03598	0.03692	0.03082	0.03082
		0.4	Present	0.03457	0.03746	0.02671	0.03311
			FSDT.[211]	0.03533	0.03772	0.02654	0.03307
			HSDT.[56]	0.03550	0.03793	0.02663	0.03372
10	SSSS	0	Present	0.09864	0.09864	0.09172	0.09172
			FSDT.[211]	0.09651	0.09651	0.08702	0.08702
			HSDT.[56]	0.09763	0.09763	0.08796	0.08796
		0.2	Present	0.09778	0.10077	0.08855	0.09317
			FSDT.[211]	0.09612	0.09864	0.08244	0.08789
			HSDT.[56]	0.09719	0.10013	0.08324	0.08859
		0.4	Present	0.09746	0.10341	0.08257	0.09492
			FSDT.[211]	0.09494	0.10112	0.07142	0.08868
			HSDT.[56]	0.09592	0.10216	0.07194	0.08991
	SCSC	0	Present	0.13622	0.13622	0.12361	0.12361
			FSDT.[211]	0.13819	0.13819	0.12475	0.12475
			HSDT.[56]	0.14069	0.14069	0.12698	0.12698
		0.2	Present	0.13447	0.13865	0.11688	0.12441
			FSDT.[211]	0.13775	0.14113	0.11845	0.12593
			HSDT.[56]	0.14031	0.14272	0.12034	0.12743
		0.4	Present	0.13250	0.14161	0.10311	0.12521
			FSDT.[211]	0.13627	0.14457	0.10310	0.12701
			HSDT.[56]	0.13869	0.14609	0.10434	0.12984
5	SSSS	0	Present	0.36304	0.36304	0.33671	0.33671
			FSDT.[211]	0.35705	0.35705	0.32249	0.32249
			HSDT.[56]	0.37112	0.37112	0.33425	0.33425
		0.2	Present	0.36188	0.36998	0.32709	0.34081
			FSDT.[211]	0.35629	0.36404	0.30681	0.32496
			HSDT.[56]	0.36981	0.37440	0.31684	0.32764
		0.4	Present	0.36269	0.37839	0.30731	0.34538
			FSDT.[211]	0.35310	0.37215	0.26825	0.32709
			HSDT.[56]	0.36558	0.37814	0.27482	0.33460
	SCSC	0	Present	0.46356	0.46356	0.42170	0.42170
			FSDT.[211]	0.48535	0.48535	0.43991	0.43991
			HSDT.[56]	0.51500	0.51500	0.45863	0.45863
		0.2	Present	0.46346	0.47058	0.40589	0.42314
			FSDT.[211]	0.48553	0.49403	0.42118	0.44286
			HSDT.[56]	0.51467	0.51878	0.43752	0.45552
		0.4	Present	0.46327	0.47868	0.36790	0.42381
			FSDT.[211]	0.48332	0.50412	0.37354	0.44539
			HSDT.[56]	0.51269	0.53345	0.38481	0.46476

Table 8.5

Natural frequencies of imperfect FGM porous plates with different porosity distributions, length-to-thickness ratios, power indexes (p = 0,0.1) and boundary conditions (SFSF, SCSF).

a/h				p = 0		p = 0.1	
	BC	*e*	Ref.	Even	Uneven	Even	Uneven
20	SFSF	0	Present	0.01425	0.01425	0.01374	0.01374
			FSDT.[211]	0.01427	0.01427	0.01372	0.01372
		0.2	Present	0.01459	0.01466	0.01401	0.01413
			FSDT.[211]	0.01471	0.01469	0.01408	0.01413
		0.4	Present	0.01515	0.01515	0.01447	0.01450
			FSDT.[211]	0.01537	0.01519	0.01462	0.01450
	SCSF	0	Present	0.01867	0.01867	0.01799	0.01799
			FSDT.[211]	0.01877	0.01877	0.01806	0.01806
		0.2	Present	0.01925	0.01924	0.01846	0.01853
			FSDT.[211]	0.01935	0.01933	0.01853	0.01859
		0.4	Present	0.02013	0.01991	0.01920	0.01918
			FSDT.[211]	0.02022	0.01999	0.01923	0.01921
10	SFSF	0	Present	0.05621	0.05621	0.05422	0.05422
			FSDT.[211]	0.05639	0.05639	0.05427	0.05427
		0.2	Present	0.05761	0.05782	0.05534	0.05576
			FSDT.[211]	0.05816	0.05806	0.05570	0.05584
		0.4	Present	0.05989	0.05974	0.05723	0.05759
			FSDT.[211]	0.06076	0.06001	0.05782	0.05769
	SCSF	0	Present	0.07290	0.07290	0.07031	0.07031
			FSDT.[211]	0.07388	0.07388	0.07110	0.07110
		0.2	Present	0.07528	0.07510	0.07228	0.07240
			FSDT.[211]	0.07619	0.07605	0.07297	0.07315
		0.4	Present	0.07883	0.07768	0.07527	0.07486
			FSDT.[211]	0.07959	0.07859	0.07576	0.07555
5	SFSF	0	Present	0.21391	0.21391	0.20664	0.20664
			FSDT.[211]	0.21576	0.21576	0.20782	0.20782
		0.2	Present	0.21992	0.21986	0.21154	0.21228
			FSDT.[211]	0.22253	0.221840	0.21331	0.21354
		0.4	Present	0.22921	0.22686	0.21937	0.21895
			FSDT.[211]	0.23245	0.22891	0.22149	0.22024
	SCSF	0	Present	0.27044	0.27044	0.26132	0.26132
			FSDT.[211]	0.27874	0.27874	0.26857	0.26857
		0.2	Present	0.28022	0.27821	0.26961	0.26868
			FSDT.[211]	0.28749	0.28643	0.27569	0.27581
		0.4	Present	0.29440	0.28723	0.28177	0.28723
			FSDT.[211]	0.30030	0.29537	0.28629	0.28427

also noteworthy that the natural frequencies of uneven-type of porosity distributions are little higher than the ones related to even-type of porosity distributions and this effect is more pronounced for SCSC FG porous plates. The reason of this phenomenon is related to the mannner of porosity distribution through the thickness direction of the FG plates: for evenly-type, the porosity is uniformly distributed through the thickness direction of the plate while for unevenly-type, the porosity spread almost around the mid-surface and vanish at the top and bottom surfaces. Consequently, the FG porous plate becomes less rigid with even-type than with uneven-type and that's why the natural frequencies have higher values in the case of uneven porosity distribution. Another remark, relying on the slope between the curves related to even and uneven types of porosity distributions which rises with the increase of the porosity coefficient. This phenomenon can be explained by the value

Table 8.6

Natural frequencies of imperfect FGM porous plates with different porosity distributions, length-to-thickness ratios, power indexes (p = 0.5,1) and boundary conditions (SFSF, SCSF).

a/h				p = 0.5		p = 1	
	BC	e	Ref.	Even	Uneven	Even	Uneven
20	SFSF	0	Present	0.01243	0.01243	0.01164	0.01164
			FSDT.[211]0.01208	0.01208	0.01089	0.01089	
		0.2	Present	0.01244	0.01275	0.01141	0.01190
			FSDT.[211]	0.01202	0.01236	0.01030	0.01100
		0.4	Present	0.01252	0.01315	0.01091	0.01222
			FSDT.[211]	0.01186	0.01269	0.00889	0.01111
	SCSF	0	Present	0.01611	0.01611	0.01488	0.01488
			FSDT.[211]	0.01590	0.01590	0.01432	0.01432
		0.2	Present	0.01616	0.01653	0.01447	0.01517
			FSDT.[211]	0.01582	0.01626	0.01355	0.01448
		0.4	Present	0.01621	0.01704	0.01344	0.01550
			FSDT.[211]	0.01561	0.01669	0.01170	0.01462
10	SFSF	0	Present	0.04907	0.04907	0.04594	0.04594
			FSDT.[211]	0.04781	0.04781	0.04309	0.04309
		0.2	Present	0.04919	0.05033	0.04510	0.04695
			FSDT.[211]	0.04760	0.04889	0.04079	0.04354
		0.4	Present	0.04955	0.05186	0.04319	0.04814
			FSDT.[211]	0.04697	0.05015	0.03526	0.04396
	SCSF	0	Present	0.06306	0.06306	0.05824	0.05824
			FSDT.[211]	0.06266	0.06266	0.05648	0.05648
		0.2	Present	0.06337	0.06467	0.05674	0.05932
			FSDT.[211]	0.06238	0.06406	0.05348	0.05706
		0.4	Present	0.06370	0.06659	0.05287	0.06054
			FSDT.[211]	0.06159	0.06570	0.04626	0.05759
5	SFSF	0	Present	0.18731	0.18731	0.17511	0.17511
			FSDT.[211]	0.18351	0.18351	0.16555	0.16555
		0.2	Present	0.18842	0.19186	0.17249	0.17851
			FSDT.[211]	0.18288	0.18738	0.15705	0.16705
		0.4	Present	0.19043	0.19729	0.16591	0.18246
			FSDT.[211]	0.18083	0.19190	0.13646	0.16839
	SCSF	0	Present	0.23518	0.23518	0.21706	0.21706
			FSDT.[211]	0.23738	0.23738	0.21422	0.21422
		0.2	Present	0.23745	0.24074	0.21266	0.22051
			FSDT.[211]	0.23667	0.24226	0.20342	0.21605
		0.4	Present	0.23993	0.24723	0.19973	0.22418
			FSDT.[211]	0.23420	0.24793	0.17711	0.21766

of ratio of mass to stiffness which can change with the increase of the porosity coefficient and the natural frequency depends hence on which part of the ratio is more reduced. As well as, the rate of porosity has also a significant effect on the diminution of the natural frequencies of FG porous plates and to control the passive vibrations, the adjustment of the percentage of porosity including in the structure constitutes an efficient way. The first five mode shapes of vibration of a simply supported (SSSS) FGM square porous plates ($a/h = 20$) are depicted in Fig. 8.11.

The influence of the thickness-to-length (h/a) ratio on the first five modes of vibration of a simply supported (SSSS) FG porous plates with two gradient indexes ($p = 0.1, 5$) and under different porosity coefficients for evenly and unevenly porosity distributions is investigated in Figs. 8.9 and 8.10, respectively. From these two figures, and for a fixed gradient index ($p = 0.1, 5$), it can be seen

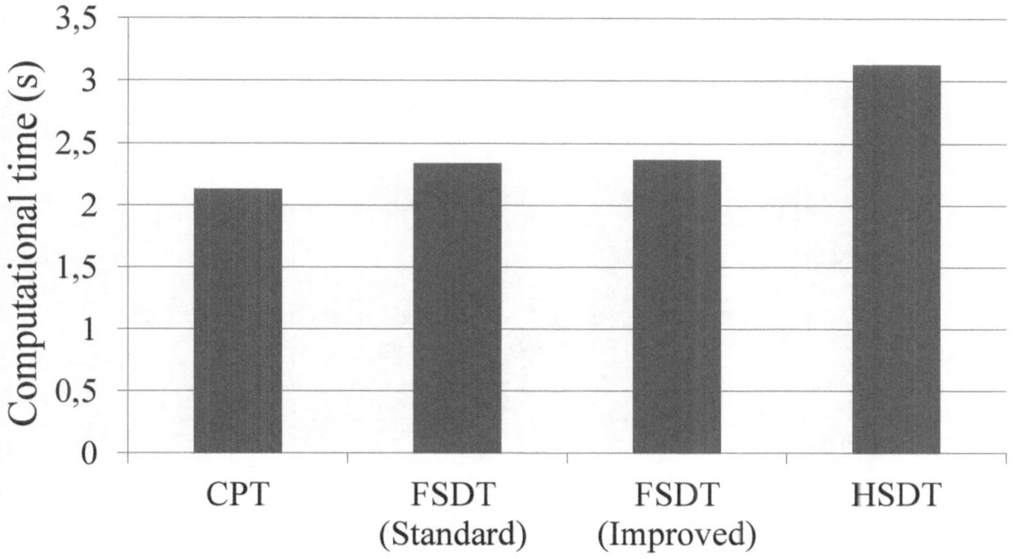

Figure 8.4 CPU time for the different theories and models.

that the non-dimensional natural frequency parameter increases as the thickness-to-length (h/a) ratio increases for all modes of vibration. In fact, with the rise of the thickness-to-length (h/a) ratio, the FG plate behaves as thick structure with high flexural rigidity and hence high values of natural frequencies. In addition, the variation of the porosity coefficient and its increase (from $(e=0)$ to $(e=0.2)$) induces a decrease in the natural frequency and this effect is more pronounced for a FG plate with a gradient index $(p=5)$ than the one with a gradient index $(p=0.1)$. This phenomenon is related to the percentage of metal in the FG plates, where the FG plates with a high portion of metal are more sensitive to the rise of the porosity coefficient due to their low rigidity. Subsequently, from Figs. 8.10(a) and 8.10(b), it can be remarked that the difference between the curves becomes large and more pointed with the increase of the porosity coefficient and with evenly type of porosity more than unevenly type of porosity distribution, and this is especially for the high modes of vibrations. This observation outlines again that the way of distribution of the pores in the whole of the structure has a significant effect on the natural frequencies of FG porous plates which must be considered in the design phase of FG porous materials.

8.3.3 CIRCULAR PLATES

Now, free vibration analysis of FGM porous circular plates are performed. First, the accuracy and validity of the proposed FEM model are verified for perfect FGM circular plates. The circular plate has a radius $(R=0.6m)$ and a total thickness $(h=0.02m)$ as shown in Fig. 8.12. The material properties of ceramic and metal portions are as mentioned in [284, 71]: $E_c=205GPa$, $\rho_c=8900Kg/m^3$, $E_m=200GPa$, $\rho_m=7800Kg/m^3$. The three first natural frequencies (in Hz) of the FGM plate for different gradient indexes and with a clamped boundary condition are presented in Tab. 8.7. Comparing the present results with those published in literature [284, 71], the maximum discrepancy is found less than (1%) which confirms the efficiency and the accuracy of the present formulation. It is noteworthy that the numbers in parentheses correspond to the circumferential wave number or vibrational mode. Another comparison is made with the results of axisymmetric perfect (Al/Al_2O_3) FGM circular plates $(R/h=20)$ on the basis of the classical plate theory using either quasi-Green's

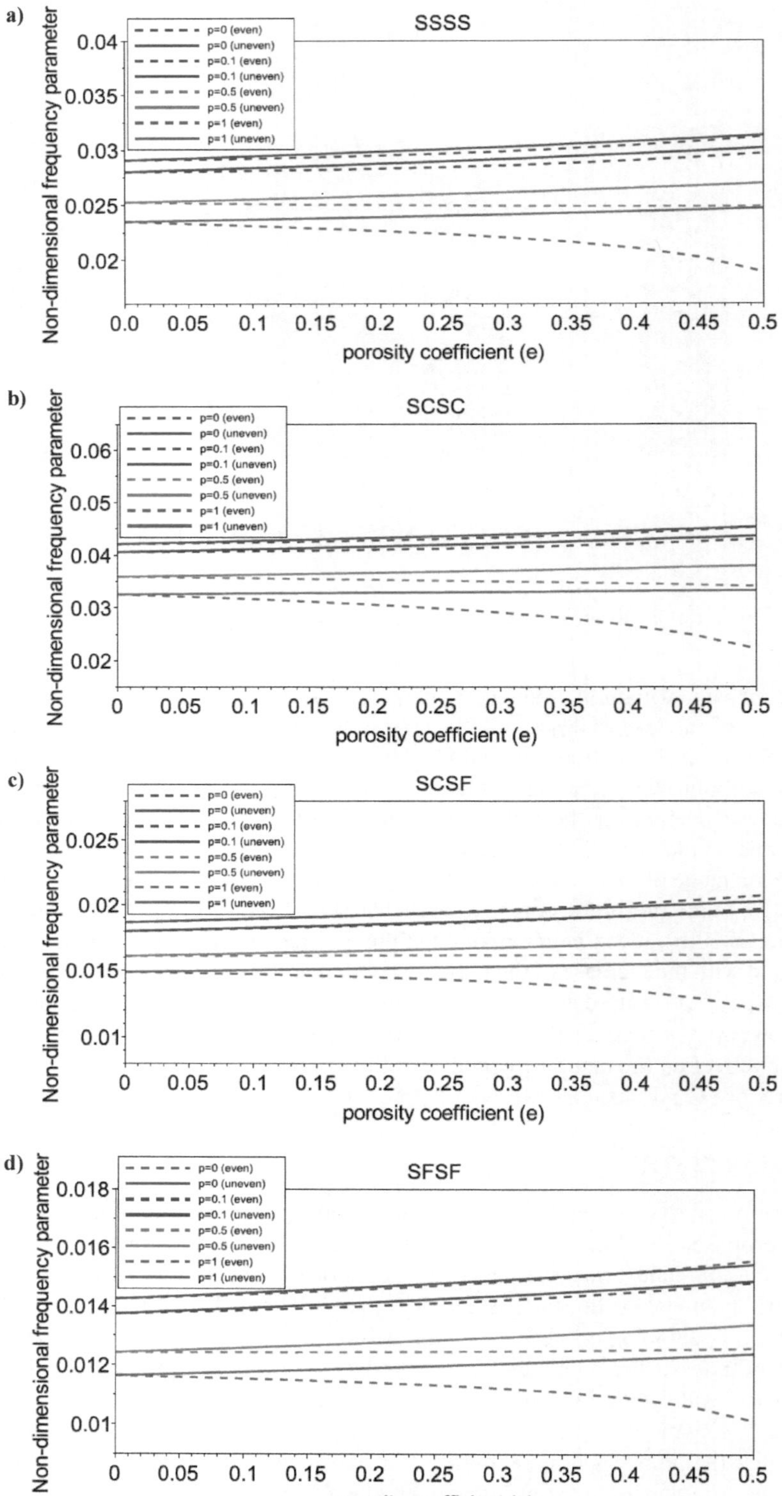

Figure 8.5 Effects of porosity distributions on natural frequencies of (a) SSSS, (b) SCSC, (c) SCSF and (d) SFSF FG square porous plates ($a/h = 20$) with different gradient indexes.

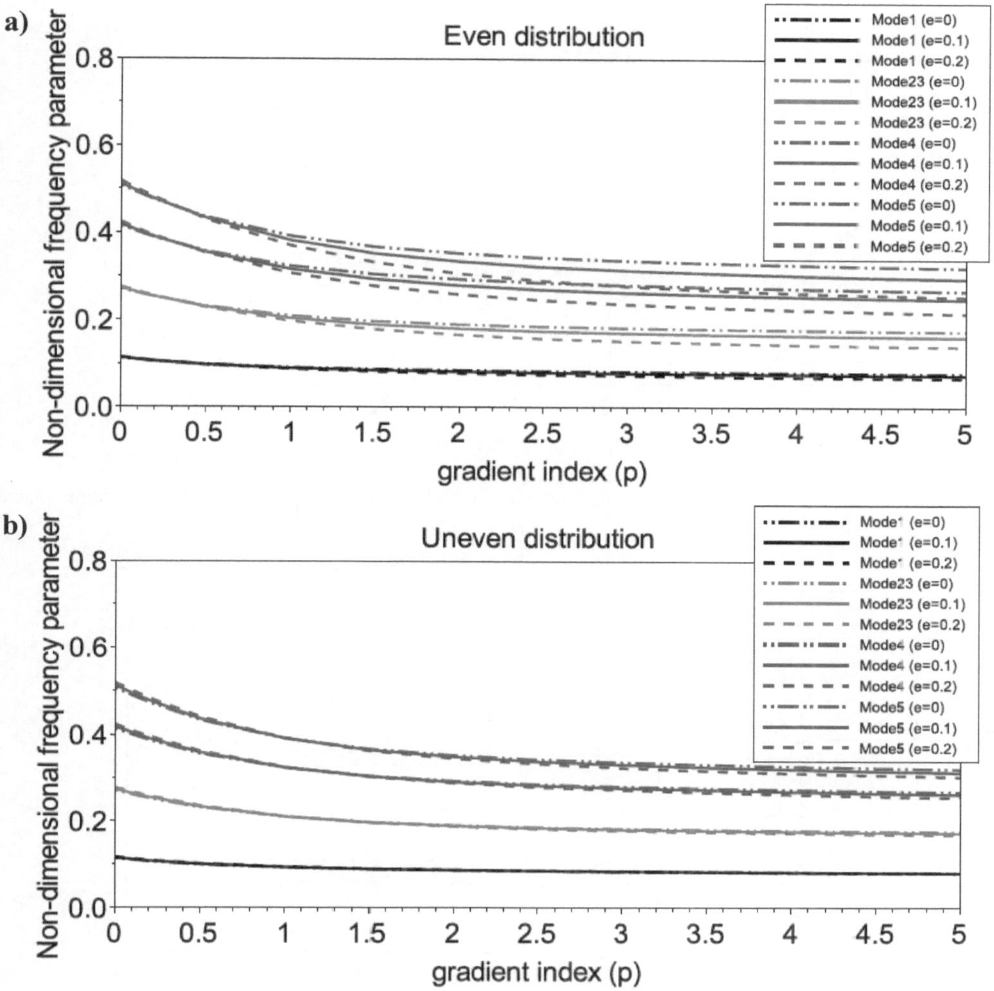

Figure 8.6 Effects of porosity coefficient on the first five modes of vibration of a simply supported (SSSS) square FG porous plates ($a/h = 10$) with (a) even and (b) uneven porosity distributions.

function [329] and analytical closed-form solutions [330]. The non-dimensional natural frequencies $\overline{\omega_c} = \omega R^2 \sqrt{\rho_c h/D}$ are computed for different gradient indexes and boundary conditions where $D = \frac{E_c h^3}{12(1-\nu_c^2)}$ is the flexural rigidity of the circular plate expressed in terms of ceramic Young's modulus E_c, thickness h and ceramic Poisson's ratio ν_c. Tables 8.8 and 8.9 show a good agreement between the results and the little difference which is about 3% for the high modes of vibration can be explained by the different theories where the FSDT-based four-nodes finite shell element is performed in this study, in contrast to classical plate theory (CPT) used in both studies [329, 330]. Moreover, it can be remarked that the present results are more closer to the analytical solutions than the ones derived from quasi-Green's function which draws the pertinence and consistency of the present formulation.

In what follows, and since the number of papers concerning free vibration analysis of FGM porous circular plates is rather scarce, we provide the first new results on vibrational behavior of imperfect FGM porous circular plates with evenly and unevenly porosity distributions. Figures 8.13

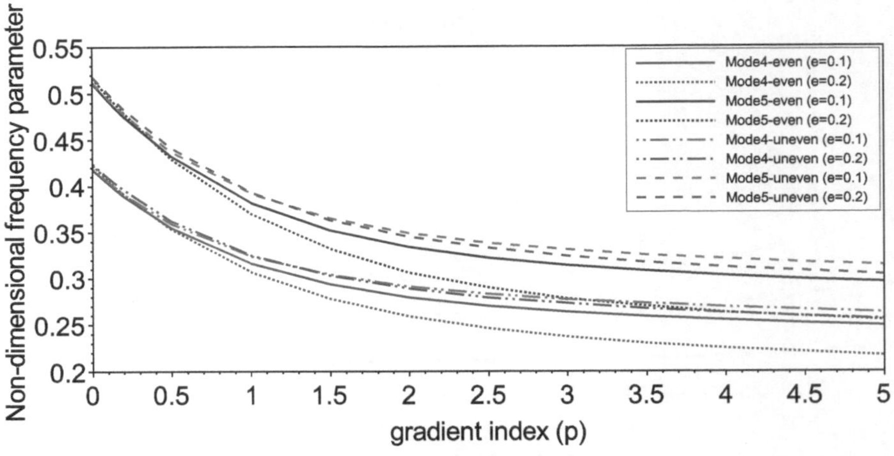

Figure 8.7 Effect of porosity distributions on fourth and fifth high modes of vibration of FG porous plates ($a/h = 10$).

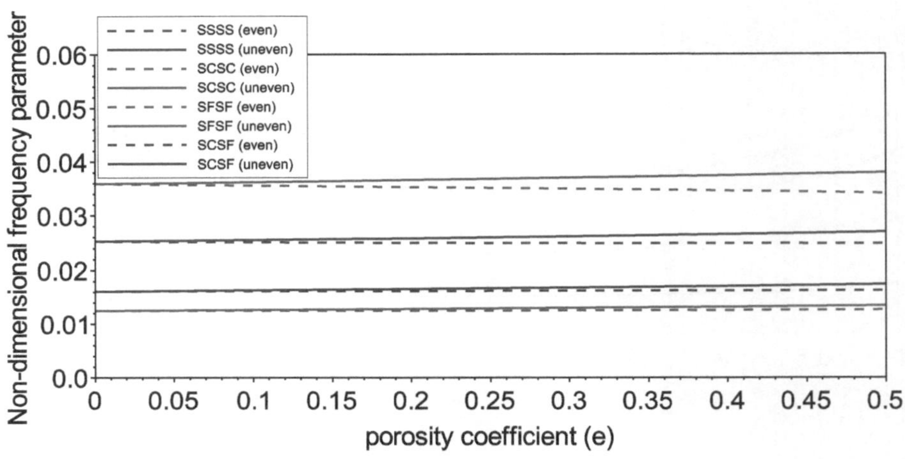

Figure 8.8 Effects of boundary conditions on natural frequency parameter of FG porous plates ($a/h = 20$, p = 0.5) with evenly and unevenly porosity distributions.

and 8.14 depict the non-dimensional natural frequencies versus the gradient index (p) for clamped (CC) and simply supported (SS) FGM porous circular plates ($R/h = 20$) for different porosity coefficients and with evenly and unevenly porosity distributions. It can be seen that as the gradient index increases (p), as the natural frequency parameter decreases and becomes constant for high values for gradient indexes. Indeed, when (p) rises, the circular plate becomes more reinforced with metallic portions which have a low stiffness and hence low natural frequencies. Moreover, when the porosity coefficient (e) increases, the natural frequencies also decreases and this effect is more prominent for high modes of vibration and with evenly porosity distributions more than unevenly distribution. Therefore, the percentage and the way by which the porosity is spread into the structure have a significant effect on vibrational behavior of FGM porous circular plates. To highlight more the effect of even and uneven porosity-types distributions, the non-dimensional natural frequencies corresponding to first mode of vibration of FGM porous circular plates ($R/h = 20$) for different porosity coefficients and under two types of boundary conditions (CC) and (SS) with even

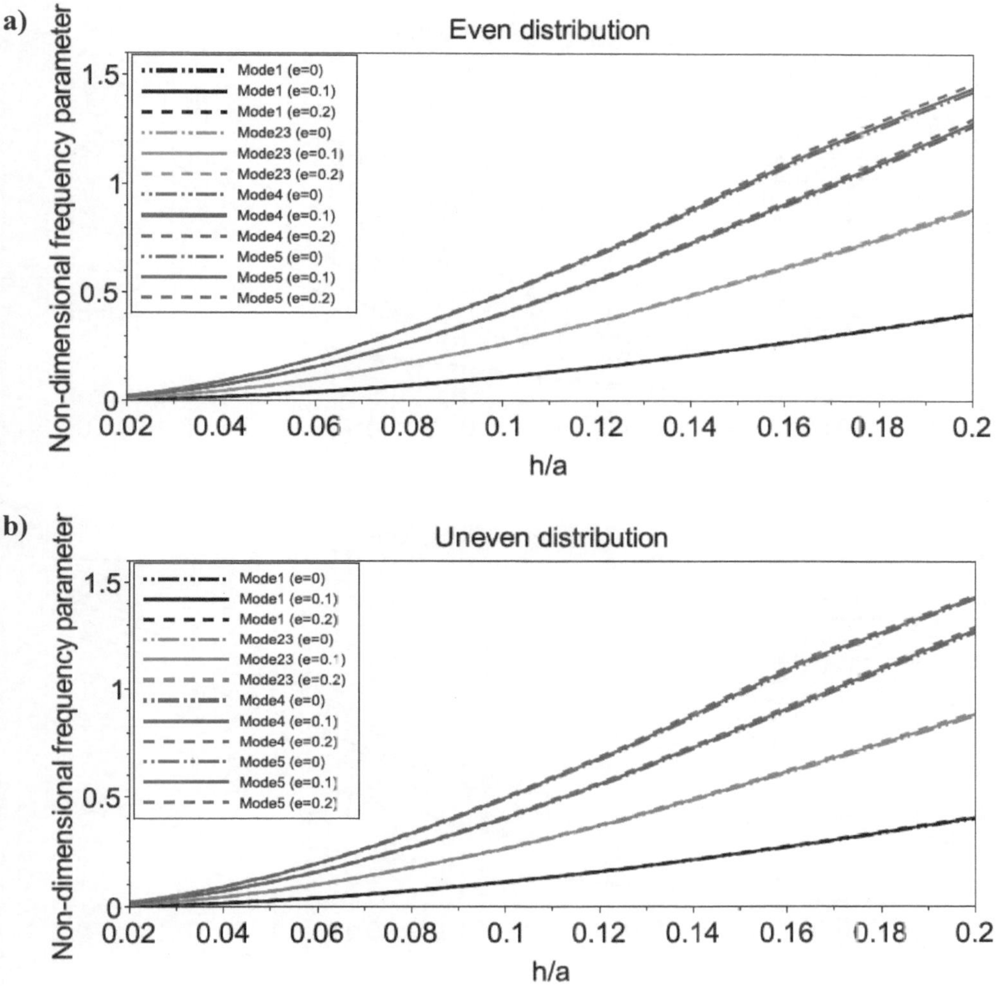

Figure 8.9 Influence of thickness-to-length (h/a) ratios on the first five modes of vibration of a simply supported (SSSS) FG porous plates ($p = 0.1$) under different porosity coefficients and for (a) evenly and (b) unevenly porosity distributions.

and uneven porosity distributions are plotted in Fig. 8.15. The non-dimensional natural frequencies of clamped (CC) FGM porous circular plates are higher than those of simply supported (SS) FGM porous circular plates due to the strong constraints imposed by the clamped (CC) boundary condition and the high stiffness of the structure in this case. Furthermore, the rise of the porosity coefficient induces a decrease in the natural frequencies which is more highlighted with the clamped (CC) boundary condition. As well as, compared the curves related to even and uneven porosity distributions, it can be remarked that the ones of even porosity-type are located blow the ones of uneven porosity-type and this fact can be related to the uniform repartition of the pores into the structure in even-type which stimulates the decrease of the rigidity of the FGM porous circular plates with low natural frequencies, whereas for uneven porosity-type the pores are concentrated in the mid-surface and vanish in the top and bottom surfaces which keep the natural frequencies a little higher. In conclusion, the variation of the natural frequencies versus the gradient index are strongly related

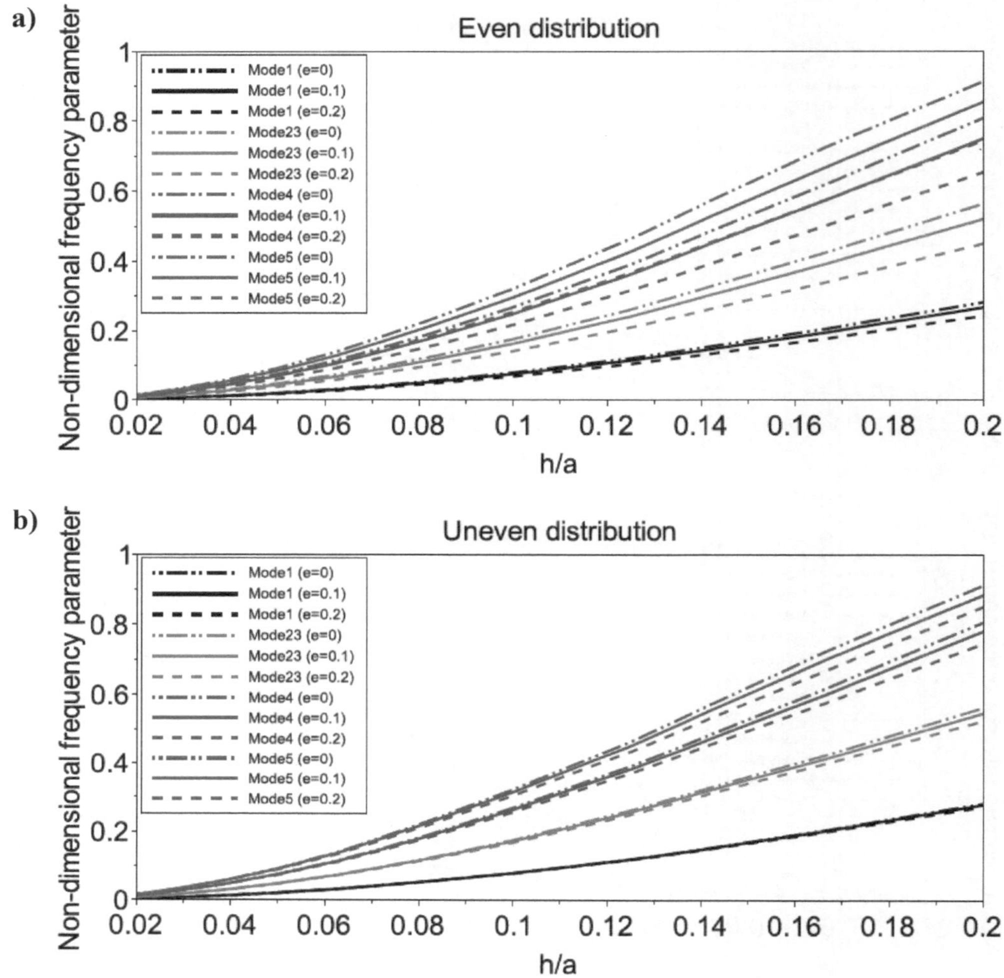

Figure 8.10 Influence of thickness-to-length (h/a) ratios on the first five modes of vibration of a simply supported (SSSS) FG porous plates ($p = 5$) under different porosity coefficients and for (a) evenly and (b) unevenly porosity distributions.

on the bending-stretching stiffness coefficients which are dependent on porosity volume fraction and types of distributions (even and uneven) as a result of the inhomogeneity of functionally graded materials. Continuously, Fig. 8.16 shows the variation of the natural frequency parameter versus the porosity coefficient for high modes of vibration of evenly and unevenly FGM porous circular plates ($R/h = 20$). This graph outlines also the fluctuations of natural frequencies for a prescribed value of gradient rely on the value of the mass to stiffness ratio at the given porosity coefficient and which part of this ratio is more reduced. Consequently, the value of the mass to stiffness ratio has a crucial role on the vibration response of FGM porous circular plates. Figures 8.17(a) and 8.17(b) show the effects of radius-to-length (R/h) ratios on natural frequencies of a fully clamped (CC) FGM porous circular plates with evenly and unevenly distributions while fixing the gradient index ($p = 5$). For both porosity distributions, the natural frequencies rise slightly in the domain ($R/h \in [5-20]$). After that and for a radius-to-length ratio ($R/h \geq 20$), they stabilized and become

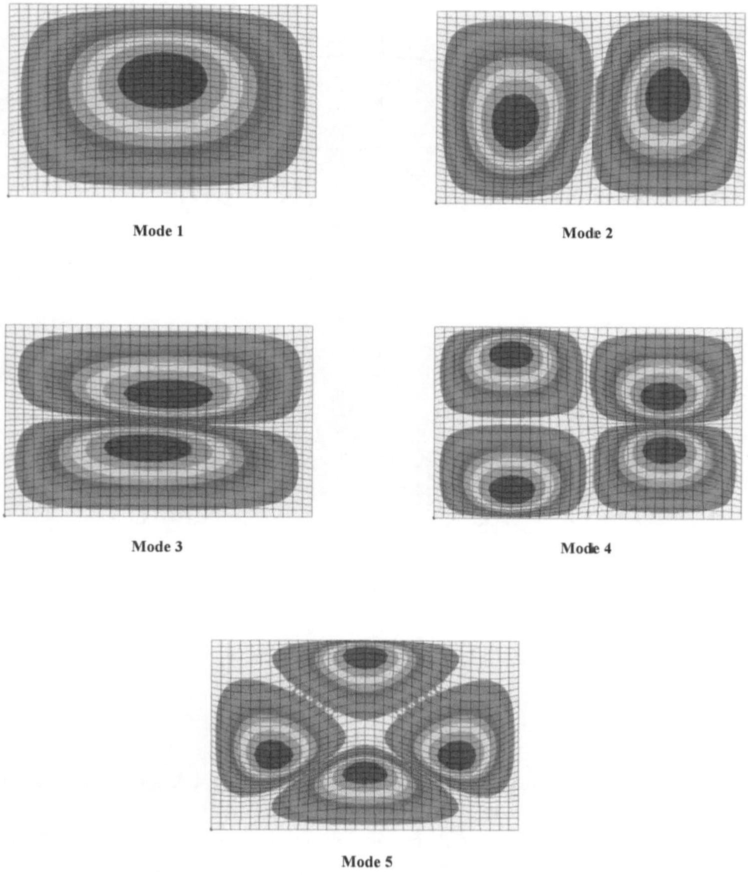

Figure 8.11 First five mode shapes of a simply supported (SSSS) FGM square porous plates $(a/h = 20)$.

constant and this is for all modes of vibration and different porosity coefficients. This effect is more prominent for the high modes of vibration in the range $(R/h \in [5-20])$ where the FGM porous circular plate behaves as thick structure and since the non-dimensional natural frequencies are proportional to (R/h) ratio, the variation of this ratio influenced significantly the vibrational behavior of FGM porous circular plates. Additionally, with the rise of porosity coefficient, the natural frequencies diminished pronouncedly for even-type of porosity more than uneven-type. To fulfill with this effect for high modes of vibration, Fig. 8.18 depicts the natural frequencies of the fourth and five modes of vibration for a fully clamped (CC) FGM porous circular plates $(p = 5)$ with different porosity distributions. It can be seen that for a fixed porosity coefficient (e), the curves related to uneven porosity distribution are higher than the ones of even porosity distribution which highlight the impact of porosity distribution manner into the FGM plate, where the concentration of the pores in the mid-surface of the FGM plate can be a good way to control passive vibration by enhancing the rigidity of the FGM circular plate in contrast to the uniform distribution of the pores related to even-type of porosity which weakens the stiffness of the plate. The first four mode shapes of vibration of a clamped (CC) FGM porous circular plates $(R/h = 20)$ are presented in Fig. 8.19.

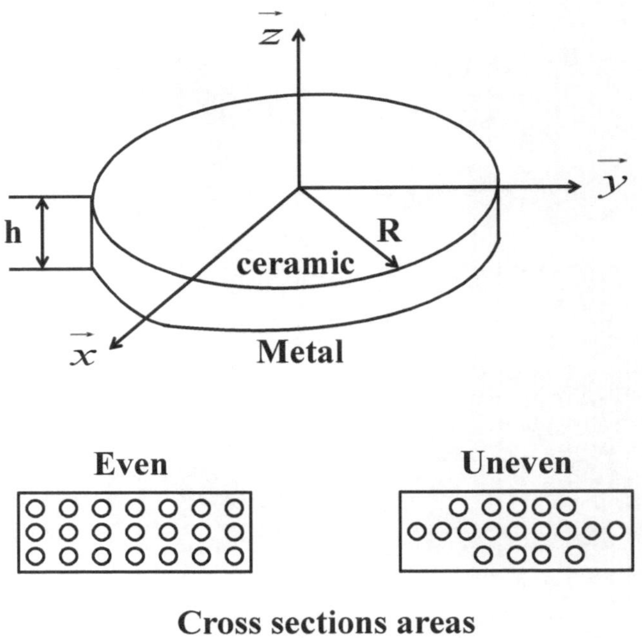

Figure 8.12 The schematic view of the FGM porous circular plate.

8.3.4 SPHERICAL CAP SHELLS

Numerical tests are also conducted on vibrational behavior of imperfect FGM spherical cap shells. Since no results are available in the open literature concerning free vibrations of imperfect FGM porous spherical caps, we start with the validation of the results with those available in the literature in the case of isotropic clamped spherical shells. The geometrical and material characteristics of the clamped spherical cap shell are taken as [246, 61]: Radius $R_s = 1m$; thickness $h = 0.05m$; Young's modulus $E = 211GPa$; Density $\rho = 7800Kg/m^3$ and Poisson's ratio $\nu = 0.3$. The spherical cap shell is modeled using 192 finite shell elements and its geometry is shown in Fig. 8.20. The dimensionless frequency parameter $\overline{\omega_s} = \omega R_s \sqrt{\rho/E}$ for the first four modes of vibration of isotropic clamped spherical cap shells $(h/R = 0.05)$ under two angles disposition $(\varphi_0 = 45^o, 60^o)$ are presented in Tab. 8.10. It is noteworthy that the numbers in brackets indicate the vibration mode (m,n) with m is the number of half-waves along a generatrix and n is the circumferential wave number. A good agreement is outlined and the little difference between the results is due to the different solution strategies used in the studies. In fact, in [246, 61] and [247] a semi-analytical-based thin shell theory and a finite element method are used, respectively, while in the present study a finite element shell model with enhanced shear distribution is implemented. Otherwise, a satisfactory correlation between the results is shown. The corresponding modes' shapes of the first four natural frequencies of the clamped spherical shells $(\varphi_0 = 45^0)$ are illustrated in Fig. 8.21. The first new results on vibrational behavior of (Al/Al_2O_3) FGM porous spherical cap shells are now presented. Figs. 8.22(a) and 8.22(b) show the variation of the dimensionless frequency parameter $\omega_s^* = \omega R_s \sqrt{\rho_c/E_c}$ versus the gradient index (p) of the clamped FGM porous spherical cap shell $(h/R_s = 0.05, \varphi_0 = 45^o)$ for different porosity coefficients and with evenly and unevenly porosity distributions, respectively. The increase of the porosity coefficient (e) induces a decrease in the natural frequency of the FGM porous spherical cap shell and this decrease is more pronounced for evenly type of porosity than unevenly type of porosity distributions. This phenomenon is related to the way of porosity distribution

Table 8.7

Comparison of the natural frequencies (Hz) of a fully clamped FGM circular plates (Numbers in parentheses () correspond to the circumferential wave number).

Gradient index (p)	Natural frequency (Hz)				
	Present	CPT-Analytical.[71]	Diff(%)	CPT-FEM.[284]	Diff(%)
0	138.05(1)	138.42(0)	0.267	138.15(1)	0.072
	469.17(2)	472.55(2)	0.720	468.72(3)	0.095
	537.74(3)	-	-	534.07(4)	0.632
1	134.25(1)	134.63(0)	0.283		
	456.27(2)	459.62(2)	0.734	-	-
	522.96(3)	-	-	-	-
3	132.34(1)	132.70(0)	0.272		
	456.27(2)	453.09(2)	0.696	-	-
	515.54(3)	-	-	-	-
5	131.75(1)	132.12(0)	0.280		
	447.79(2)	451.06(2)	0.730	-	-
	513.23(3)	-	-	-	-
7	131.48(1)	131.85(0)	0.281		
	446.86(2)	450.13(2)	0.731	-	-
	512.17(3)	-	-	-	-
9	131.32(1)	131.69(0)	0.281		
	446.33(2)	449.60(2)	0.732	-	-
	511.57(3)	-	-	-	-
10	131.27(1)	131.64(0)	0.281		
	446.33(2)	446.15(2)	0.040	-	-
	511.36(3)	-	-	-	-

through the thickness direction of the FGM porous spherical cap which is uniform in evenly type and random in unevenly type. As well as, the variation of the natural frequencies with respect to the gradient index can be explained by the effect of bending-stretching coupling coefficients which depends on porosity volume fraction and types of distributions. Therefore, the present improved (FSDT) model constitutes a general finite element model that can describe accurately the buckling responses of FGM porous shells. Equally, the effect of the variation of the angle ($\varphi_0(^o)$) geometrical parameter on natural frequencies of the first four modes of vibration of (Al/Al_2O_3) FGM porous spherical cap shells ($p = 0.5, h/R_s = 0.05$) with evenly and unevenly porosity distributions is investigated as shown in Figs. 8.23(a) and 8.23(b), respectively. It can be seen that the increase of the values of the angle ($\varphi_0(^o)$) leads to a decrease in the natural frequencies for the first four modes of vibration. Beyond ($\varphi_0 = 45^0$), the variation of the natural frequencies become constant which highlight that the FGM porous spherical cap shells are more sensitive to the variation of the angle geometrical parameter in the range ($\varphi_0 \in [5-45]$). Otherwise, the porosity coefficient (e) has not any significant effect on natural frequencies of the FGM porous spherical shells with the change of the angle geometrical parameter. The variation of the dimensionless frequency parameter ω_s^* versus the radius-to-thickness (R/h) ratio of a clamped FGM porous spherical cap shells ($p = 5, \varphi_0 = 45^o$) for different porosity coefficients and with evenly and unevenly porosity distributions is plotted in Figs. 8.24(a) and 8.24(b), respectively. From these figures, it is shown that the increase of the radius-to-thickness (R/h) ratio leads to a decrease in the natural frequencies of the imperfect FGM porous spherical cap shells which can be divided into two ranges: $R/h \in [5-25]$ and $R/h \in [25-100]$. In the first range, the FGM spherical shell is more sensitive to the variation of the (R/h) ratio than the second range. Indeed, in the second range of (R/h) ratio, the FGM spherical shell behaves as thin structure with low rigidity and the natural frequencies become hence as steady state motion.

Table 8.8

Non-dimensional natural frequency parameter $\overline{\omega_c}$ of perfect FGM clamped circular plates $(R/h = 20)$ for power indexes (p = 1,2,3) (Numbers in parentheses () correspond to the circumferential wave number).

BC	*Theories*	Gradient index (p)		
		1	2	3
CC	Present	7.748(1)	7.040(1)	6.812(1)
	CPT-Analytical.[330]	7.797(0)	7.090(0)	6.867(0)
	CPT-Axisymmetric.[329]	8.498(0)	8.123(0)	7.911(0)
	Present	29.788(3)	27.023(3)	26.098(3)
	CPT-Analytical.[330]	30.357(1)	27.604(1)	26.737(1)
	CPT-Axisymmetric.[329]	33.086(1)	31.625(1)	30.798(1)
	Present	65.382(6)	59.186(6)	57.002(6)
	CPT-Analytical.[330]	68.012(2)	61.845(2)	59.902(2)
	CPT-Axisymmetric.[329]	74.127(2)	70.855(2)	69.002(2)
SS	Present	3.758(1)	3.416(1)	3.307(1)
	CPT-Analytical.[330]	3.767(0)	3.425(0)	3.317(0)
	CPT-Axisymmetric.[329]	4.105(0)	3.924(0)	3.821(0)
	Present	22.422(3)	20.359(3)	19.687(3)
	CPT-AnalyticalS.[330]	22.685(1)	20.628(1)	19.980(1)
	CPT-Axisymmetric.[329]	24.724(1)	23.633(1)	23.015(1)
	Present	55.022(6)	49.867(6)	48.117(6)
	CPT-Analytical.[330]	56.602(2)	51.470(2)	49.853(2)
	CPT-Axisymmetric.[329]	61.692(2)	58.968(2)	57.426(2)

Table 8.9

Non-dimensional natural frequency parameter $\overline{\omega_c}$ of perfect FGM clamped circular plates $(R/h = 20)$ for power indexes (p = 4,5,∞) (Numbers in parentheses () correspond to the circumferential wave number).

BC	*Theories*	Gradient index (p)		
		4	5	∞
CC	Present	6.717(1)	6.660(1)	5.160(1)
	CPT-Analytical.[330]	6.777(0)	6.724(0)	5.199(0)
	CPT-Axisymmetric.[329]	7.733(0)	7.573(0)	5.199(0)
	Present	25.691(3)	25.439(3)	19.803(3)
	CPT-Analytical.[330]	26.386(1)	26.177(1)	20.243(1)
	CPT-Axisymmetric.[329]	30.107(1)	29.485(1)	20.243(1)
	Present	55.684(6)	55.339(6)	43.366(6)
	CPT-Analytical.[330]	59.116(2)	58.649(2)	45.352(2)
	CPT-Axisymmetric.[329]	67.453(2)	66.059(2)	45.352(2)
SS	Present	3.263(1)	3.237(1)	2.504(1)
	CPT-Analytical.[330]	3.274(0)	3.248(0)	2.512(0)
	CPT-Axisymmetric.[329]	3.736(0)	3.658(0)	2.512(0)
	Present	19.402(3)	19.203(3)	14.927(3)
	CPT-AnalyticalS.[330]	19.717(1)	19.562(1)	15.127(1)
	CPT-Axisymmetric.[329]	22.498(1)	22.033(1)	15.127(1)
	Present	47.334(6)	44.273(6)	36.569(6)
	CPT-Analytical.[330]	49.199(2)	48.810(2)	37.744(2)
	CPT-Axisymmetric.[329]	56.137(2)	54.977(2)	37.744(2)

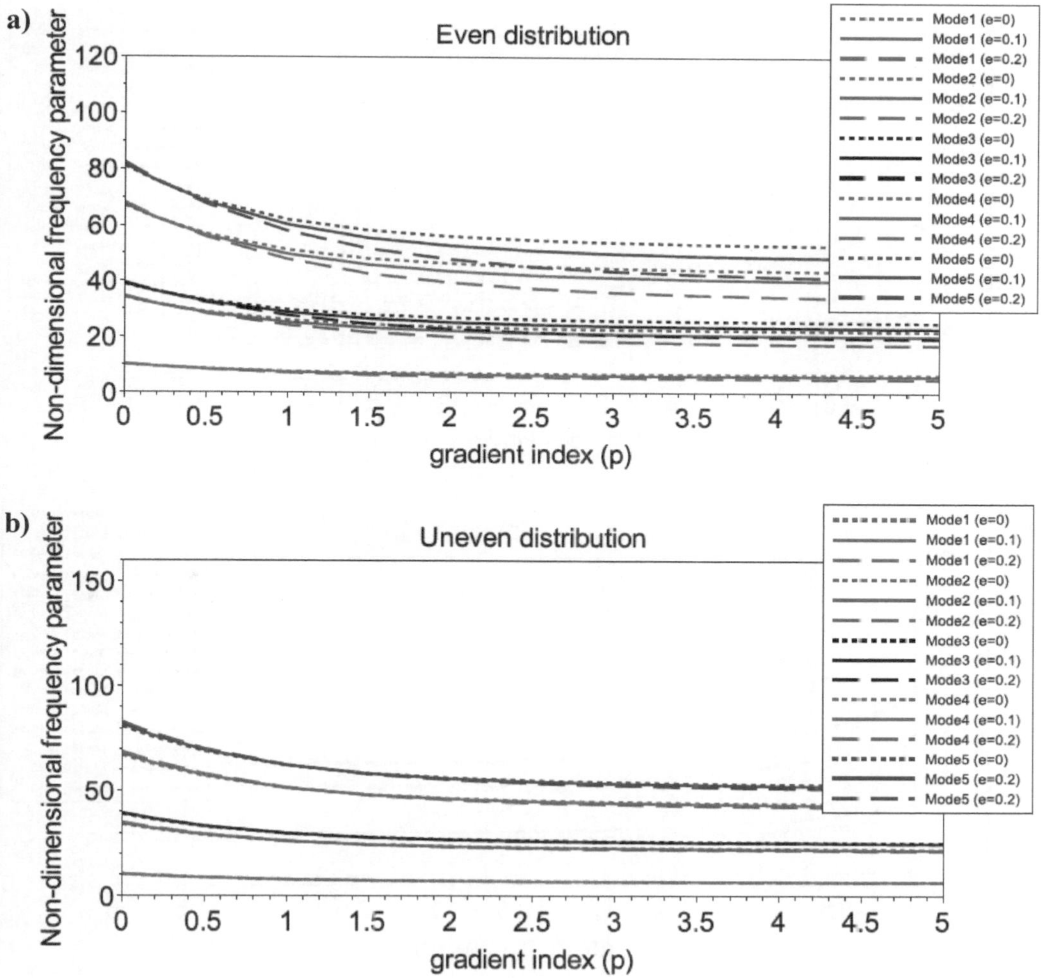

Figure 8.13 Non-dimensional natural frequencies versus the gradient index (p) of a clamped (CC) circular FGM porous plates ($R/h = 20$) for different porosity coefficients and with (a) evenly and (b) unevenly porosity distributions.

8.4 TIME-DEPENDENT DEFLECTION RESPONSES OF FG POROUS STRUCTURES

This part of the chapter deals with the time-deflection responses of functionally graded (FG) porous plates and spherical shells subjected to different external pulse excitations by means of an improved first-order shear deformation theory (Im-FSDT) with improved transverse shear deformations. The governing equations for the dynamic transient response is systematically derived using a variational principle and finite element method-based (I-FSDT) theory and they are solved numerically using the constant average acceleration Newmark's integration algorithm. The effective material properties are evaluated using a modified power-law distribution with a porosity parameter as described in section 8.2. The pores are assumed to be dispersed uniformly or randomly along the thickness of the shell leading to even or uneven porosity distributions. Furthermore, it should be mentioned that the porosity parameter for dynamic analysis is denoted by (ξ) in all numerical tests. Otherwise, this

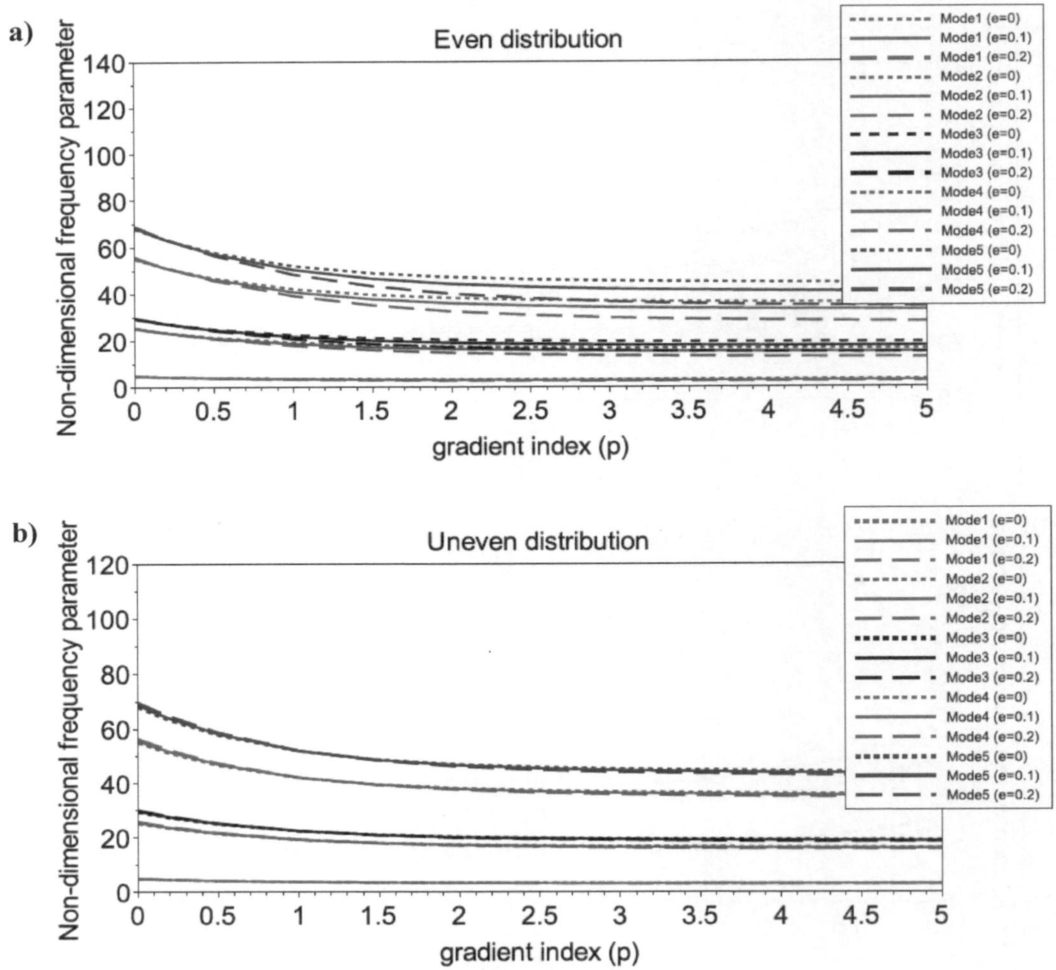

Figure 8.14 Non-dimensional natural frequencies versus the gradient index (p) of a simply supported (SS) circular FGM porous plates $(R/h = 20)$ for different porosity coefficients and with (a) evenly and (b) unevenly porosity distributions.

coefficient is denoted (α). The model accuracy is checked by comparing the present time-deflection responses with the available data. Further, the effects of porosity volume fraction, types of distributions, external pulse loads, geometrical and material parameters are explored and discussed. It is found that the porosity rate and the type of dispersion have a significant effect on the dynamic transient responses of FG porous plates and shells. In such a way, the dynamic behavior of aerospace, automobile and naval structural components can be controlled passively by the percentage and types of added porosity within such structure.

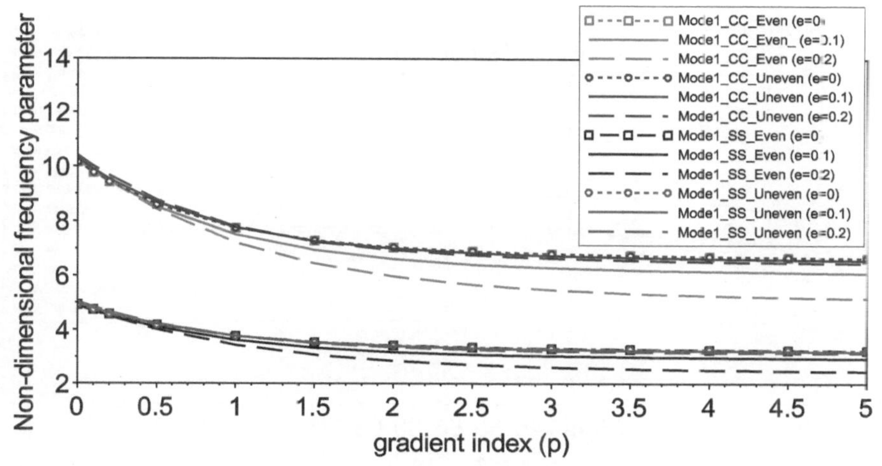

Figure 8.15 Non-dimensional natural frequencies of FGM porous circular plates ($R/h = 20$) for different porosity coefficients and boundary conditions and with evenly and unevenly porosity distributions.

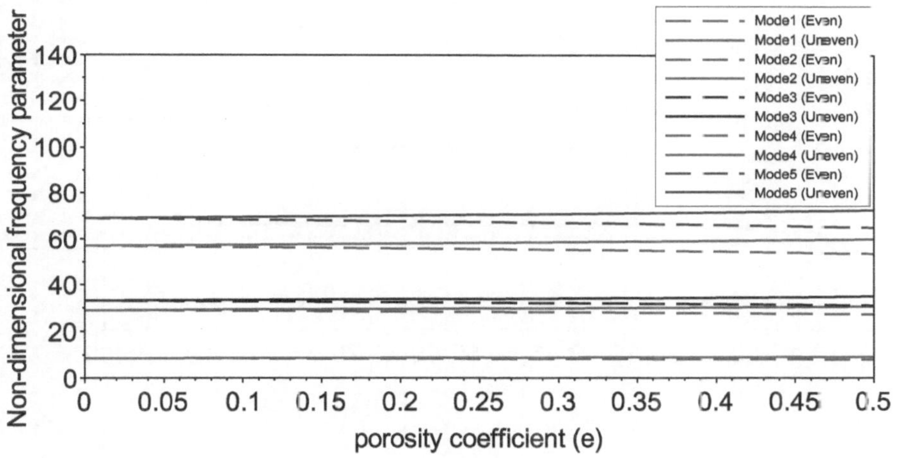

Figure 8.16 Variation of natural frequency parameter versus the porosity coefficient for high modes of vibration of evenly and unevenly FGM porous circular plates ($R/h = 20$).

8.4.1 DYNAMIC PROBLEM AND DISCRETE APPROXIMATION IN TIME DOMAIN

8.4.1.1 Dynamic problem

For elastodynamic problems, the corresponding equation of motion of the FG porous structure can be written, in time domain, as follows:

$$\mathbf{M}\ddot{U}_n(t) + \mathbf{K}U_n(t) = F(t). \tag{8.4}$$

where M, K, $U_n(t)$, $\ddot{U}_n(t)$ and $F(t)$ represent the mass, the stiffness matrices; the displacement and the acceleration vectors and the external load vector, respectively. The mass and stiffness matrices are obtained via the discretization of the kinetic and potential energy and they can be expressed as

$$\begin{cases} M^I = \begin{bmatrix} m_{1I} & m_{1I} & m_{1I} & m_{2I} & m_{2I} \end{bmatrix}_{diag} \\ K = \int_A B^T H_T B dA. \end{cases} \tag{8.5}$$

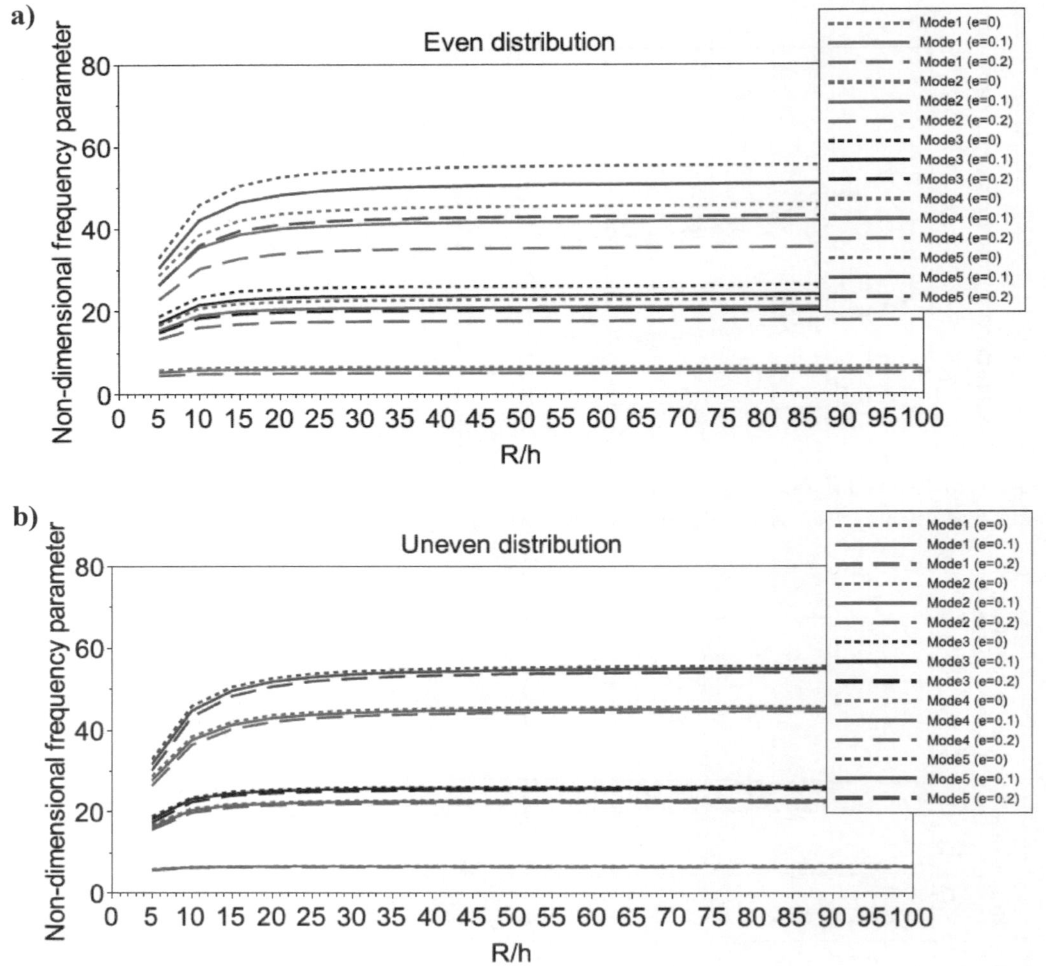

Figure 8.17 Effects of radius-to-length (R/h) ratio on natural frequencies of a fully clamped (CC) FGM porous circular plates ($p = 5$) with (a) evenly and (b) unevenly porosity distributions.

with

$$m_{kI} = \frac{\int_A \overline{\rho}_k dA \int_A \overline{\rho}_k N^I N^I dA}{\int_A (\sum \overline{\rho}_k N^I N^I) dA}, \quad k = 1, 2 \tag{8.6}$$

and

$$(\overline{\rho}_1, \overline{\rho}_2) = \int_A \rho(z, \xi)(1, f^2) dA. \tag{8.7}$$

8.4.1.2 Newmark integration method

Newmark method [188, 25] is used in this study since some members of this family of integration methods are widely applied for dynamic problems. This method allows the direct solution of the second-order differential equation (Eq. (8.4)) and its algorithm operates, considering step-by-step solution procedure, as shown in Tab. 8.11.

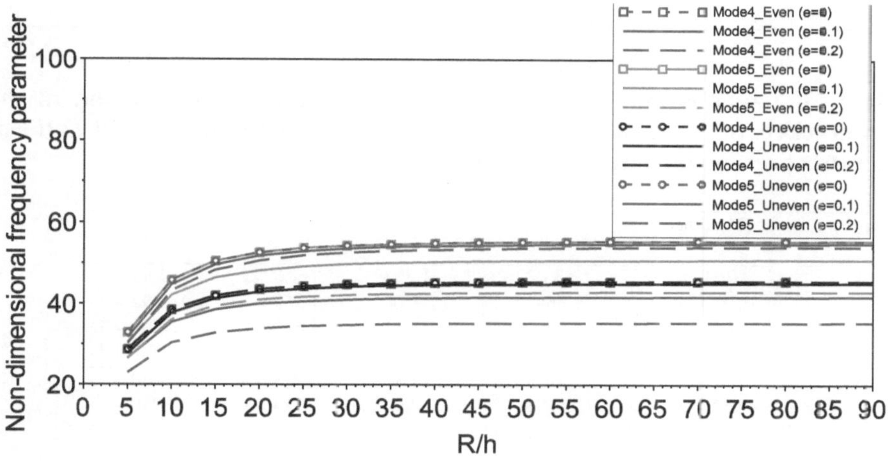

Figure 8.18 Effects of radius-to-length (R/h) ratio on high modes of vibration of a fully clamped (CC) FGM porous circular plates ($p = 5$) with different porosity distributions.

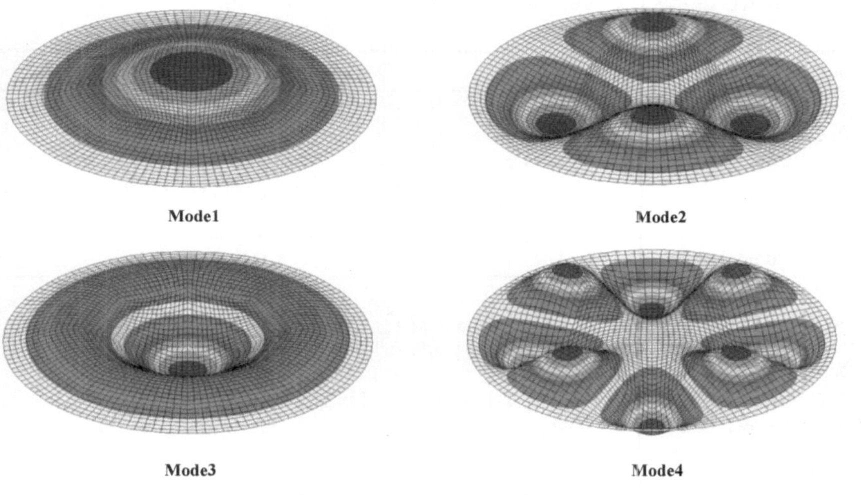

Figure 8.19 First four mode shapes of vibration of a clamped (CC) FGM porous circular plates ($R/h = 20$).

In Tab. 8.11, β and γ are the Newmark constants which control the stability and accuracy of the response. To obtain stable and convergent results, the constant average acceleration method is used in this study for which $\gamma = 1/2$ and $\beta = 1/4$. At this stage we recall that four well-known members of the Newmark method can be used based on the values of β and γ parameters, stability and order of accuracy of the method [254, 45]. The numerical properties of these members of the Newmark family of methods are summarized in Tab. 8.12.

Briefly, the algorithm of the adopted implicit Newmark integration scheme of constant average acceleration type ($\beta = 1/4$, $\gamma = 1/2$), neglecting the damping effect, is demonstrated in Fig. 8.25.

Table 8.10
Dimensionless frequency parameter $\overline{\omega_s}$ of clamped isotropic ($\nu = 0.3$) spherical shells (Numbers in brackets indicate the vibration mode (m,n). m is the number of half-waves along a generatrix and n is the circumferential wave number).

$\varphi_0\,(^o)$	Reference	Mode			
		1	2	3	4
45	Present	1.0820	1.2942	1.6015	1.9340
	FEM.[247]	1.0688(1,1)	1.2930(2,1)	1.5696(3,1)	1.9292(4,1)
	Semi-analytical.[246]	1.0690(1,1)	1.2931(2,1)	1.5700(3,1)	1.9300(4,1)
	Semi-analytical.[61]	1.0698(1,1)	1.3064(2,1)	1.5802(3,1)	1.9533(4,1)
60	Present	0.8987	1.0800	1.3715	1.4024
	FEM.[247]	0.8990(1,1)	1.0790(2,1)	1.2140(3,1)	1.3930(4,1)
	Semi-analytical.[246]	0.8990(1,1)	1.0791(2,1)	1.2140(3,1)	1.3931(4,1)
	Semi-analytical.[61]	-	-	-	-

Table 8.11
Solution procedure using the Newmark integration technique.

Initial calculations ($t = 0$)
(1) Calculating the values of U_0, $\dot{U}_0$ and $\ddot{U}_0$; (2) Choosing a step time Δt and the parameters β, γ (3) calculation of the integral constants $c_0 = \frac{1}{\beta \Delta t^2},\ c_1 = \frac{1}{\beta \Delta t},\ c_2 = \frac{1}{2\beta} - 1,\ c_3 = \frac{\gamma}{\beta \Delta t},\ c_4 = \frac{\gamma}{\beta} - 1\ c_5 = \Delta t \left(\frac{\gamma}{2\beta} - 1\right) c_6 = \Delta t\,(1-\gamma)\ c_7 = \gamma \Delta t$ (4) Forming the effective stiffness matrix: $\boxed{\mathbf{K}} = [\mathbf{K}] + c_0\,[\mathbf{M}] + c_3\,[\mathbf{C}]$
For each time step ($t + \Delta t$)
(5) Calculating the residual term: $\{R_{t+\Delta t}\} = \{F_{t+\Delta t}\} + [\mathbf{M}]\left(c_0\{U_t\} + c_1\{\dot{U}_t\} + c_2\{\ddot{U}_t\}\right) + [\mathbf{C}]\left(c_3\{U_t\} + c_4\{\dot{U}_t\} + c_5\right)\{\ddot{U}_t\}$ (6) calculating the displacement $\{U_{t+\Delta t}\}$ at time $(t + \Delta t)$ by the resolution of Eq. (7.23) $\boxed{\mathbf{K}}\,\{U_{t+\Delta t}\} = \{R_{t+\Delta t}\}$ (7) Calculation of the acceleration $\{\ddot{U}_{t+\Delta t}\}$ and velocity $\{\dot{U}_{t+\Delta t}\}$ at time $(t + \Delta t)$: $\{\ddot{U}_{t+\Delta t}\} = c_0\left(\{U_{t+\Delta t}\} - \{U_t\}\right) - c_1\{\dot{U}_t\} - c_2\{\ddot{U}_t\}$ $\{\dot{U}_{t+\Delta t}\} = \{\dot{U}_t\} + c_6\{\ddot{U}_t\} + c_7\{\ddot{U}_{t+\Delta t}\}$

8.4.2 ACCURACY OF THE (IM-FSDT) MODEL FOR DYNAMIC BEHAVIOR OF ISOTROPIC, LAMINATED AND FG PERFECT PLATES

In this section, three case studies for dynamic analysis of rectangular plates subjected to impulse sudden loads are presented to illustrate the efficiency and accuracy of the present formulation.

Case Study 1: Isotopic homogeneous plates

The time response of a square isotropic simply supported plate subjected to a sudden pulse load of amplitude $F_0 = 10N/cm^2$ is compared first with HSDT available response. The following geometrical assumptions are made: $a = 25cm$; $h = 5cm$; $E = 2.1 \times 10^6 N/cm^2$; $\nu = 0.25$; $\rho = 8 \times 10^{-6} N - sec^2/cm^4$ [115]. The comparison is presented in Fig. 8.26 and an excellent performance among the present (Im-FSDT) deflection recovery with the HSDT model is shown. It can be

Table 8.12
Numerical properties of the Newmark's family of methods [254, 45].

Method	β	γ	Type	Stability	Order of accuracy
Constant average acceleration	1/4	1/2	Implicit	Unconditional	2
Linear average acceleration	1/6	1/2	Implicit	Conditional	2
Fox-Goodwin	1/12	1/2	Implicit	Conditional	3
Central difference	0	1/2	Explicit	Conditional	2

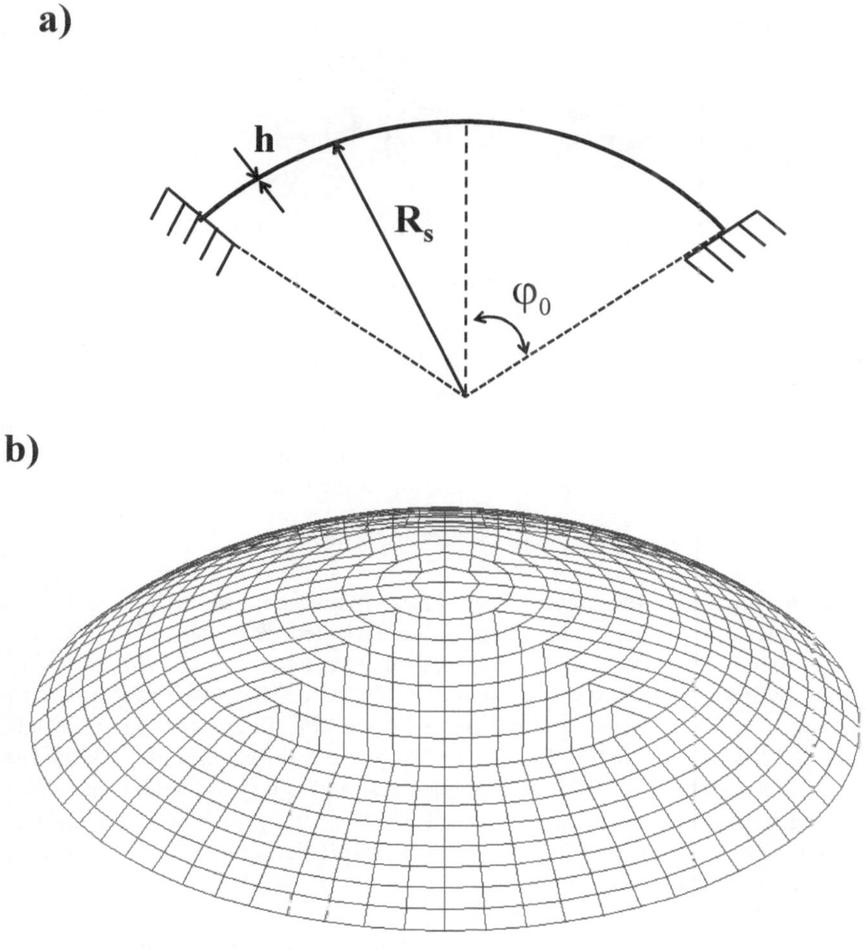

Figure 8.20 Geometry of the spherical cap with (a) system coordinates and (b) mesh configuration.

stated that first-order shear deformable model with a quadratic function update of the shear part can give best estimates for the transient response of homogeneous plates.

Case Study 2: Laminated orthotropic plats

The second assessment consists of a four-layer laminated orthotropic $[0/90/90/0^o]$ simply supported plate subjected to a sudden pulse load of amplitude $F_0 = 10N/cm^2$ with the following

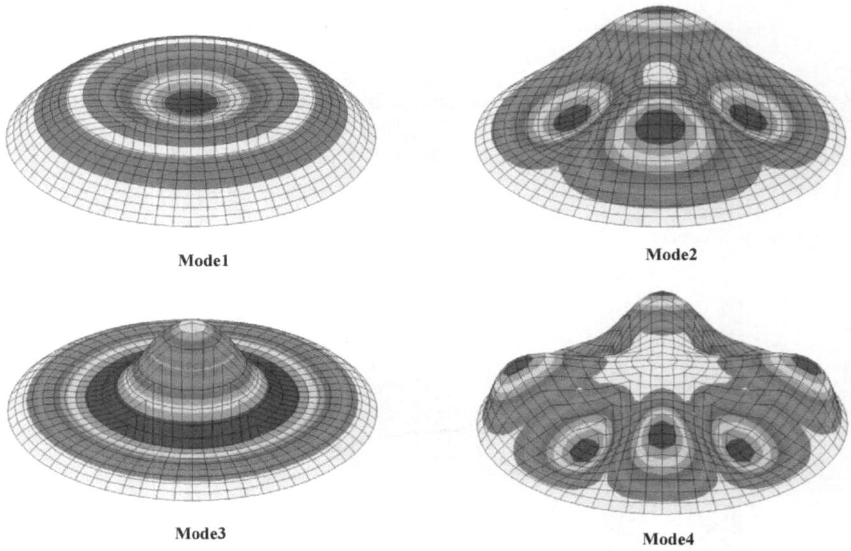

Figure 8.21 First four mode shapes of clamped spherical cap shells ($h/R_s = 0.05, \varphi_0 = 45^o$).

geometrical properties [115]: $a = 25cm$; $h = 5cm$; $E_1/E_2 = 25$; $E_2 = 2.1 \times 10^6 N/cm^2$; $\nu_{12} = 0.25$; $G_{12} = G_{13} = G_{23} = 0.5E_2$; $\rho = 8 \times 10^{-6} N - sec^2/cm^4$. Figure 8.27 shows a comparative plot of the present (Im-FSDT) time response with the (HSDT) solution for a laminated orthotropic plate. A good correlation is observed leading to trace the effectiveness of the present (Im-FSDT) finite element model for the dynamic analysis of laminated orthotropic plates.

Case Study 3: FG perfect plats

In this case, to further verify the present results for dynamic analysis of graded FG perfect (without porosities) plates, the obtained results of a simply supported square FG plate ($a = b = 1, h = 0.01, p = 2$) under a uniformly distributed load of magnitude $F_0 = 1N/m^2$ are compared with those reported by Ramteke et al. [201] as shown in Fig. 8.28. The elastic properties of metal (Al-Aluminium) and ceramic (Al_2O_3-Alumina) phases of the FG plate, used in this study, are taken as indicated in [201]: $E_m = 70GPa$; $E_c = 3800GPa$; $\rho_m = 2700(kg/m^3)$; $\rho_c = 3800(kg/m^3)$; $\nu_m = \nu_c = 0.3$. As illustrated in Fig. 8.28, the obtained results are in good agreement with those given in [201] using a nine-node finite element-based HSDT theory which proves again the consistency and capability of the current (Im-FSDT) model for predicting transient responses of FG perfect plates with high accuracy.

8.4.3 NATURAL FREQUENCY RESPONSE OF FG POROUS RECTANGULAR PLATES

A quick check of the formulation for the prediction of natural frequencies of FG porous plate made of Aluminium and Alumina phases (see, section 8.3.1) is performed in Tabs. 8.13 and 8.14. The FG porous plate is simply supported in all its edges (SSSS) with a geometrical ratio ($a/h = 20$) and two types of porosity, namely even and uneven distributions. The natural frequency parameter is computed as: $\overline{\omega} = \omega h\sqrt{\rho_m/E_m}$, where E_m and ρ_m are the material properties of Aluminium metal phase. It is worth observing that there is a good agreement among results, which indicates the performance of the present (Im-FSDT) (noted Imp) for the determination of the frequency response of FG porous plates. However, few differences detected in some cases can be explained by the different theories and strategies solutions applied in the two works. In Ref.[211], indeed, the

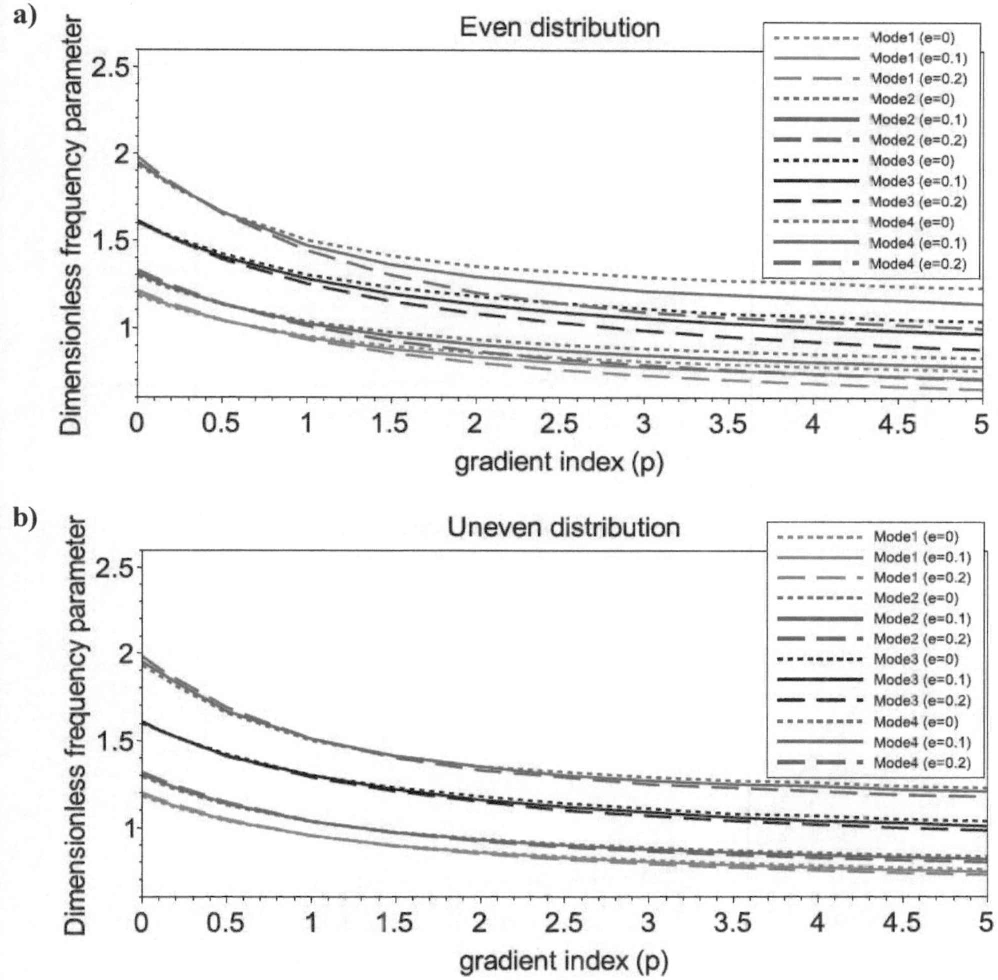

Figure 8.22 Variation of the dimensionless frequency parameter ω_s^* versus the gradient index (p) of a clamped FGM porous spherical cap shells $(h/R_s = 0.05, \varphi_0 = 45^o)$ for different porosity coefficients and with (a) evenly and (b) unevenly porosity distributions.

authors use a standard (FSDT) theory involving a shear correction factor, which is otherwise not required by the present Im-FSDT-based formulation. Otherwise, an acceptable correlation is provided. Next, the first seven mode shapes of the FG porous vibrating plates $(SSSS, p = 1)$ are plotted in Fig. 8.29.

8.4.4 DYNAMIC RESPONSE OF FG POROUS RECTANGULAR PLATES UNDER DIFFERENT EXTERNAL PULSE LOADS

This section provides the first new results concerning the dynamic response of FG porous rectangular plates subjected to different external pulse loads while varying the porosity coefficient, porosity distribution, type of applied pulse load and different geometrical and material characteristics. Fig. 8.30 illustrates the FG porous plate subjected to a pulse load, while its material properties are as

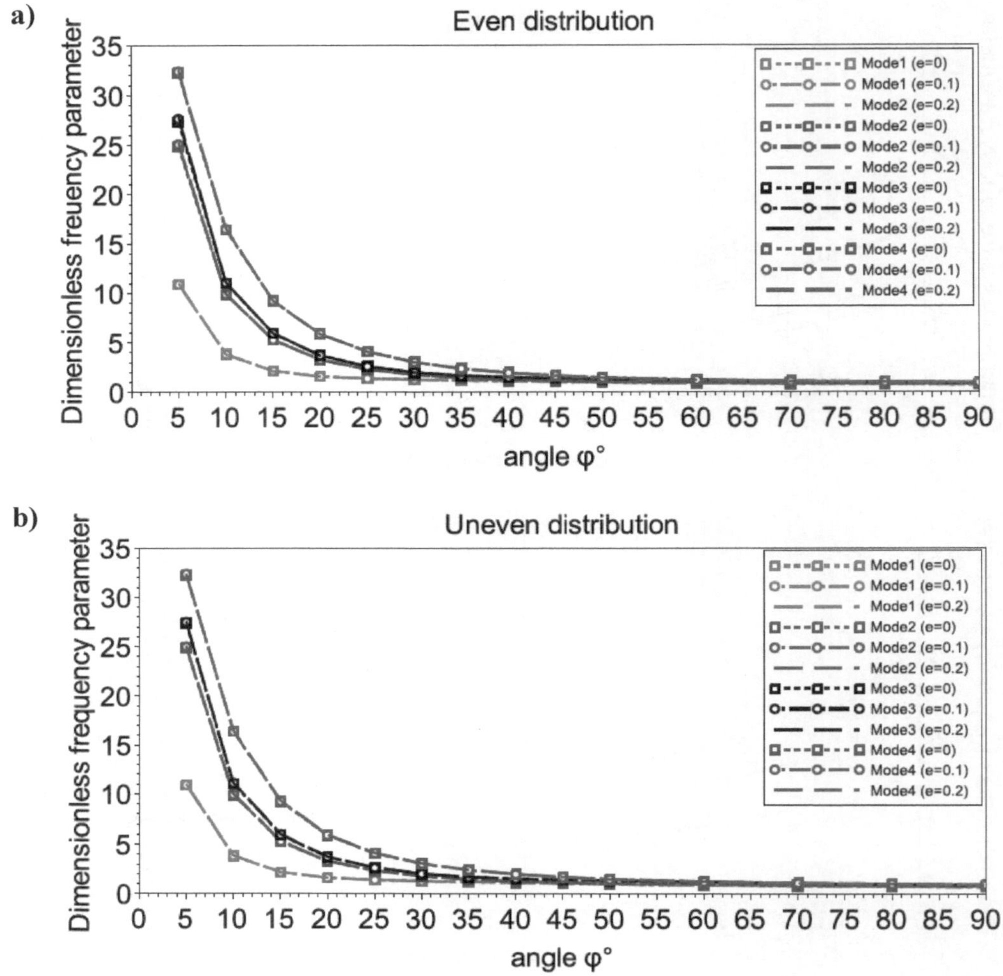

Figure 8.23 Variation of the dimensionless frequency parameter ω_s^* versus the angle $\varphi_0(^o)$ of a clamped FGM porous spherical cap shells $(p = 0.5, h/R_s = 0.05)$ for different porosity coefficients and with (a) evenly and (b) unevenly porosity distributions.

given in section 8.3.1. Two types of external pulse loads can be applied at the point (x_0, y_0) on the upon surface of the FG porous plate as illustrated in Fig. 8.31. The time response of a simply supported FG porous plate $(a/h = 20,\ p = 1)$ under a transverse harmonic excitation of the form $F(t) = F_0 sin(\Omega t)$ with: $F_0 = 100N$ and $\Omega = 5600 rad/s$ and for various porosity volume fractions $(\xi = 0, 0.2, 0.4)$ is illustrated in Fig. 8.32. It can be seen that, as the magnitude of porosity volume fraction increases, the deflection magnitude becomes larger for both porosity distributions. This observation relies on the reduction of the structural stiffness for an increased porosity volume fraction, such that the FG porous plate bends more easily. In Fig. 8.33, we plot the effect of porosity distributions on the transient response of FG porous plate $(a/h = 20,\ p = 1,\ SSSS)$ under a transverse harmonic excitation for two sets of porosity volume fractions (a): $(\xi = 0.2)$ and (b): $(\xi = 0.4)$. It is well shown that even porosity distribution yields larger oscillation amplitudes than in the uneven

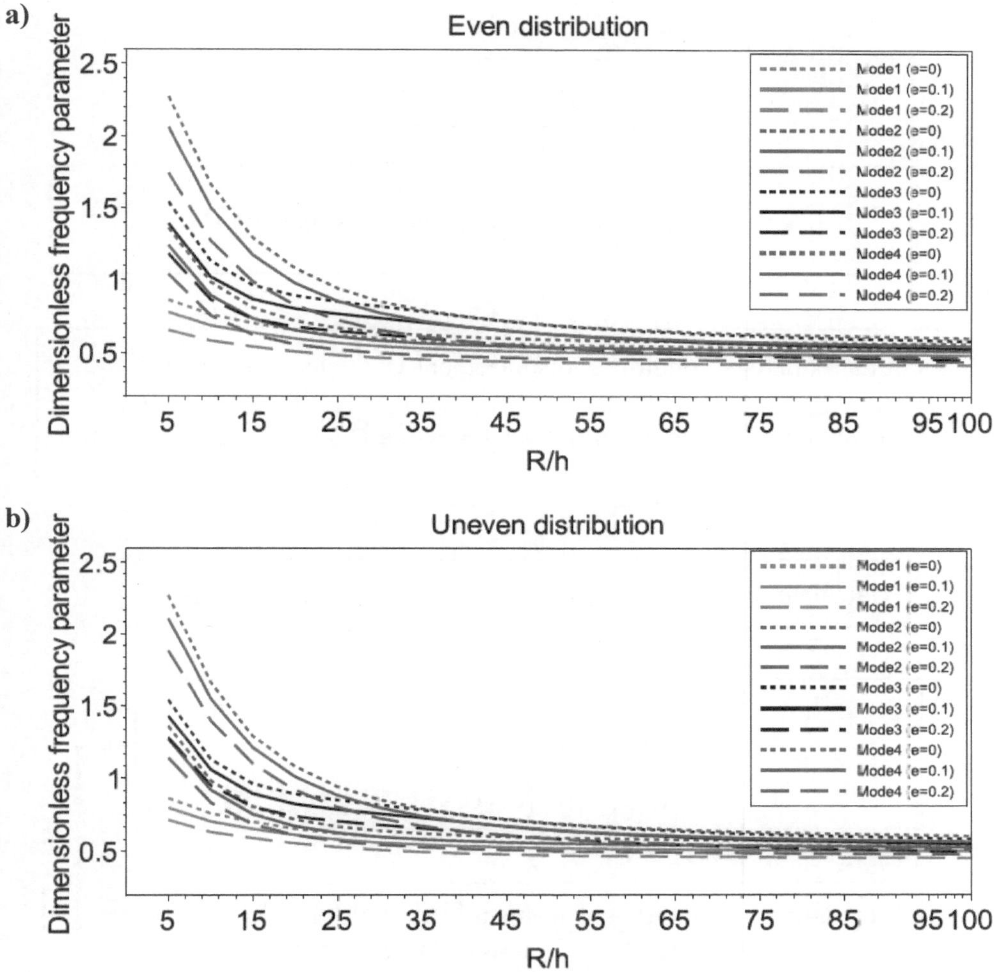

Figure 8.24 Variation of the dimensionless frequency parameter ω_s^* versus the radius-to-thickness (R/h) ratio of a clamped FGM porous spherical cap shells $(p = 5, \varphi_0 = 45^o)$ for different porosity coefficients and with (a) evenly and (b) unevenly porosity distributions.

type. The reason is due to the uniform distribution of porosity across the section area of the FG porous plate, in even case, whereas there is no porosity in the corners of the plate for an uneven dispersion. Therefore, the reduction in structural stiffness is more pronounced for even porosities. The effect of variation of the gradient index (p) on transient response of evenly and unevenly FG porous plate $(a/h = 20,\ \xi = 0.2,\ SSSS)$ subjected to transverse harmonic excitation is illustrated in Fig. 8.34. It is deduced that, as the gradient index (p) increases, the amplitude oscillations increases and this can be attributed to the decrease of the strength of the FG porous plate with the reinforcement of the metal constituent. Furthermore, the effect of the applied boundary conditions on the transient response of FG porous plates $(a/h = 20,\ \xi = 0.2)$ is examined in Fig. 8.35. It can be seen that a fully clamped (CCCC) FG porous plate features the lowest deflections while a fully simply supported (SSSS) FG porous plate reaches the highest oscillation magnitudes. Note also that for combined constraints (SCSC), the transient response of the FG porous plate is located between

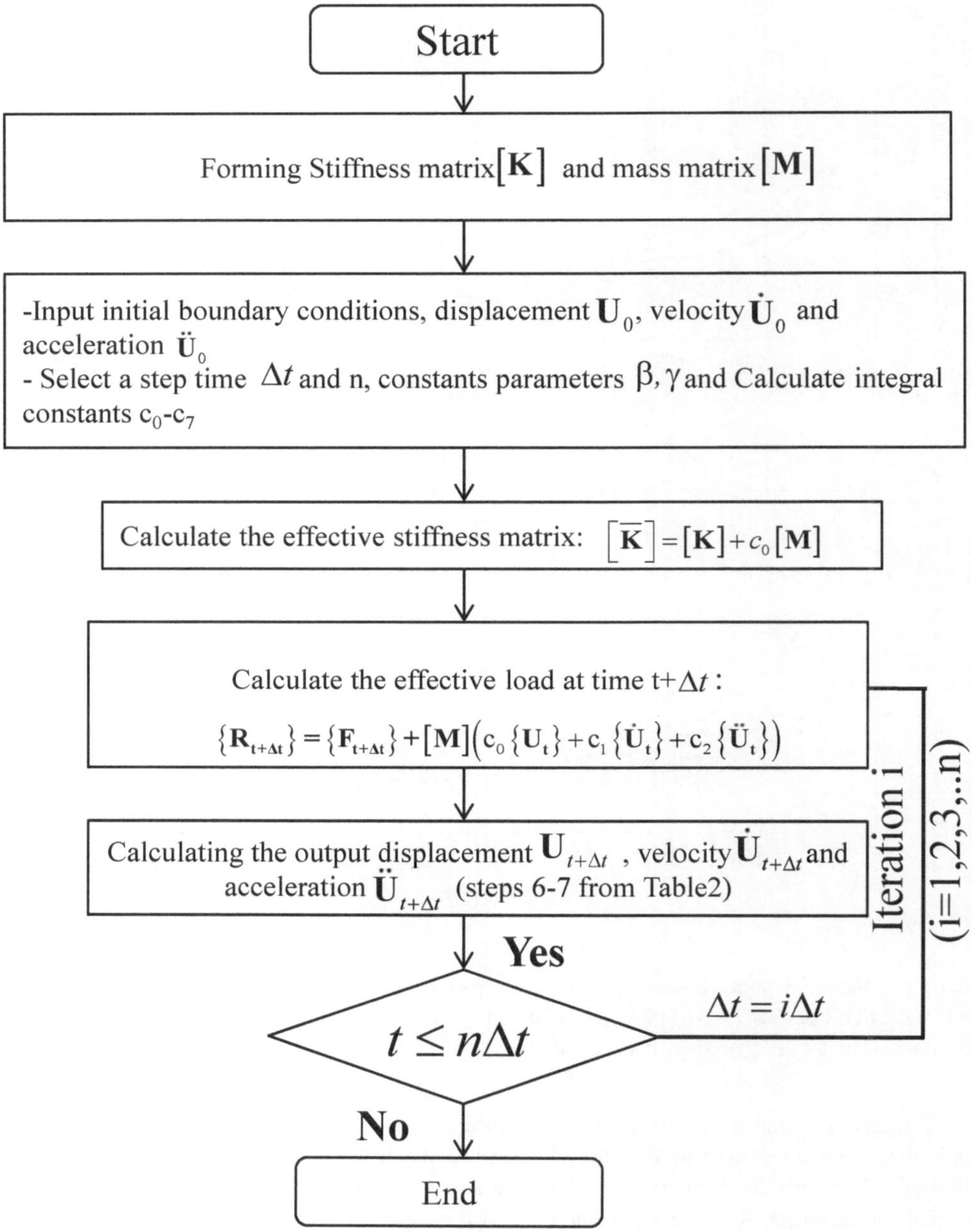

Figure 8.25 The algorithm of the constant average acceleration Newmark's integration technique for transient analysis of FG porous structures.

these extreme cases. Consequently, the type of applied boundary condition plays a significant role on the structural stiffness of the plate and its dynamic response.

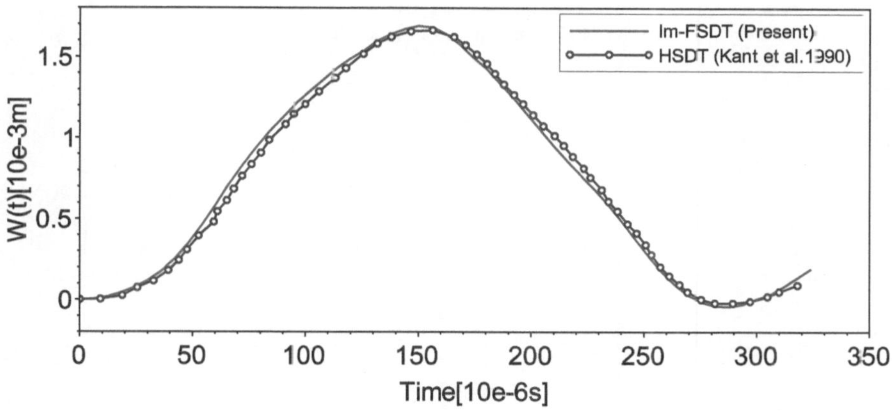

Figure 8.26 Time-deflection response of isotropic square plate under sudden pulse load.

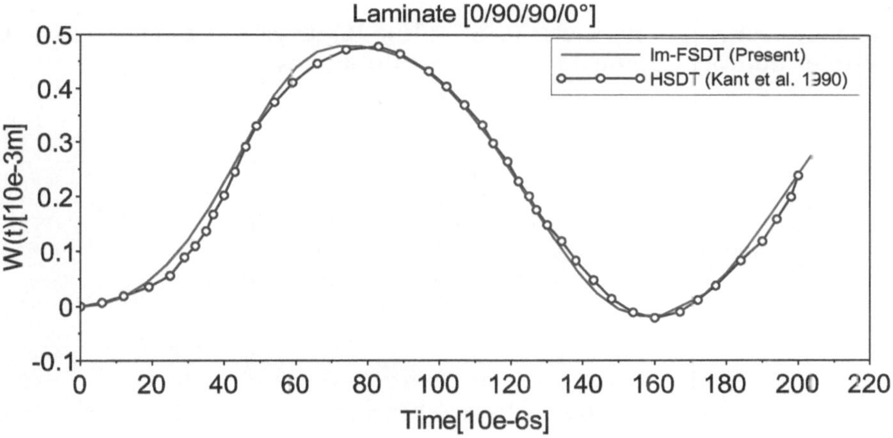

Figure 8.27 Time-deflection response of laminated orthotropic $[0/90/90/0^{\circ}]$ plate under sudden pulse load.

Now, we provide the first new dynamic responses of FG porous plates ($a/h = 20$, $SSSS$, $p = 1$) under a suddenly applied uniform load of intensity $F_0 = -10^{-6} N/m^2$, see Fig. 8.36 for even and uneven porosity distributions. From this figure, it is found that the maximum value and period of transverse displacement $W(t)$ occurs for imperfect FG plates with a porosity volume fraction ($\xi = 0.4$) while the minimum occurs for perfect FG plates with a porosity volume fraction ($\xi = 0$). In addition, the amplitude oscillations of even FG porous plates are larger than uneven types of porosity, just because of the different pores distributions. The uniform distribution, indeed, is reached for an even type of porosity, whereas a concentration of pores is observed in the middle surface for uneven types of porosity. It can be concluded that FG porous plates with concentrated pores in the mid-surface are more rigid than FG porous plates with uniformly distributed pores. To highlight better this effect, Fig. 8.38 depicts the transient response of evenly and unevenly FG porous plates, for two porosity volume fractions ($\xi = 0.2$, $\xi = 0.4$). Even FG porous plates are more flexible

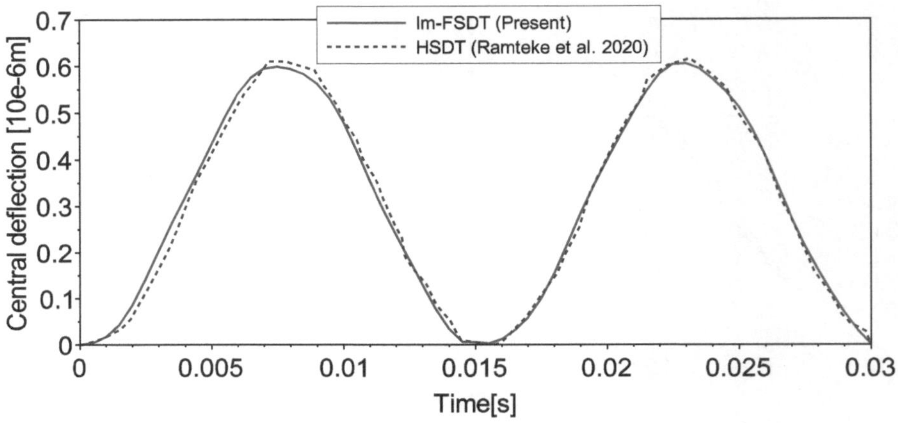

Figure 8.28 Time-deflection response of FG perfect plates $(a/h = 100, p = 2)$ under uniformly distributed load.

Table 8.13

Natural frequency $\overline{\omega}$ of FG porous plates (SSSS, a/h = 20) for gradient indexes (p = 0,0.1) and under various porosity volume fractions and distributions.

Porosity	ξ	Gradient index (p)			
		0		0.1	
		Imp	Ref.[211]	Imp	Ref.[211]
Even	0	0.02908	0.02912	0.02804	0.02802
	0.2	0.02949	0.03003	0.02831	0.02875
	0.4	0.03042	0.03137	0.02904	0.02985
Uneven	0	0.02908	0.02912	0.02804	0.02802
	0.2	0.02984	0.02999	0.02876	0.02884
	0.4	0.03077	0.03101	0.02964	0.02980

than uneven ones, as visible from the larger deflection and period due to the reduced flexural rigidity of the FG plates for uniform pores distributions. Figure 8.37 also shows the effect of the geometrical parameter (a/h) on the transient response of even FG porous plates $(SSSS,\ p = 1,\ \xi = 0.2)$. Based on these plots, note that the increase of (a/h) ratio makes the FG porous plate thinner and more flexible with a lower load capacity. Next, the examination of the effect of gradient index (p) on the dynamic behavior of FG clamped (CCCC) porous plates is carried out in Fig. 8.39. It can be seen that, as the gradient index (p) increases, the deflections and period of oscillations become larger for both even and uneven porosity distributions. This phenomenon is related to the decreased structural stiffness of the FG porous plate with the addition of metal portions. Therefore, the graded property constitutes an efficient way to tailor the vibration oscillations of structural components with a good solution for passive control of vibrations.

8.4.5 DYNAMIC RESPONSE OF FG POROUS SPHERICAL SHELLS SUBJECTED TO DIFFERENT EXTERNAL PULSE LOADS

The second assessment consists of the investigation of the dynamic response of FG porous vibrating spherical shells subjected to different external pulse loads. Since, no results are available in the

Table 8.14

Natural frequency $\overline{\omega}$ of FG porous plates $(SSSS, a/h = 20)$ for gradient indexes (p = 0.5,1) and under various porosity volume fractions and distributions.

Porosity	ξ	gradient index (p)			
		0.5		1	
		Imp	Ref.[211]	Imp	Ref.[211]
Even	0	0.02526	0.02467	0.02350	0.02224
	0.2	0.02499	0.02456	0.02265	0.02104
	0.4	0.02487	0.02423	0.02107	0.01817
Uneven	0	0.02526	0.02467	0.02350	0.02224
	0.2	0.02582	0.02524	0.02390	0.02247
	0.4	0.02653	0.02590	0.02440	0.02269

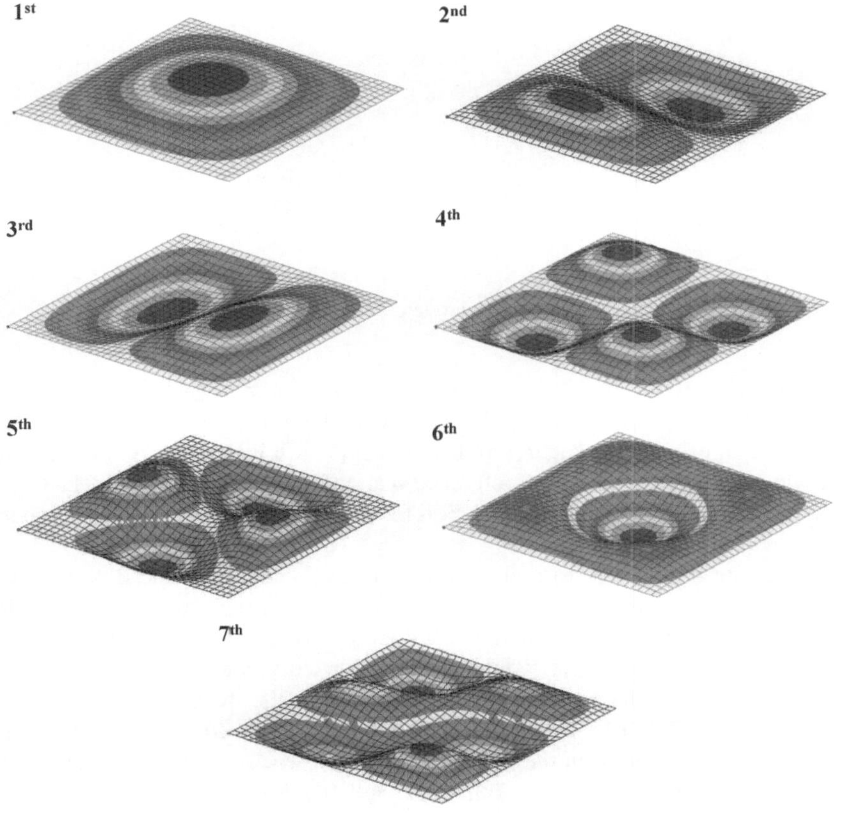

Figure 8.29 Seven mode shapes of simply supported FG porous vibrating plates.

open literature about the dynamic analysis of FG porous spherical shells, a comparative study is carried out for the isotropic case. Thus, a clamped spherical shell under a concentrated apex load is considered, where the structural characteristics of the shell and the concentrated load are presented in Fig. 8.40. The dynamic response of the isotropic spherical structure is subjected to a sudden pulse

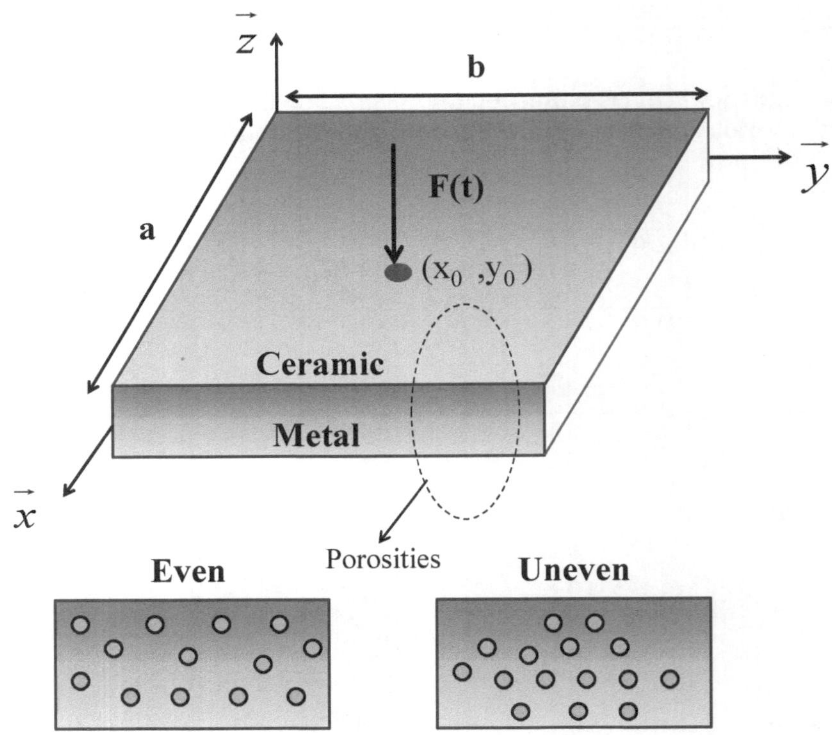

Figure 8.30 Configuration of FG porous plate under a pulse point load.

load of magnitude $F = 100$ in the time interval $[0-280]\mu s$, see Fig. 8.41. The present (Im-FSDT) normalized vertical displacement at shell apex is compared to the one obtained by Duarte Filho and Armando using an eight-node hexahedral finite element [77]. It is pointed out that the present (Im-FSDT) dynamic curve follows well the reference one and these close agreements serve to verify the present approach and establish the foundation for its application for FG porous spherical shells where no comparative solution is available.

To further validate the present formulation, a comparative study of the dimensionless natural frequency parameter $\overline{\omega_s} = \omega R_s \sqrt{\rho/E}$ for isotropic clamped spherical shells is also conducted. The structural properties of the isotropic spherical shell are given in Tab. 8.15. The comparative results are reported in Tab. 8.16, where it is visible a satisfactory correlation among the results with few discrepancies related to the different solution strategies used in the studies. Indeed, a semi-analytical method is employed in [246, 61], a finite element method is utilized in [247] while an improved (FSDT) finite shell model is used in the present work. Otherwise, a good agreement is shown. The first seven mode shapes of the vibrating spherical shell ($R_s/h = 20$; $\beta = 60^0$) are plotted in Fig. 8.42.

Now, the first new elasto-dynamic response of FG porous spherical shells is performed. The spherical shell is subjected to a sudden pulse load of magnitude $F_0 = -1 \times 10^6 N/m^2$. The geometrical characteristics of the spherical shell are plotted in Fig. 8.40 and its mechanical properties are mentioned in section 8.3.1. The transient response of the FG porous spherical shell under different porosity volume fractions and for even and uneven porosity distributions is given in Fig. 8.43. It is pointed out that an increased porosity volume fraction from ($\xi = 0$) to ($\xi = 0.4$) intensifies the oscillations amplitudes of the FG porous spherical shell and the period of oscillations becomes larger

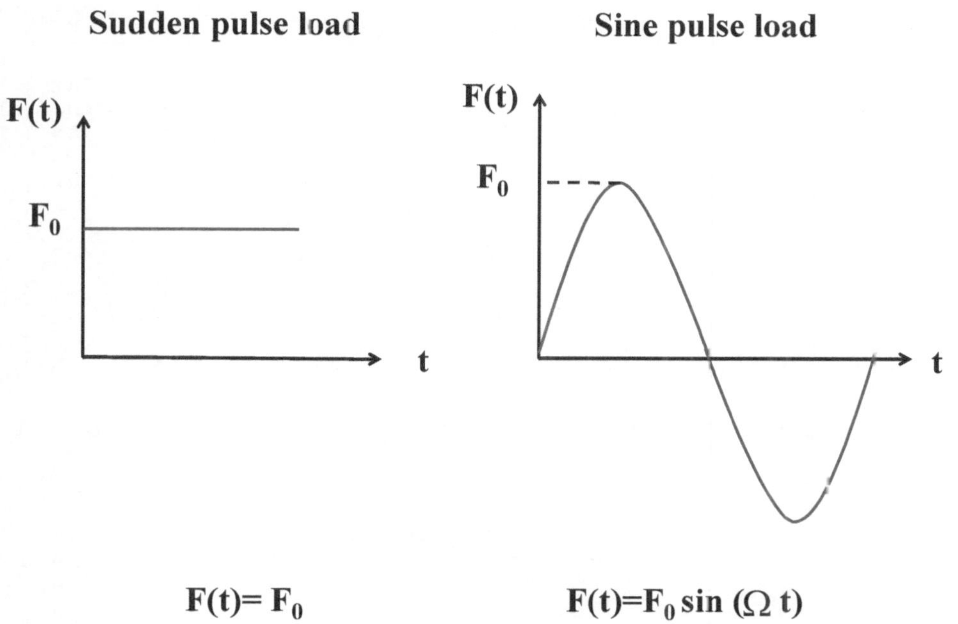

Figure 8.31 Different applied external pulse loads.

in the case of even porosity distributions. therefore, a high percentage of porosity volume fraction included in the structure makes the FG spherical shell more flexible and bends easily. On the other hand, the density and the Young's modulus of the FG porous spherical shell are further reduced for a high rate of porosity which stimulates the deflections of the shell. To a better comprehension of this phenomenon, Fig. 8.44 shows the effect of types of porosity distributions on the transient dynamic response for FG porous spherical shells under a sudden pulse load, with a comparative study between even and uneven deflection responses. For the same porosity volume fraction, by comparing even and uneven transient responses, it can be remarked that even porosity deflections are higher than uneven deflections. Such deflections increase for an increased porosity rate. This can be attributed to different pores distributions in the whole spherical structure where even type of porosity induces a uniform distribution of pores across the section of the shell, while an uneven porosity reflects a concentration of pores in the mid-surface and a vanishing distribution in the upper and lower surfaces. Thereby, the concentration of pores in the mid-surfaces of the shell makes the spherical structure more rigid against vibration oscillations, in contrast to uniformly distributed pores which stimulates further vibration oscillations. Consequently, the vibrations of FG spherical shells can be adjusted by controlling the percentage and type of porosity included into the structure. This method constitutes an efficient way to control passively the structural vibrations of plates and shells. Next, Fig. 8.45 depicts the effect of a varying gradient index (p) on the transient transient dynamic response of even and uneven FG porous spherical shells (CC) subjected to a sudden pulse load. From these plots, it can be noted that an increased gradient index (p) leads to large deflections and period for both even and uneven FG porous spherical shells. It is related to the added metal portions with the rise of gradient index and the decreased stiffness of the spherical shell which features larger oscillations and periods. The effect of a varying (R_s/h) ratio on the transient dynamic response for both even and unevenly FG porous spherical shells is plotted in Fig. 8.46. Note that an increased (R_s/h) ratio induces larger deflections especially for even porosity dispersions. Indeed,

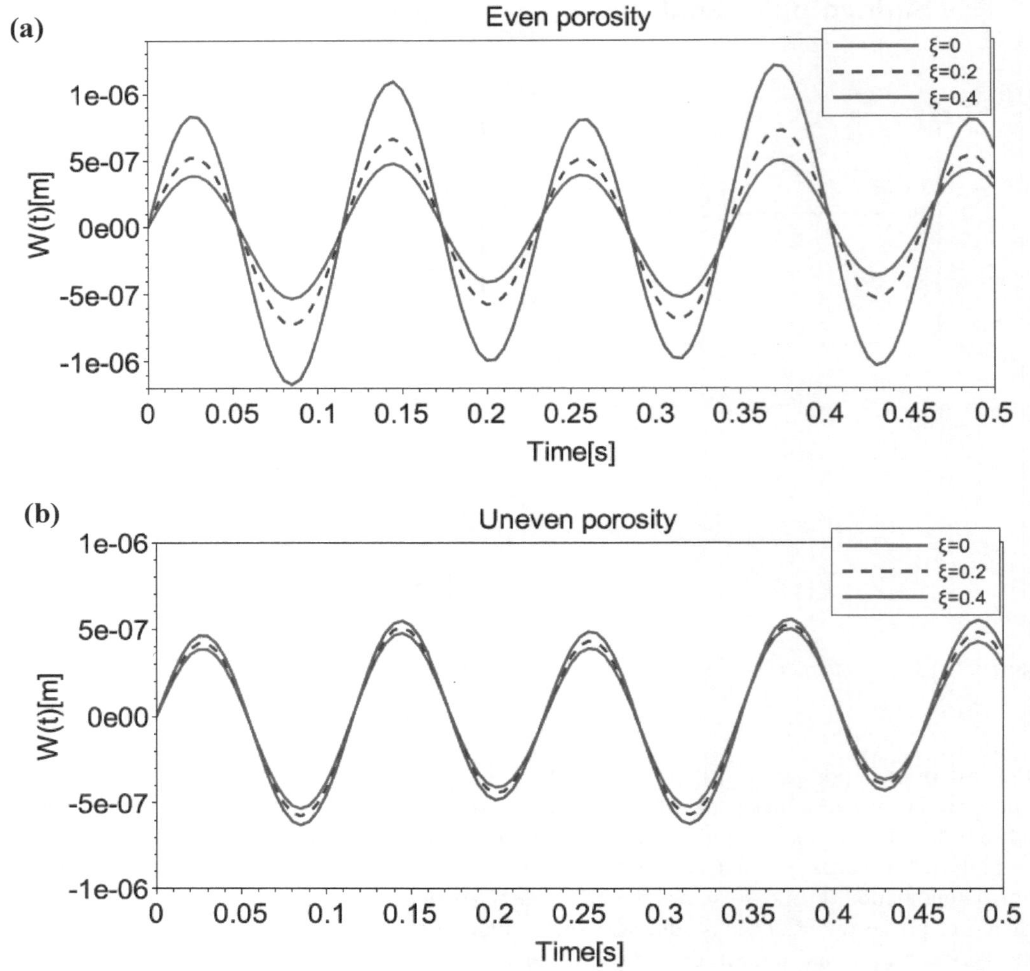

Figure 8.32 Time-deflection response of (a) evenly and (b) unevenly FG porous plates $(a/h = 20, p = 1, SSSS)$ under a transverse harmonic excitation and for different porosity volume fractions.

for an increased (R_s/h) ratio, the FG porous spherical shell becomes thinner and exposed to larger bending effects. In addition, the uniform dispersion of porosity emphasizes this phenomenon due ti the decreased flexural stiffness of the FG porous spherical shell. Consequently, the type of porosity dispersion, its volume fraction, material and geometrical characteristics of the shell represent some crucial parameters that must be considered in the design and control of vibrations for such structures.

Finally, Fig. 8.47 represents the harmonic response of FG porous spherical shells subjected to an harmonic excitation of magnitude $F_0 = 40N$ and pulsation $\Omega = 1800rad/s$. Both even and uneven porosity dispersions are considered. As stated, the FG imperfect spherical shell with a porosity volume fraction $(\xi = 0.4)$ yields the largest vibration amplitude, while the perfect FG spherical shell with a porosity volume fraction $(\xi = 0)$ has the lowest vibration amplitude. This can be related to the reduced stiffness terms with the increased porosity volume fraction. Besides, this effect is more pronounced for even FG spherical shells than uneven ones. As mentioned earlier, this is mainly

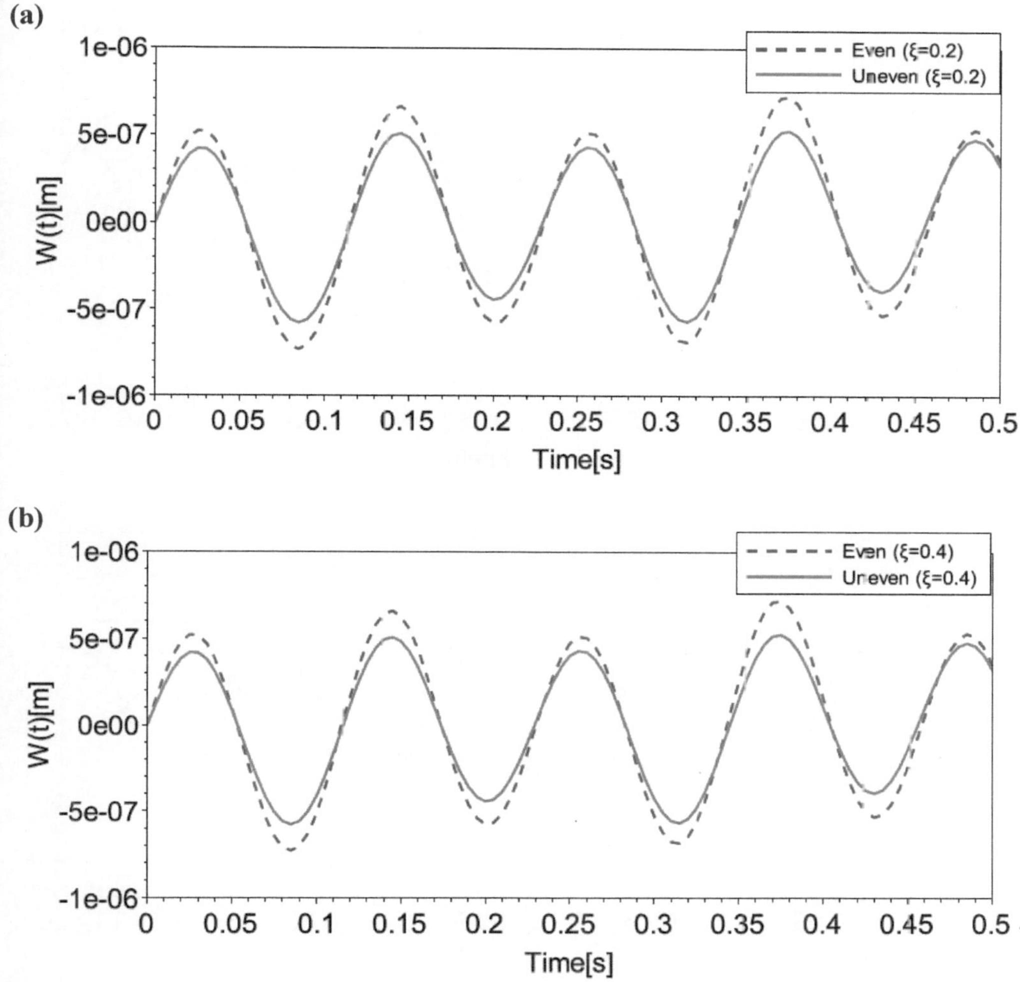

Figure 8.33 Time-deflection response of FG porous plate ($a/h = 20$, $p = 1$, $SSSS$) under a transverse harmonic excitation for two sets of porosity volume fractions (a) ($\xi = 0.2$) and (b) ($\xi = 0.4$) and for different porosity distributions.

related to the uniform dispersion of pores through the thickness direction of the shell which reduces the structural stiffness with an overall decrease in the vibration amplitudes.

8.5 BUCKLING RESPONSES OF FG POROUS PLATES AND SHELLS

In this section, the buckling responses of FG porous plates and shells under different edges compression loads are investigated and presented.

8.5.1 BUCKLING EIGENVALUE PROBLEM FOR FG POROUS STRUCTURES

Herein, the used steps for the determination of the critical buckling load of FG porous plates and shells is presented and highlighted. It should be mentioned that the governing equations are obtained

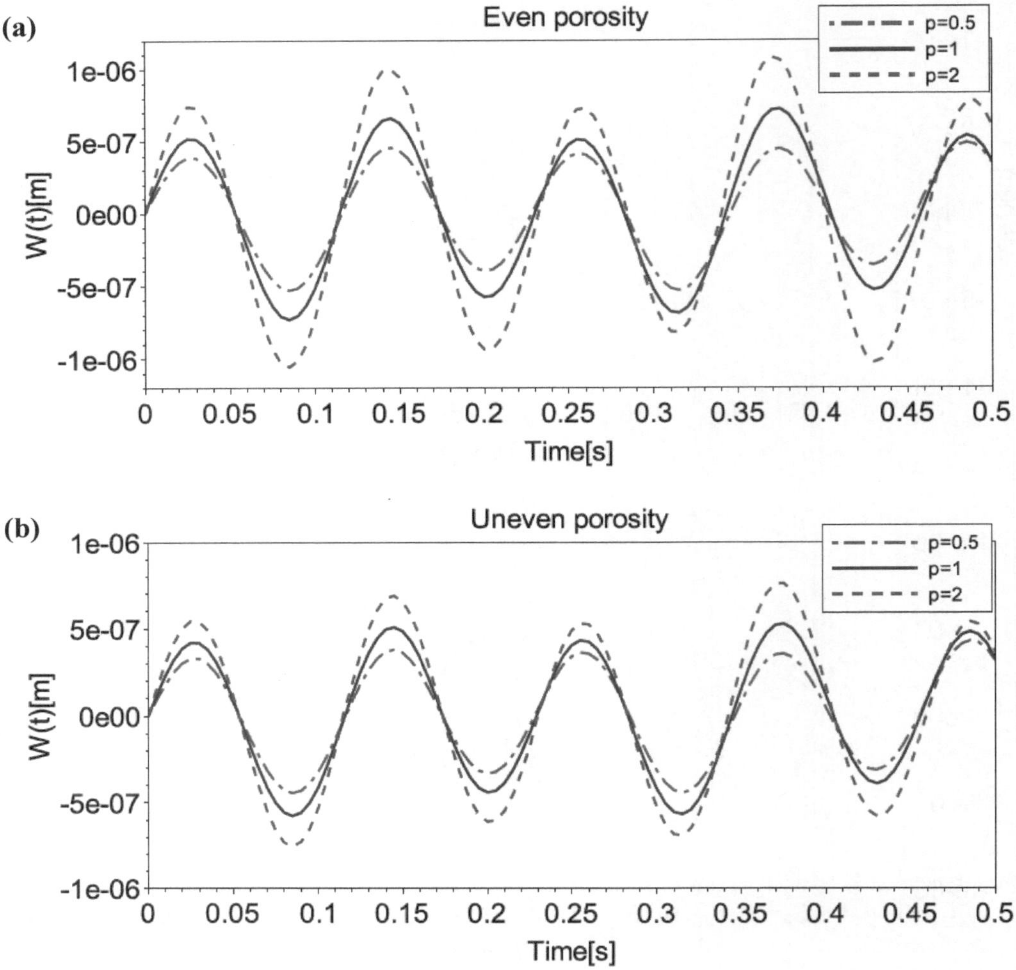

Figure 8.34 Effect of the variation of gradient index (p) on Time-deflection response of (a) even and (b) uneven FG porous plate ($a/h = 20$, $\xi = 0.2$, $SSSS$) under a transverse harmonic excitation.

using the (I-FSDT) model where its detailed steps are as indicated in previous Chapter 1. The resolution procedure can be summarized as follows:

Prebuckling step

This step consists of the computation of the in-plane stresses by solving the static in-plane bending problem as given in the following equation:

$$k_m U_n = F_n. \tag{8.8}$$

where k_m is the linear stiffness matrix, U_n is the global displacement vector and F_n is the external

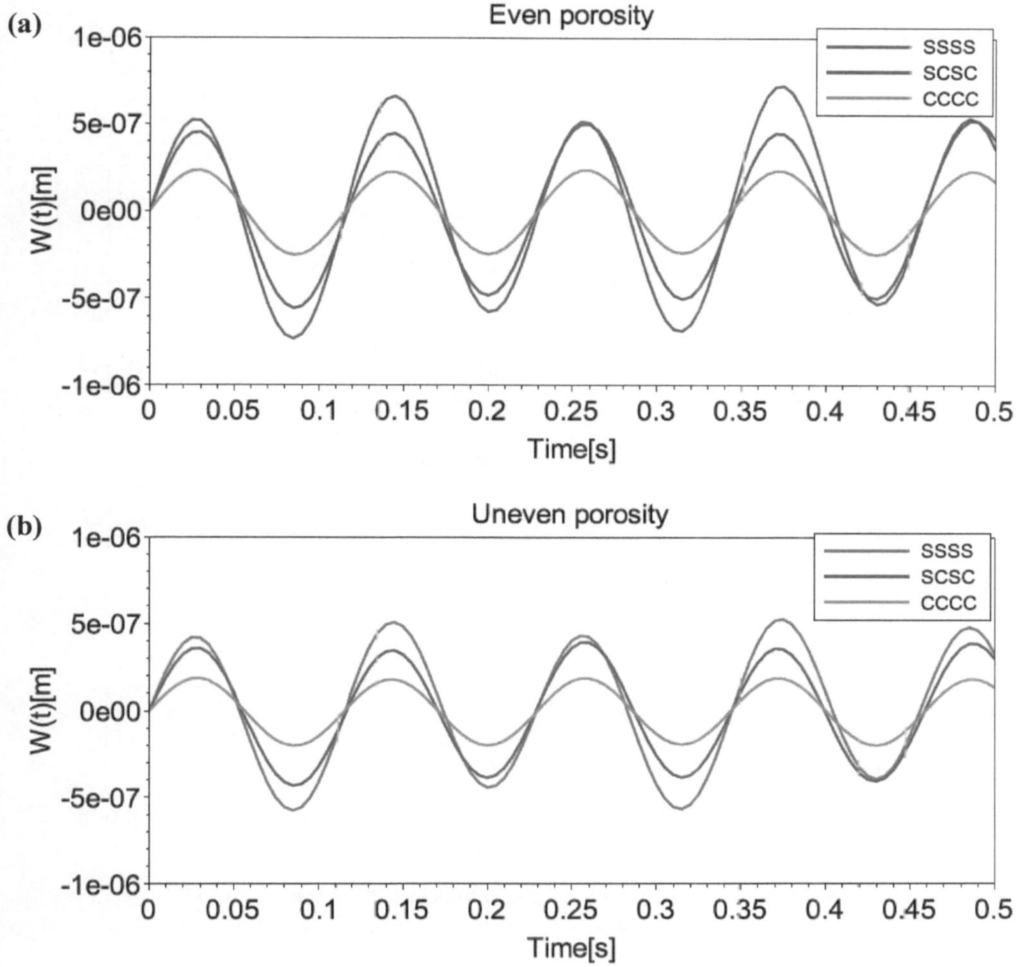

Figure 8.35 Effect of boundary conditions on transient response of (a) even and (b) uneven FG porous plate ($a/h = 20$, $\xi = 0.2$) under a transverse harmonic excitation.

force vector. The expression of the linear stiffness matrix k_m is given as:

$$k_m = \int_A B^T H_T B dA. \tag{8.9}$$

where $B = \begin{bmatrix} B_m \\ B_b \\ B_s \end{bmatrix}$. is the global strain-displacement vector.

On another hand, the boundary conditions constitute an important aspect for prebuckling analysis in the finite element formulation. As a result, the definition of the used boundary conditions is given as:

Simply supported (S): $\begin{matrix} \text{At } y = 0, b: & v = w = \Theta_1 = 0, \\ \text{At } x = 0, a: & u = w = \Theta_2 = 0, \end{matrix}$

Clamped (C): $\begin{matrix} \text{At } y = 0, b: & u = v = w = \Theta_1 = \Theta_2 = 0, \\ \text{At } x = 0, a: & u = v = w = \Theta_1 = \Theta_2 = 0, \end{matrix}$

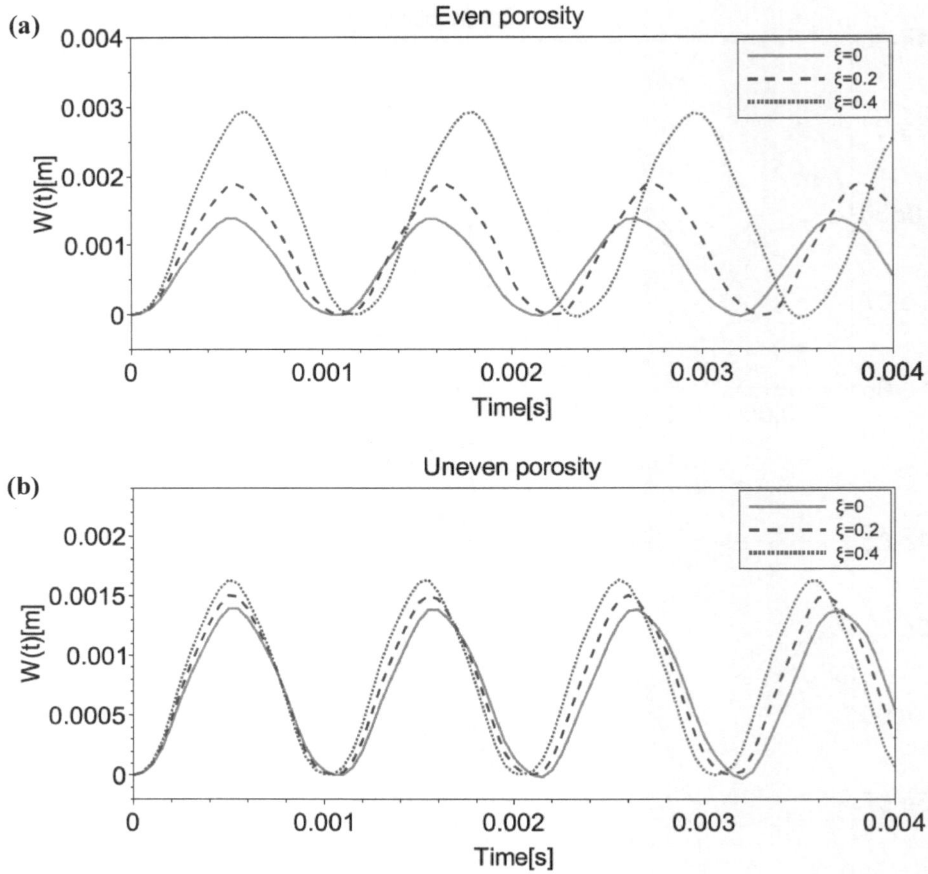

Figure 8.36 Time-deflection response of (a) even and (b) uneven FG porous plate ($a/h = 20$, $SSSS$, $p = 1$) under a suddenly applied uniform load of intensity $F_0 = -10^{-6}N/m^2$ and for different porosity volume fractions.

Buckling step

After carrying out a standard linear static analysis, the membrane stresses in the elements are obtained and the critical buckling load can be evaluated by solving the following eigenvalue problem:

$$(k_m + \lambda_{cr} k_g)U_n = 0. \tag{8.10}$$

where k_g is the geometric stiffness matrix given by:

$$k_{g_{ij}} = \begin{bmatrix} UUM_{ij}I & (UBF_{ij} + UBC_{ij})I \\ (BUF_{ij} + BUC_{ij})I & (BBF_{ij} + BBC_{ij})I \end{bmatrix}, \tag{8.11}$$

The terms UUM, (UBF, BBF) and (UBC, BBC), for a couple of nodes (i, j) are given as:

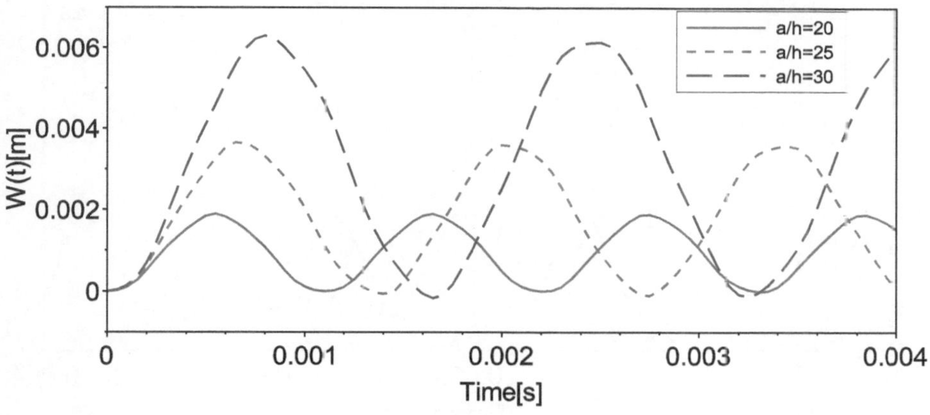

Figure 8.37 Effect of length-to-thickness ratio (a/h) on time-deflection response of FG porous plate ($SSSS$, $p=1$, $\xi=0.2$) under a suddenly applied uniform load of intensity $F_0=-10^{-6}N/m^2$.

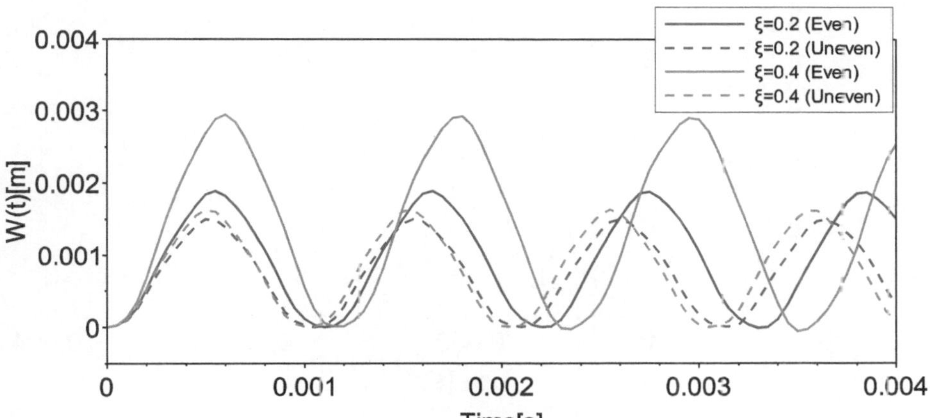

Figure 8.38 Effect of porosity distributions on time-deflection response of FG porous plate ($a/h=20$, $SSSS$, $p=1$) under a suddenly applied uniform load of intensity $F_0=-10^{-6}N/m^2$.

$$\begin{cases} UUM_{ij}=\int_A\left(\overline{N}^i_{,1}\left(N^{11}\overline{N}^j_{,1}+N^{12}\overline{N}^j_{,2}\right)+\overline{N}^i_{,2}\left(N^{12}\overline{N}^j_{,1}+N^{22}\overline{N}^j_{,2}\right)\right)dA \\ UBF_{ij}=\int_A\left(\overline{N}^i_{,1}\left(M^{11}\overline{N}^j_{,1}+M^{12}\overline{N}^j_{,2}\right)+\overline{N}^i_{,2}\left(M^{12}\overline{N}^j_{,1}+M^{22}\overline{N}^j_{,2}\right)\right)dA;\ BUF_{ij}=UBF_{ji} \\ BBF_{ii}=-\int_A\left(\overline{N}^i_{,1}\left(a_1M^{11}+a_2M^{12}\right)+\overline{N}^i_{,2}\left(a_1M^{12}+a_2M^{22}\right)\right)\ d_i\ dA;\ BBF_{ij}=0\ \ if\ \ i\neq j \\ UBC=\frac{1}{8}\begin{bmatrix} -\alpha-\beta & -\beta & 0 & -\alpha \\ \beta & \beta-\gamma & -\gamma & 0 \\ 0 & \gamma & \gamma+\delta & \delta \\ \alpha & 0 & -\delta & \alpha-\delta \end{bmatrix},\quad \begin{cases} \alpha=\int_A(1-\xi)T_2dA\ ,\ \beta=\int_A(1-\eta)T_1dA \\ \gamma=\int_A(1+\xi)T_2dA\ ,\ \delta=\int_A(1+\eta)T_1dA \end{cases} \\ BUC_{ij}=UBC_{ji},\qquad BBC_{ii}=-\frac{1}{4}\begin{Bmatrix} \left(\alpha a_2^A+\beta a_1^B\right)\cdot d_1 \\ \left(\beta a_1^B+\gamma a_2^C\right)\cdot d_2 \\ \left(\gamma a_2^C+\delta a_1^D\right)\cdot d_3 \\ \left(\delta a_1^D+\alpha a_2^A\right)\cdot d_4 \end{Bmatrix};\ BBC_{ij}=0\ \ if\ \ i\neq j \end{cases} \tag{8.12}$$

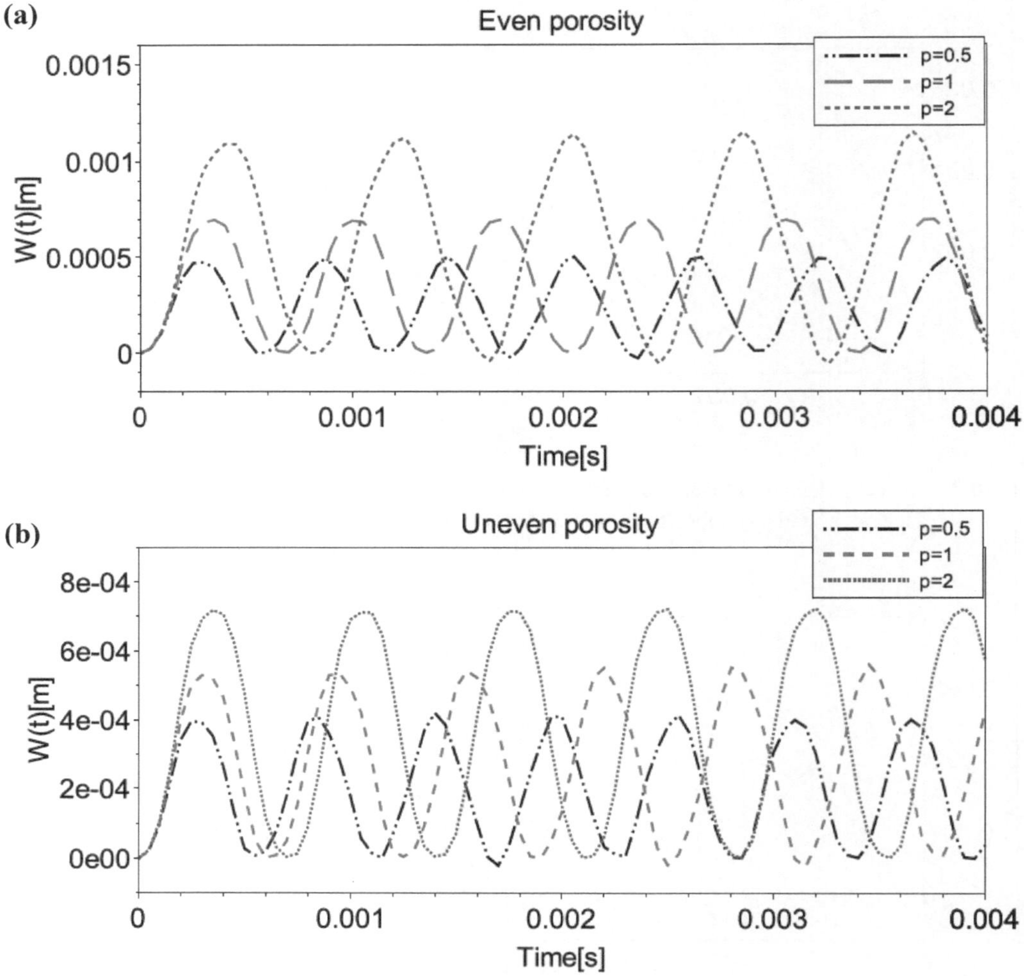

Figure 8.39 Time-deflection response of (a) even and (b) uneven FG porous plate ($a/h = 20$, $CCCC$, $\xi = 0.2$) under a suddenly applied uniform load of intensity $F_0 = -10^{-6} N/m^2$ and for different gradient indexes.

It should be mentioned that the lowest eigenvalue of Eq. (B.10) is called the critical buckling load and the vector U_n provides the buckling mode shape of the FGM structure. These steps can be summarized in the flowchart illustrated in Fig. 8.48.

It is also noteworthy, that the proposed improved first-order shear deformation theory (I-FSDT) constitutes a suitable choice for thin to moderately thick structures and it can hence bring up two levels of analysis: the prediction of buckling behavior of porous FGM structures taking into account geometrical nonlinearities and the addition of a parabolic function $f(z)$ in the shear part of deformations which reflect the amelioration in the FSDT theory without an addition of the computational time. Moreover, it can be used for further analysis such as vibration, thermal buckling and post-buckling of beams, plates and shells.

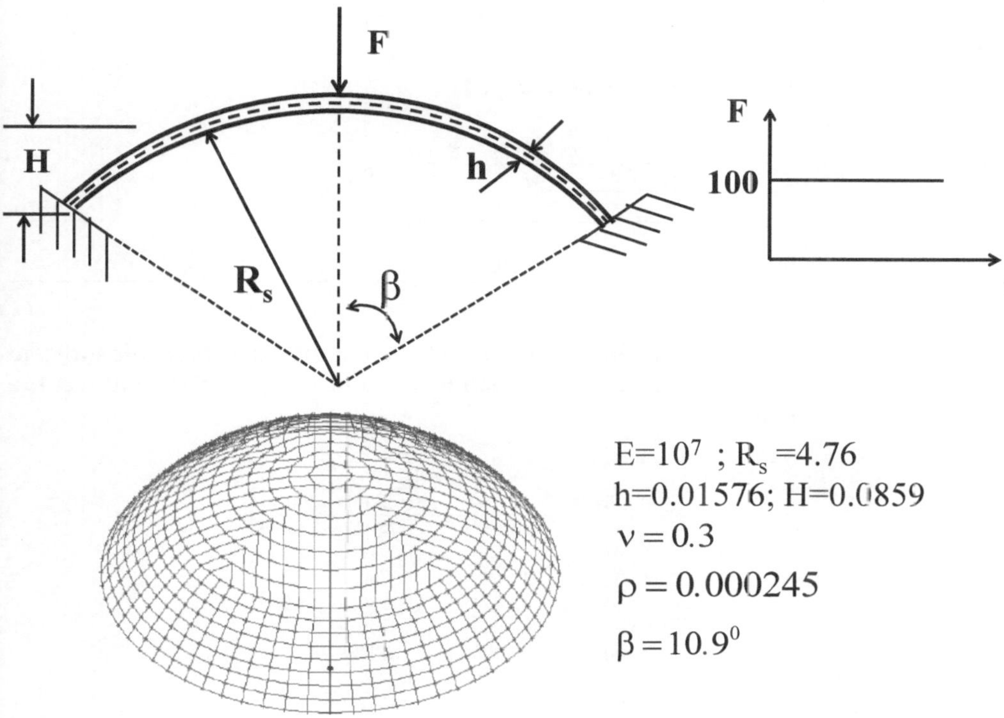

Figure 8.40 Geometric sketch, characteristics and finite element mesh of the vibrating spherical shell.

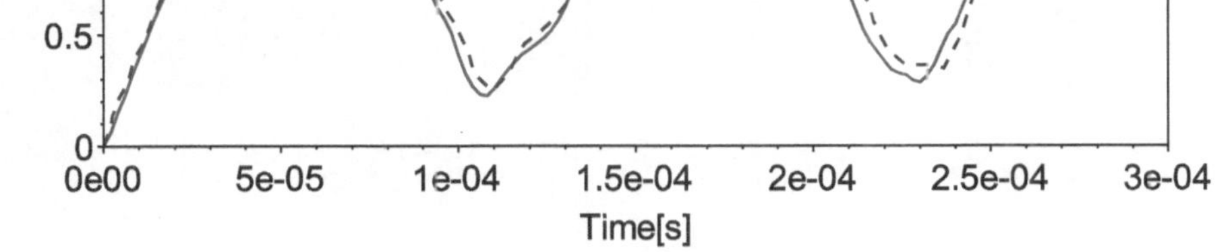

Figure 8.41 Time-deflection dynamic response of the isotropic spherical shell.

8.5.2 POWER-BASED FUNCTIONALLY GRADED PLATES (P-FGM)

The buckling responses of perfect P-FGM plates are computed and analyzed. The obtained results are presented in the next sections.

Table 8.15
Structural properties of the spherical shell [246, 61].

$E(GPa)$	ν	$\rho(kg/m^3)$	$R_s(m)$	$h(m)$
211	0.3	7800	1	0.05

Table 8.16
Natural frequencies of isotropic clamped spherical shells (Numbers in brackets indicate the vibration mode (m,n). m is the number of half-waves along a generatrix and n is the circumferential wave number).

$\beta(^o)$	(m,n)	Natural frequency $\overline{\omega_s}$			
		(Im-FSDT)	Ref.[247]	Ref.[246]	Ref.[61]
45	(1,1)	1.0820	1.0688	1.0690	1.0698
	(2,1)	1.2942	1.2930	1.2931	1.3064
	(3,1)	1.6015	1.5696	1.5700	1.5802
	(4,1)	1.9340	1.9292	1.9300	1.9533
60	(1,1)	0.8987	0.8990	0.8990	-
	(2,1)	1.0800	1.0790	1.0791	-
	(3,1)	1.2715	1.2140	1.2140	-
	(4,1)	1.4024	1.3930	1.3931	-

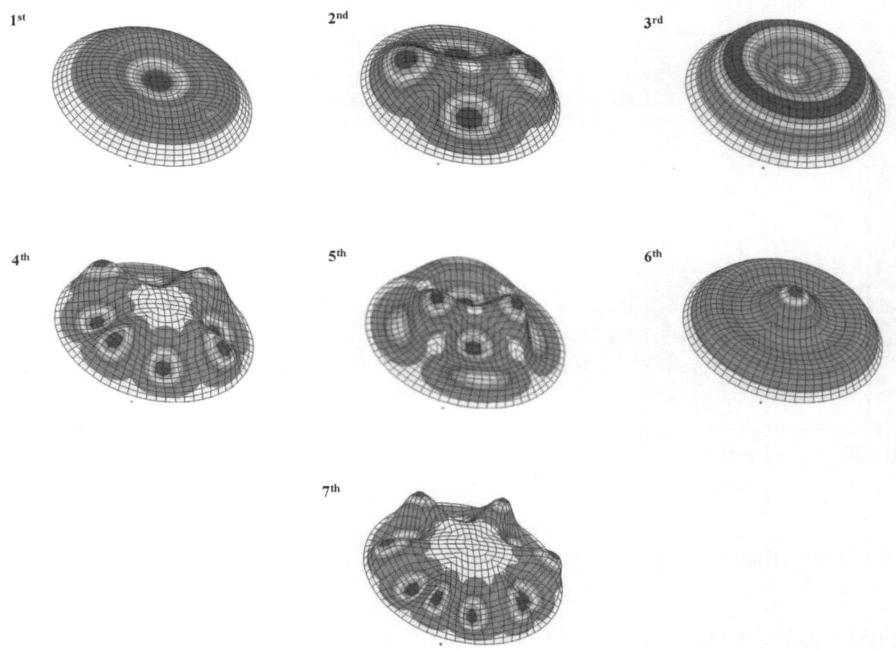

Figure 8.42 Seven mode shapes of vibrating spherical shell.

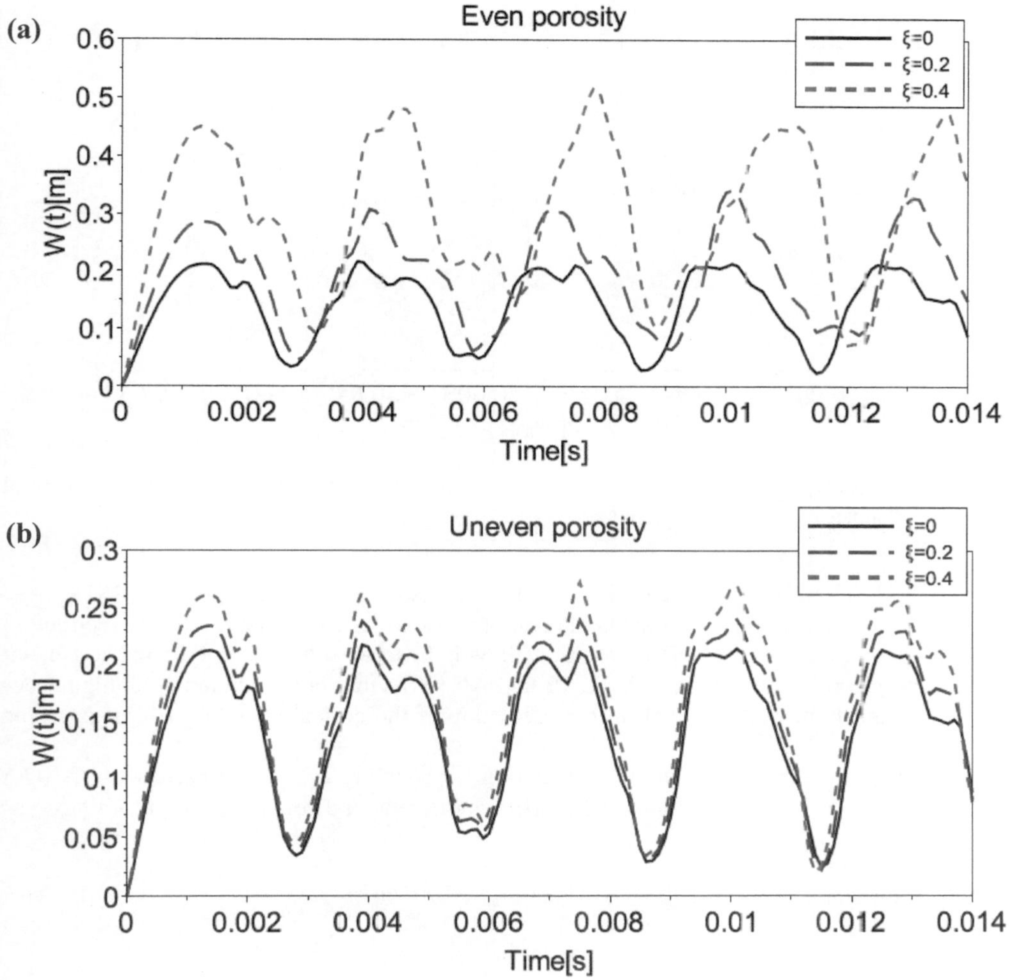

Figure 8.43 Transient response of (a) even and (b) uneven FG porous spherical shell $(CC, p = 1)$ under different porosity volume fractions and subjected to a sudden pulse load.

8.5.2.1 Convergence and validation studies

At first, the convergence rate of an isotropic simply supported (SSSS) square plate subjected to uniform compression load is examined in Fig. 8.49. The critical buckling load for the isotropic plate is expressed in terms of dimensionless value: $\overline{N}_{cr} = N_{cr}a^2/D_0$, where $a = 10$ in is the length side of the plate, $D_0 = E_0h^3/12(1-\nu^2)$, $E_0 = 3 \times 10^6 psi$, $h = 0.2$ in and $\nu = 0.3$ are the rigidity, the Young's modulus, the thickness and the Poisson's ratio of the isotropic plate, respectively [291, 135]. From Fig. 8.49, it is observed that the current finite shell element presents a good convergence rate with the mesh refinement for the first four modes of buckling of the isotropic plates where (18×18) finite elements per side are sufficient to determine the critical buckling load. Furthermore, a quick check of the formulation in regard to standard (FSDT) theory and analytical approach is presented in Tab. 8.17. As can be seen, the present formulation provides better results toward analytical buckling response in comparison to Standard (FSDT) theory which outlines that the present finite shell element is simple and efficient in the prediction of the critical buckling load of isotropic plates.

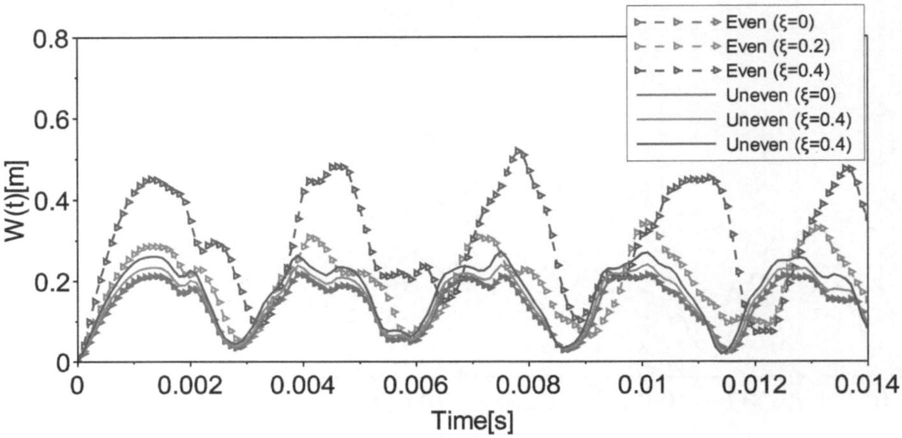

Figure 8.44 Effect of types of porosity distributions on transient dynamic response of the FG porous spherical shell $(CC, p = 1)$ subjected to a sudden pulse load.

Another check out of the formulation is carried out for isotropic simply supported (SSSS) rectangular plates subjected to uniaxial and biaxial compression loads where the results are reported in Tab. 8.18. As can be seen, the present results fit well with the exact solutions and a good agreement is revealed toward others' analytical, MATLAB and ANSYS solutions which draw the high accuracy of the present finite element shell in the prediction of the critical buckling load of isotropic plates.

At second step, the consistency of the formulation for buckling analysis of perfect (Al/Al_2O_3) P-FGM plates is also checked. The material properties of metal and ceramic portions are taken as [258]:

metal (Aluminum, Al): $E_m = 70 \times 10^9 N/m^2$; $\rho_m = 2702 kg/m^3$; $\nu = 0.3$

ceramic (Alumina, Al_2O_3): $E_c = 380 \times 10^9 N/m^2$; $\rho_c = 3960 kg/m^3$; $\nu = 0.3$

The dimensionless critical buckling load of the (Al/Al_2O_3) FGM plates subjected to different types of compression loads and for different length-to-thickness (a/h) and aspect (a/b) ratios is evaluated relative to the metal portion as $\overline{N}_{cr} = N_{cr}a^2/E_m h^3$ and the obtained results are reported in Tabs. 8.19, 8.20, 8.21 and 8.22. A good correlation between the present evaluated critical buckling loads and those derived from the literature is obtained and the difference, in some cases, notably when the length-to-thickness ratio (a/h) is equal to 5 or 10, is related in one hand to the different theories and on other hand, to the aspect of the plate which is considered thick in the range of [5–10] of the length-to-thickness ratio (a/h). Otherwise, the present finite element model constitutes a simple and efficient model for buckling analysis of perfect FGM plates.

Now, we provide the first known results for buckling analysis of porous FGM plates under different types of applied edge compression. In fact, in the literature there is a lack of availability of buckling analysis of porous FGM structures and few performed works are generally for isotropic or sandwich composite structures. So, a parametric study for porous FGM plates varying several design parameters, coefficient of porosity, types of porosity distributions, gradient indexes, boundary conditions and types of applied loadings are investigated in the following parts.

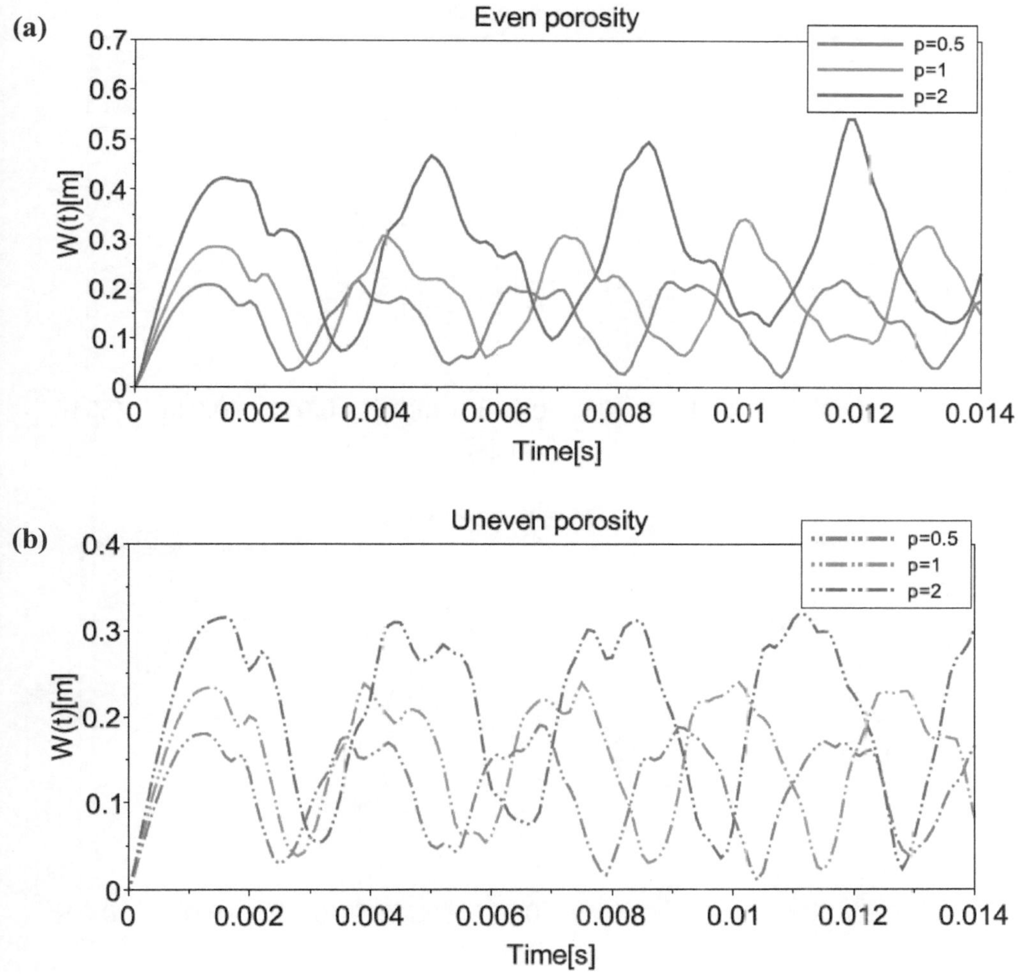

Figure 8.45 Effect of gradient indexes on transient dynamic response of (a) even and (b) uneven FG porous spherical shell (CC) subjected to a sudden pulse load.

8.5.2.2 Parametric analysis

The second assessment covers a parametric study on the effects of geometrical and material parameters on critical buckling load of power-based functionally graded (P-FGM) plates. The first examination concerns the tracing of the dimensionless critical buckling load $\overline{N}_{cr} = N_{cr}a^2/E_m h^3$ in the function of the gradient index (p) for the first four modes of a simply supported (SSSS) square (Al/Al_2O_3) P-FGM plates ($a/h = 20$) under uniaxial compression load varying the porosity parameter (α) and for two porosity patterns as shown in Fig. 8.50. In one hand, it is observed that the critical buckling load for a pure ceramic P-FGM plate has the maximum value and the buckling load diminishes as the gradient index (p) value rises due to the increase in the metal constituent in the P-FGM plate with the rise of the gradient index (p). On the other hand, moving from perfect ($\alpha = 0$) to imperfect ($\alpha = 0.1, 0.2, 0.3$) P-FGM plates, the critical buckling load decreases, in a pronounced way, with the high buckling modes. Concerning the porosity patterns, the difference between the buckling curves is more highlighted with uneven porosity pattern compared to even pattern. This

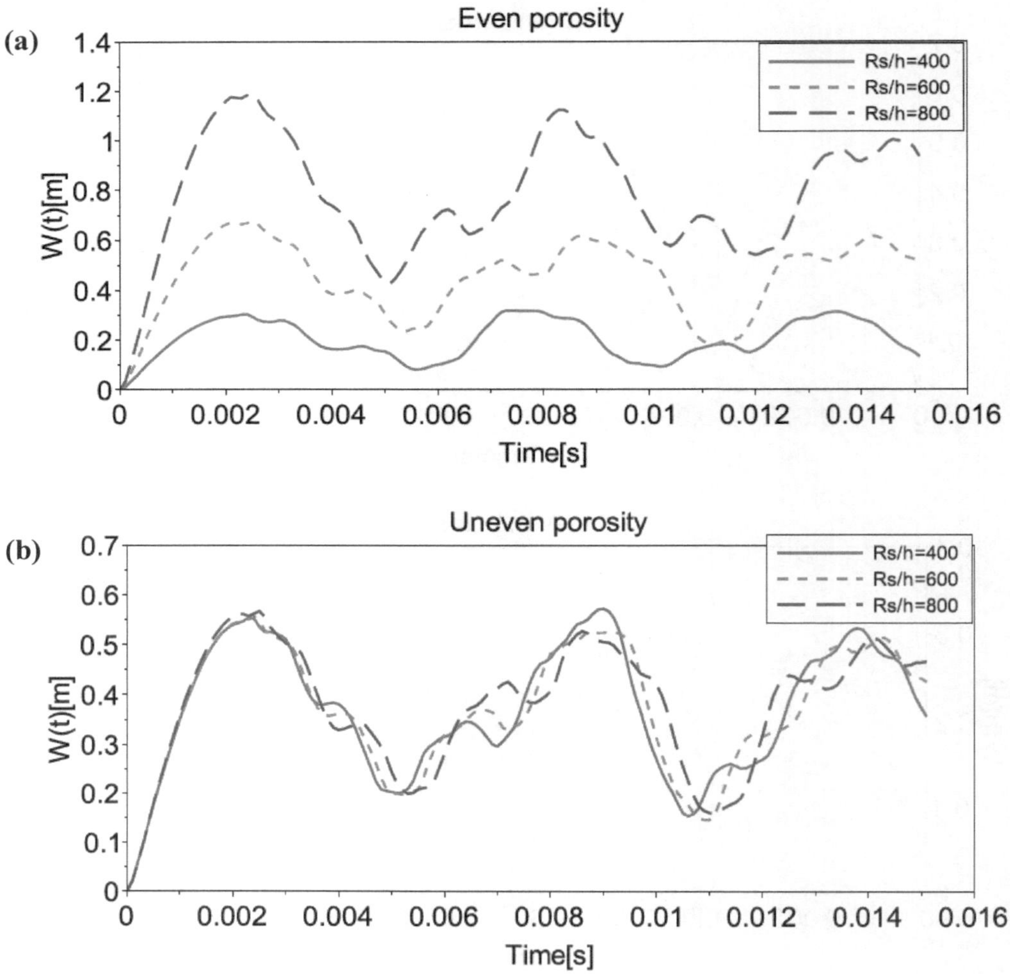

Figure 8.46 Effect of variation of (R_s/h) ratio on transient dynamic response of (a) even and (b) uneven FG porous spherical shell $(CC, p=1, \xi=0.2)$ subjected to a sudden pulse load.

observation can be explained by the manner of porosity distribution through the whole of the structure where concentrated pores in the mid-surface of the plate influence significantly the buckling response more than uniform distributed pores. Therefore, the profile of pores distributions is an important parameter on buckling response of P-FGM plates which must be considered in the stability analysis of such structural components. It should be noted that all computations of the critical buckling load is carried out adimensionally with the aforementioned expression: $\overline{N}_{cr} = N_{cr}a^2/E_m h^3$.

The second examination consists of the exploration of the effect of porosity parameter (α) on buckling response of a simply supported (SSSS) (Al/Al_2O_3) P-FGM plates $(a/h = 20, p = 1)$ under different edge compression loads as plotted in Fig. 8.51. As can be seen, the increase of the porosity parameter (α) leads to a decrease in the critical buckling load for all applied loads and this effect is more prominent for even than uneven porosity distributions. Indeed, the percentage of porosity parameter coupled with the pattern of distributions accentuated the rigidity of the plate which manifest by a pronounced down in the buckling load. Furthermore, it can be noted that for uniaxial

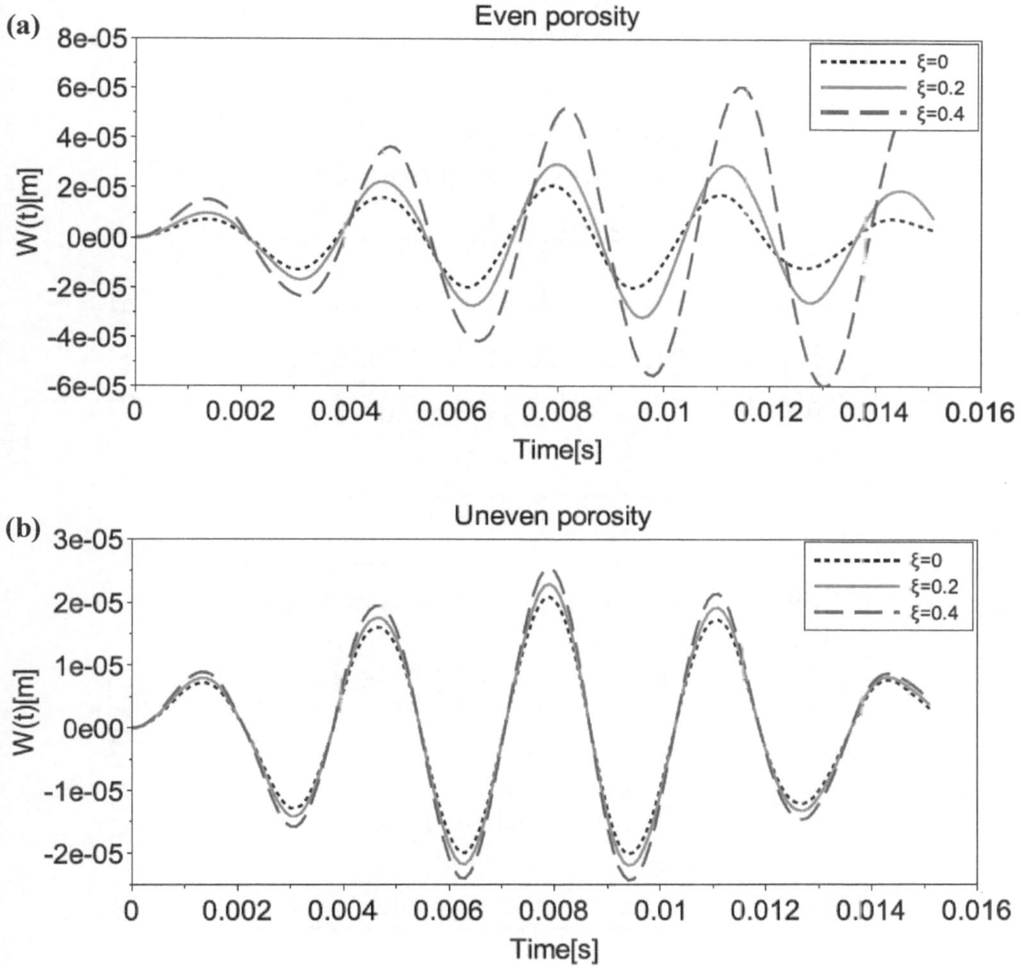

Figure 8.47 Harmonic transient dynamic response of (a) even and (b) uneven FG porous spherical shell $(CC, p = 1,)$ subjected to an harmonic excitation.

compression load, the second and third modes are decoupled while for pure biaxial compression or combined biaxial compression tension loads, these buckling modes are coupled which trace the influence of applied edge load on buckling modes of the P-FGM plates. To better point this effect, Fig. 8.52 traces the buckling response of the first four modes of a simply supported (SSSS) (Al/Al_2O_3) evenly and unevenly P-FGM plates $(a/h = 20, p = 1)$ subjected to uniaxial compression, biaxial compression and biaxial compression tension loads. It is clear that the type of applied load plays a crucial role on buckling behavior of P-FGM plates where the biaxial compression tension load induces the highest critical buckling load while the uniaxial compression has the lowest one. The biaxial compression load is located between these two extreme cases for the first and fourth buckling modes, in contrast to the second and third modes where coupled tensile and compressive stresses in x and y directions result in an overlap between uniaxial and biaxial loaded edges. Hence, the type of applied load stimulates the buckling behavior of P-FGM plates which should be taken into account in buckling stability analysis of such advanced materials.

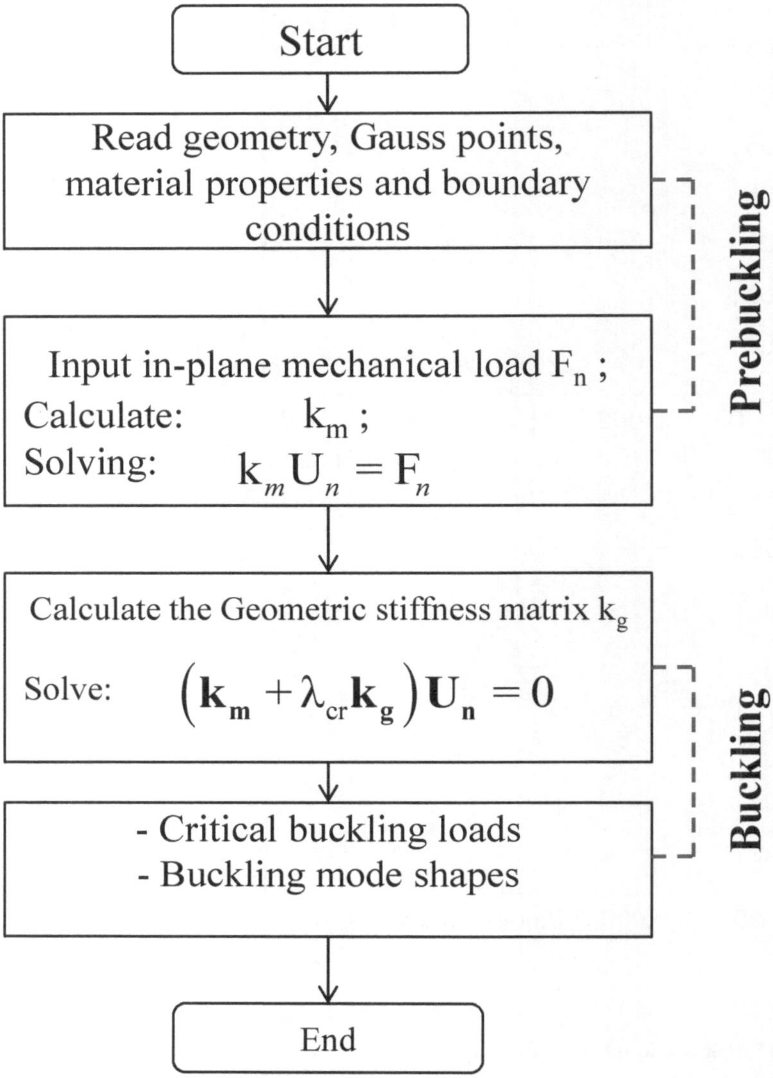

Figure 8.48 Flowchart of solving buckling problem of FGM porous structures.

The impact of porosity distribution patterns on the critical buckling behavior of P-FGM square plates $(SSSS, b/h = 10, p = 1)$ subjected to uniaxial compression load is presented in Fig. 8.53. For perfect P-FGM plates without porosities $(\alpha = 0)$, the first four buckling modes are conserved. However, when the porosity parameter changes, the effect of porosity pattern (even or uneven distributions) becomes more prounonced. In fact, as discussed earlier, even porosity type induces less buckling load than uneven porosity type. This observation is due to the way of pores distribution into the structure where uniform distributed pores weakened more the stiffness of the plate than concentrated pores in the mid-surface. Therefore, the manner of pores distribution affects earnestly the elastic modulus of the P-FGM plates and as a result the buckling behavior of these structural parts. Next, the impact of geometrical ratio (b/h) on the buckling modes of evenly and unevenly (Al/Al_2O_3) P-FGM plates $(SSSS, p = 1)$ under uniaxial compression is explored in Fig. 8.54. It is shown that

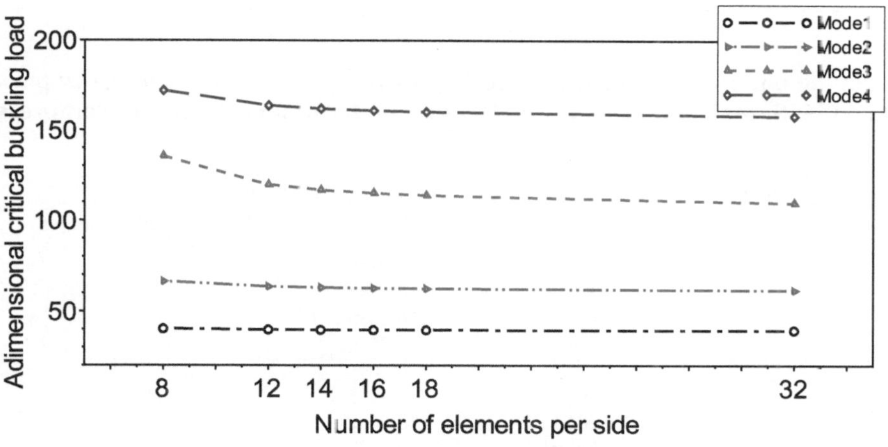

Figure 8.49 Convergence rate of dimensionless critical buckling load for an isotropic square plate (SSSS) subjected to uniform compression load.

Table 8.17
Comparison of the dimensionless critical buckling load $\overline{N}_{cr} = N_{cr}a^2/D_0$ of the first four buckling modes of an isotropic square plates (SSSS) under uniform compression load.

Reference	Mode			
	1	2	3	4
Current	39.5740	62.4624	113.6881	160.1272
Standard-FSDT[291]	39.2040	60.3722	103.9630	151.987
%difference	0.9437	3.3463	9.3500	5.0835
Analytical-approach[262]	39.4761	61.6814	109.656	157.904
%difference	0.2473	1.2661	3.6770	1.4079

the increase of the width-to-thickness ratio (b/h) has a significant effect on the third and fourth buckling modes more than the first and second modes. Indeed, for low values of $(b/h = 10, 20)$, the FGM plates behave as thick structure which resists to the compressive stresses, while for high values of $(b/h = 50, 100)$, the FGM plate is slender and buckles hence easily with the compression loaded edges. This effect is more pronounced with evenly than unevenly distributed pores which outlines again that the manner of pores distributions can intensify the buckling response of P-FGM plates. This parameter is really crucial and must be considered in both design and experiments procedure of aeroelastic, automobile and reactors FGM portions.

8.5.3 SIGMOID FUNCTIONALLY GRADED PLATES (S-FGM)

This section is devoted to sigmoid functionally graded plates, known as S-FGM, where some validations are performed and presented in tabular form. To this end, the critical buckling load $\widetilde{N_{cr}} = N_{cr}\eta_0 a^2/E_m h^3$ of a simply supported square (Al/Zr_O2) S-FGM plates $(a/h = 100)$ subjected to uniaxial compression load is computed and reported in Tab. 8.23. η_0 is the magnitude of in-plane buckling load applied in the x direction and the Aluminium and Zirconia material properties are given as indicated in [248]:

Table 8.18

Comparison of the dimensionless critical buckling load $\overline{N}_{cr} = N_{cr}b^2/\pi^2 D_0$ for isotropic simply supported (SSSS) rectangular plates ($b/h = 100$) under different types of compression loadings.

Aspect ratio (a/b)	Reference	Type of loading	
		Uniaxial compression	Biaxial compression
0.5	Current	6.2578	5.0059
	MATLAB[4]	6.2418	4.9935
	ANSYS[4]	6.1855	4.9484
	Analytical solution[327]	6.2380	–
	Exact solutions[262]	6.2500	–
1	Current	4.0095	2.0046
	MATLAB[4]	3.9978	1.9990
	ANSYS[4]	3.9602	1.9802
	Analytical solution[327]	3.9970	–
	Exact solutions[262]	4.0000	–
1.5	Current	4.3786	1.4489
	MATLAB[4]	4.3374	1.4439
	ANSYS[4]	4.3056	1.4342
	Analytical solution[327]	4.3370	–
	Exact solutions[262]	–	–
2	Current	4.0095	1.2546
	MATLAB[4]	3.9981	1.2496
	ANSYS[4]	3.9696	1.2439
	Analytical solution[327]	–	–
	Exact solutions[262]	–	–
3	Current	4.0095	1.1158
	MATLAB[4]	3.9993	1.1108
	ANSYS[4]	3.9727	1.1083
	Analytical solution[327]	–	–
	Exact solutions[262]	–	–

metal (Aluminum, Al): $E_m = 70 \times 10^9 N/m^2$; $\rho_m = 2702 kg/m^3$; $\nu = 0.3$

ceramic (Zirconia, ZrO_2): $E_c = 151 \times 10^9 N/m^2$; $\rho_c = 3000 kg/m^3$; $\nu = 0.3$

The relative errors between the current results and those of analytical [248] and numerical [94] solutions rise with added metal portions (11%) due to different adopted kinematic assumptions and solutions strategies. Indeed, in Ref. [248], a hyperbolic shear deformation theory with Von-karman hypothesis is introduced into the formulation and a corrective in-plane buckling factor is included which generates the observed difference. As well as, in Ref. [94] a quasi-conforming shell element with rotary inertia is considered and the solutions are performed using Navier's method. So, the different theories induces the difference between the critical buckling results. Otherwise, a good correlation is revealed. Continuously, a quick check out of the formulation for (Al/Al_2O_3) perfect and imperfect S-FGM plates $(p = 1)$ subjected to uniaxial compression in-plane load with even and uneven porosity types and for different boundary conditions is examined in Tab. 8.24. The results are in good agreement with an average error (3%) when compared with analytical solutions [248] and the slight difference $[8-11]\%$ in some cases is related to different theories and methodology performed in these two works. In particular, the used corrective in-plane buckling factor η_0 presented in the work of Singh and Harsha [248] add to the induced difference. Therefore, to provide a general definition of the critical buckling load, the expression of the buckling is the same as mentioned in the previous section of P-FGM plates, $\overline{N}_{cr} = N_{cr}a^2/E_m h^3$.

Table 8.19

Dimensionless critical buckling load $\overline{N}_{cr} = N_{cr}a^2/E_mh^3$ of a simply supported (SSSS) (Al/Al_2O_3) FGM plates subjected to uniaxial compression load for different aspect ratios (a/b), length-to-thickness ratios $(a/h = 5, 10)$ and power indexes.

a/b	Power index (p)	Reference	a/h	
			Uniaxial compression load	
			5	10
0.5	0	Current	6.3284	7.1976
		High-order-based FE [4]	6.7203	7.4054
		HSDT[203]	6.7140	7.3970
		RPT[258]	6.7203	7.4050
	0.5	Current	4.1823	4.7011
		High-order-based FE [4]	4.4234	4.8206
		HSDT[203]	4.4100	4.8100
		RPT[258]	4.4235	4.8200
	1	Current	3.2190	3.6153
		High-order-based FE [4]	3.4154	3.7108
		HSDT[203]	3.3900	3.7000
		RPT[258]	3.4164	3.7100
	5	Current	1.9821	2.3269
		High-order-based FE [4]	2.1447	2.4152
		HSDT[203]	2.1240	2.1240
		RPT[258]	2.1484	2.4100
1	0	Current	15.2323	17.3749
		High-order-based FE [4]	16.0210	13.5786
		HSDT[203]	16.0000	13.5400
		RPT[258]	16.0200	13.5700
	0.5	Current	9.4757	11.4082
		High-order-based FE [4]	10.6250	12.1229
		HSDT[203]	10.5700	12.0800
		RPT[258]	10.6200	12.1200
	1	Current	7.9278	8.7771
		High-order-based FE [4]	8.2210	9.3380
		HSDT[203]	8.1460	9.2990
		RPT[258]	8.2200	8.1460
	5	Current	4.4413	5.5630
		High-order-based FE [4]	5.0408	6.0305
		HSDT[203]	4.9700	5.9900
		RPT[258]	5.0500	6.0300
1.5	0	Current	28.0012	38.2041
		High-order-based FE [4]	28.1960	40.7488
		HSDT[203]	28.1500	40.5800
		RPT[258]	28.1900	40.7400
	0.5	Current	19.1200	25.4531
		High-order-based FE [4]	19.2475	26.9092
		HSDT[203]	19.0900	26.7200
		RPT[258]	19.2500	26.9000
	1	Current	14.9896	19.7140
		High-order-based FE [4]	15.0205	20.7959
		HSDT[203]	14.7600	20.5700
		RPT[258]	15.0300	20.8000
	5	Current	8.3356	12.0163
		High-order-based FE [4]	8.4399	12.9238
		HSDT[203]	8.2500	12.7400
		RPT[258]	8.4700	12.9500

Table 8.20

Dimensionless critical buckling load $\overline{N}_{cr} = N_{cr}a^2/E_mh^3$ of a simply supported (SSSS) (Al/Al_2O_3) FGM plates subjected to uniaxial compression load for different aspect ratios (a/b), length-to-thickness ratios $(a/h = 50, 100)$ and power indexes.

a/b	Power index (p)	Reference	a/h	
			Uniaxial compression load	
			50	100
0.5	0	Current	7.6702	7.6933
		High-order-based FE [4]	7.6555	7.6636
		HSDT[203]	7.6400	7.6500
		RPT[258]	7.6500	7.6600
	0.5	Current	4.9778	4.9909
		High-order-based FE [4]	4.9635	4.9681
		HSDT[203]	4.9500	4.9600
		RPT[258]	4.9600	4.9680
	1	Current	3.8253	3.8353
		High-order-based FE [4]	3.8166	3.8201
		HSDT[203]	3.8100	3.8100
		RPT[258]	3.8100	3.8200
	5	Current	2.5198	2.5297
		High-order-based FE [4]	2.5171	2.5205
		HSDT[203]	2.5100	2.5100
		RPT[258]	2.5100	2.2900
1	0	Current	19.5380	19.6650
		High-order-based FE [4]	19.5815	19.6147
		HSDT[203]	19.5400	19.5700
		RPT[258]	19.5800	19.6100
	0.5	Current	12.6880	12.7590
		High-order-based FE [4]	12.6971	12.7159
		HSDT[203]	12.6700	12.6900
		RPT[258]	12.6900	12.7100
	1	Current	9.75163	9.8051
		High-order-based FE [4]	9.7637	9.7776
		HSDT[203]	9.7430	9.7500
		RPT[258]	9.7630	9.7700
	5	Current	6.4105	6.4636
		High-order-based FE [4]	6.4371	6.4507
		HSDT[203]	6.4200	6.4300
		RPT[258]	6.4300	6.4500
1.5	0	Current	47.6340	48.2057
		High-order-based FE [4]	47.5809	47.8322
		HSDT[203]	47.2900	47.5300
		RPT[258]	47.5700	47.8200
	0.5	Current	31.0520	31.3789
		High-order-based FE [4]	30.8705	31.0135
		HSDT[203]	30.6700	30.8200
		RPT[258]	30.8600	31.0100
	1	Current	23.9720	24.2195
		High-order-based FE [4]	23.7421	23.8481
		HSDT[203]	23.5900	23.6900
		RPT[258]	23.7400	23.8400
	5	Current	15.7540	15.9986
		High-order-based FE [4]	15.6228	15.7259
		HSDT[203]	15.5100	15.6200
		RPT[258]	15.6280	15.7200

Table 8.21

Dimensionless critical buckling load $\overline{N}_{cr} = N_{cr}a^2/E_m h^3$ of a simply supported (SSSS) (Al/Al_2O_3) FGM plates subjected to biaxial compression load for different aspect ratios (a/b), length-to-thickness ratios $(a/h = 5, 10)$ and power indexes.

a/b	Power index (p)	Reference	a/h	
			Biaxial compression load	
			5	10
0.5	0	Current	5.0617	5.7577
		High-order-based FE [4]	5.3763	5.9243
		HSDT[203]	5.3710	5.9180
		RPT[258]	5.3760	5.9260
	0.5	Current	3.4362	3.7609
		High-order-based FE [4]	3.5387	3.8565
		HSDT[203]	3.5270	3.8500
		RPT[258]	3.5390	3.8570
	1	Current	2.5765	2.8926
		High-order-based FE [4]	2.7323	2.9687
		HSDT[203]	2.7150	2.9610
		RPT[258]	2.7330	2.9690
	5	Current	1.7587	1.8619
		High-order-based FE [4]	1.7158	1.9322
		HSDT[203]	1.7000	1.9250
		RPT[258]	1.7190	1.9330
1	0	Current	7.1178	8.6879
		High-order-based FE [4]	8.0105	9.2893
		HSDT[203]	8.0010	9.2730
		RPT[258]	8.0110	9.2890
	0.5	Current	4.7512	5.7045
		High-order-based FE [4]	5.3125	6.0614
		HSDT[203]	5.2880	6.0450
		RPT[258]	5.3130	6.0620
	1	Current	3.6665	4.3925
		High-order-based FE [4]	4.1105	4.6690
		HSDT[203]	4.0730	4.6500
		RPT[258]	4.1120	4.6700
	5	Current	2.1930	2.7822
		High-order-based FE [4]	2.5204	3.0152
		HSDT[203]	2.4870	2.9998
		RPT[258]	2.5270	3.0180
1.5	0	Current	10.4700	13.6150
		High-order-based FE [4]	11.6817	14.6085
		HSDT[203]	11.6650	14.5710
		RPT[258]	11.6820	14.6080
	0.5	Current	7.0483	8.9693
		High-order-based FE [4]	7.8293	9.5684
		HSDT[203]	7.7820	9.5280
		RPT[258]	7.8300	9.5630
	1	Current	5.8932	6.9077
		High-order-based FE [4]	6.0763	7.3779
		HSDT[203]	6.0000	7.3310
		RPT[258]	6.0800	7.3790
	5	Current	3.1512	4.3199
		High-order-based FE [4]	3.6060	4.7065
		HSDT[203]	3.5440	4.6660
		RPT[258]	3.6180	4.7120

Table 8.22

Dimensionless critical buckling load $\overline{N}_{cr} = N_{cr}a^2/E_m h^3$ of a simply supported (SSSS) (Al/Al_2O_3) FGM plates subjected to biaxial compression load for different aspect ratios (a/b), length-to-thickness ratios $(a/h = 50, 100)$ and power indexes.

a/b	Power index (p)	Reference	a/h	
			Biaxial compression load	
			50	100
0.5	0	Current	6.1361	6.1546
		High-order-based FE [4]	6.1244	6.1309
		HSDT[203]	6.1170	6.1230
		RPT[258]	6.1240	6.1310
	0.5	Current	3.9822	3.9927
		High-order-based FE [4]	3.9708	3.9745
		HSDT[203]	3.9660	3.9700
		RPT[258]	3.9710	3.9740
	1	Current	3.0603	3.0681
		High-order-based FE [4]	3.0533	3.0560
		HSDT[203]	3.0490	3.0520
		RPT[258]	3.0530	3.0560
	5	Current	2.0158	2.0237
		High-order-based FE [4]	2.0137	2.0164
		HSDT[203]	2.0110	2.0140
		RPT[258]	2.0140	2.0160
1	0	Current	9.7692	9.8323
		High-order-based FE [4]	9.7908	9.8073
		HSDT[203]	9.7720	9.7880
		RPT[258]	9.7910	9.8070
	0.5	Current	6.3438	6.3796
		High-order-based FE [4]	6.3485	6.3580
		HSDT[203]	6.3360	6.3450
		RPT[258]	6.3490	6.3450
	1	Current	4.8757	4.9026
		High-order-based FE [4]	4.8818	4.8888
		HSDT[203]	4.8720	4.8790
		RPT[258]	4.8820	4.8890
	5	Current	3.2053	3.2318
		High-order-based FE [4]	3.2185	3.2254
		HSDT[203]	3.2120	3.2190
		RPT[258]	3.2190	3.2250
1.5	0	Current	15.8220	15.9630
		High-order-based FE [4]	15.8876	15.9313
		HSDT[203]	15.8370	15.8800
		RPT[258]	15.8880	15.9310
	0.5	Current	10.2780	10.3590
		High-order-based FE [4]	10.3037	10.3285
		HSDT[203]	10.2700	10.2950
		RPT[258]	10.3360	10.3280
	1	Current	7.8999	7.9604
		High-order-based FE [4]	7.9236	7.9419
		HSDT[203]	7.8970	7.9160
		RPT[258]	7.9240	7.9420
	5	Current	5.1865	5.2453
		High-order-based FE [4]	5.2210	5.2389
		HSDT[203]	5.2030	5.2220
		RPT[258]	5.2210	5.2390

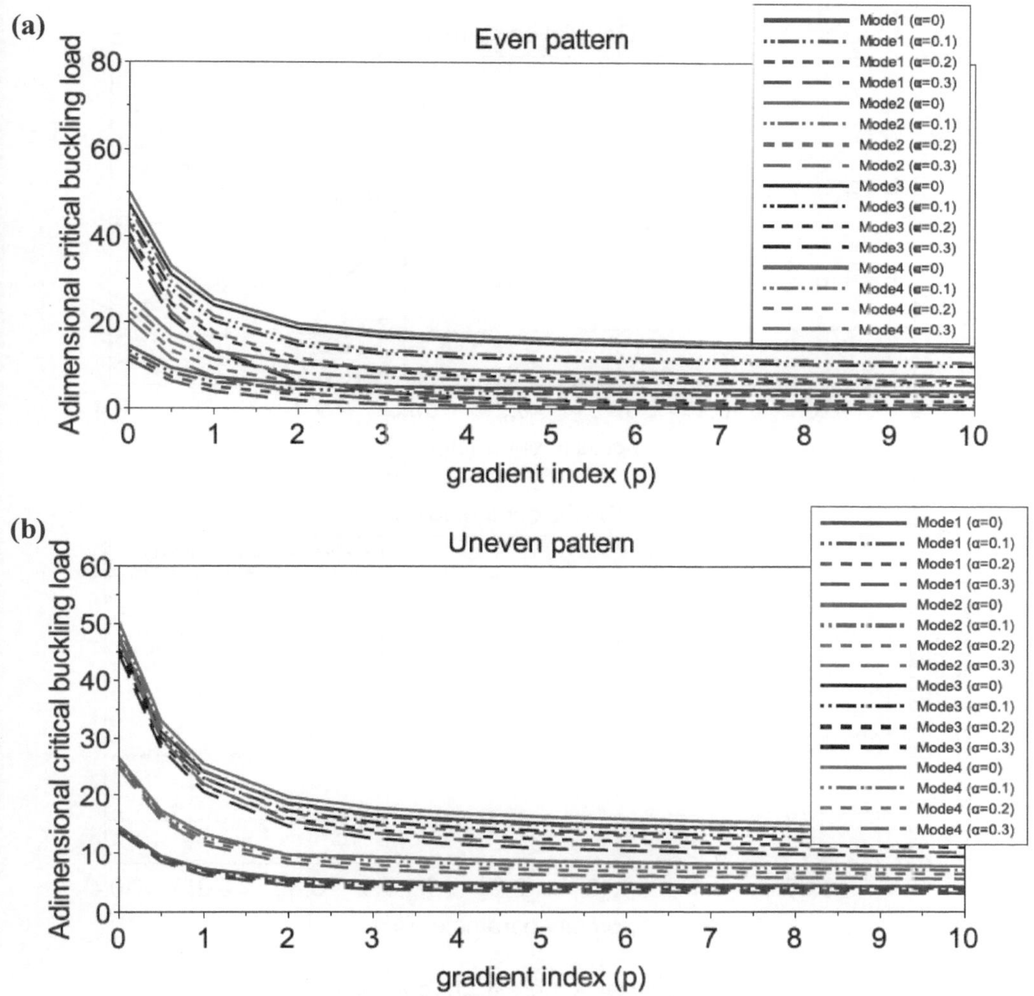

Figure 8.50 Dimensionless critical buckling load versus the gradient index (p) for the first four buckling modes of a simply supported (SSSS) (Al/Al_2O_3) P-FGM plates $(a/h = 20)$ under uniaxial compression with variation of the porosity coefficient and for (a) even and (b) uneven patterns.

Now, we provide newly buckling results on porous S-FGM plates using a simple and efficient finite element model. The critical buckling load of a simply supported (SSSS) square (Al/Al_2O_3) S-FGM plate $(a/h = 100)$ subjected to various edge compression loadings (uniaxial, biaxial compressions and biaxial compression and tension) is evaluated adimensionally for different values of gradient index (p) and the obtained curves are plotted in Fig. 8.55. Three levels of analysis can be derived from this figure: (a) the rise of the gradient index (p) leads to a decrease in the critical buckling load and when (p) reaches high values, the critical buckling load becomes stabilized due to the reinforcement of the structure by metal portions, (b) the increase of porosity parameter from $(\alpha = 0)$ to $(\alpha = 0.3)$ induces a reduction in the values of the buckling load and this phenomenon is more prominent for even than uneven porosity type. This fact is expected since uniform distributed pores weaken further the stiffness of the structure compared to non-uniform spread pores. (c) Uneven buckling curves are located above even buckling curves which mark the crucial impact

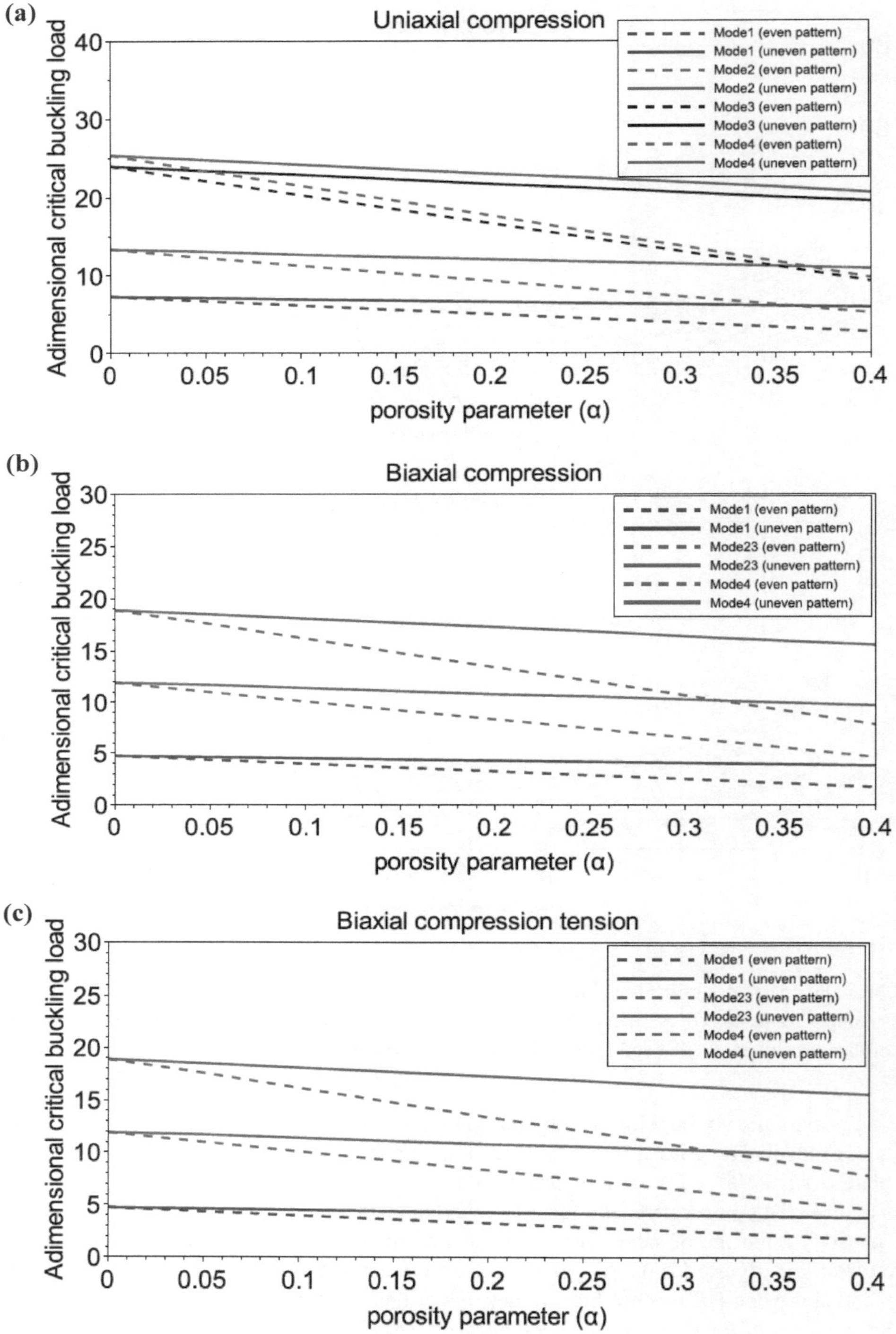

Figure 8.51 Dimensionless critical buckling load versus the porosity parameter (α) for the first four buckling modes of a simply supported (SSSS) (Al/Al_2O_3) P-FGM plates $(a/h = 20, p = 1)$ under different edge compression loads.

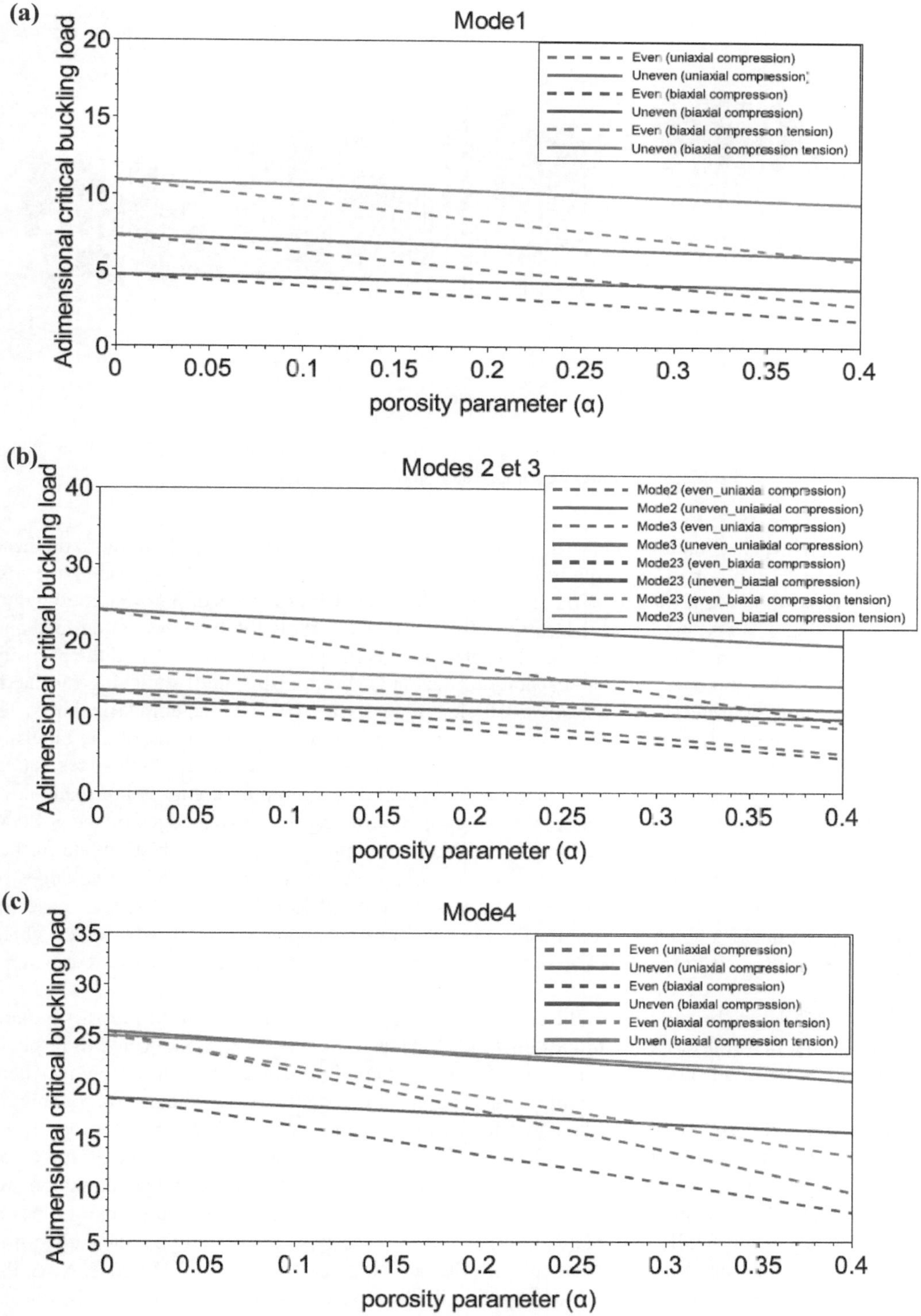

Figure 8.52 Influence of applied edge loads on the first four buckling modes of a simply supported (SSSS) (Al/Al_2O_3) P-FGM plates $(a/h = 20, p = 1)$.

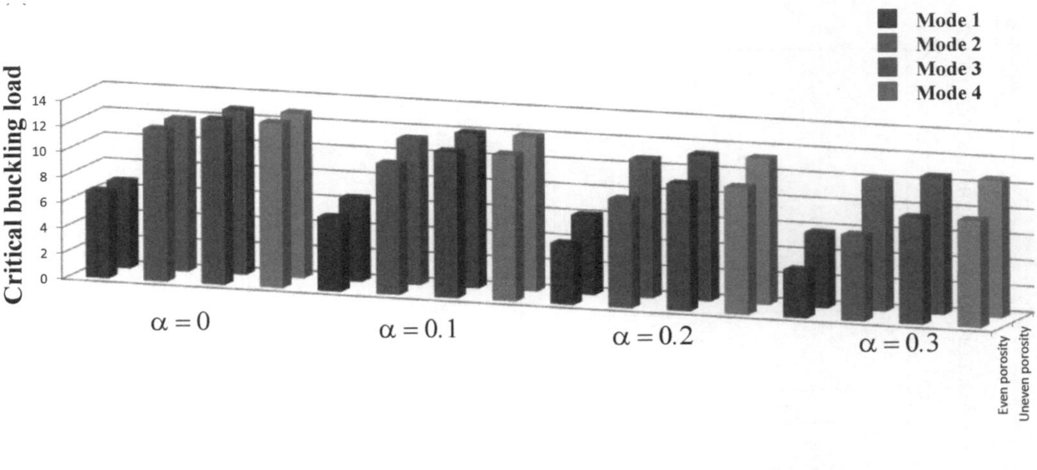

Figure 8.53 Histogram of repartition of buckling modes of evenly and unevenly (Al/Al_2O_3) P-FGM plates $(SSSS, b/h = 10, p = 1)$ subjected to uniaxial compression load.

of type of porosity distribution on buckling responses of S-FGM plates. To highlight more on these insightful points, and in particular the impact of applied edge loadings, Fig. 8.56 presents the histogram of repartition of the buckled (Al/Al_2O_3) S-FGM plates with four sets of porosity coefficients $(\alpha = 0, 0.1, 0.2, 0.3)$ and gradient indexes $(p = 0, 1, 5, 10)$ by varying the type of applied load and for even and uneven porosity distributions. It is clear that as the gradient index (p) and the porosity coefficient (α) rise, the buckling load capacity of the S-FGM diminishes until vanishing for maximum value of gradient index and porosity coefficient $(p = 10, \alpha = 0.3)$ which emphasis that metal portions are more sensitive to the variation of these material parameters. Furthermore, the intensity of the load is maximum for biaxial compression and tension edge loaded where high stress values are developed in the two edges of the plate, whereas, uniaxial or biaxial compression loads have less buckling intensity. In fact, the compressive applied stresses on the two edges of the S-FGM plate are compensated by the tensile stresses applied to the opposite edges which stimulate further the buckling response of the plate compared to pure compressive stresses applied in two edges or four edges of the plate. Therefore, biaxial compression type of load is the most serious case for which the plate buckles quickly, in contrast to biaxial compression and tension load type for which the plate gets sufficient time to buckle and can provide a marge of security for buckled structural components.

Figure 8.57 shows the variation of the critical buckling load in the function of the porosity parameter (α) for the first four buckling modes of S-FGM plates $(a/h = 20, p = 2)$ subjected to in-plane axial compression load. Four combinations of boundary conditions are taken into account where (C), (S) and (F) refer to clamped, simply supported and free edges. As expected, a fully clamped S-FGM induces the highest critical buckling load while Clamped-free S-FGM presents the lowest ones. The others two configurations (SSSS, CSCS) are situated between these two extreme cases which highlight that clamped edges present the highest resistance against buckling phenomenon due to the strong applied constraints. Furthermore the slop between even and uneven porosity types is more pronounced for a fully clamped S-FGM plates. So, the type of applied boundary condition combined with the type of porosity can enhance the buckling behavior of S-FGM plates. Next, the coupled effect of boundary conditions and applied load on critical buckling response of the porous S-FGM plates is explored in Fig. 8.58. As can be seen, for a simply supported (S) or free (F) edges, the applied compressive stresses notably in the case of biaxial compression load and with a high rate of porosity coefficient $(\alpha = 0.3)$ stimulate further the carrying load capacity of the S-FGM

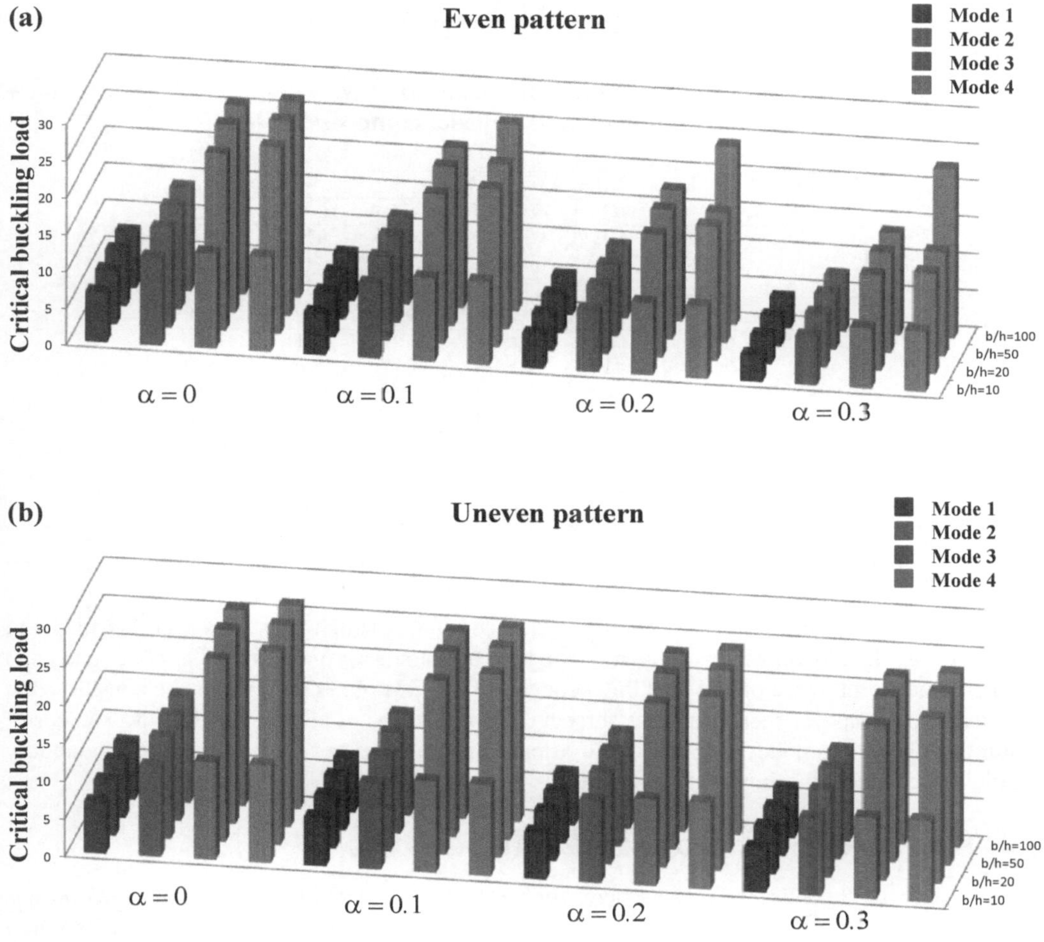

Figure 8.54 Histogram of repartition of the first four buckling modes versus different width-to-thickness ratio (b/h) of (a) evenly and (b) unevenly (Al/Al_2O_3) P-FGM plates $(SSSS, p=1)$ under uniaxial compression load.

Table 8.23

Computation of the dimensionless critical buckling load $\widetilde{N_{cr}} = N_{cr}\eta_0 a^2/E_m h^3$ for a simply supported (SSSS) square (Al/ZrO_2) S-FGM plates $(a/h = 100)$ subjected to uniaxial compression load.

Gradient index	Models				
p	Current	Analytical.[248]	Numerical.[94]	%(error1)	%(error2)
Ceramic	7.900	7.794	7.828	1.34	0.91
1	5.523	5.448	5.484	1.35	0.70
2	5.063	5.304	5.346	4.54	5.29
5	4.967	5.182	5.229	4.14	5.01
10	4.614	5.146	5.195	10.33	11.18
Metal	3.662	3.613	3.629	1.33	0.90

Table 8.24
Computation of the dimensionless critical buckling load $\widetilde{N_{cr}} = N_{cr}\eta_0 a^2/E_m h^3$ of square (Al/Al_2O_3) S-FGM plates $(p = 1)$ subjected to uniaxial compression load.

BC	a/h	Model	Perfect	Even		Uneven	
			$\alpha = 0$	$\alpha = 0.1$	$\alpha = 0.4$	$\alpha = 0.1$	$\alpha = 0.4$
SSSS	10	Current	4.4113	3.7373	1.7026	4.2052	3.5674
		Ref.[248]	4.6696	4.0254	1.8944	4.4794	3.8738
	100	Current	4.9172	4.1527	1.8489	4.6948	4.0142
		Ref.[248]	4.8888	4.2060	1.9537	4.6952	4.0781
CCCC	10	Current	11.488	10.111	5.213	11.116	9.821
		Ref.[248]	11.58	10.012	4.8068	11.089	9.5284
	100	Current	13.767	11.982	5.839	13.343	11.877
		Ref.[248]	13.026	11.208	5.207	12.51	10.865
CSCS	10	Current	8.430	7.208	3.393	8.054	6.833
		Ref.[248]	8.5616	7.3958	3.5295	8.2031	7.0623
	100	Current	9.508	8.055	3.621	9.092	7.808
		Ref.[248]	9.4232	8.1074	3.7664	9.0499	7.8601

plates which buckle more faster, in contrast to clamped plates which possesses high resistance to compressive applied loadings. Consequently, the type of boundary condition and applied load edge have a significant influence on the buckling behavior of S-FGM plates which must be considered in the experiments phase of such structures subject to various types of applied edge loads. The corresponding six buckling modes shapes of S-FGM plates including porosities under various boundary conditions are depicted in Figs. 8.59 and 8.60.

8.5.4 FUNCTIONALLY GRADED POROUS SPHERICAL CAPS

To deal with curved shells, we present here, for the first time, buckling analysis of FGM porous spherical caps. In fact, according to the open literature, there is a remarkable lack in investigation of buckling behavior of spherical shell-type geometry. So, we provide new results on buckling responses of FGM porous spherical caps. At first, we go through a quick check of the formulation for perfect FGM spherical caps subjected to uniform radial pressure q (Pascal) as shown in Fig. 8.61. The spherical cap is made of Aluminium and Alumina constituents and the pores are uniformly or randomly distributed along the thickness direction with even and uneven porosity patterns. Besides, the applied boundary condition is of type simply supported. The computation of the critical buckling load q_{cr} (MPa) for different radius-to-thickness (R/h) ratios and gradient indexes (p) is performed and presented in Tab. 8.25. In all computations, the value of the angle is taken as: $\varphi = 60^0$. It should be noted that the numbers in brackets indicate the buckling mode (m,n). The present results are in good correlation with those given by Bich et al. using analytical solutions [29, 2, 179] which highlight the efficiency and the reliability of the present formulation for buckling analysis of FGM spherical caps.

Now, we provide the first new results on buckling responses of FGM porous spherical caps subjected to uniform pressure load. Figures 8.62 and 8.63 present the critical buckling load q_{cr} (MPa) versus the gradient index (p) of (Al/Al_2O_3) FGM porous spherical cap for two sets of radius-to-length $(R/h = 400, 800)$ ratios. The rise of the gradient index (p) from ceramic to metal portions leads to decrease in the critical buckling load q_{cr} for all porosity parameters. Furthermore, for a high value of porosity coefficient $\alpha = 0.3$, the FGM spherical cap has the low capacity carrying load. This phenomenon occurs for high rate of porosity which weakens the rigidity of the FGM spherical cap and induces hence low values of the critical buckling load for both even or uneven

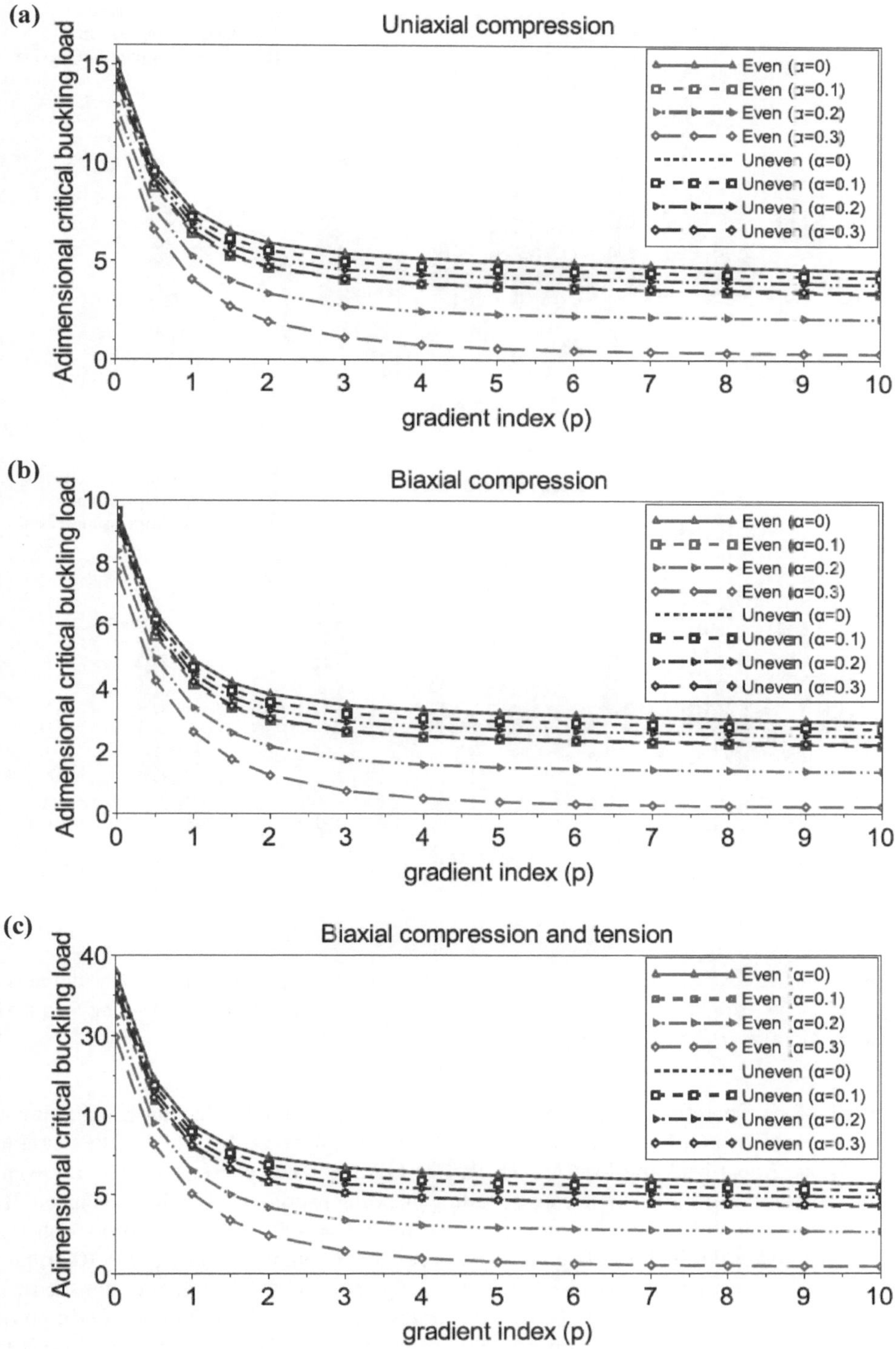

Figure 8.55 Dimensionless critical buckling load of a simply supported (SSSS) (Al/Al_2O_3) S-FGM plates $(a/h = 100)$ subjected to (a) uniaxial compression, (b) biaxial compression and (c) biaxial compression and tension for different values of gradient indexes (p).

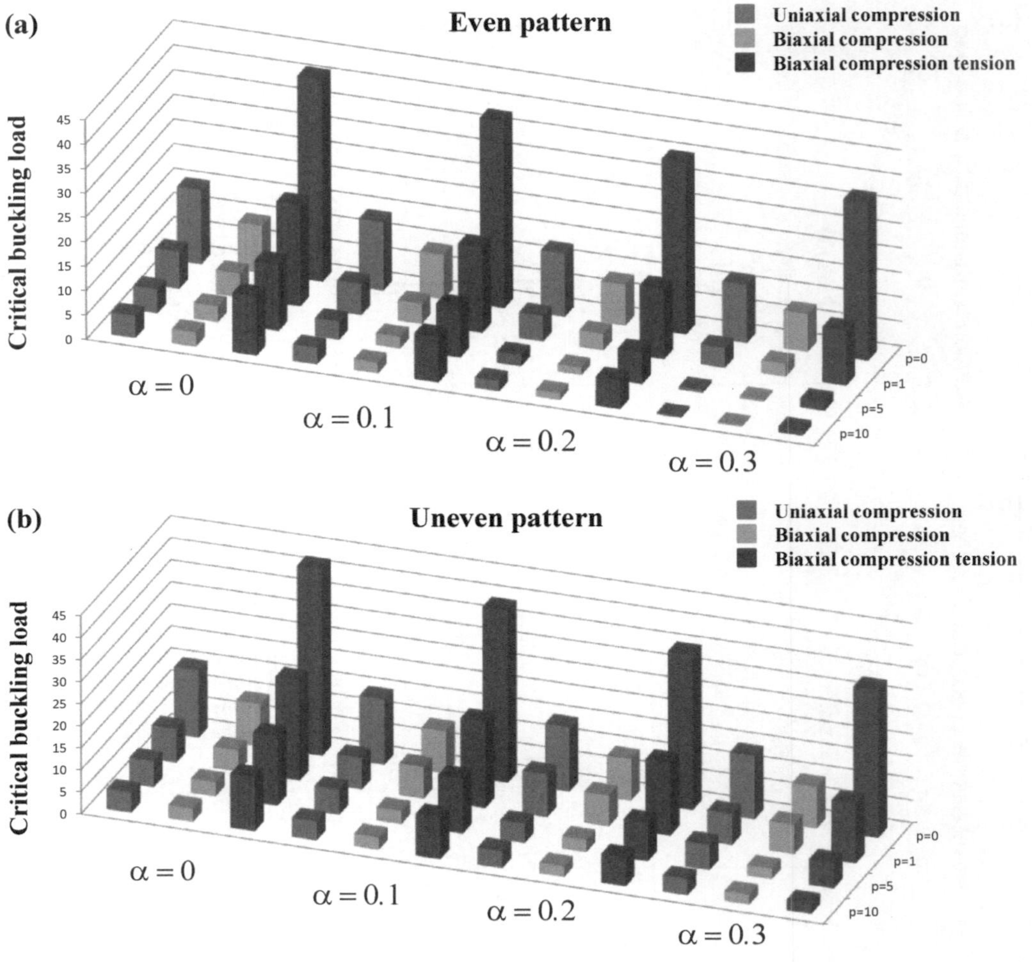

Figure 8.56 Histogram of repartition of the buckled (Al/Al_2O_3) S-FGM plates $(a/h = 100)$ with four sets of porosity coefficients $(\alpha = 0, 0.1, 0.2, 0.3)$, four gradient indexes $(p = 0, 1, 5, 10)$ and varying the type of applied load and for (a) even and (b) uneven porosity distributions.

porosity distributions. Therefore, the percentage of porosity spread into the shell has a major impact on buckling response of FGM spherical shell caps. To outline in detail the effect of the variation of (R/h) ratio on critical buckling load (q_{cr} (MPa)) of FGM porous spherical shells with evenly and unevenly porosity distributions, the corresponding buckling histogram (Fig. 8.64) is given. It is clear that for a low value of the radius-to-thickness ratio $(R/h = 400)$, the FGM porous spherical cap possesses the highest critical buckling pressure, in contrast to high value of $(R/h = 1000)$ ratio, the FGM spherical cap presents the lowest critical buckling pressure which further diminished with high rate of porosity coefficient $(\alpha = 0.3)$. Indeed, for low values of (R/h) ratio, the FGM porous spherical cap behaves as thicker structure with high rigidity, while for high values of (R/h), the spherical cap is more slender and it buckles hence easily. So, the variation of the radius-to-thickness ratio has an important impact on buckling responses of FGM porous spherical shells which must be considered in the design phase of such structures.

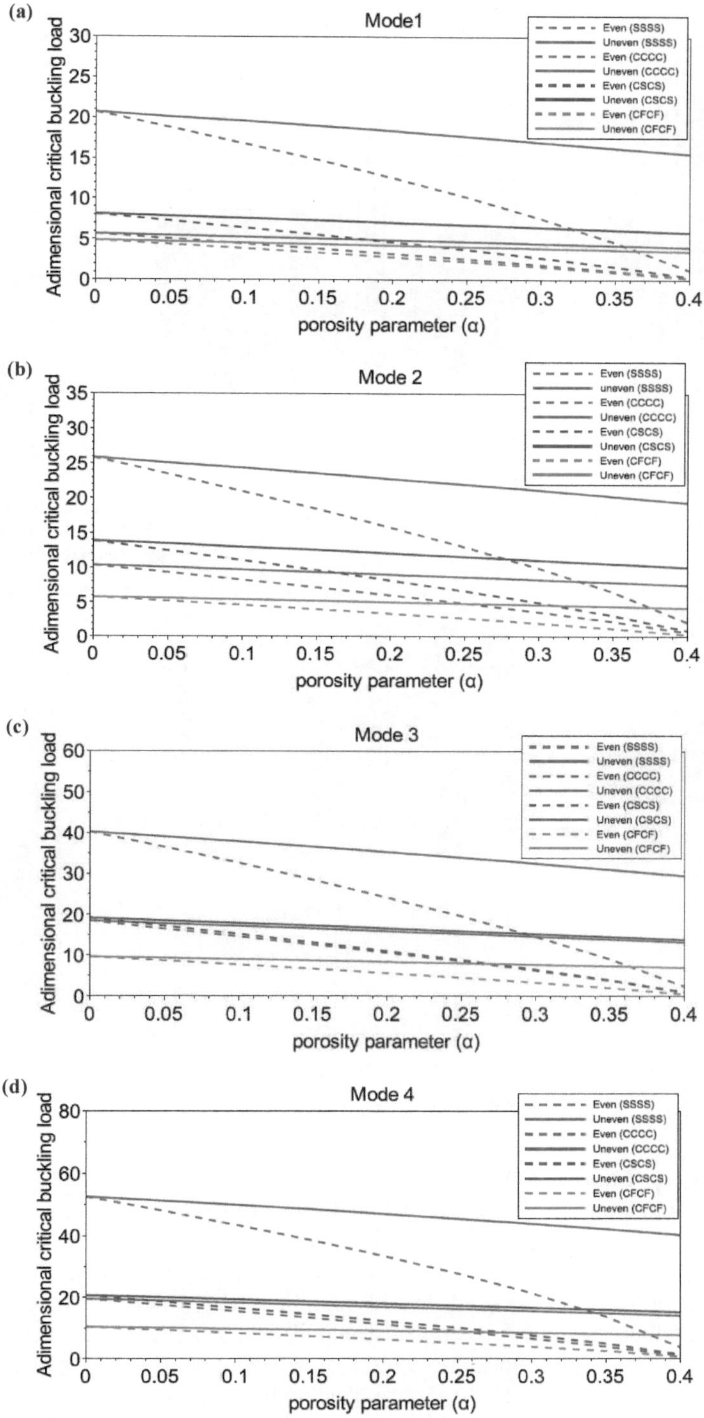

Figure 8.57 Variation of the critical buckling load in the function of the porosity parameter (α) for the first four buckling modes of S-FGM plates ($a/h = 20, p = 2$) subjected to in-plane axial compression load.

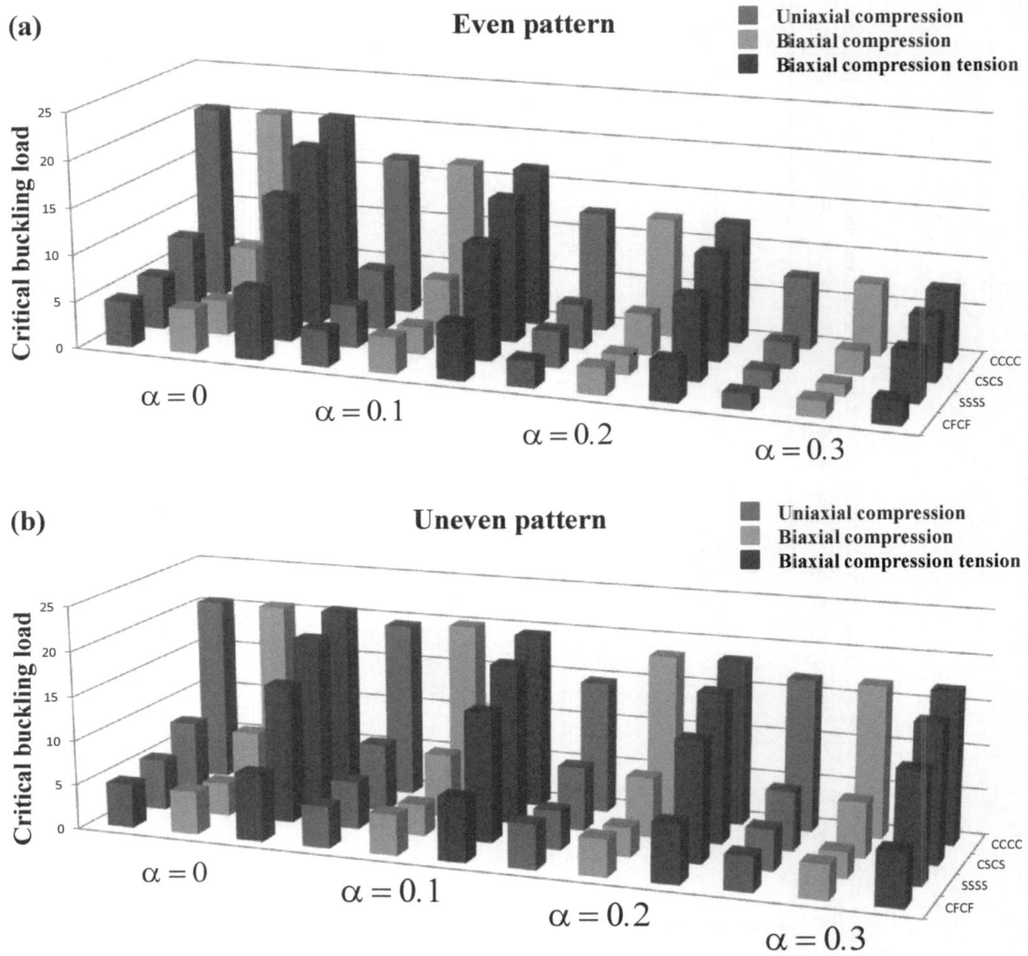

Figure 8.58 Coupled effect of boundary conditions and applied loadings on the critical buckling responses of (Al/Al_2O_3) S-FGM plates $(a/h = 20, p = 2)$ for different porosity coefficients $(\alpha = 0, 0.1, 0.2, 0.3)$ and with (a) even and (b) uneven porosity distributions.

Figure 8.65 plots q_{cr} (MPa) as a function of porosity parameter (α) of an SS (Al/Al_2O_3) FGM porous spherical cap under uniform pressure load for various (R/h) ratios. As observed, the increase in porosity parameter induces a significant reduction in the critical bucking pressure for both evenly and unevenly FGM porous spherical caps. The reason is that the rise of porosity coefficient reduces the Young's modulus of the spherical shell. In addition, the effect is more prominent for the evenly distributed porosity rather than the unevenly distributed one. Another observation reveals that a minimum value of $(R/h = 400)$ ratio induces high values of critical buckling pressure compared to maximum values of $(R/h = 800, 1000)$. So, thicker shells have the capability/ability to support further stresses than slender shells and provide hence more resistance against buckling phenomenon.

Figures 8.66 and 8.67 illustrate the effect of boundary conditions on critical buckling load q_{cr} (MPa) of S-FGM spherical caps $(R/h = 400, p = 2)$ subjected to uniform pressure load and for various values of porosity parameter. As noticed, clamped edges perform high critical buckling

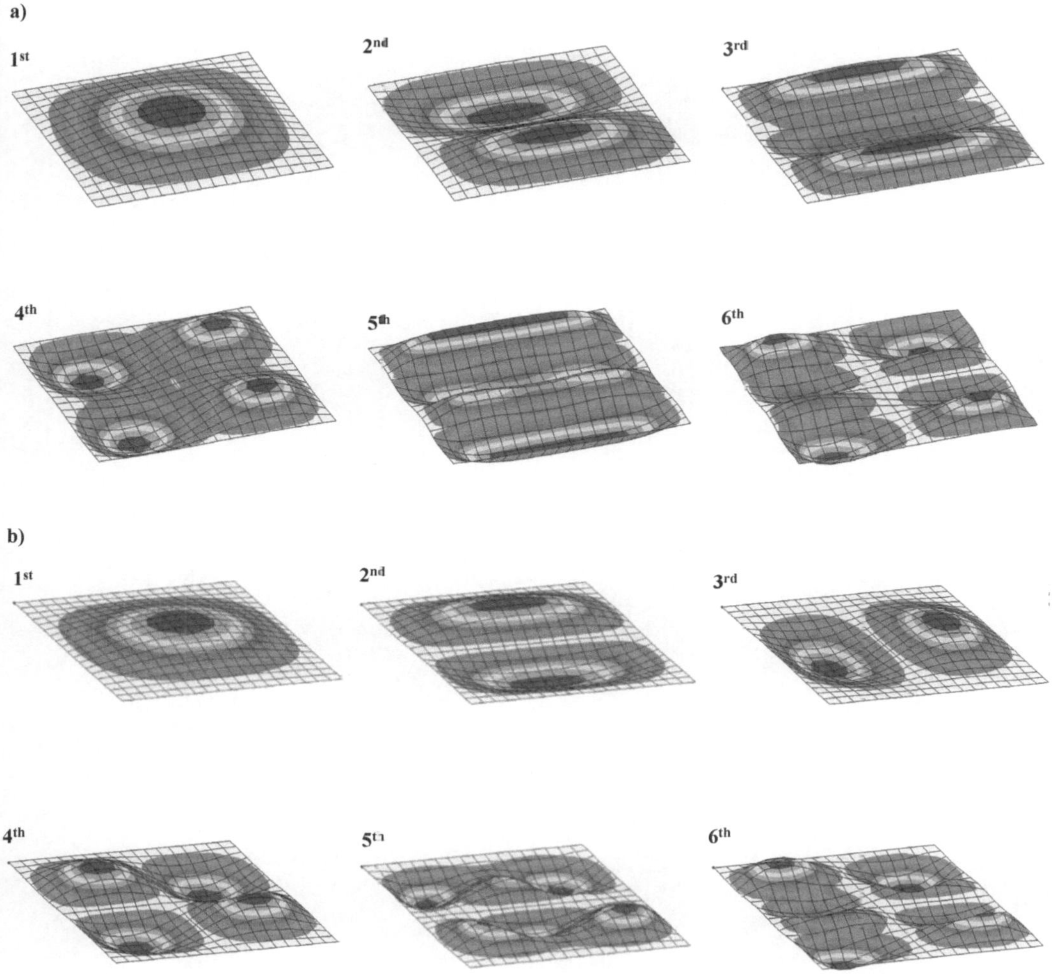

Figure 8.59 First six modes shape of porous S-FGM plates under various boundary conditions (a) SSSS, (b) CSCS and subjected to in-plane axial compression load

pressure compared to simply supported edges. Subsequently, a sudden drop emerges for even type of porosity compared to uneven type of porosity. This behavior is associated to the manner of porosity distribution through the thickness of S-FGM spherical cap which is uniform in the case of even-type and random in the case of uneven-type. So, with high rate of porosity, the spherical cap has weaker buckling resistance capacity and it starts to deflect more rapidly. The corresponding buckled mode shape of the SS porous FGM spherical cap is given in Fig. 8.68.

8.6 THERMOELASTIC BUCKLING ANALYSIS OF FG POROUS PLATES AND SHELLS

In this section, the thermoelastic buckling analysis of FG porous plates and shells made of different combinations of metal/ceramic materials is investigated using the (I-FSDT) model developed as done in Chapter 1. Herein, a metal-ceramic structure with porosity and temperature dependent

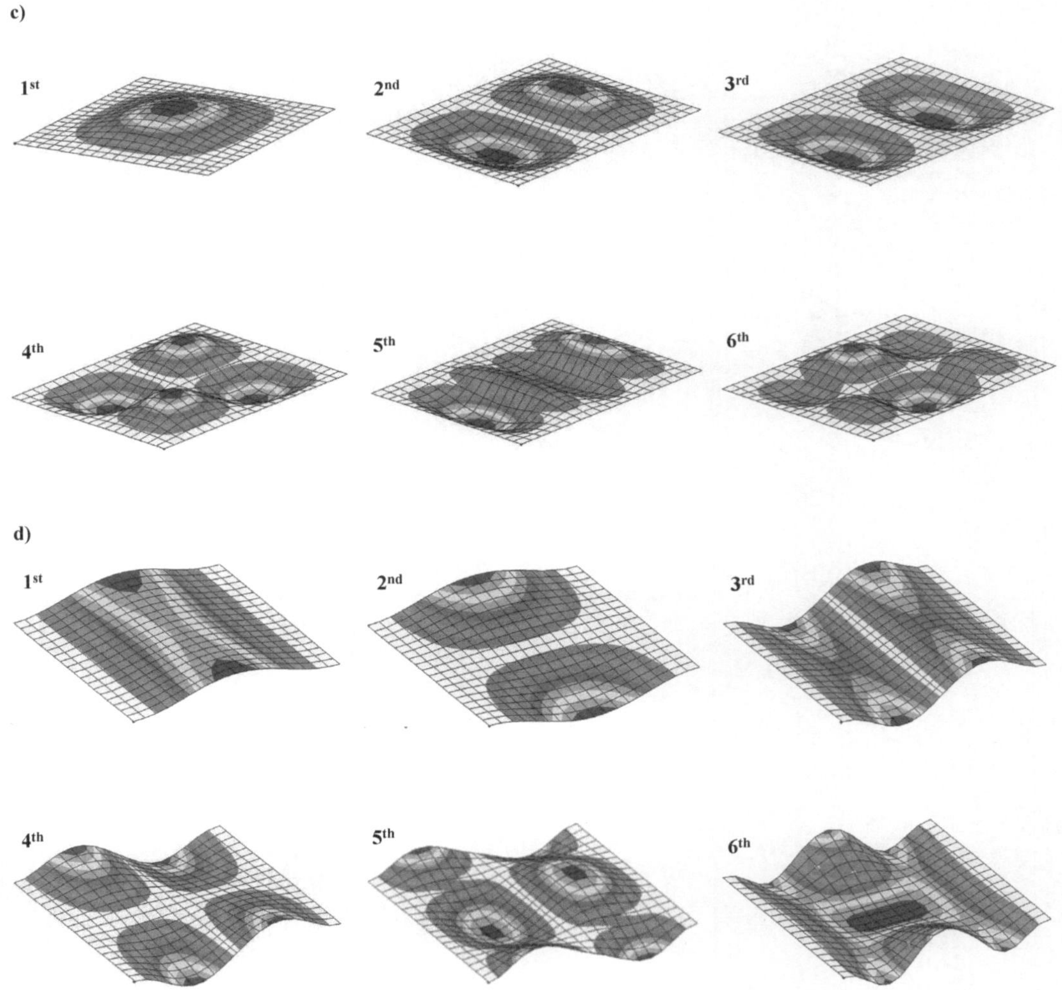

Figure 8.60 First six modes shape of porous S-FGM plates under various boundary conditions (c) CCCC, (d) CFCF and subjected to in-plane axial compression load.

materials is considered, where pores can be uniformly or randomly distributed along the thickness direction of the structure, see Fig. 8.69. A coordinate system (x, y, z) located on the mid-surface of the functionally graded porous (FGP) structure is established. The structure is exposed to uniform thermal load along its thickness h and two kinds of FG pore distributions, including uniform (or even distribution) and random (or uneven distribution) forms are introduced as depicted in Fig. 8.69. The temperature-dependant material properties (such as Young's modulus E, coefficient of thermal expansion α, coefficient of thermal conductivity K, density ρ) of metal and ceramic materials can be obtained by the Touloukian model as a polynomial function [303, 126]:

$$P_{eff}(T) = P_0\left(P_{-1}T^{-1} + 1 + P_1 T + P_2 T^2 + P_3 T^3\right) , \tag{8.13}$$

In which:

$P_{eff} : (E_m, E_c, \alpha_m, \alpha_c, K_m, K_c, \rho_m, \rho_c)$

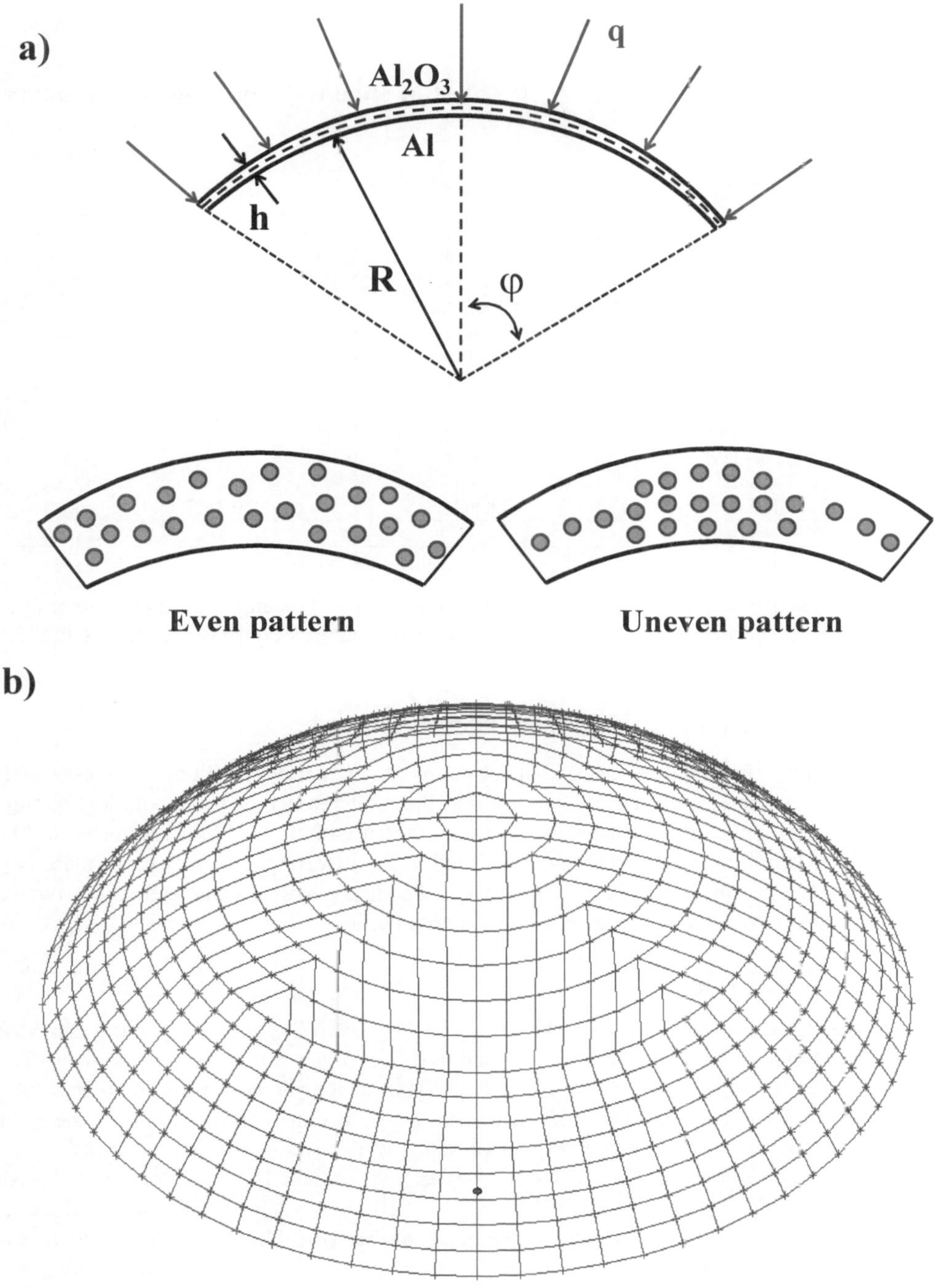

Figure 8.61 FGM porous spherical shell cap subjected to (a) uniform radial pressure and (b) mesh configuration.

Table 8.25

Critical buckling load q_{cr} (MPa) of (Al/Al_2O_3) FGM spherical cap subjected to uniform pressure load.

R/h	Theory	Gradient index (p)			
		0	1	2	5
400	Current	2.8503	1.5911	1.2296	0.9272
	Ref.[29]	2.8748	1.5618	1.2109	0.9329
	Ref.[2]	2.8808(12,1)	1.5617(12,1)	1.2111(12,1)	-
	Ref.[179]	2.8732(9,1)	1.566(10,1)	–	0.9336(9,1)
600	Current	1.2555	0.6901	0.5531	0.4265
	Ref.[29]	1.2777	0.6941	0.5382	0.4146
	Ref.[2]	1.2771(7,1)	0.6944(15,1)	0.5386(15,1)	-
	Ref.[179]	1.278(11,1)	0.6944(12,1)	–	0.415(11,1)
800	Current	0.7103	0.4181	0.3069	0.2330
	Ref.[29]	0.7187	0.3904	0.3027	0.2332
	Ref.[2]	0.7190(16,1)	0.3904(17,1)	0.3027(17,1)	-
	Ref.[179]	0.7182(13,1)	0.3909(14,1)	–	0.2332(13,1)

P_0, P_{-1}, P_1, P_2 and P_3 are the temperature coefficients and T is the temperature parameter in K.

The definition of the thermal loads, as well as porosity patterns distributions are as indicated previously in section 8.2.

8.6.1 THERMOELASTIC BUCKLING OF FG POROUS PLATES

In this part of study, numerical results and illustrations about thermal buckling responses of FG porous plates are presented to trace the validity and the applicability of the theoretical formulation based on an efficient computational finite element model within the (FSDT) framework. First, Comparisons of the present model with models found in the literature review is performed. Next, a parametric study covering the effects of FG gradient index, porosity volume fraction, types of distributions, aspect ratios, thermal loadings and material compositions is presented.

8.6.1.1 Convergence test

For the convergence test of the critical buckling thermal load of ΔT_{cr} (in K) of a simply supported (Al/Al_2O_3) FGM plate subjected to uniform temperature rise $(b/h = 80)$, a sets a mesh sizes (8×8, 12×12, 14×14, 16×16, 18×18, 32×32 and 100×100) is considered for different gradient indexes as shown in Table 8.26. The material properties of FGM constituents which are the Young modulus, the coefficient of the thermal expansion and the Poisson's ratio are given as follows: [286]: for Aluminium (Al) phase: $E_m = 70 \times 10^9$ (N/m^2), $\alpha_m = 23 \times 10^{-6}/°C$, $\nu_m = 0.3$; for Alumina (Al_2O_3) phase, $E_c = 380 \times 10^9 (N/m^2)$, $\alpha_c = 7.4 \times 10^{-6}/°C$, $\nu_c = 0.3$. From Tab. 8.26, it is observed that there is a rapid and excellent convergence of the critical buckling thermal load as the mesh size rises for the different gradient indexes. Furthermore, (32×32) mesh size is sufficient to obtain accurate thermal buckling results. Thus, this mesh size is adopted to analyze thermal buckling behavior of the FGM plates in the rest of numerical examples.

8.6.1.2 Validation and comparison studies

In this validation part, the exactness of the present method is established by comparing the present results with the available models in the literature for isotropic and FGM plates. The following numerical tests are conducted for the validation purpose.

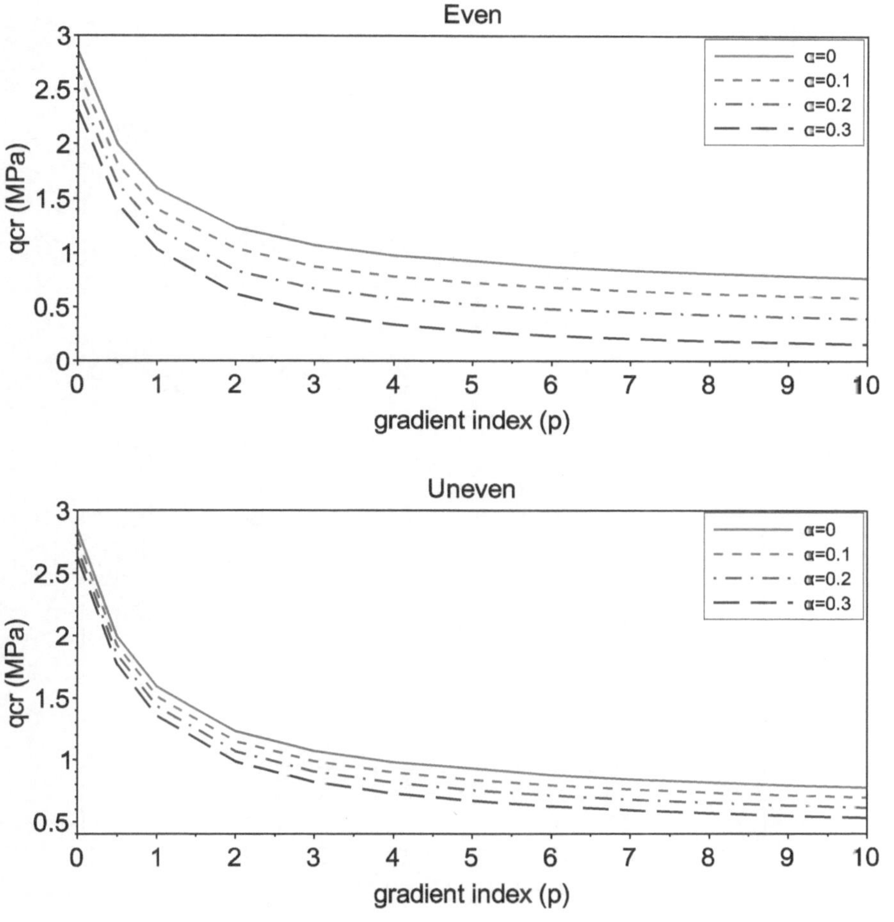

Figure 8.62 Critical buckling load q_{cr} (MPa) of (Al/Al_2O_3) FGM porous spherical cap $(R/h = 400)$ subjected to uniform pressure load.

Test 1: Isotropic plates

To illustrate the efficiency of the present model, the critical buckling temperature $(\widehat{\Delta T_{cr}} = \alpha_0 \Delta T_{cr})$ of a simply supported (SSSS) isotropic plate (a/b = 1, $\alpha_0 = 1 \times 10^{-6}/K$, $E = 1 \times 10^{-6} N/m^2$, $\nu = 0.3$) is computed and compared with a refined plate theory (RPT) [36] and a global higher-order deformation theory [158] as mentioned in Table 8.27. It is shown that the present model is in excellent agreement with other approaches which draw its efficiency for thermal analysis in the isotropic case. Furthermore, it should be noted that in [36, 158] analytical solutions are provided while in our work a numerical solution via finite element method is proposed. Nevertheless, the analytical method is limited when the geometry model, boundary conditions, or types of the load becomes more complex. As an alternative, a simple and an efficient finite element model is proposed in the present work to analyze thermoelastic behavior of FG porous plates and shells. Note that $Error(\%)$ $= 100 \times \frac{\|\Delta T_{crp} - \Delta T_{cr-ref}\|}{\|\Delta T_{cr-ref}\|}$, where ΔT_{crp} and ΔT_{cr-ref} denote the critical buckling temperature of the proposed formulation and the references, respectively.

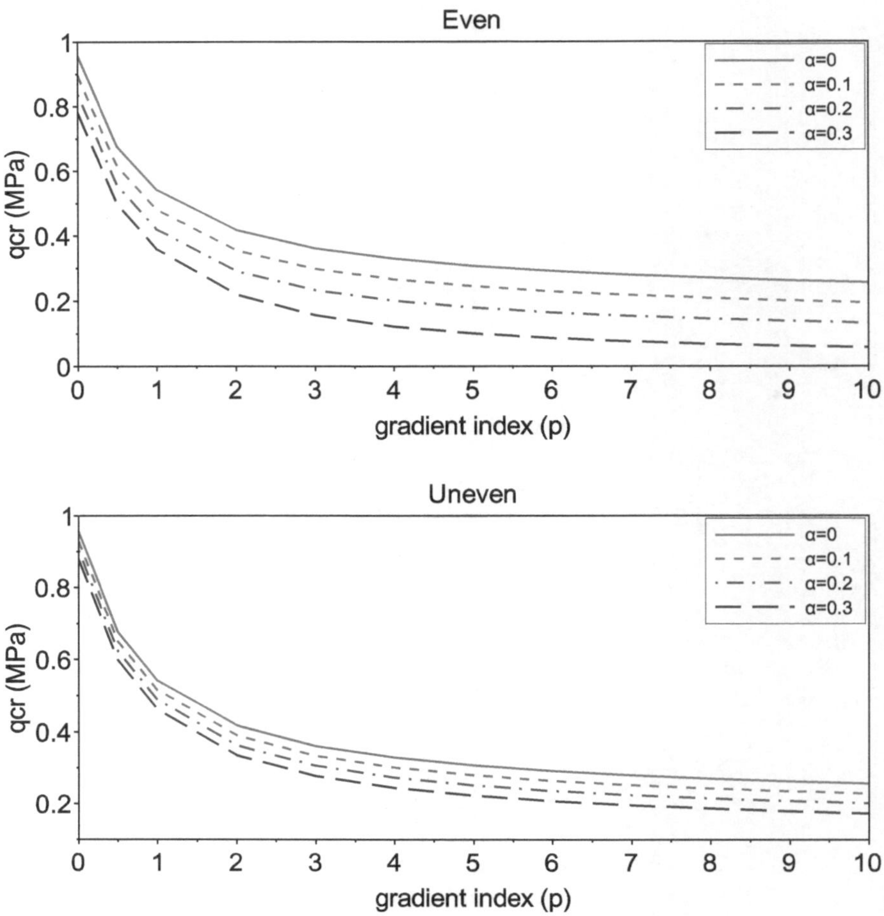

Figure 8.63 Critical buckling load q_{cr} (MPa) of (Al/Al_2O_3) FGM porous spherical cap $(R/h = 800)$ subjected to uniform pressure load.

Test 2: FGM plates

Continuously, the reliability of the model for (FGM) structures is checked via a comparison study of the critical thermal buckling load (ΔT_{cr}, in K) of a simply supported (Al/Al_2O_3) FGM plates subjected to uniform thermal load and for different width-to-thickness (b/h) ratios. The material properties of the FGM plate are as mentioned previously. From Tab. 8.28, it can be seen that the relative error between the results is around 3% in maximum with [286] whereas it can reach 15% compared to [110]. Thus, it can be explained by the different theories and also to the definition of simply boundary conditions. In fact, in [286] a numerical solution is provided based on improved (TSDT) theory while an analytical solution is provided in [110] in terms of essential and natural boundary conditions. Furthermore, the effect of bending-coupling stiffness components, which vanishes in [286], in the case of pure metal (Al) and ceramic (Al_2O_3) cases, can explain the difference between the results in these extreme cases.

Additionally, it should be mentioned that the plates which are thicker than the limitations could be evaluated by failure analysis due to thermal stresses, because their critical buckling temperatures

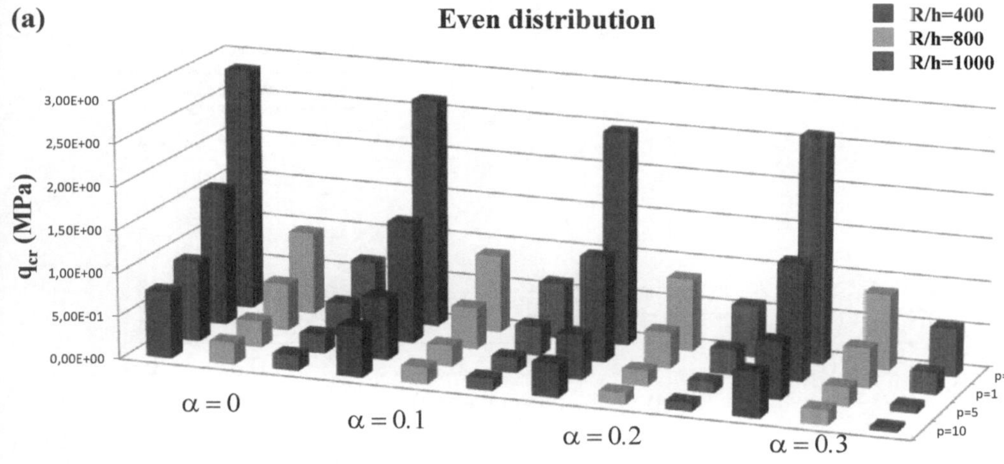

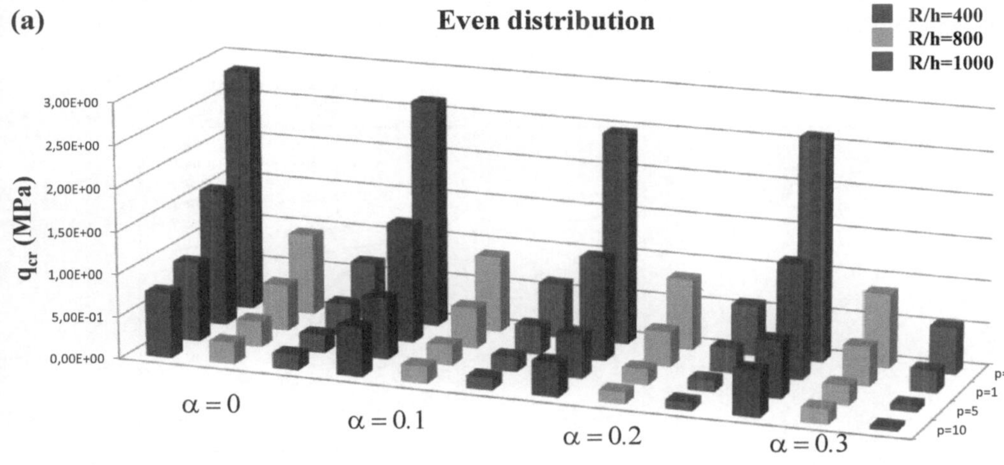

Figure 8.64 Critical buckling histogram variation of SS (Al/Al_2O_3) FGM porous spherical cap subjected to uniform pressure load for different sets of radius-to-thickness ratios (R/h) with (a) evenly and (b) unevenly porosity distributions.

are too high and close to the melting point temperature of the materials. So, the present model constitutes a more general finite element with the enhancement of the shear part of transverse shear strains instead of shear correction factors used in [110] which overestimates the critical buckling temperature. Another observation from Table 8.28, can be made on the variation of the width-to-thickness (b/h) ratio which reaches its maximum for low values of (b/h) ratio and minimum for high values of (b/h) ratio. Indeed, for low value of (b/h) ratio, the FGM plate is considered as thick plate with high stiffness and can hence resist to high thermal field by high value of critical buckling thermal load.

8.6.1.3 Effects of material composition and aspect ratios

This section presents the results of several numerical examples, in which the thermal buckling behavior of FG porous plates subjected to various types of thermal loadings is involved. The effects

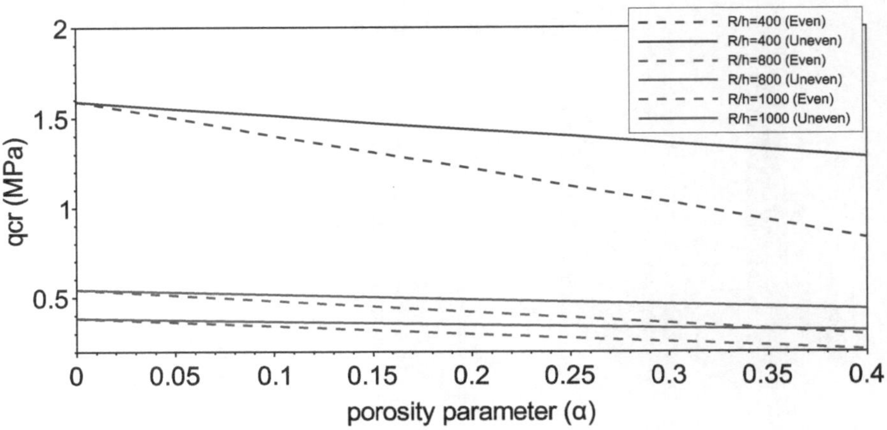

Figure 8.65 q_{cr} (MPa) as a function of porosity parameter (α) of SS (Al/Al_2O_3) FGM porous spherical cap under uniform pressure load for various (R/h) ratios.

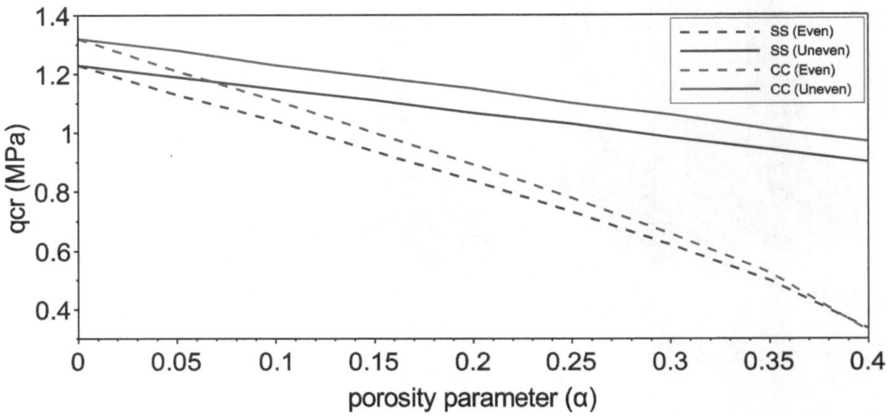

Figure 8.66 q_{cr} (MPa) as a function of porosity parameter (α) of S-FGM porous spherical cap $(R/h = 400, p = 2)$ under uniform pressure load.

of material and geometrical parameters such as the porosity parameter, types of porosity distributions, FG gradient index, aspect ratios and thermal loads are examined and investigated. To this end, an FG porous plate of a width (b), a length (a), and a thickness (h) is considered. The material composition at the top surface $(z = h/2)$ is assumed to be the ceramic-rich and it is varying continuously to the metal-rich surface at the opposite side $(z = -h/2)$. The pores can be uniformly or randomly distributed in the thickness direction of the plate as shown in Fig. 8.69. To explore the thermal buckling analysis of a variety of FGM materials, three types of ceramics, Al_2O_3, Si_3N_4 and ZrO_2 are chosen to mix with stainless steel $SUS304$ metal. The temperature dependency of these materials is provided in Tab. 8.29.

The first investigation concerns the effect of power-law index (p) on the critical buckling temperature ΔT_{cr} for three types of materials of a simply supported (SSSS) FG porous plates $(a/h = 50, e = 0.2)$ subjected to uniform thermal load, while varying the type of porosity

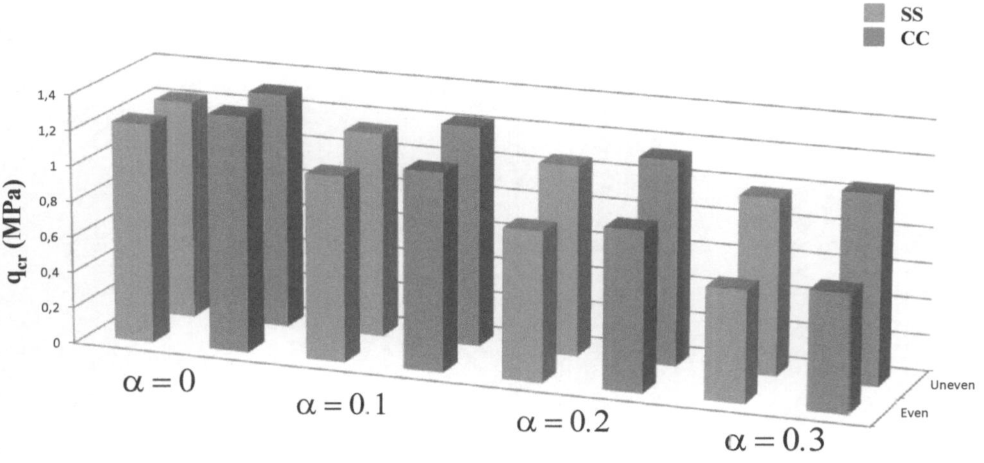

Figure 8.67 Critical buckling q_{cr} (MPa) Histogram variation of evenly and unevenly S-FGM porous spherical cap $(R/h = 400, p = 2)$ under uniform pressure load.

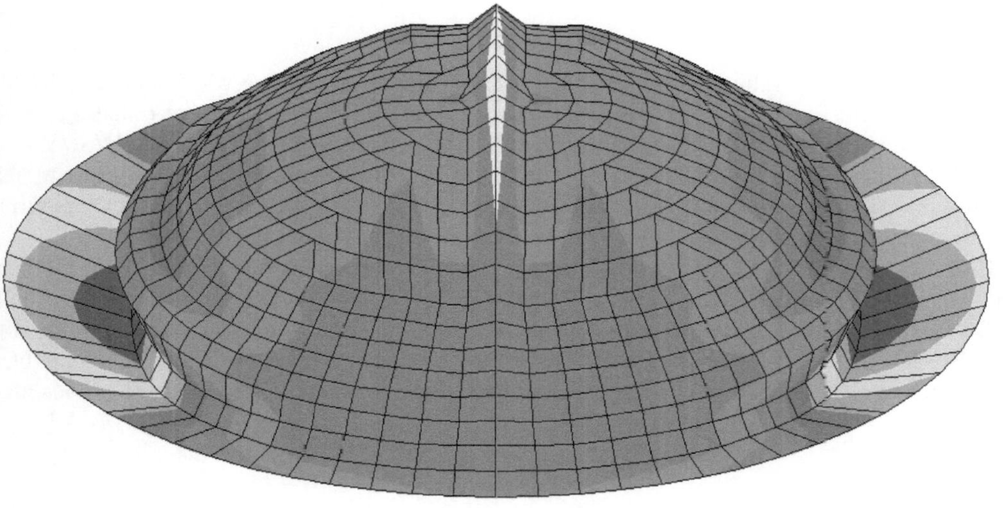

Figure 8.68 Buckled mode shape of porous S-FGM spherical cap $(R/h = 400, p = 2)$ under uniform pressure load.

distributions (Even or Uneven), temperature dependant (TD) or independant (TID) materials as shown in Figs. 8.70 and 8.71.

In one hand, it is remarked that the critical buckling temperature of $Si_3N_4/SUS304$ FG porous plate has the top value for even and uneven porosity distributions and for temperature dependent (TD) and independent (TID) configurations, while the $ZrO_2/SUS304$ FG porous plate has the lowest value of the critical buckling temperature among the pairs of material combinations. $Al_2O_3/SUS304$ FG porous plate is classed between these extreme cases and its critical buckling temperature values are close to those of $Si_3N_4/SUS30$ FG porous plate in the case of (TID) material. This can be

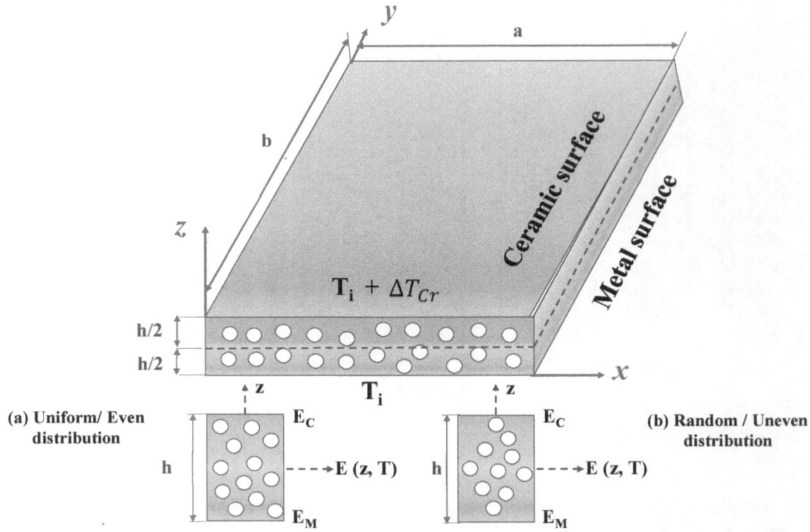

Figure 8.69 Geometry and cross-section areas of metal-ceramic structure with graded elastic pores and exposed to uniform thermal load.

explained by its high value of Young's modulus and therefore stiffness which manifest by high values of the critical buckling temperatures. On the other hand, it can be observed that the critical buckling temperature ΔT_{cr} decreases as the power index (p) rises for the pairs $Si_3N_4/SUS30$ and $Al_2O_3/SUS30$ but not for $Zr_{O_2}/SUS30$ FG porous plate, for which the buckling response remains almost constant during the increase for the power index (p). This phenomenon is attributed to the elevation in the metal constituent with the rise of the power index increment.

Continuously, (TD) FG porous plates induce higher critical buckling temperature than (TID) ones and the buckling response of $(Al_2O_3/SUS304)$ and $(Si_3N_4/SUS304)$ FG porous plates are sticked for (TID) case. Therefore, the temperature dependency of FG materials plays an important role in the stimulation of the thermal buckling behavior of FG porous plates by high values of critical buckling temperatures. The critical buckling temperature of simply supported (SSSS) FG porous plates

Table 8.26

Convergence test of the critical buckling thermal load ΔT_{cr} (in K) for (Al/Al_2O_3) FGM plate subjected to uniform temperature rise.

b/h	Mesh size	Al_2O_3	p=1	p=5	p=10	Al
80	8 × 8	27.279	14.688	14.811	13,783	8.777
	12 × 12	26.951	14.479	14.592	13.595	8.6712
	14 × 14	26.882	14.435	14.546	13.555	8.4910
	16 × 16	26.838	14.406	14.517	13.529	8.6348
	18 × 18	26.808	14.387	14.496	13.512	8.6251
	32 × 32	26.730	14.366	14.444	13.466	8.6000
	100 × 100	26.731	14.366	14.444	13.466	8.6000
	[286]	26.687	14.385	13.982	13.126	8.6000

Table 8.27
Critical buckling temperature $\widehat{\Delta T_{cr}}$ of simply supported isotropic plates.

a/h	Present	Source		*difference*(%)	
		[158]	[36]	*ER*1	*ER*2
10	0.1199×10^{-1}	0.1183×10^{-1}	0.1198×10^{-1}	1.35	0.083
20	0.3123×10^{-2}	0.3109×10^{-2}	0.3119×10^{-2}	0.45	0.128
100	0.1266×10^{-3}	0.1264×10^{-3}	0.1265×10^{-3}	0.158	0.079

Table 8.28
Critical buckling thermal load ΔT_{cr} (in K) of (Al/Al_2O_3) FGM plate subjected to uniform temperature rise.

Materials	Source	b/h=10	b/h=20	b/h=40	b/h=60	b/h=80
	Present	1620.700	422.090	106.64	47.506	26.730
Al_2O_3	[286]	1595.148	419.856	106.386	47.402	26.687
	%Error	1.60	0.532	0.238	0.2194	0.161
	[110]	1617.484	421.516	106.492	47.424	26.693
	%Error	0.198	0.136	0.138	0.172	0.138
	Present	869.180	226.04	57.104	25.456	14.336
p=1	[286]	862.044	226.464	57.353	25.552	14.385
	%Error	0.827	0.187	0.434	0.375	0.340
	[110]	757.891	196.257	49.500	22.037	12.402
	%Error	14.684	15.17	15.36	15.51	15.594
	Present	846.210	225.17	57.274	25.597	14.444
p=5	[286]	809.660	217.448	55.405	24.713	13.982
	%Error	4.514	3.389	3.373	3.901	3.304
	[110]	678.926	178.528	45.213	20.144	11.340
	%Error	24.639	26.125	26.675	27.07	270336
	Present	786.960	210.100	53.468	23.889	13.466
p=10	[286]	761.634	204.929	52.243	23.304	13.126
	%Error	3.325	2.52	2.344	0.107	2.590
	[110]	692.519	183.141	46.455	20.703	11.657
	%Error	13.637	14.720	13.116	15.389	15.347
Al	Present	521.440	135.8	34.309	15.285	8.600
	[286]	513.222	135.084	34.228	15.251	8.586
	%Error	1.601	0.530	0.236	0.222	0.018

obtained from three pairs of materials, $(Si_3N_4/SUS304)$,$(Al_2O_3/SUS304)$ and $(ZrO_2/SUS304)$, while varying the type of porosity patterns and temperature dependency is shown in Fig. 8.71. Increasing the value of (p) leads to a decrease in the value of buckling temperature of $(Al_2O_3/SUS304)$ and $(Si_3N_4/SUS304)$ FG plates; while, $(ZrO_2/SUS304)$ FG plate shows the opposite behavior. Therefore, it is reasonable to conclude that ZrO_2 is the only type of ceramic, which is not recommended to mix with $(SUS304)$ to make FG plates owing to its inferior material property.

Secondly, the inspection of the effect of the slenderness ratio (b/a) on critical buckling temperature of simply supported (SSSS) TD and TID $(Al_2O_3/SUS304)$, $(ZrO_2/SUS304)$ and $(Si_3N_4/SUS304)$ FG porous plates under uniform and non-uniform thermal loads is examined in Figs. 8.72, 8.73, 8.74 and 8.75, while varying the porosity parameter and types of distributions.

In one hand, from Fig. 8.72, it is detected that the increase of the slenderness ratio (b/a) induces a progressive increase in the critical buckling temperature for the three sets of porosity

Table 8.29

Temperature dependent coefficients of Young's modulus $E(P_a)$ thermal expansion coefficient α(1/K), Poisson's ratio ν and mass density ρ (kg/m3) for various materials [286].

Materials	P_0	P_{-1}	P_1	P_2	P_3	P, @300K
SUS304						
E	$201.04e^{+9}$	0	$3.079e^{-4}$	$-6.534e^{-7}$	0	$207.79e^{+9}$
α	$12.330e^{-6}$	0	$8.086e^{-4}$	0	0	$15.321e^{-6}$
ν	0.3262	0	$-2.002e^{-4}$	$3.797e^{-7}$	0	0.318
ρ	8166	0	0	0	0	8166
Al_2O_3						
E	$349.55e^{+9}$	0	$-3.853e^{-4}$	$4.027e^{-7}$	$-1.673e^{-10}$	$320.24e^{+9}$
α	$6.8269e^{-6}$	0	$1.838e^{-4}$	0	0	$7.203e^{-6}$
ν	0.26	0	0	0	0	0.260
ρ	3800	0	0	0	0	3800
Si_3N_4						
E	$348.43e^{+9}$	0	$-3.070e^{-4}$	$2.160e^{-7}$	$-8.946e^{-11}$	$322.27e^{+9}$
α	$5.8723e^{-6}$	0	$9.095e^{-4}$	0	0	$7.475e^{-6}$
ν	0.24	0	0	0	0	0.240
ρ	2370	0	0	0	0	2370
ZrO_2						
E	$244.27e^{+9}$	0	$-1.371e^{-3}$	$1.214e^{-6}$	$-3.681e^{-10}$	$168.06e^{+9}$
α	$12.766e^{-6}$	0	$-1.491e^{-3}$	$1.006e^{-5}$	$-6.778e^{-11}$	$18.591e^{-6}$
ν	0.2882	0	$1.133e^{-4}$	0	0	0.298
ρ	3657	0	0	0	0	3657

parameters ($e = 0.1; e = 0.2; e = 0.3$). Furthermore, it is noted that the material ($Si_3N_4/SUS304$) exposed to uniform change of dependent temperature (TD UT), has the highest value of critical temperature for both porosity patterns (Even and Uneven) compared to ($Al_2O_3/SUS304$) and ($ZrO_2/SUS304$) FG porous plates. Besides, it is detected that the value of critical buckling temperature of ($Al_2O_3/SUS304$) plate is not far from the value of ($Si_3N_4/SUS304$) material, but the last material ($ZrO_2/SUS304$) has the lowest value of critical temperature. This outlines that the Silicon nitride (Si_3N_4) and alumina (Al_2O_3) ceramics are more efficient to stimulate the thermal buckling behavior of FG porous plates than Zirconia (Zr_{O_2}) material due to their high values of Young's moduli. As well as, it can be remarked that the difference between the curves becomes more pronounced with the increase of porosity volume fraction (from $e = 0.1$ to $e = 0.3$). Similarly for non-uniform change of dependent temperature (TD NUT), the ($Si_3N_4/SUS304$) induces the highest values of buckling temperature in both even and uneven porosity distributions, while (b/a) ratio is varying and for all sets of porosity coefficients ($e = 0.1; e = 0.2; e = 0.3$) compared to others material pairs ($Al_2O_3/SUS304$) and ($ZrO_2/SUS304$) as shown in Fig. 8.73. In conclusion, the type of material mixing between the ceramic and metal phases is of great importance on buckling response of FG porous plates under thermal loadings.

8.6.2 EFFECTS OF POROSITY VOLUME FRACTION, TYPES OF DISTRIBUTIONS AND BOUNDARY CONDITIONS

In the next investigations, the effect of porosity parameter, types of distributions, boundary conditions and types of thermal loadings on critical buckling temperature load of ($Si_3N_4/SUS304$); ($Al_2O_3/SUS304$) and ($ZrO_2/SUS304$) FG porous plates with temperature depended (TD) and independed (TID) materials are performed in Figs. 8.76, 8.77,8.78,8.79, 8.80 and 8.81.

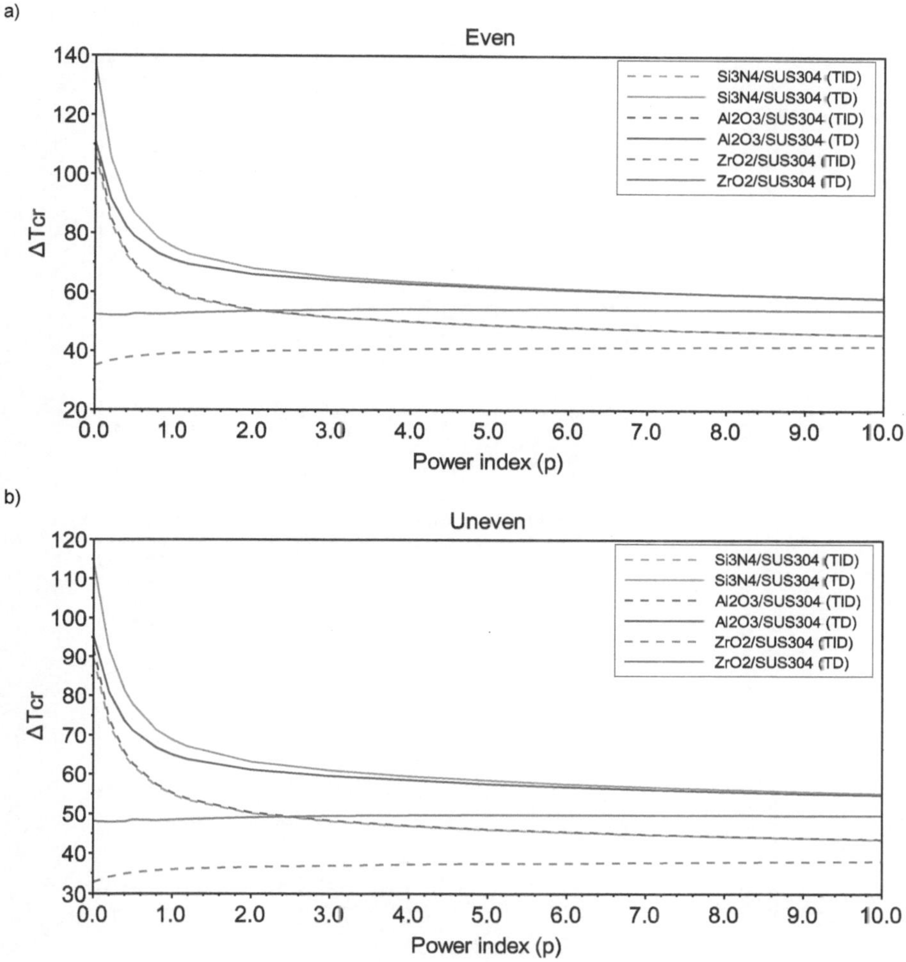

Figure 8.70 Critical buckling temperature (K) versus the power index (p) of (a) Even (b) Uneven FG porous plates $(a/h = 50, e = 0.2)$ made out of different pairs of materials and subjected to uniform thermal load.

First, the evaluation of the impact of the boundary conditions on even and uneven and for uniform and non-uniform FG plate with $(Si_3N_4/SUS304)$ material are portrayed in Figs. 8.76 and 8.77, respectively where the critical buckling temperature is plotted versus the porosity parameter for different boundary conditions in both cases: uniform and non-uniform FG plates and with even and uneven porosity distribution types while imposing a length-to-thickness ratio $(a/h = 20)$. The curves can be organized from the maximum to minimum of the critical temperature in the following way:CC (Clamped-Clamped) – CF (Clamped-Free) – CS (Clamped-Supported) – SS (Simply supported). Actually, the more the plate is subjected to critical buckling, the more it becomes rigid and thus its critical temperature increases.

Moreover, the effect of porosity distribution type can be well interpreted through the plotted figures. It can be noticed that the critical buckling temperature is higher with even porosity distribution than uneven type of porosity and with depended temperature (TD) than independed temperature (TID) type of materials. Well, for even-type the porosity is spread uniformly along the plate section, whereas, for uneven-type, it is spread mostly around the mid-plane and vanishes at the top and bottom surfaces. Thus, the plate becomes more rigid with even-type than with uneven-type and that's

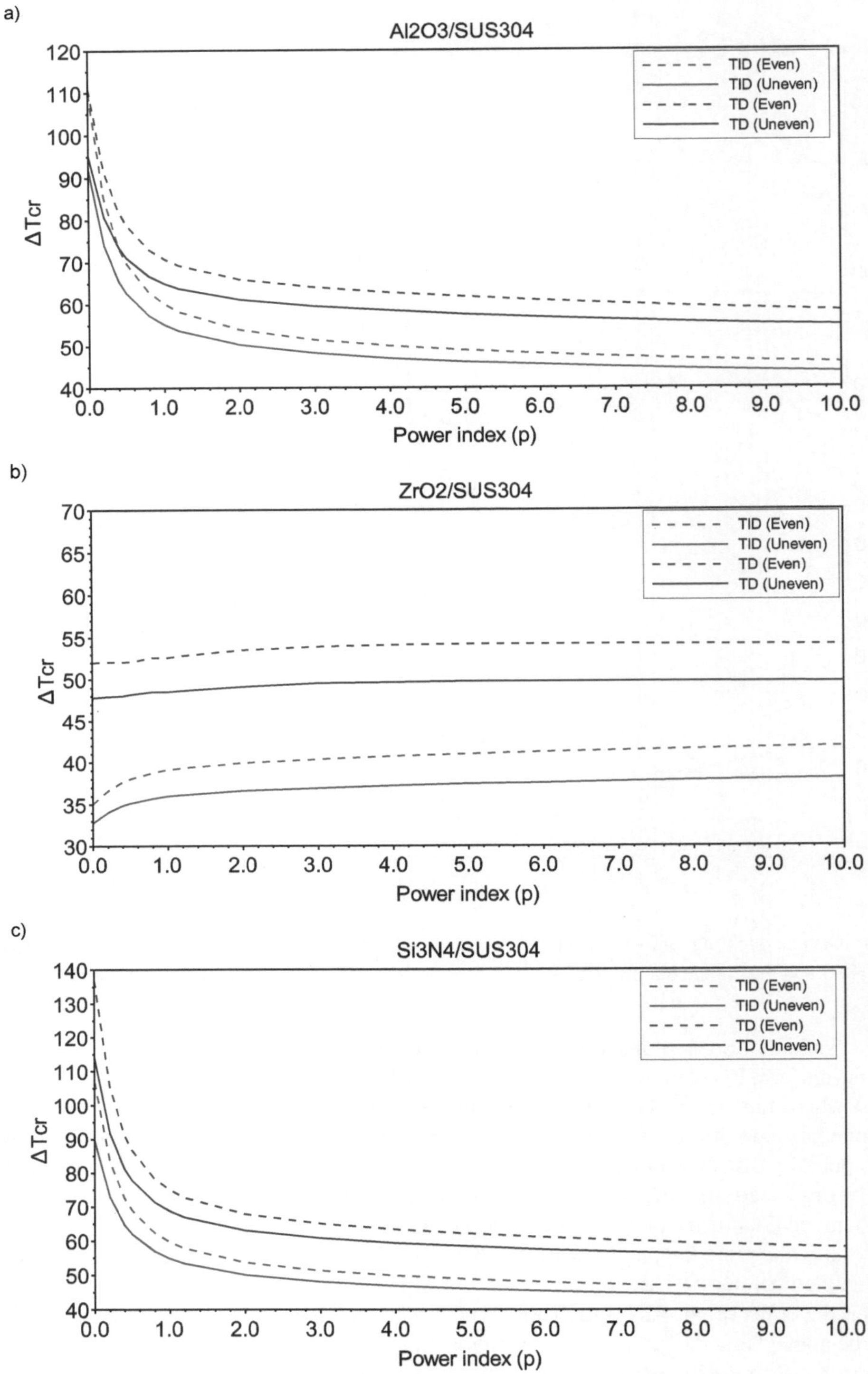

Figure 8.71 Critical buckling temperature (K) versus the power index (p) of FG porous plates $(a/h = 50, e = 0.2)$ made out of different pairs of materials (a) $Al_2O_3/SUS304$, (b) $ZrO_2/SUS304$ and (c) $Si_3N_4/SUS304$ under uniform thermal load.

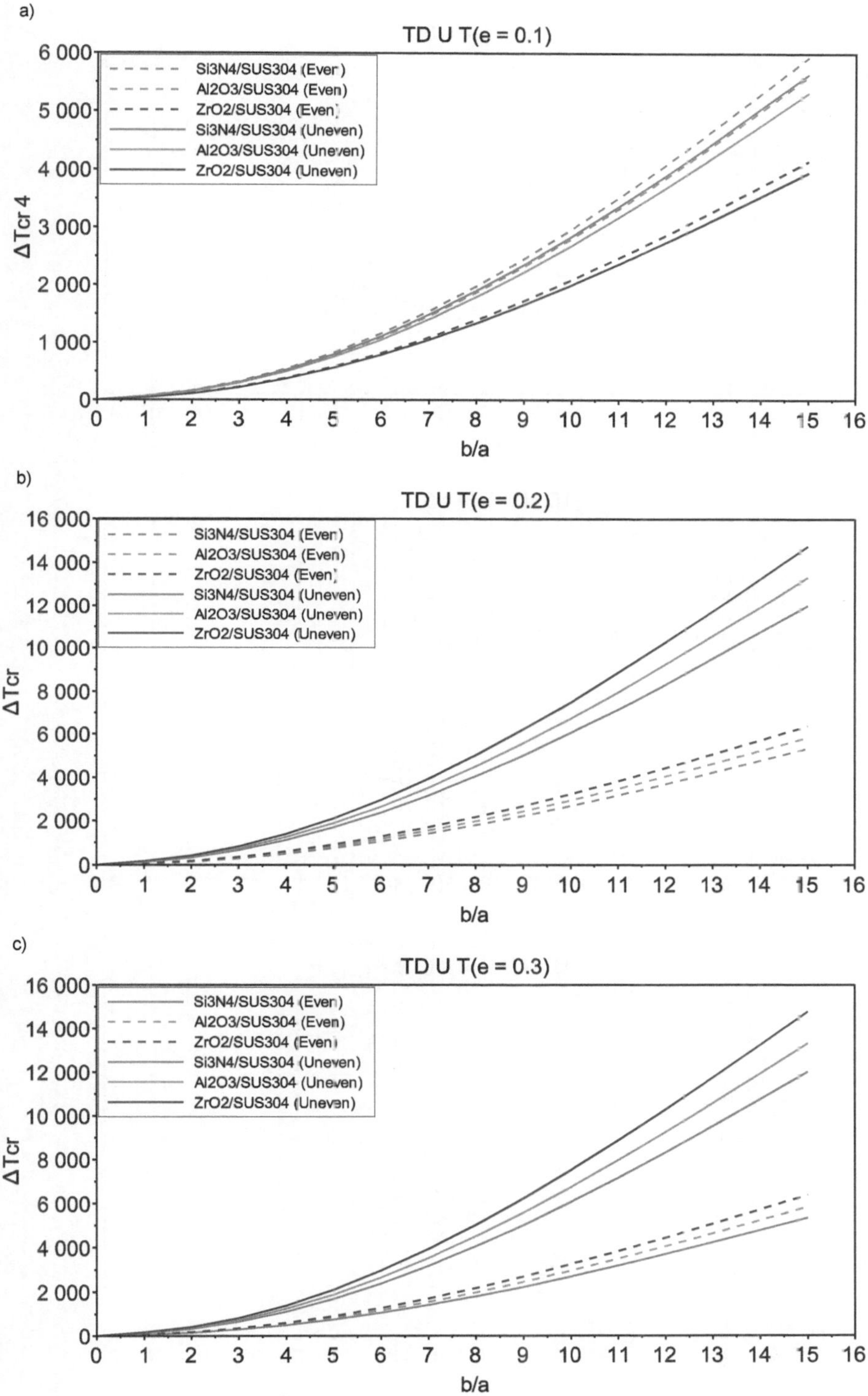

Figure 8.72 Critical buckling temperature (K) in the function of the slenderness ratio (b/a) of a simply supported (SS) (TD) FG porous plate $(p = 1)$ for different porosity coefficients $(e = 0.1; e = 0.2; e = 0.3)$ and with even and uneven porosity distributions under uniform thermal load.

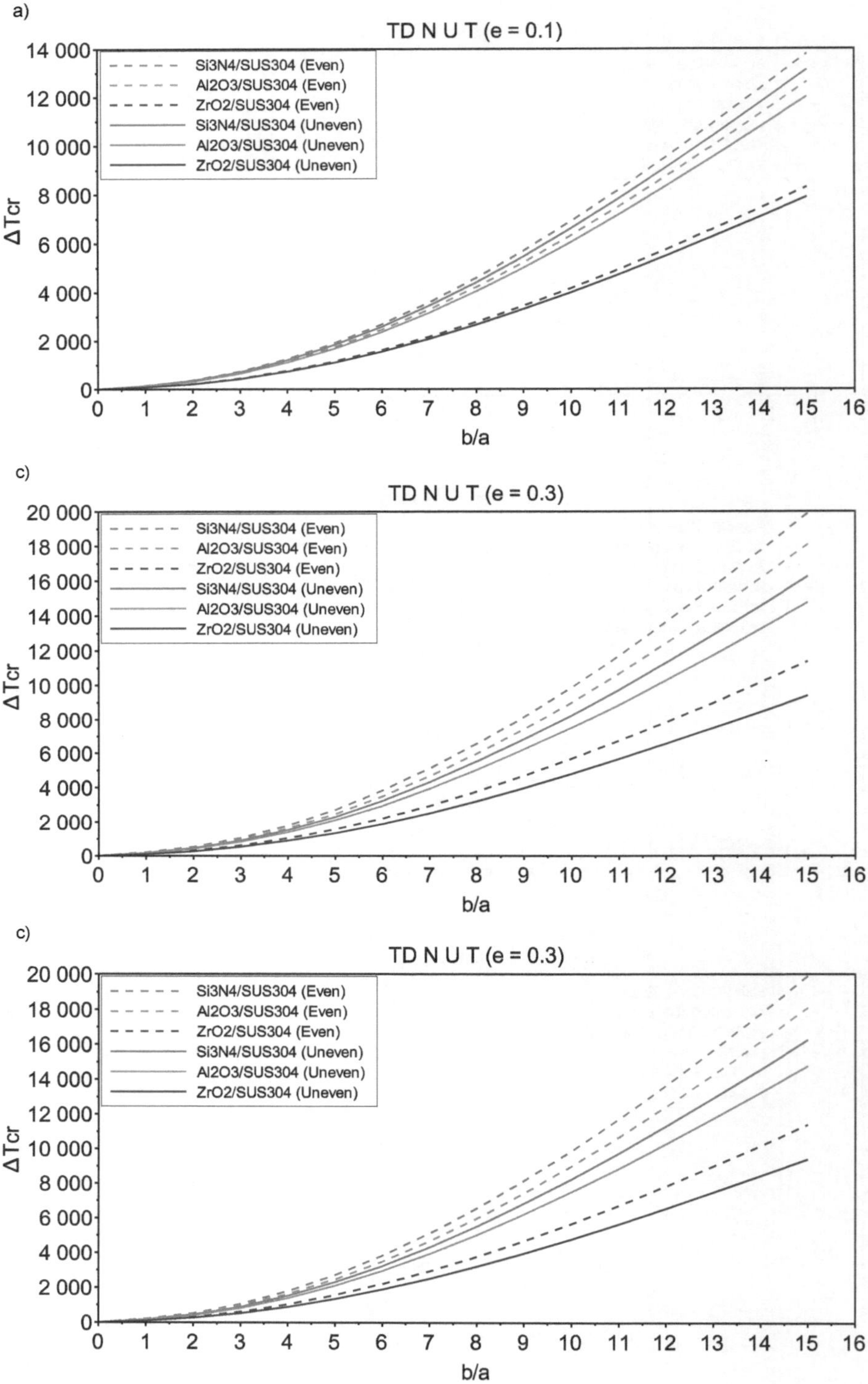

Figure 8.73 Critical buckling temperature (K) in the function of the slenderness ratio (b/a) of a simply supported (SS) (TD) FG porous plate $(p = 1)$ for different porosity coefficients $(e = 0.1; e = 0.2; e = 0.3)$ and with even and uneven porosity distributions under non-uniform thermal load.

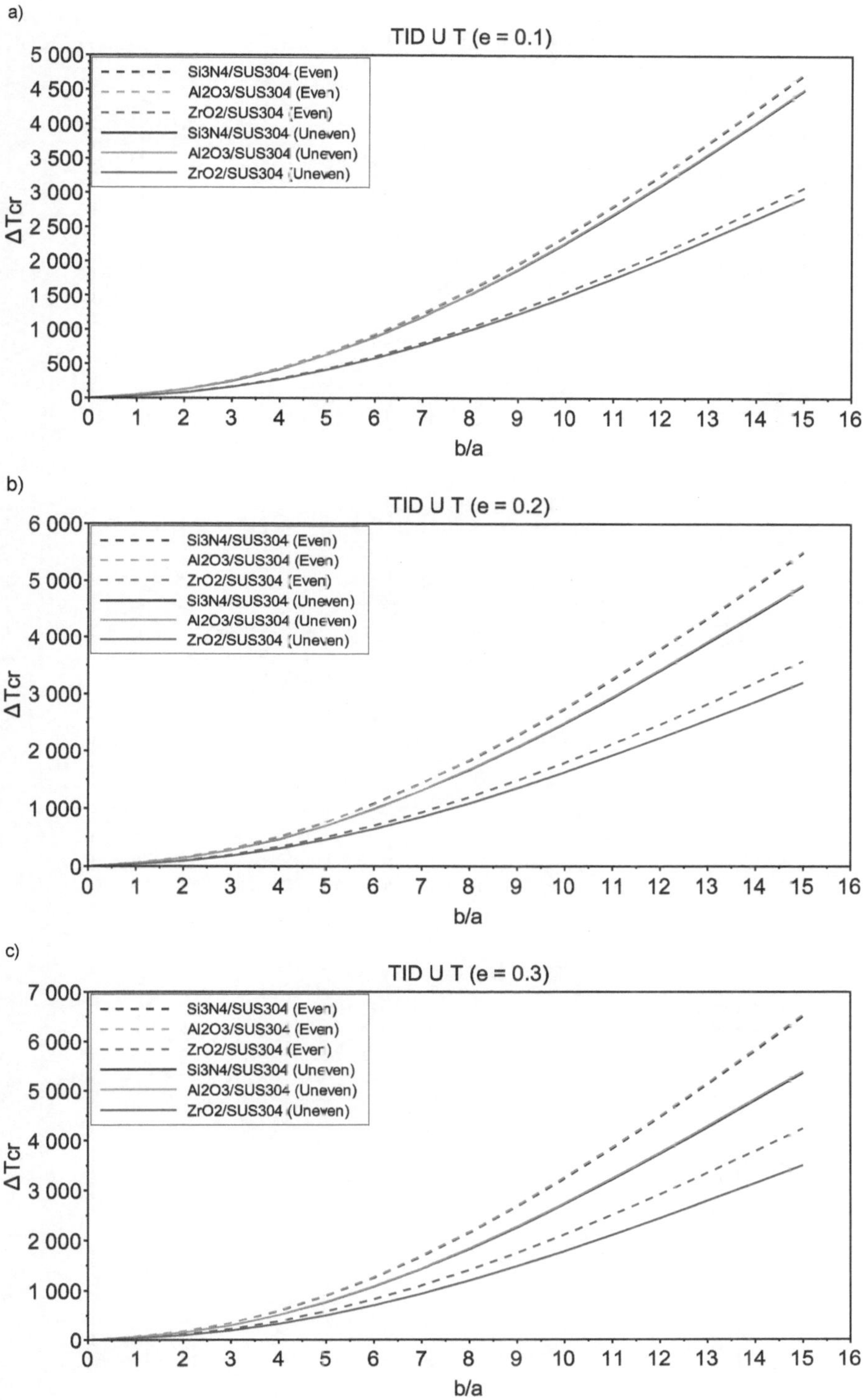

Figure 8.74 Critical buckling temperature (K) in the function of the slenderness ratio (b/a) of a simply supported (SS) (TID) FG porous plate $(p = 1)$ for different porosity coefficients $(e = 0.1; e = 0.2; e = 0.3)$ and with even and uneven porosity distributions under uniform thermal load.

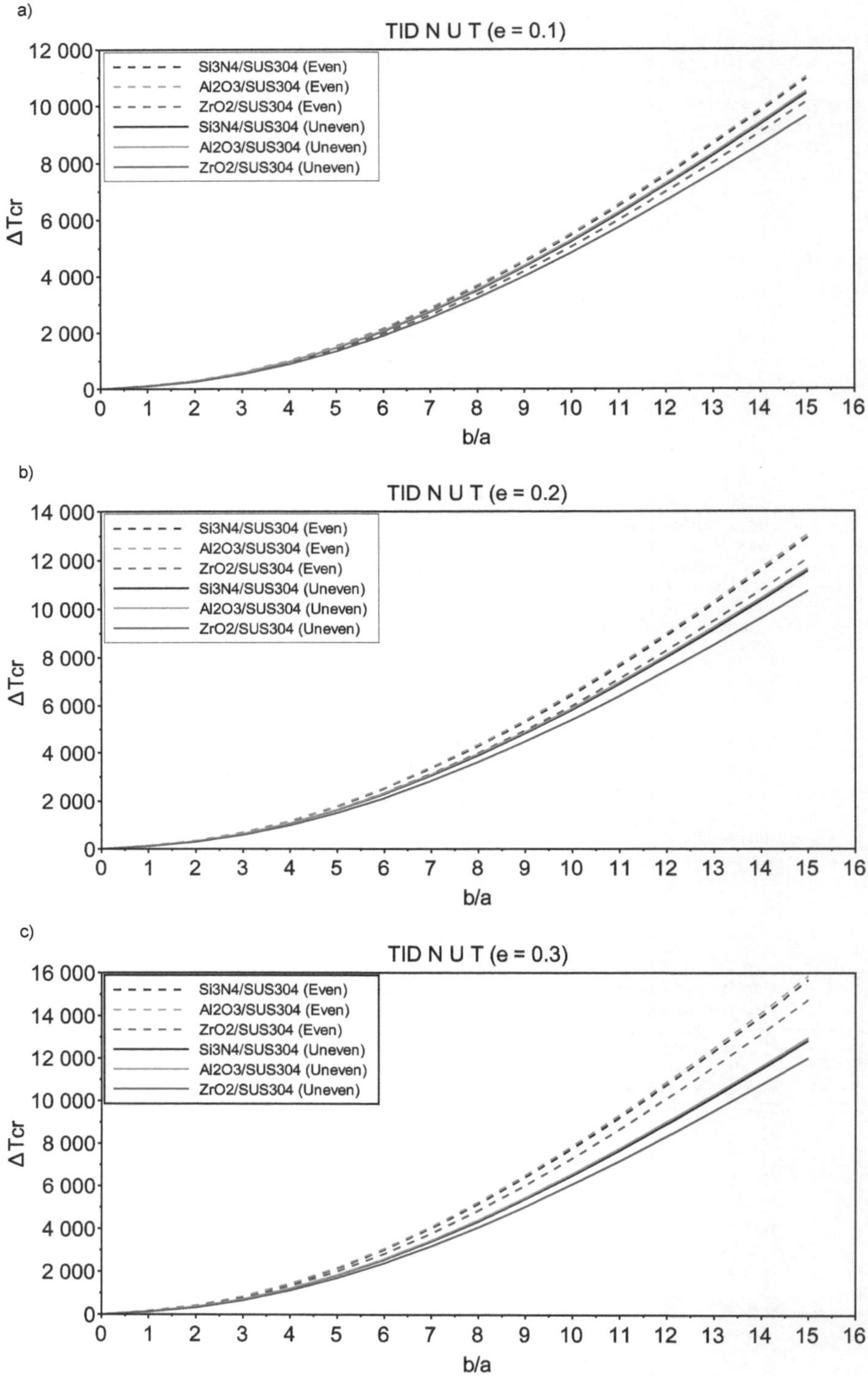

Figure 8.75 Critical buckling temperature (K) in the function of the slenderness ratio (b/a) of a simply supported (SS) (TID) FG porous plate $(p = 1)$ for different porosity coefficients $(e = 0.1; e = 0.2; e = 0.3)$ and with even and uneven porosity distributions under non-uniform thermal load.

a)

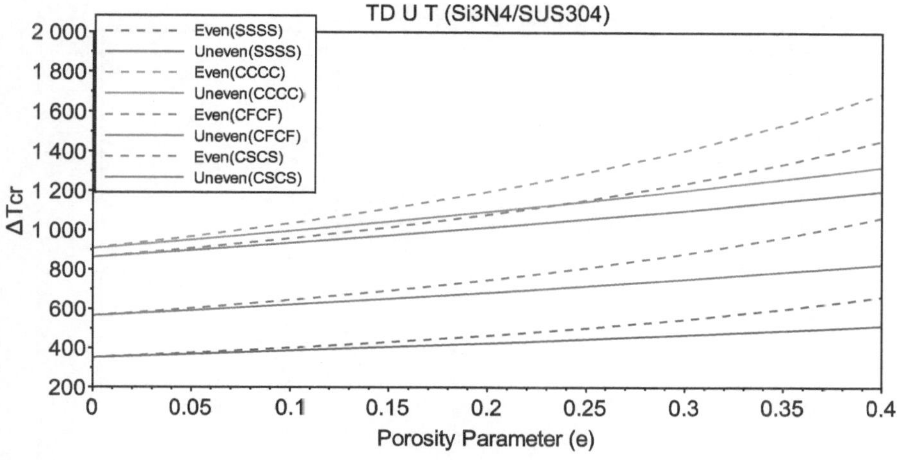

a)

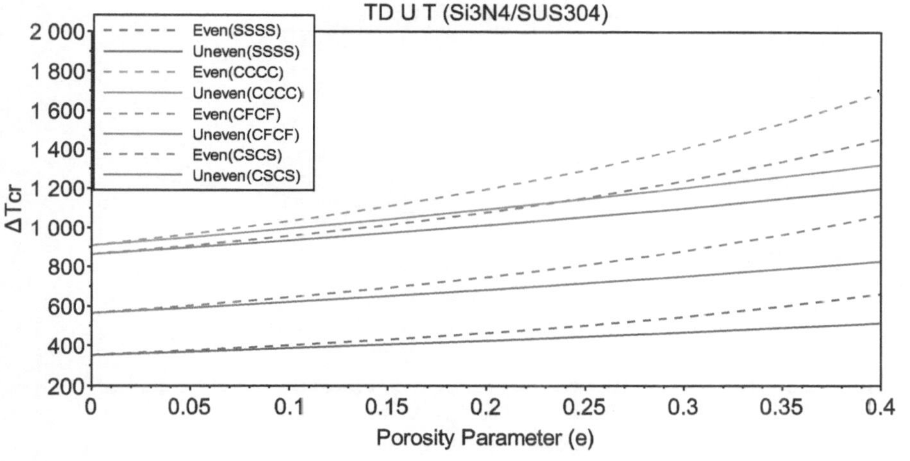

Figure 8.76 Critical buckling temperature (K) in the function of porosity parameter of ($Si_3N_4/SUS304$) FG porous plate ($a/h = 20$) under different boundary conditions and porosity types and for uniform and non-uniform temperature depended loads.

why the critical buckling temperature has higher values in the case of even porosity distribution. Additionally, it can be seen that with higher porosity rates, the difference between even and uneven types of porosity distributions becomes more pronounced which present the effect of porosity rate in the augmentation of the critical buckling temperature of FG porous plates.

As the same as in case of ($Si_3N_4/SUS304$) material, it is observed that the values of critical buckling temperature in the case of non-uniform temperature depended (TD NUT) are more higher than in case of uniform temperature depended (TD UT), as the same way with non-uniform temperature independent (TID NUT) is more than in case of uniform temperature independent (TID UT), with both materials $ZrO_2/SUS304$ and $Al_2O_3/SUS304$ as shown in Figs. 8.78, 8.79 and Figs. 8.80, 8.81, respectively. It can be also remarked that the value of the critical buckling temperature in the case of even clamped clamped (CC) is more than all the other boundary conditions of different pairs of materials. Consequently, the type of thermal loads, temperature dependency of materials, type of porosity distributions and boundary conditions play an important role in the determination of the critical thermal buckling load and it provides the best possible combination in terms of material

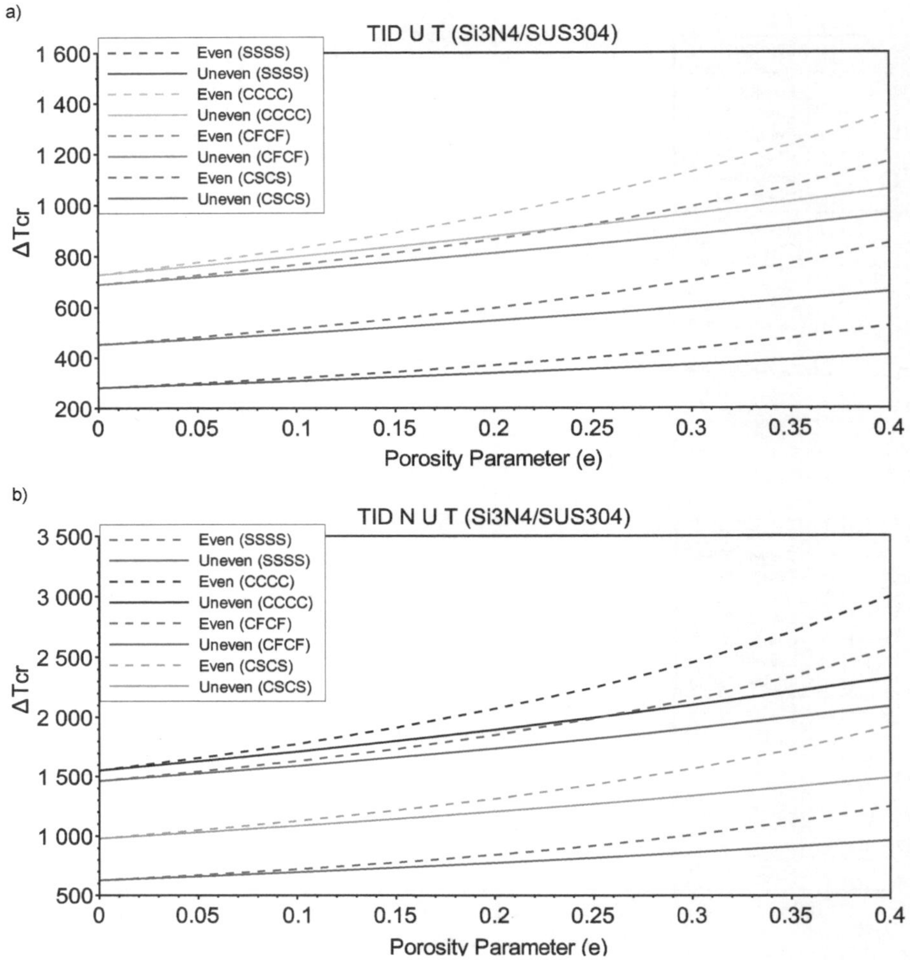

Figure 8.77 Critical buckling temperature (K) in the function of porosity parameter of $(Si_3N_4/SUS304)$ FG porous plate $(a/h = 20)$ under different boundary conditions and porosity types and for uniform and non-uniform temperature independent loads.

constituents, percentage of porosity included in the structure and thermal and boundary conditions to ensure the safety of structural components against the buckling failure mode.

Figure 8.82 displays the histogram repartition of the critical buckling temperature of $(Si_3N_4/SUS304)$ FG plate $(p = 2, a/h = 20)$ for four types of boundary conditions, under uniform and non-uniform depended and independed temperatures (TD UT, TD NUT, TID UT, TID NUT) and with evenly and unevenly types of porosity distributions. It is clear that as the porosity coefficient rise as the critical buckling temperature rise especially with the buckled $Si_3N_4/SUS304$ FG plate under clamped clamped (CCCC) boundary condition and in the case of non-uniform temperature depended (TD NUT) and with the maximum value of porosity coefficient (e = 0.4). Furthermore, it can be seen that evenly FG porous plates induces the highest values of critical buckling temperature compared to unevenly ones. From gained results, we can see that the increase in the power-law index (p) results in a decrease in the critical temperature buckling, conversely, the increase of the porosity factor (e) increases the buckling strength of the FG porous plates. This proves that the percentage of added porosity volume fraction, as well as the manner of pores distributions

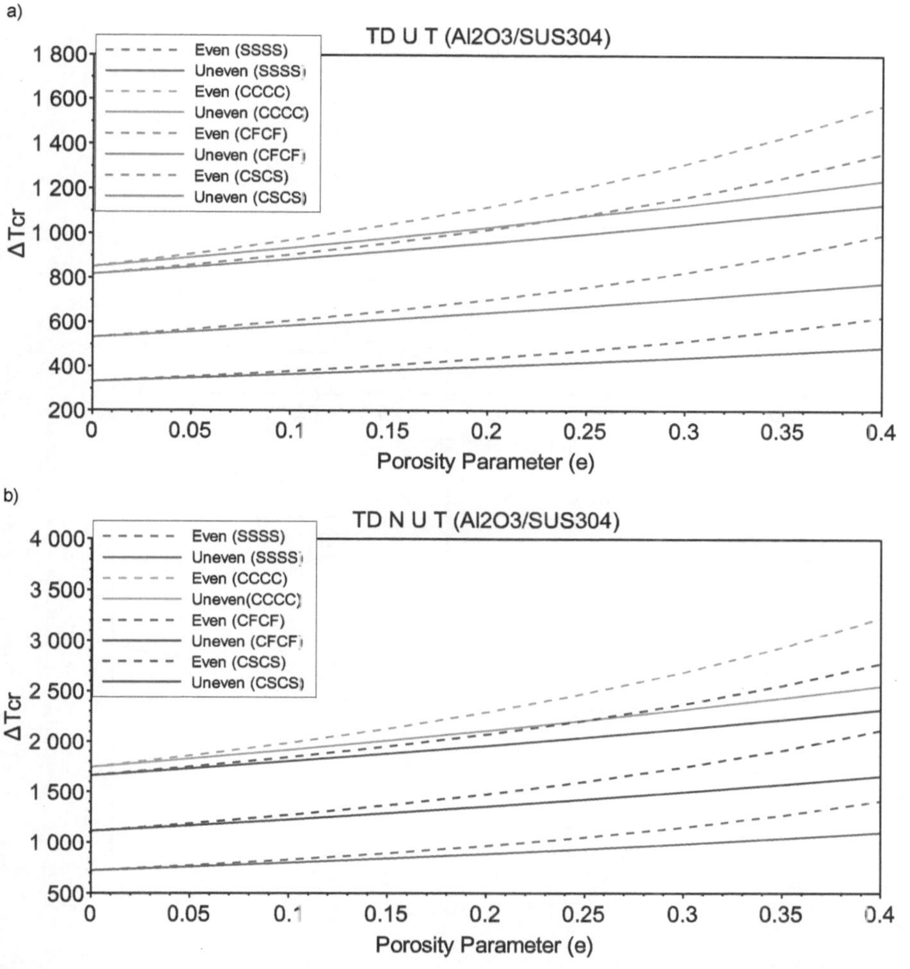

Figure 8.78 Critical buckling temperature (K) in the function of porosity parameter of ($Al_2O_3/SUS304$) FG porous plate ($a/h = 20$) under different boundary conditions and porosity types and for uniform and non-uniform temperature depended loads.

have a significant impact on the strength of the FG porous plates against elevated gradient of temperatures. In conclusion, from these parametric illustrations, it is found that mixing the Silicon nitride (Si_3N_4) and alumina (Al_2O_3) with stainless steel ($SUS304$) materials is more convenient to mixing Zirconia (ZrO_2) with ($SUS304$) and this is due to high flexural rigidity of these pairs of materials compared to last one and since the buckling capacity of the structure depends on material nature, it is crucial to choose the best combination of mixing materials to ensure the safety of the engine against buckling and later post-buckling failure modes.

The thermal buckling mode shapes of FG porous plates under uniform temperature rise and for different boundary conditions are presented in Fig. 8.83. It is shown that as the (FGP) plate is subjected to strong constraints as clamped or simply supported conditions at its edges, as the critical buckling temperature is higher. Therefore, the type of applied constraints has a significant effect on buckling response of FG porous plates and its bearing-load strength against instability problems.

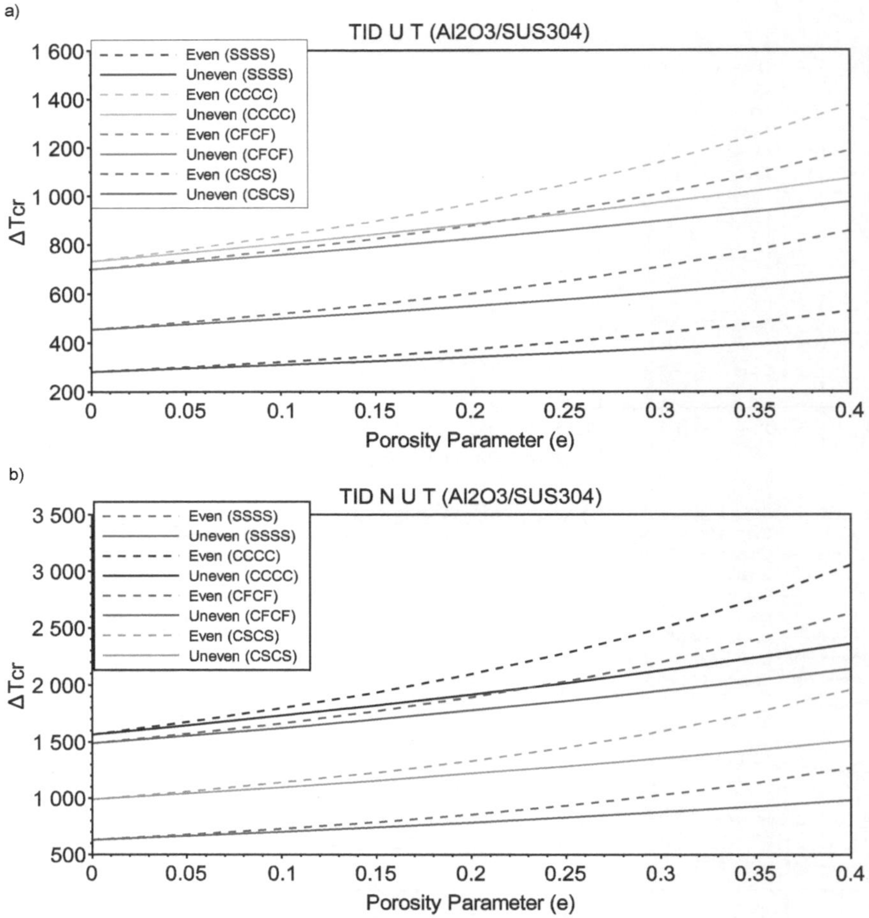

Figure 8.79 Critical buckling temperature (K) in the function of porosity parameter of $(Al_2O_3/SUS304)$ FG porous plate $(a/h = 20)$ under different boundary conditions and porosity types and for uniform and non-uniform temperature independed loads.

8.6.3 THERMOELASTIC BUCKLING OF FG POROUS SPHERICAL SHELLS

This section is intended to present for the first time the thermal buckling analysis of FG porous spherical shells through an entailed parametric study including the effects of FG power index, porosity volume fraction, types of graded distributions, applied constraints and thermal loads.

8.6.3.1 Accuracy and reliability verification of the numerical approach

First of all, the accuracy and reliability of the proposed finite element numerical approach are verified in the case of isotropic spherical shells. Since there are no works related to thermal buckling analysis of FG porous spherical shells, we perform here the comparison of the present results to the ones given by Eslami et al. [73]. The critical thermal buckling strains $\varepsilon_T = \alpha_T \times \Delta_T$ for a simply supported (SS) isotropic spherical shell $(\nu = 0.3)$ with latitude angle $\varphi = 90^0$, subjected to uniform thermal load and for various ratios of h/R are reported in Tab. 8.30. It should be noted that the thermal expansion is taken as $\alpha_T = 1 \times 10^{-6}$. The geometry and mesh configuration of the spherical shell under thermal load is shown in Fig. 8.84.

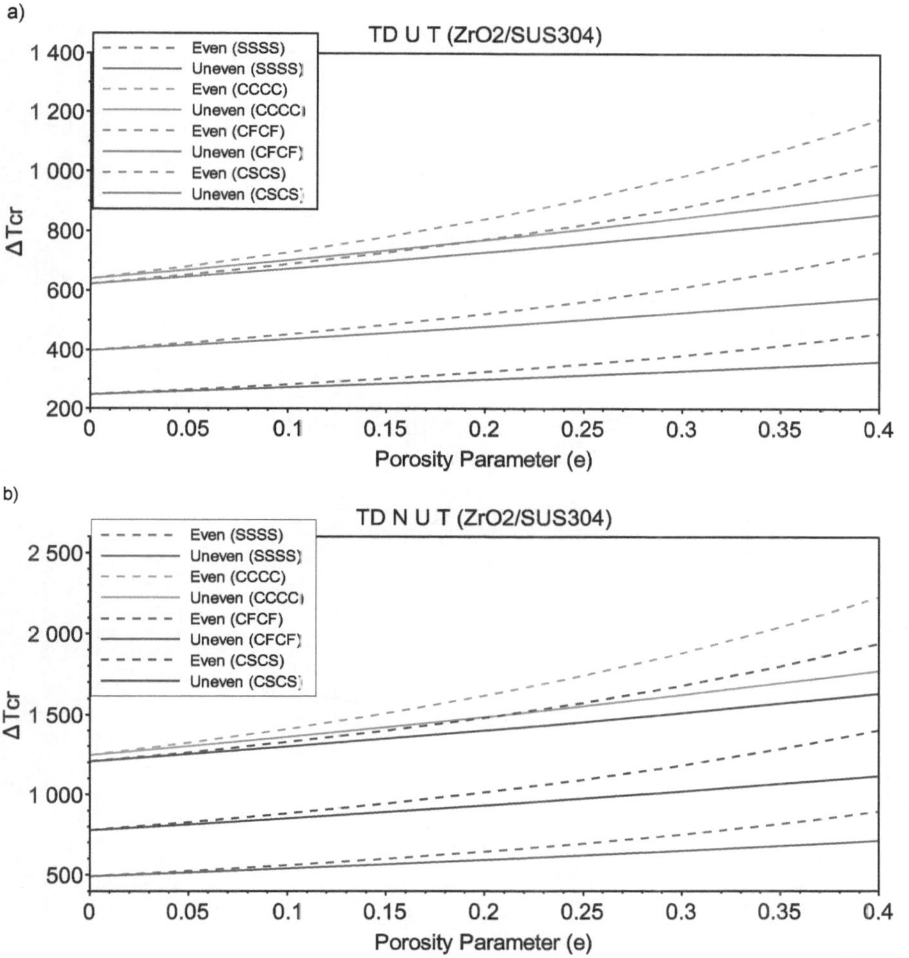

Figure 8.80 Critical buckling temperature (K) in the function of porosity parameter of ($ZrO_2/SUS304$) FG porous plate ($a/h = 20$) under different boundary conditions and porosity types and for uniform and non-uniform temperature depended loads.

Table 8.30 compares the lowest thermal buckling strains obtained by analytical thin shell theory using either a deep shell theory (DST) or shallow shell theory (SST) theory [73]. The relative errors (ER1 and ER2) evaluated between the present results and those from the literature increases with the increase of h/R ratio and this is expected since different shell theories are used and with the increase of h/R, the spherical shell behaves as thick structure. Furthermore, the deep shell theory (DST) predicts buckling load for m-values (high values) whereas the shallow shell theory (SST), predicts the buckling load for m = 1 (low values) and the present results hold well with the (DST) theory more than (SST) theory and the slight difference is due to the improvement of stability in the (DST) compared to (SST) theory. Hence, it is reasonable to perform the thermal buckling behavior of spherical shells under thermal loadings within the present formulation, which is numerically locking-free and doesn't overestimate the critical buckling values.

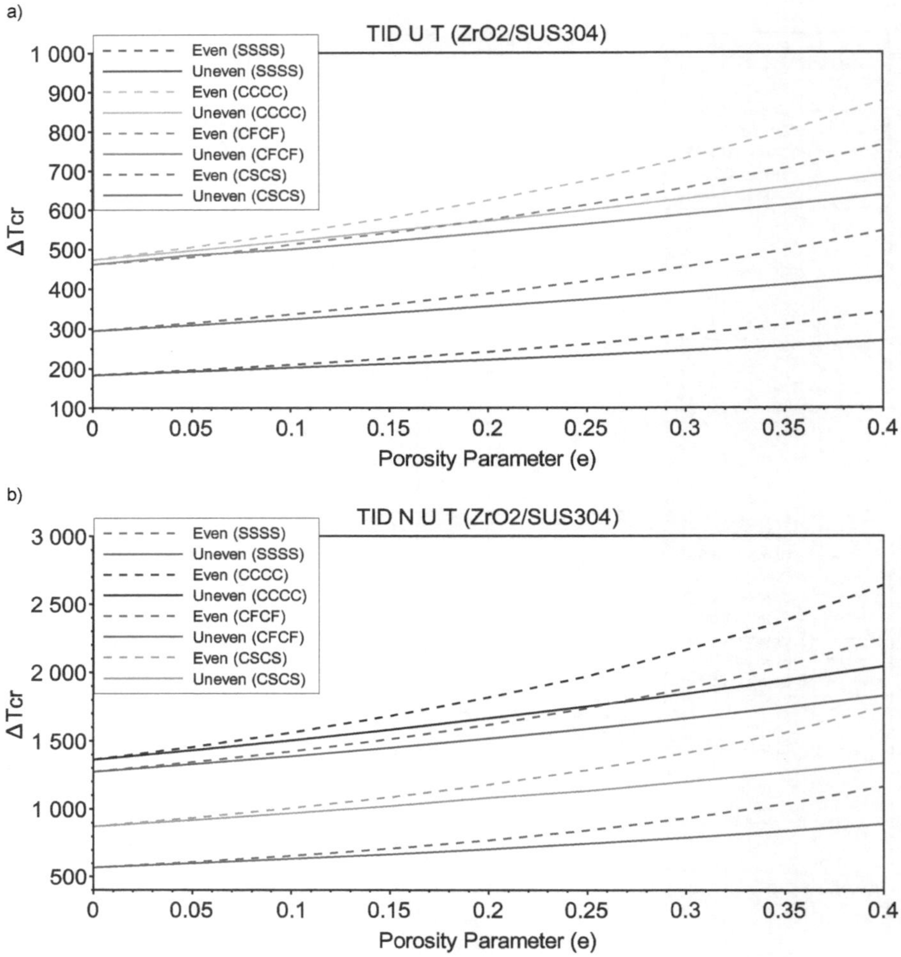

Figure 8.81 Critical buckling temperature (K) in the function of porosity parameter of $(ZrO_2/SUS304)$ FG porous plate $(a/h = 20)$ under different boundary conditions and porosity types and for uniform and non-uniform temperature independed loads.

8.6.3.2 Buckling temperature responses of FG porous spherical shells

Now, we provide the first new results on buckling temperature responses of FG porous spherical shells. For the sake of brevity and convenience, the pair of material $Al_2O_3/SUS304$ is considered, where the spherical shell is composed from Alumina Al_2O_3 and Stainless Steel $SUS304$ components. The material properties of these components are as mentioned, in Tab. 8.29. In these analyses, the latitude angle is taken as $\varphi = 60^0$ and the porosity can be distributed either evenly or unevenly along the thickness direction. The type of applied boundary condition is simply supported (SS). The critical buckling temperature is evaluated relative to the metal thermal expansion as: $\Delta T_{cr-Adim-m} = \alpha_m \times \Delta T_{cr}$. The first study aims to analyze the critical buckling temperature $\Delta T_{cr-Adim-m}$ in the function to the power index (p) FG porous spherical caps subjected to uniform and non-uniform temperature depended and independed loads for two sets of radius-to-thickness ratios $(R/h = 400, 800)$.

In Figs. 8.85 and 8.86 we can show that the increase of the power index (p) from metal to ceramic portions leads to a decrease in the critical buckling temperature $\Delta T_{cr-Adim-m}$ for all porosity

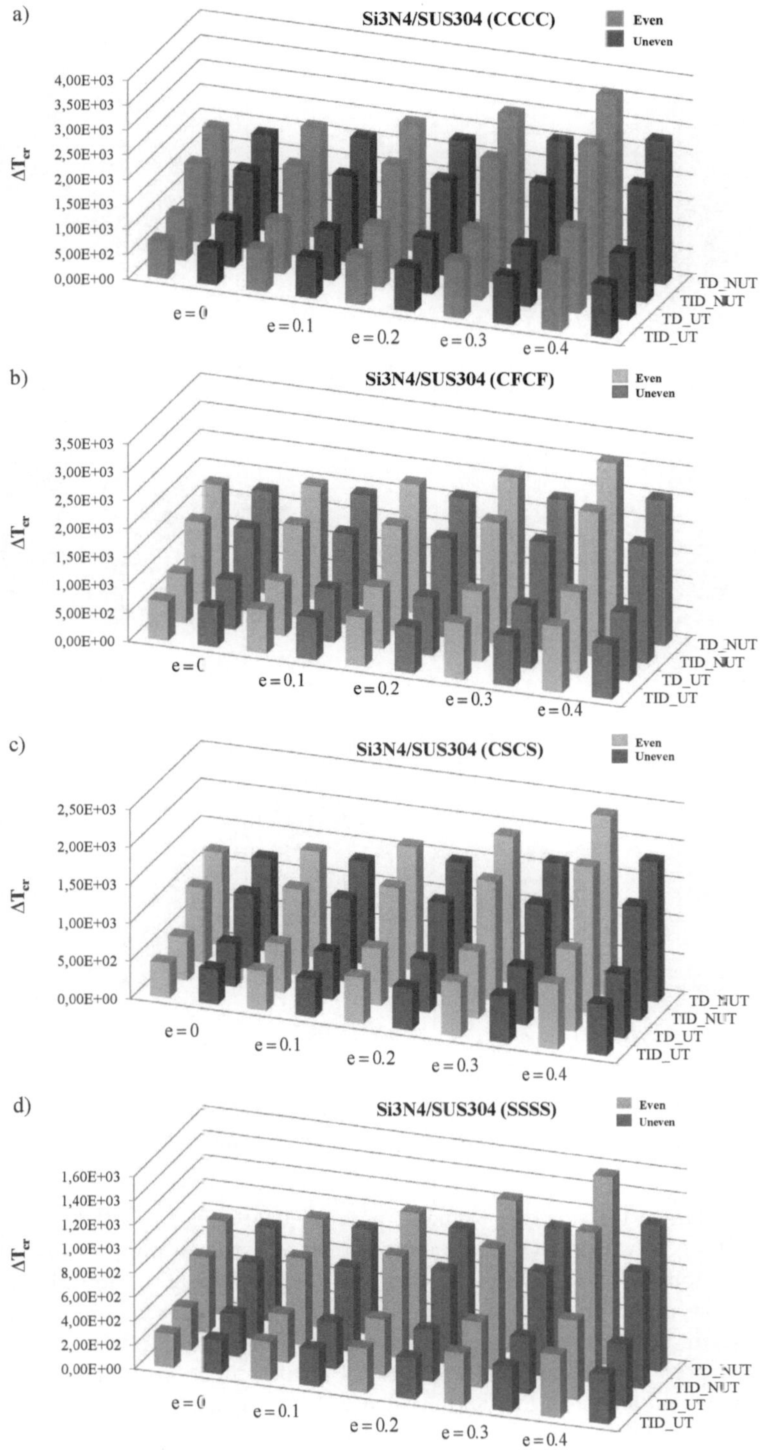

Figure 8.82 Histogram of repartition of the buckled $Si_3N_4/SUS304$ FG porous plates ($a/h = 20, p = 2$) for five sets of porosity coefficients and under different temperature dependency and thermal loads while varying the type of Boundary conditions (a) (CCCC), (b) (CFCF), (c) (CSCS) and (d) (SSSS).

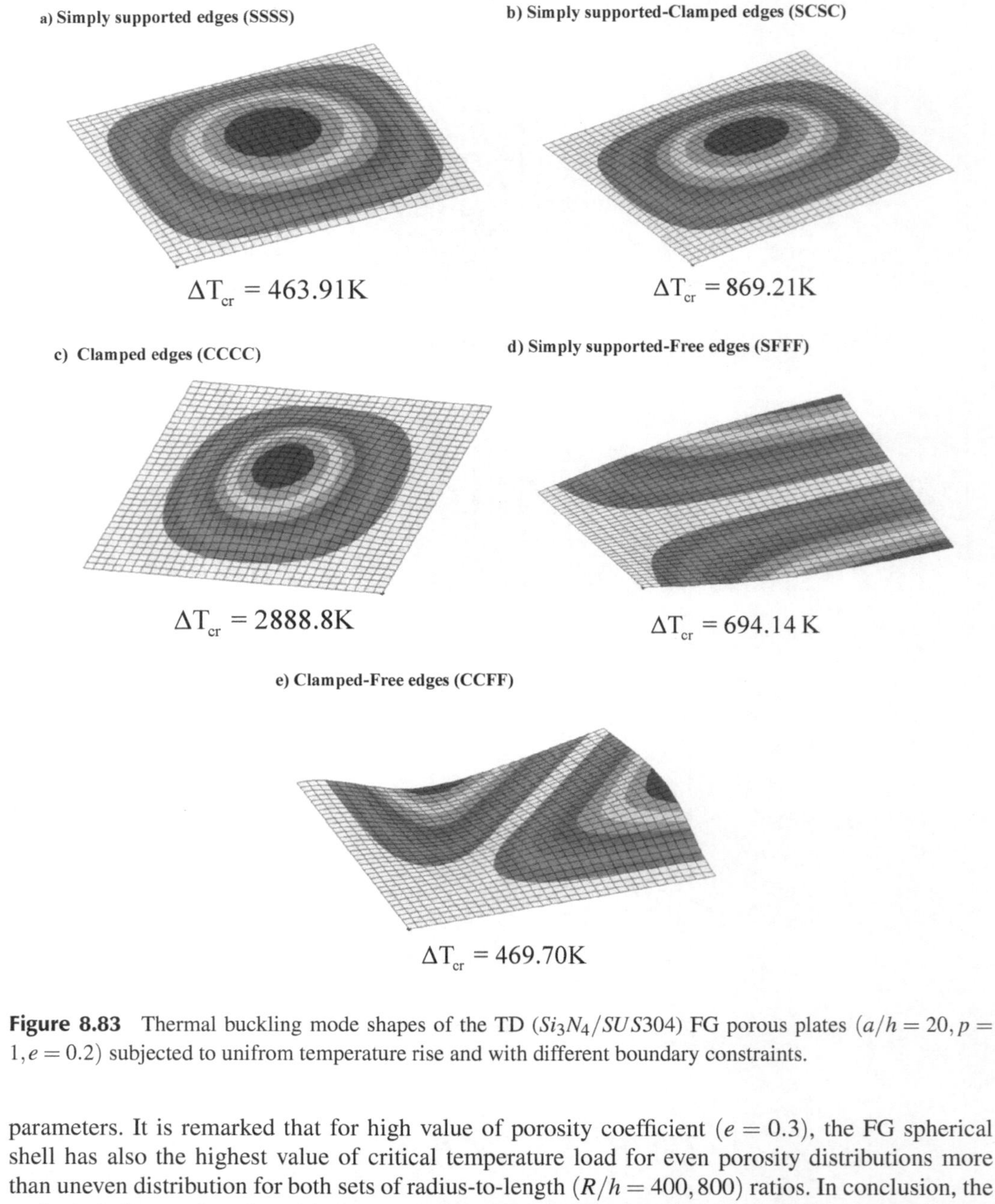

Figure 8.83 Thermal buckling mode shapes of the TD ($Si_3N_4/SUS304$) FG porous plates ($a/h = 20, p = 1, e = 0.2$) subjected to unifrom temperature rise and with different boundary constraints.

parameters. It is remarked that for high value of porosity coefficient ($e = 0.3$), the FG spherical shell has also the highest value of critical temperature load for even porosity distributions more than uneven distribution for both sets of radius-to-length ($R/h = 400, 800$) ratios. In conclusion, the analysis indicates that the critical temperature change, $\Delta T_{cr-Adim-m}$, decreases in terms of power index and increases in terms of porosity volume fraction. This is due to the variation of flexural stiffness in the FG porous spherical shell, as well as to the reduction in effective coefficient of thermal expansion, which increases the thermal load-bearing capacity, unlike, the critical buckling load which decreases with the increase in porosity volume fraction due to a reduction in effective Young's modulus.

Figure 8.87 plots ΔTcr as a function of porosity parameter (e) of an SS ($Al_{2O3}/SUS304$) FG porous spherical shell under uniform temperature load for various temperature material configura-

Table 8.30
Values of critical thermal buckling strains ε_T of simply supported spherical shell ($\nu = 0.3, \varphi = 90^0$) subjected to uniform thermal load.

h/R	Present	[73]		*Difference*(%)	
		DST	SST	*ER*1	*ER*2
0.01	4.1018E-03	4.130E-03	4.133E-03	0.68	0.75
0.02	8.5557E-03	8.435E-03	8.350E-03	1.43	2.46
0.03	1.2268E-02	1.272E-02	1.255E-02	3.55	2.24
0.04	1.7428E-02	1.726E-02	1.708E-02	0.97	2.03
0.05	2.2070E-02	2.149E-02	2.107E-02	2.69	4.74
0.06	2.6313E-02	2.614E-02	2.571E-02	0.66	2.33
0.07	3.1734E-02	3.162E-02	3.061E-02	0.36	3.67
0.08	3.6019E-02	3.574E-02	3.442E-0	0.87	4.64
0.09	4.0256E-02	3.991E-02	3.859E-02	0.86	4.30
0.1	4.4480E-02	4.4570E-02	4.325E-02	0.2	2.84

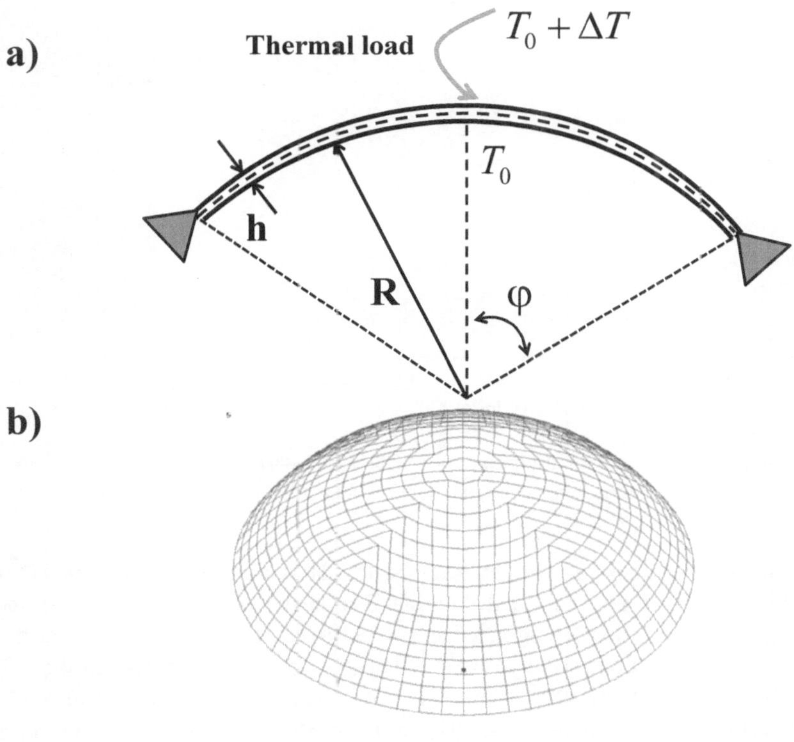

Figure 8.84 Sketch (a) and meshed (b) configurations of a simply supported spherical shell under a temperature gradient field.

tions (TD and TID). We can distinguish that if the porosity coefficient rises, the critical buckling temperature increases. It is noted that when the FG spherical shell is made of temperature depended (TD) material, the critical buckling temperature has the highest value more than when the FG spherical shell is made of temperature independed (TID) material and this is for both even and uneven porosity distributions. Furthermore, it can be observed that the increase of porosity parameter (e)

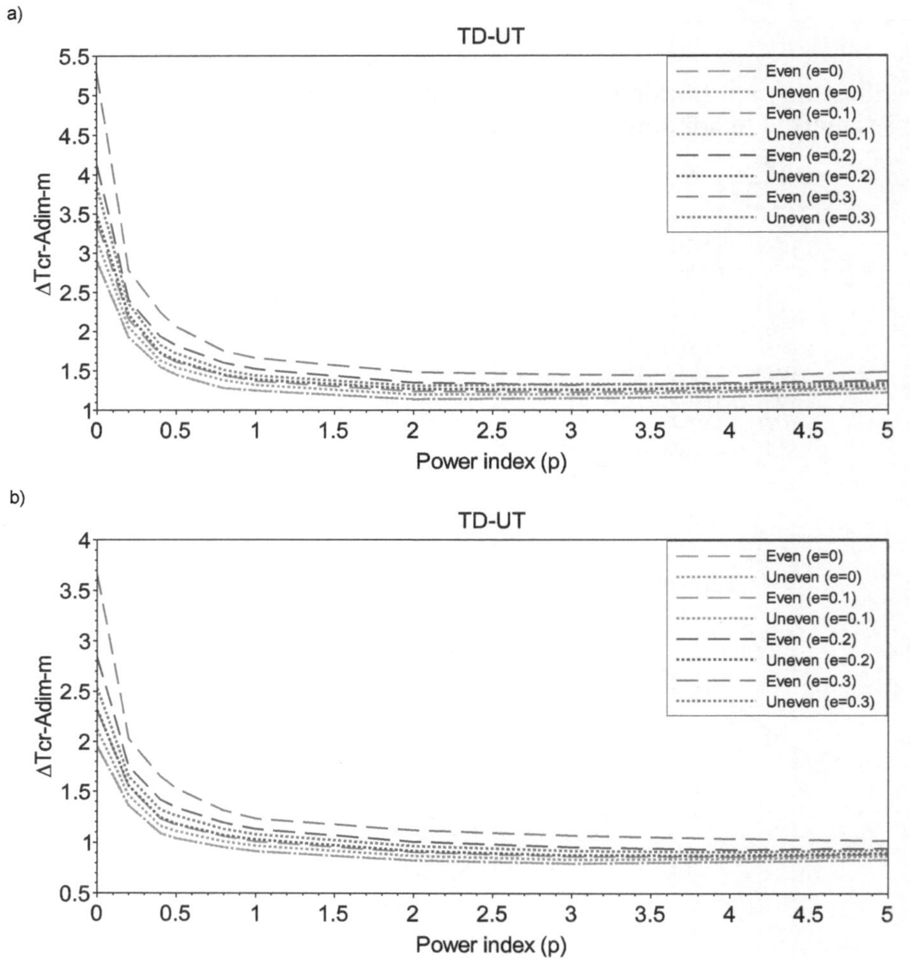

Figure 8.85 Adimensional critical buckling temperature of ($Al_2O_3/SUS304$) FG porous spherical cap shell with (a) $R/h = 400$ and (b) $R/h = 800$ subjected to uniform depended (TD UT) temperature load.

induces an increase in the critical buckling temperature and this is more pronounced with TD than TID materials and with evenly than unevenly distributions. Similarly observations can be made in the case of non-uniform temperature load as depicted in Fig. 8.88. Another observation concerning the radius-to-thickness ratio, for which the increase of this ratio leads to a decrease in the values of the critical thermal buckling loads. In fact, when this ratio increases, the FG spherical shell behaves as thin shell structure, with low flexural rigidity and becomes more sensitive to the thermal loadings. In conclusion, the type of temperature material dependency, type of porosity distributions, thermal loadings and aspect ratios have a significant impact on thermal buckling behavior of such shell structures.The effect of gradient index (p) on thermal buckling responses of FG TID porous spherical shells is giving in Fig. 8.89.

Figures 8.90 and 8.91 present the curves of FG porous spherical shell as a function of porosity parameter (e) under uniform and non-uniform temperature loads subjected to even or uneven porosity distribution and for various (R/h) ratios. Herein, we can observe that when the porosity coefficient (e) rises, also the critical buckling temperature increases especially when the porosity equal to 0.3 and for even temperature depended load. It is remarked that the highest values of critical buckling

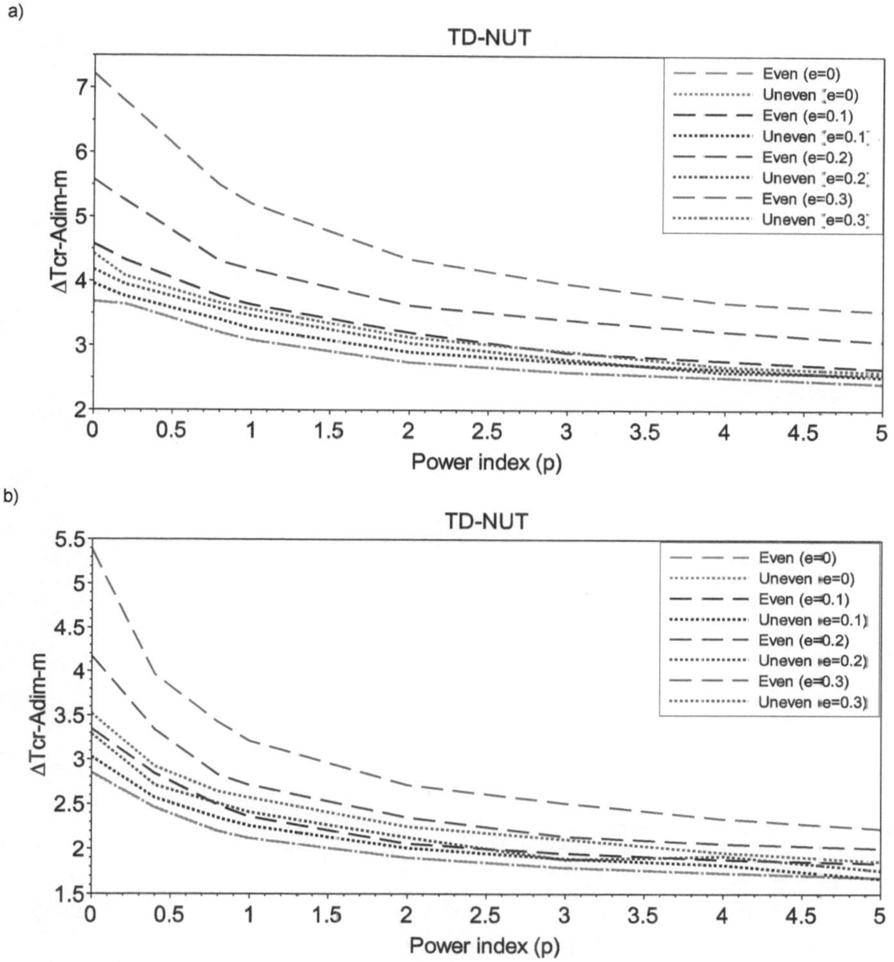

Figure 8.86 Adimensional critical buckling temperature of ($Al_2O_3/SUS304$) FG porous spherical cap shell with (a) $R/h = 400$ and (b) $R/h = 800$ subjected to non-uniform depended (TD NUT) temperature load.

temperature are for radius-to-thickness ratio ($R/h = 400$) with even porosity distribution in the both case of temperature load (depended and independed).

Also, it is clear that when the FG spherical cap subjected to non-uniform temperature depended load, the FG porous spherical shell possessess the highest value of critical buckling temperature. So, more the FG porous spherical shell is thicker and contains a rate of porosity superior to 15%, more its resistance to thermal buckling is also superior. This can be explained by the bending-stretching components which are temperature and porosity dependent and offers hence the maximum stiffness, thus yielding in high critical values.

Finally, the effect of variation of the latitude angle on thermal buckling response of FG porous spherical shell is examined in Fig. 8.92 for three curved shell angles (ϕ=30°, 60°, 90°) and for TD and TID materials under uniform temperature rise. The curves are plotted while maintaining $e = 0.2$ and $p = 1$. It is shown that the high value of the critical buckling temperature is obtained for an angle equal to ϕ = 30° whereas, the lowest value is obtained for ϕ = 90°. This is attributed to the energy contribution due to the membrane and bending deformations which is enhanced with more curved shells. Furthermore, more the R/h ratio increases, the thermal buckling response decreases

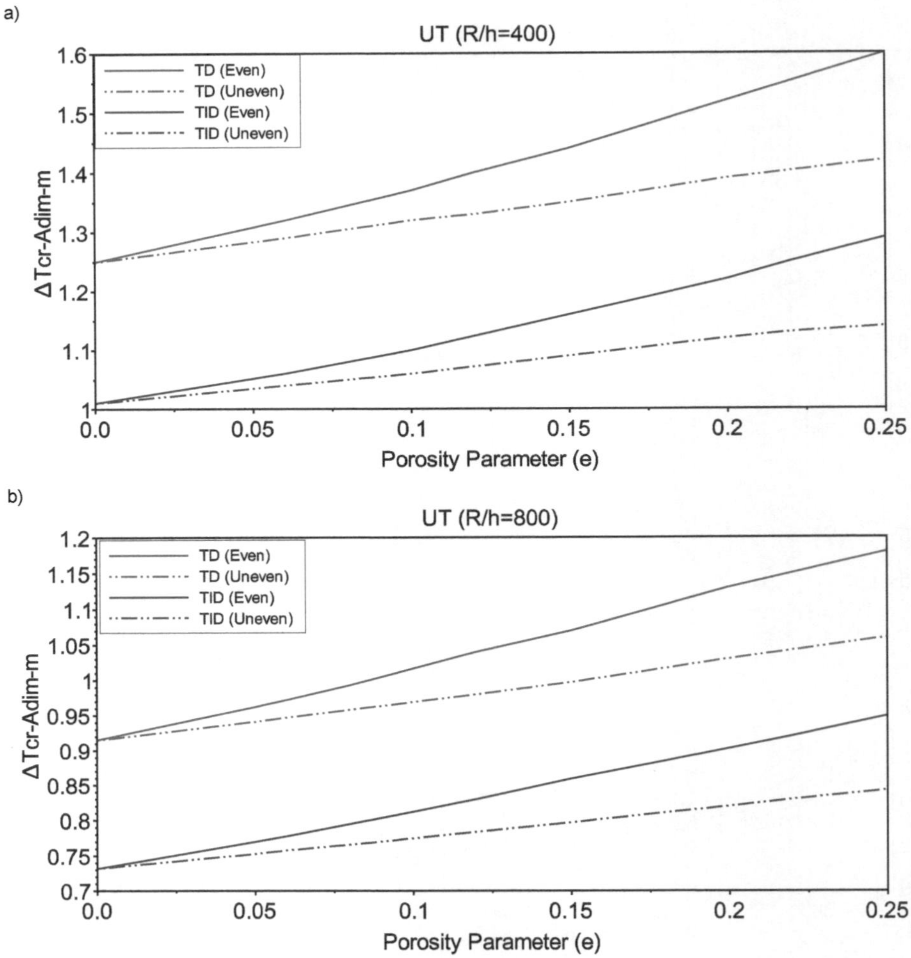

Figure 8.87 $\Delta T cr\text{-}adim\text{-}m$ as a function of porosity parameter (e) of $(Al_2O_3/SUS304)$ FG porous spherical cap with (a) $R/h = 400$ and (b) $R/h = 800$ under uniform temperature (UT) load.

drastically and this is for both TD and TID FG porous spherical shells. Therefore, the angle $\phi = 30^\circ$ and the radius-to-thickness ratio $R/h = 100$ is the best configuration to obtain higher buckling load and the FG shell can hence provide more resist to thermal fields. In addition, it can be seen that evenly type of porosity gives more buckling capacity than uneven type. Thus, the type of porosity distribution patterns, aspect ratios and curved angles have an important influence on thermal buckling behavior of FG porous spherical shells.

The buckling mode shape of CC $(Si_3N_4/SUS304)$ FG porous spherical shell $(R/h = 400, p = 1, e = 0.2, \phi = 30\circ)$ under unifrom thermal load is plotted in Fig. 8.93.

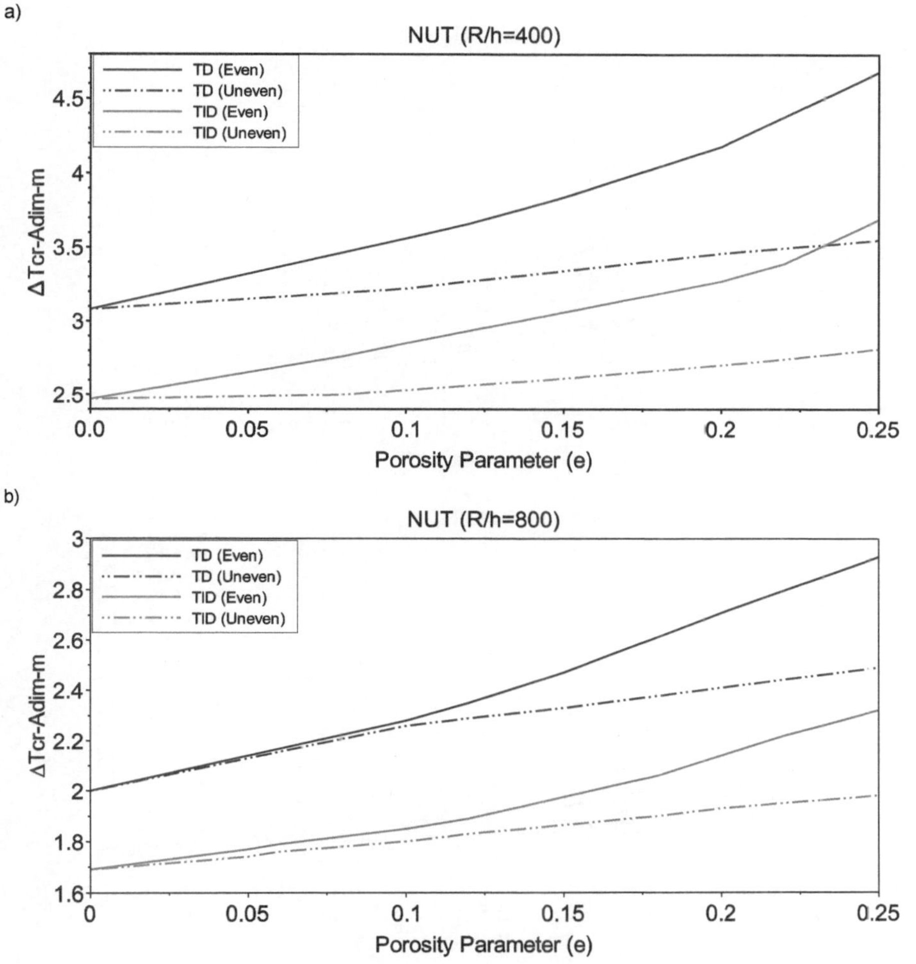

Figure 8.88 $\Delta Tcr\text{-}adim\text{-}m$ as a function of porosity parameter (e) of $(Al_2O_3/SUS304)$ FG porous spherical cap with (a) $R/h = 400$ and (b) $R/h = 800$ under non-uniform temperature (NUT) load.

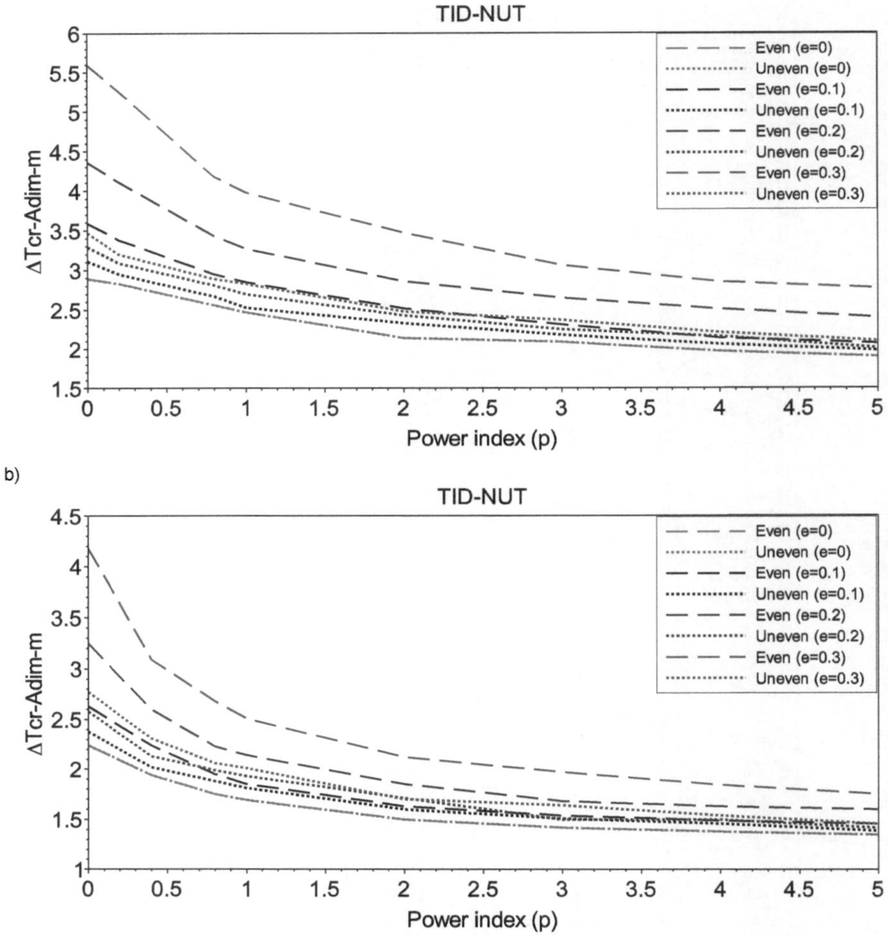

Figure 8.89 $\Delta Tcr\text{-}adim\text{-}m$ as a function of gradient index (p) of TID ($Al_2O_3/SUS304$) FG porous spherical cap with (a) $R/h = 400$ and (b) $R/h = 800$ under non-uniform temperature (NUT) load.

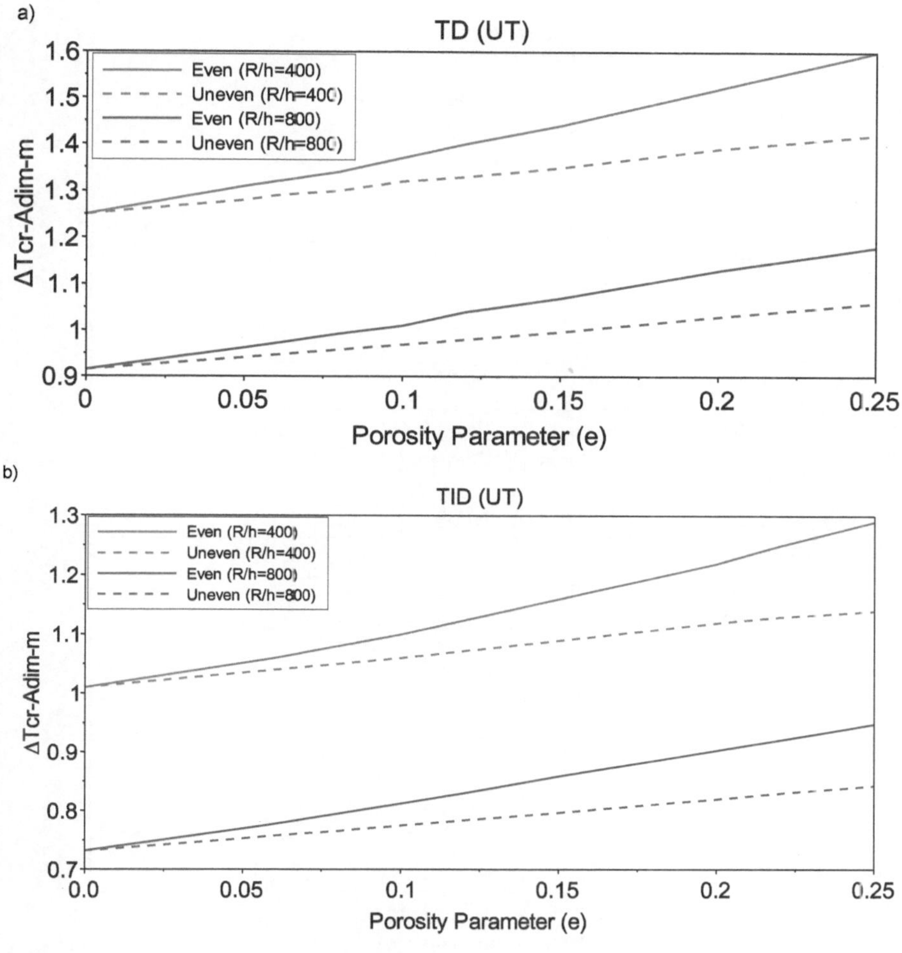

Figure 8.90 $\Delta Tcr\text{-}adim\text{-}m$ as a function of porosity parameter (e) of SS ($Al_2O_3/SUS304$) FG porous spherical cap under uniform temperature load for various (R/h) ratios.

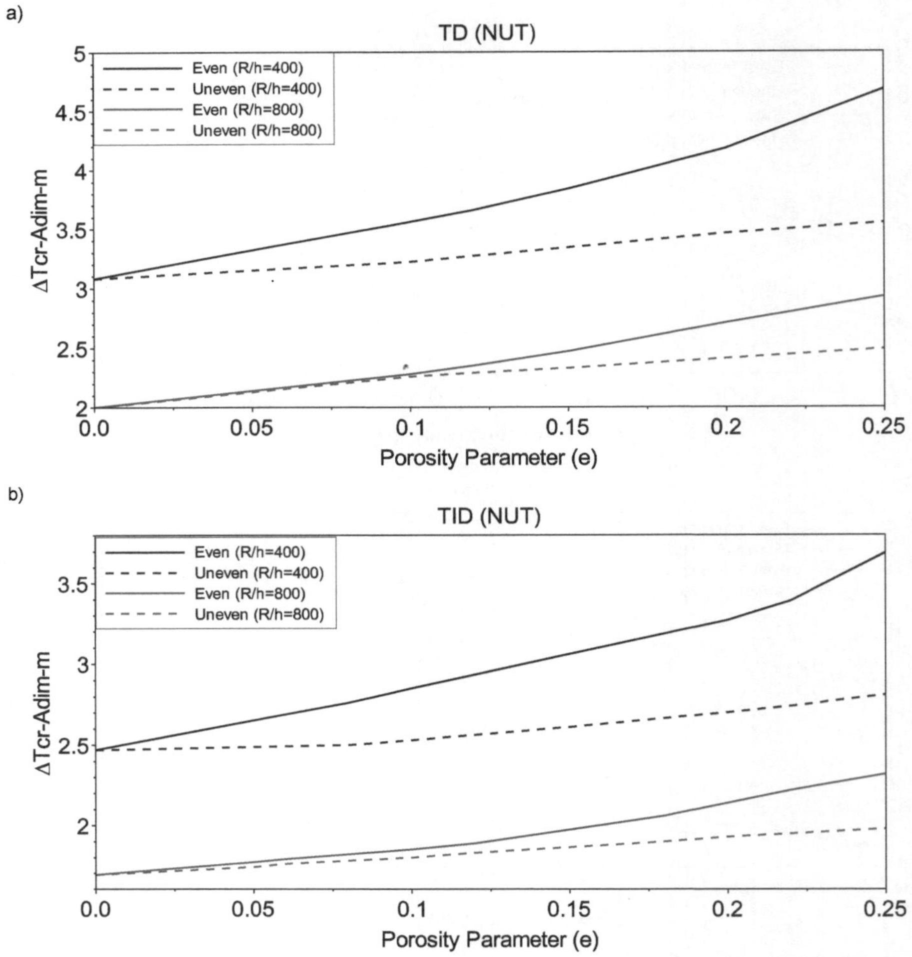

Figure 8.91 $\Delta Tcr\text{-}adim\text{-}m$ as a function of porosity parameter (e) of SS ($Al_2O_3/SUS304$) FG porous spherical cap under non-uniform temperature load for various (R/h) ratios.

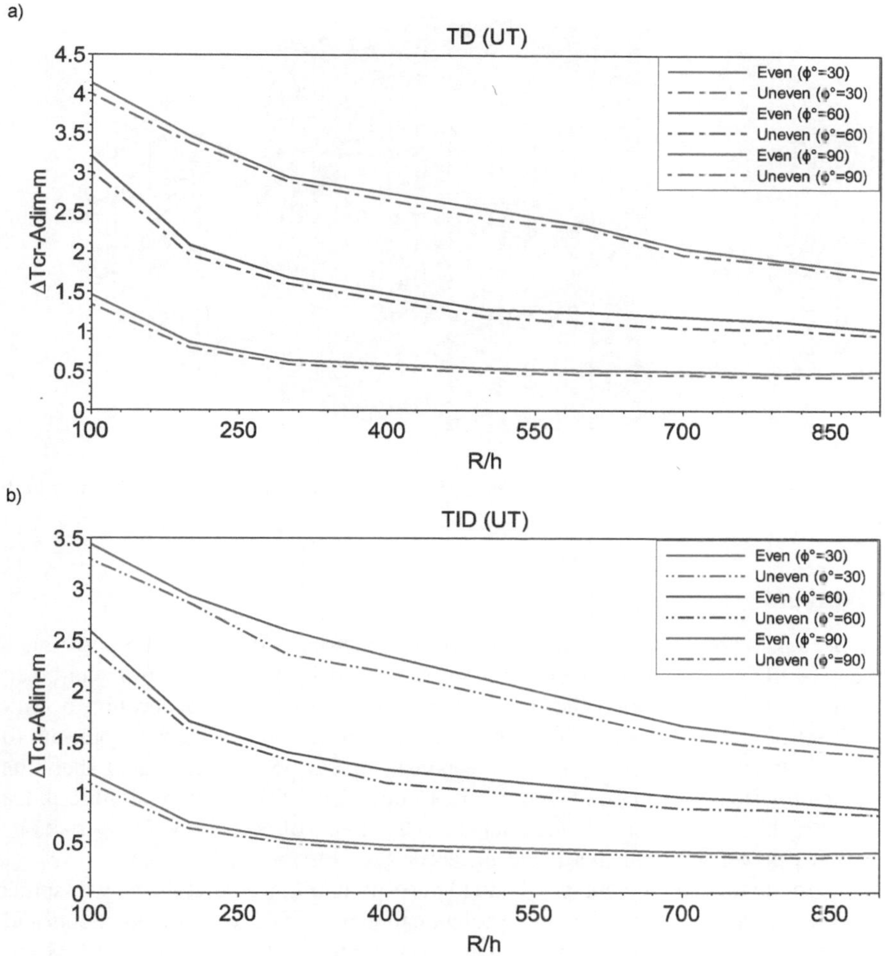

Figure 8.92 $\Delta Tcr\text{-}adim\text{-}m$ as a function of (R/h) ratio of SS $(Al_2O_3/SUS304)$ FG porous (TD) and (TID) spherical cap ($e = 0.2$ and $p = 1$) under uniform temperature load and for various (ϕ) angles.

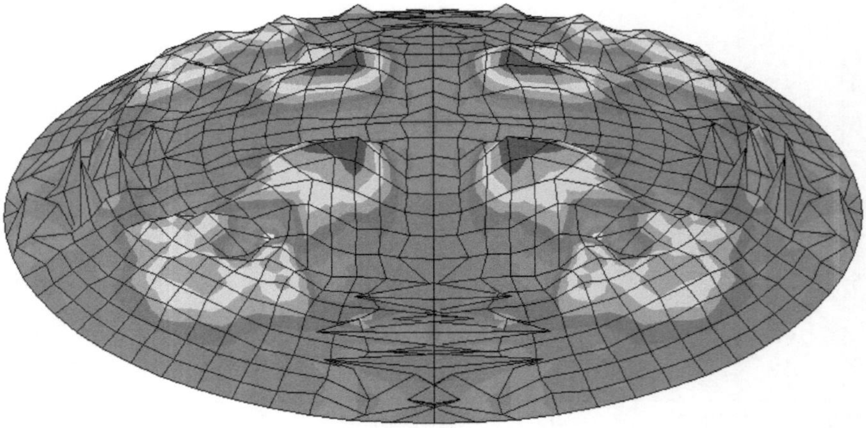

Figure 8.93 Thermal buckling mode shape of CC ($Si_3N_4/SUS304$) FG porous spherical shell under uniform temperature rise.

8.7 CONCLUSION

In this chapter, mechanical and thermoelastic responses of FG porous structures are performed using the improved (I-FSDT) finite element shell model. The porosity volume fraction can be either evenly or unevenly distributed through the thickness direction of the shell according to a modified power-law function. Different sets of numerical illustrations are provided for vibration, time-transient deflections, buckling and thermoelastic responses of FG porous plates and shells under different mechanical/thermal solicitations. Temperature dependent (TD) and temperature independent (TID) FG materials are considered in the analysis. The effects of porosity volume fraction, the type of distributions, power gradient index and different geometrical parameters are investigated and examined. It is found that the percentage of added porosity in the structure, the way of spread of pores imperfections have a significant effect on mechanical responses of structural elements and for a percentage of porosity volume fraction higher than 50%, the flexural rigidity of the FGM porous structure becomes more weak and the strength of FGM porous plate/shell becomes less rigid which affect drastically the mechanical response of such structures.

A Geometrical stiffness matrix components-Kirchhoff model

Herein, the geometrical stiffness matrix components related to the Kirchhoff-Love shell model are given in detail. The expression of these component, presented in Eq.(7.39), are given as:

$$UUM_{IJ} = \int_A \left(N_{I,1}\left(N^{11}N_{J,1} + N^{12}N_{J,2}\right) + N_{I,2}\left(N^{12}N_{J,1} + N^{22}N_{J,2}\right)\right) dA. \tag{A.1}$$

$$\begin{aligned} UUF_{IJ} = & \int_A \left(N_{Ju,1}\left(N_{I,1}M^{11} + N_{I,2}M^{12}\right) + N_{Ju,2}\left(N_{I,1}M^{12} + N_{I,2}M^{22}\right)\right) dA + \\ & \int_A \left(N^T_{Iu,1}\left(N_{J,1}M^{11} + N_{J,2}M^{12}\right) + N^T_{Iu,2}\left(N_{J,1}M^{12} + N_{J,2}M^{22}\right)\right) dA. \end{aligned} \tag{A.2}$$

The terms N_{Iu} are expressed as follows:

$$N_{Iu} = P^I_k td^I_k - P^I_m td^I_m \ , \ td^I_m = \frac{2}{2L_k} t_m \otimes d_m, \tag{A.3}$$

where t_m and d_m are the unit vectors defined by the segment of the element with center m and the normal in this point, respectively. The coupling terms are expressed as follows:

$$\begin{cases} UBF_{IJ} = \int_A \left(N_{Jd,1}\left(N_{I,1}M^{11} + N_{I,2}M^{12}\right) + N_{Jd,2}\left(N_{I,1}M^{12} + N_{I,2}M^{22}\right)\right) dA \\ BUF_{IJ} = UBF_{JI} \end{cases} . \tag{A.4}$$

where N_{Id} are expressed as follows:

$$N_{Id} = N_I I - P^I_k tt^I_k - P^I_m tt^I_m \ , \ tt^I_m = \frac{3}{4} t_m \otimes t_m. \tag{A.5}$$

The rotation terms are given by:

$$\begin{cases} BBF_{IJ} = 0 \ , \ I \neq J \\ BBF_{IJ} = -BBF_{IJ} I \ , \ I = J \end{cases} , \tag{A.6}$$

where BBF_{IJ} is defined by:

$$BBF_{IJ} = \int_A \left(\left(M^{11}a_1 + M^{12}a_2\right) \cdot N_{Id,1} + \left(M^{12}a_1 + M^{22}a_2\right) \cdot N_{Id,2}\right) d_I dA. \tag{A.7}$$

DOI: 10.1201/9781003483786-A

B Geometrical stiffness matrix components-FSDT model

The geometrical stiffness matrix components related to the first-order shear deformation (FSDT) shell model, presented in Eq.(1.82), are given as:

$$UUM_{IJ} = \int_A \left(\overline{N}^I_{,1} \left(N^{11}\overline{N}^J_{,1} + N^{12}\overline{N}^J_{,2} \right) + \overline{N}^I_{,2} \left(N^{12}\overline{N}^J_{,1} + N^{22}\overline{N}^J_{,2} \right) \right) dA. \quad \text{(B.1)}$$

$$UBF_{IJ} = \int_A \left(\overline{N}^I_{,1} \left(M^{11}\overline{N}^J_{,1} + M^{12}\overline{N}^J_{,2} \right) + \overline{N}^I_{,2} \left(M^{12}\overline{N}^J_{,1} + M^{22}\overline{N}^J_{,2} \right) \right) dA. \quad \text{(B.2)}$$

$$BUF_{IJ} = UBF_{JI}. \quad \text{(B.3)}$$

$$BBF_{II} = -\int_A \left(\overline{N}^I_{,1} \left(a_1 M^{11} + a_2 M^{12} \right) + \overline{N}^I_{,2} \left(a_1 M^{12} + a_2 M^{22} \right) \right) d_I \, dA. \quad \text{(B.4)}$$

$$BBF_{IJ} = 0 \ \ if \ \ I \neq J \quad \text{(B.5)}$$

$$UBC = \frac{1}{8} \begin{bmatrix} -\alpha-\beta & -\beta & 0 & -\alpha \\ \beta & \beta-\gamma & -\gamma & 0 \\ 0 & \gamma & \gamma+\delta & \delta \\ \alpha & 0 & -\delta & \alpha-\delta \end{bmatrix} \quad \text{(B.6)}$$

$$\begin{cases} \alpha = \int_A (1-\xi) T_2 dA \ , \ \beta = \int_A (1-\eta) T_1 dA \\ \gamma = \int_A (1+\xi) T_2 dA \ , \ \delta = \int_A (1+\eta) T_1 dA \end{cases} \quad \text{(B.7)}$$

$$BUC_{IJ} = UBC_{JI} \quad \text{(B.8)}$$

$$BBC_{II} = -\frac{1}{4} \begin{Bmatrix} \left(\alpha a_2^A + \beta a_1^B \right) \cdot d_1 \\ \left(\beta a_1^B + \gamma a_2^C \right) \cdot d_2 \\ \left(\gamma a_2^C + \delta a_1^D \right) \cdot d_3 \\ \left(\delta a_1^D + \alpha a_2^A \right) \cdot d_4 \end{Bmatrix} \quad \text{(B.9)}$$

$$BBC_{IJ} = 0 \ \ if \ \ I \neq J \quad \text{(B.10)}$$

DOI: 10.1201/9781003483786-B

References

1. Alijani A, Darvizeh M, Darvizeh A, and Ansari R. On nonlinear thermal buckling analysis of cylindrical shells. *Thin Wall Struct*, 95:170–182, 2015.
2. D.H. Bich abd N.T. Phuong. Buckling analysis of functionally graded annular spherical shells and segments subjected to mechanic loads. *VNU Journal of Mathematics – Physics*, 29(3):14–31, 2013.
3. F.Y. Addou, M. Meradjah, A.A. Bousahla, A. Benachour, F. Bourada, A. Tounsi, and S.R. Mahmoud. Influences of porosity on dynamic response of fg plates resting on winkler/pasternak/kerr foundation using quasi 3d hsdt. *Computers and Concrete*, 24(4):347–367, 2019.
4. H. Adhikari, P. Dash, and B.N. Singh. Buckling analysis of porous fgm sandwich plates under various types nonuniform edge compression based on higher order shear deformation theory. *Composite Structures*, 251(1):112597, 2020.
5. H. Akbari, M. Azadi, and H. Fahham. Free vibration analysis of thick sandwich cylindrical panels with saturated fg-porous core. *Mechanics Based Design of Structures and Machines*, doi:10.1080/15397734.2020.1748051, 2020.
6. M. Akbari, Y. Kiani, and M.R. Eslami. Thermal buckling of temperature-dependent fgm conical shells with arbitrary edge supports. *Acta Mechanica*, 226:897–915, 2014.
7. S. Alimirzaei, M. Mohammadimehr, and A. Tounsi. Nonlinear analysis of viscoelastic micro-composite beam with geometrical imperfection using fem: Msgt electro-magneto-elastic bending, buckling and vibration solutions. *Structural Engineering and Mechanics*, 71(5):485–502, 2019.
8. M. Bouazza and A. Tounsi, E.A. Adda-Bedia and A. Megueni. Thermoelastic stability analysis of functionally graded plates: An analytical approach. *Computers and Material Sciences*, 49:865–870, 2010.
9. U. Andelfinger and E. Ramm. Eas-elements for two-dimensional, three-dimensional plate and shell structures and their equivalence to hr-elements. *International Journal for Numerical Methods in Engineering*, 36:1311–1337, 1993.
10. R. Ansari, E. Hasrati, A.H. Shakouri, M. Bazdid-Vahdati, and H. Rouhi. Nonlinear large deformation analysis of shells using the variational differential quadrature method based on the six-parameter shell theory. *International Journal of Non-Linear Mechanics*, 106:130–143, 2018.
11. R. Ansari and J. Torabi. Numerical study on the buckling and vibration of functionally graded carbon nanotube-reinforced composite conical shells under axial loading. *Composites Part B: Engineering*, 95:196–208, 2016.
12. R. Ansari, J. Torabi, M. Faghih Shojaei, and E. Hasrati. Buckling analysis of axially-loaded functionally graded carbon nanotube-reinforced composite conical panels using a novel numerical variational method. *Composite Structures*, 157:398–411, 2016.
13. H. A. Atmane, A. Tounsi, and F. Bernard. Effect of thickness stretching and porosity on mechanical response of a functionally graded beams resting on elastic foundations. *International Journal of Mechanics and Materials in Design*, 13(1):71–84, 2017.
14. Mirzavand B, Eslami MR, and Reddy JN. Dynamic thermal postbuckling analysis of shear deformable piezoelectric-fgm cylindrical shells. *J therm stresses*, 36:189–206, 2013.
15. Mirzavand B, Eslami MR, and Shakeri M. Dynamic thermal postbuckling analysis of piezoelectric functionally graded cylindrical shells. *J therm stresses*, 33:646–660, 2010.
16. E. Bagherizadeh, Y. Kiani, and M.R. Eslami. Mechanical buckling of functionally graded material cylindrical shells surrounded by pasternak elastic foundation. *Composite Structures*, 93:3063–3071, 2011.
17. M. Balubaid, A. Tounsi, B. Dakhel, and S.R. Mahmoud. Free vibration investigation of fg nanoscale plate using nonlocal two variables integral refined plate theory. *Computers and Concrete*, 24(6):579–586, 2019.
18. Shariat BAS and Eslami MR. Thermal buckling of imperfect functionally graded plates. *Int J Solids Struct*, 43:4082–4096, 2006.
19. Y. Basar, Y. Ding, and R. Schultz. Refined shear-deformation models for composite laminates with finite rotations. *International Journal of Solids and Structures*, 30:2611–2638, 1993.

20. K. J. Bathe and E. Dvorkin. A four-node plate bending element based on mindlin/reissner plate theory and a mixed interpolation. *International Journal for Numerical Methods in Engineering*, 21:367–383, 1985.
21. K.J Bathe. *Finite Element Procedures in Engineering Analysis*. Prentice-Hail, Engiewood Cliffs, N J, 1st edition, 1982.
22. R.H. Bathe and F. Brezzi. On the convergence of a four-node plate bending element based on mindlin-reissner plate theory and a mixed interpolation. *Proceedings Conference on Mathematics of Finite Elements and Applications V (Academic Press, New York)*, 99:491–503, 1985.
23. J.L. Batoz and G. Dhatt. *Modelisation des structures par elements finis*. Hermès-Lavoisier, 1st, 2nd, 3rd edition, 1990.
24. W.A. Bedia, M.S.A. Houari, A. Bessaim, A.A. Bousahla, A. Tounsi, T. Saeed, and M.S. Alhodaly. A new hyperbolic two-unknown beam model for bending and buckling analysis of a nonlocal strain gradient nanobeams. *Journal of Nano Research*, 57:175–191, 2019.
25. T. Belytschko and T. J. R. Hughes. *Computational Methods for Transient Analysis*. Elsevier Science Publishers B.V., North-Holland, 1983.
26. A. Benachour, H.D. Tahar, H.A. Atmane, A. Tounsi, and M. Sid Ahmed. A four variable refined plate theory for free vibrations of functionally graded plates with arbitrary gradient. *Composites: Part B*, 42:1386–1394, 2011.
27. H. Berghouti, E.A.A. Bedia, A. Benkhedda, and A. Tounsi. Vibration analysis of nonlocal porous nanobeams made of functionally graded material. *Advances in Nano Research*, 7(5):351–364, 2019.
28. V. Bhagat, P. Jeyaraj, and S.M. Murigendrappa. Buckling and free vibration characteristics of a uniformly heated isotropic cylindrical panel. *Procedia Engineering*, 144:474–481, 2016.
29. D.H. Bich. Nonlinear buckling analysis of fgm shallow spherical shells. *Vietnam Journal of Mechanics*, 31:17–30, 2009.
30. W.B. Bickford. A consistent higher order beam theory. *Developments in theoretical and Applied Mechanics*, 11:137–150, 1982.
31. E.M. Blandzi, K. Magnucki, and W. Stawecki. Bending and buckling of a circular plate with symmetrically varying mechanical properties. *Applied Mathematical Modelling*, 89:1198–1205, 2021.
32. M. Bodaghi and A.R. Saidi. Levy-type solution for buckling analysis of thick functionally graded rectangular plates based on the higher-order shear deformation plate theory. *Applied Mathematical Modelling*, 34:3659–3673, 2010.
33. Z. Boukhlif, M. Bouremana, F. Bourada, A.A. Bousahla, M. Bourada, A. Tounsi, and M.A.A. Osta. A simple quasi-3d hsdt for the dynamics analysis of fg thick plate on elastic foundation. *Steel and Composite Structures*, 31(5):503–516, 2019.
34. L. Boulefrakh, H. Hebali, A. Chikh, A.A. Bousahla, A. Tounsi, and S.R. Mahmoud. The effect of parameters of visco-pasternak foundation on the bending and vibration properties of a thick fg plate. *Geomechanics and Engineering*, 18(2):161–178, 2019.
35. F. Bourada, A.A. Bousahla, M. Bourada, A. Azzaz, A. Zinata, and A. Tounsi. Dynamic investigation of porous functionally graded beam using a sinusoidal shear deformation theory. *Wind and Structures*, 28(1):19–30, 2019.
36. M. Bourada, A. Tounsi, M.S.A. Houari, and E.A.E. Bedia. A new four-variable refined plate theory for thermal buckling analysis of functionally graded sandwich plates. *Journal of Sandwich Structures and Materials*, 14(1):5–33, 2012.
37. B. Brank, D. Peric, and FB. Damjanic. On implementation of a nonlinear four node shell finite element for thin multilayered elastic shells. *Computational Mechanics*, 16:341–359, 1995.
38. B. Brendel and E. Ramm. Linear and nonlinear stability analysis of cylindrical shells. *Computational Mechanics*, 12:549–558, 1980.
39. N. Buchter and E. Ramm. Shell theory versus degeneration-a comparison in large rotation finite element analysis. *International Journal for Numerical Methods in Engineering*, 34:39–59, 1992.
40. N. Buechter and E. Ramm. Shell theory versus degeneration - a comparison in large rotation finite element analysis. *International Journal for Numerical Methods in Engineering*, 50:39–59, 1992.
41. T.Q. Bui, T.V. Do, T.T. Yu, D.T. Pham, and C.T. Nguyen. Role of material combination and new results of mechanical behavior for fg sandwich plates in thermal environment. *Journal of Computers Sciences*, 21:164–181, 2017.

42. E. Carrera, S. Brischetto, M. Cinefra, and M. Soave. Refined and advanced models for multilayered plates and shells embedding functionally graded material layers. *Mechanics of Advanced Materials and Structures*, 17(8):603–621, 2010.
43. E. Carrera, S. Brischetto, and P. Nali. *Plates and Shells for Smart Structures: classical and Advanced Theories for Modeling and Analysis*. John Wiley & Sons, 1st edition, 2011.
44. L.A. Chaabane, F. Bourada, M. Sekkal, S. Zerouati, F.Z. Zaoui, A. Tounsi, A. Derras, A.A. Bousahla, and A. Tounsi. Analytical study of bending and free vibration responses of functionally graded beams resting on elastic foundation. *Structural Engineering and Mechanics*, 71(2):185–196, 2019.
45. Y.S. Chang. Accuracy evaluation of Newmark method referring to theoretical solutions. *Earthquake Engineering*, 12:1–16, 2008.
46. D. Chen, J. Yang, and S. Kitipornchai. Elastic buckling and static bending of shear deformable functionally graded porous beam. *Composite Structures*, 133:54–61, 2015.
47. D. Chen, J. Yang, and S. Kitipornchai. Free and forced vibrations of shear deformable functionally graded porous beams. *International Journal of Mechanical Sciences*, 108-109:14–22, 2016.
48. W.Q. Chen, C.F. Lu, and Z.G. Bian. A mixed method for bending and free vibration of beams resting on a pasternak elastic foundation. *Applied Mathematical Modelling*, 28:877–890, 2004.
49. JR. Cho and HJ. Park. High strength fgm cutting tools: finite element analysis on thermoelastic characteristics. *Journal of Material Processing Technology*, 130:351–356, 2002.
50. M. Cinefra, E. Carrera, L. Della Croce, and C. Chinosi. Refined shell elements for the analysis of functionally graded structures. *Composite Structures*, 94:415–422, 2012.
51. Hirwani CK and Panda SK. Numerical nonlinear frequency analysis of pre-damaged curved layered composite structure using higher-order finite element method. *Int J Non Linear Mech*, 102:14–24, 2018.
52. Hirwani CK, Panda SK, Mahapatra SS, Mandal SK, and De AK. Dynamic behaviour of delaminated composite plate under blast loading. *Proceedings of the ASME 2017 Gas Turbine India Conference*, 2017.
53. Hirwani CK, Panda SK, and Mahapatra TR. Thermomechanical deflection and stress responses of delaminated shallow shell structure using higher-order theories. *Compos Struct*, 184:135–145, 2018.
54. F. Dammak, S. Abid, A. Gakwaya, and G. Dhatt. A formulation of the non linear discrete kirchhoff quadrilateral shell element with finite rotations and enhanced strains. *Revue Européenne des Éléments*, 14:7–31, 2005.
55. D. Debowski, K. Magnucki, and M. Malinowski. Dynamic stability of a metal foam rectangular plate. *Steel and Composite Structures*, 10(2):151–168, 2010.
56. P.A Demirhan and V. Taskin. Bending and free vibration analysis of levy-type porous functionally graded plate using state space approach. *Composites Part B*, 160:661–676, 2019.
57. G. Dhatt and G. Touzot. *Une presentation de la methode des elements finis*. Maloine S.A. Paris et Les Presses de l'Université de Laval Québec, 1st edition, 1981.
58. J. Ding, L. Chu, L. Xin, and JD. Dui. Nonlinear vibration analysis of functionally graded beams considering the influences of the rotary inertia of the cross section and neutral surface position. *Mechanics Based Design of Structures and Machines*, 46(2):225–237, 2018.
59. D. Draiche, A.A. Bousahla, A. Tounsi, A.S. Alwabli, A. Tounsi, and S.R. Mahmoud. Static analysis of laminated reinforced composite plates using a simple first-order shear deformation theory. *Computers and Concrete*, 24(4):369–378, 2019.
60. A. Draoui, M. Zidour, A. Tounsi, and B. Adim. Static and dynamic behavior of nanotubes-reinforced sandwich plates using (FSDT). *Journal of Nano Research*, 57:117–135, 2019.
61. Y. Du, R. Huo, F. Pang, S. Li, Y. Huang, and H. Zhang. Free vibration of spherical cap subjected to various boundary conditions. *Advances in Mechanical Engineering*, 11(9):1–12, 2019.
62. N.D. Duc. Nonlinear static and dynamic stability of functionally graded plates and shells. *Vietnam National University Press, Hanoi*, 2014.
63. N.D. Duc and P.H. Cong. Nonlinear postbuckling of symmetric s-fgm plates resting on elastic foundations using higher order shear deformation plate theory in thermal environments. *Composite Structures*, 100:566–574, 2013.
64. N.D. Duc and P.H. Cong. Nonlinear thermo-mechanical dynamic analysis and vibration of higher order shear deformable piezoelectric functionally graded material sandwich plates resting on elastic foundations. *International Journal of Sandwich and Structure Materials*, 20:191–218, 2016.

65. N.D. Duc, P.T. Thang, and N.T. Daoand H.V. Tac. Nonlinear buckling of higher deformable s-fgm thick circular cylindrical shells with metal-ceramic-metal layers surrounded on elastic foundations in thermal environment. *Composite Structures*, 121:134–141, 2015.
66. N.D. Duc and H.V. Tung. Mechanical and thermal postbuckling of higher order shear deformable functionally graded plates on elastic foundations. *Composite Structures*, 93(11):2874–2881, 2011.
67. D.V. Dung, L.K. Hoa, B.T. Thuyet, and N.T. Nga. Buckling analysis of functionally graded material (fgm) sandwich truncated conical shells reinforced by fgm stiffeners filled inside by elastic foundations. *Applied Mathematics and Mechanics*, 37(7):879–902, 2016.
68. D.V. Dung and N.T. Nga. Buckling and postbuckling nonlinear analysis of imperfect fgm plates reinforced by fgm stiffeners with temperature-dependent properties based on tsdt. *Acta Mechanica*, 227:2377–2401, 2016.
69. D.V. Dung, N.T. Nga, and L.K. Hoa. Mechanical and thermal postbuckling of fgm thick circular cylindrical shells reinforced by fgm stiffener system using higher-order shear deformation theory. *Applied Mathematics and Mechanics*, 38(1):73–98, 2017.
70. D.V. Dung and H.T. Thiem. Mechanical and thermal postbuckling of fgm thick circular cylindrical shells reinforced by fgm stiffener system using higher-order shear deformation theory. *Applied Mathematics and Mechanics*, 38(1):73–98, 2017.
71. F. Ebrahimi and A. Rastgo. An analytical study on the free vibration of smart circular thin fgm plate based on classical plate theory. *Thin-Walled Structures*, 46:1402– 1408, 2008.
72. A.M.K. Esawi and M.M. Farag. Carbon nanotube reinforced composites: potential and current challenges. *Materials and Design*, 28:2394–2401, 2007.
73. M.R. Eslami, H.R. Ghorbani, and M. Shakeri. Thermoelastic buckling of thin spherical shells. *Journal of Thermal Stresses*, 24:1177–1198, 2001.
74. M. Estili and Y. Sakka. Recent advances in understanding the reinforcing ability and mechanism of carbon nanotubes in ceramic matrix composites. *Science and Technology of Advanced Materials*, 15:1–25, 2014.
75. A.J.M. Ferreira, G.E. Fasshauer, R.C. Batra, and J.D. Rodrigues. Static deformations and vibration analysis of composite and sandwich plates using a layerwise theory and rbf-ps discretizations with optimal shape parameter. *Composite Structures*, 86:328–343, 2008.
76. L. A. Duarte Filho and M. A. Armando. Geometrically nonlinear static and dynamic analysis of shells and plates using the eight-node hexahedral element with one-point quadrature. *Finite Elements in Analysis and Design*, 40:1297–1315, 2004.
77. L.A. Duarte Filho and M.A. Armando. Geometrically nonlinear static and dynamic analysis of shells and plates using the eight-node hexahedral element with one-point quadrature. *Finite Element in Analysis and Design*, 40:1297–1315, 2004.
78. D.D. Fox and J.C. Simo. A drill rotation formulation for geometrically exact shells. *Computer Methods in Applied Mechanics and Engineering*, 98:329–343, 1992.
79. A. Frikha and F. Dammak. Geometrically non-linear static analysis of functionally graded material shells with a discrete double directors shell element. *Computer Methods in Applied Mechanics and Engineering*, 315:1–24, 2017.
80. A. Frikha and F. Dammak. Geometrically non-linear static analysis of functionally graded material shells with a discrete double directors shell element. *Computer Methods in Applied Mechanics and Engineering*, 315:1–24, 2017.
81. A. Frikha, M. Wali, A. Hajlaoui, and F. Dammak. Dynamic response of functionally graded material shells with a discrete double directors shell element. *Composite Structures*, 154:385–395, 2016.
82. A. Frikha, S. Zghal, and F. Dammak. Dynamic analysis of functionally graded carbon nanotubes-reinforced plate and shell structures using a double directors finite shell element. *Aerospace Science and Technology*, 78:438–451, 2018.
83. A. Frikha, S. Zghal, and F. Dammak. Finite rotation three and four nodes shell elements for functionally graded carbon nanotubes-reinforced thin composite shells analysis. *Computer Methods in Applied Mechanics and Engineering*, 329:289–311, 2018.
84. E. Garcia-Macias, R. Castro-Triguero, E. I. S. Flores, M.I. Friswell, and R. Gallego. Static and free vibration analysis of functionally graded carbon nanotube reinforced skew plates. *Composite Structures*, 140:473–490, 2016.

85. E. Garcia-Macias, L. Rodriguez-Tembleque, R. Castro-Triguero, and A. Saez. Buckling analysis of functionally graded carbon nanotube-reinforced curved panels under axial compression and shear. *Composites Part B: Engineering*, 108:243–256, 2017.
86. E. Garcia-Macias, L. Rodriguez-Tembleque, R. Castro-Triguero, and A. Saez. Eshelby-mori-tanaka approach for post-buckling analysis of axially compressed functionally graded cnt/polymer composite cylindrical panels. *Composites Part B: Engineering*, 128:208–224, 2017.
87. O. Gohardani, M.C. Elola, and C. Elizetexea. Potential and prospective implementation of carbon nanotubes on next generation aircraft and space vehicles: A review of current and expected applications in aerospace sciences. *Progress in Aerospace Sciences*, 3:42–68, 2014.
88. M. Gorji. On large deflection of symmetric composite plates under static loading. *Journal of Mechanical Engineering Science*, 200:13–19, 1986.
89. R.M.S. Gowda and K.A.V.Pandalai. Thermal buckling of orthotropic plates. *Studies in Structural Mechanics*, K. A. V. Pandalai (ed.):9–44, 1970.
90. A. Gupta and M. Talha. Influence of initial geometric imperfections and porosity on the stability of functionally graded material plates. *Mechanics Based Design of Structures and Machines*, 46(6):693–711, 2018.
91. Ounis H, Tati A, and Benchabane A. Thermal buckling behavior of laminated composite plates: a finite-element study. *Front Mech Eng*, 9(1):41–49, 2014.
92. A. Hajlaoui, A. Jarraya, K. El Bikri, and F. Dammak. Buckling analysis of functionally graded materials structures with enhanced solid-shell elements and transverse shear correction. *Composite Structures*, 132:87–97, 2015.
93. A. Hajlaoui, E. Triki, A. Frikha, M. Wali, and F. Dammak. Nonlinear dynamics analysis of fgm shell structures with a higher order shear strain enhanced solid-shell element. *Latin American Journal of Solids and Structures*, 14:72–91, 2017.
94. S.C. Han, G.R. Lomboy, and K.D. Kim. Mechanical vibration and buckling analysis of fgm plates and shells using a four-node quasi-conforming shell element. *International Journal of Structural Stability and Dynamics*, 8:203–229, 2008.
95. Y. Han and J. Elliot. Molecular dynamics simulations of the elastic properties of polymer/carbon nanotube composites. *Computational Materials Science*, 39:315–323, 2007.
96. S.H.H. Hashemi, H.R.D. Taher, H. Akhavan, and M. Omidi. Free vibration of functionally graded rectangular plates using first-order shear deformation plate theory. *Applied Mathematical Modelling*, 34:1276–1291, 2010.
97. M. Haskul. Elastic state of functionally graded curved beam on the plane stress state subject to thermal load. *Mechanics Based Design of Structures and Machines*, doi:10.1080/15397734.2019.1660890, 2019.
98. F.R. Heyliger and J.N. Reddy. A higher order beam finite element for bending and vibration problems. *Journal of Sound and Vibration*, 126:309–326, 1988.
99. Ovesy HR, Ghannadpour SAM, and Nassirnia M. Post-buckling analysis of rectangular plates comprising functionally graded strips in thermal environments. *Comput Struct*, 147:209–215, 2015.
100. M.M. Hrabok and T.M. Hrudey. A review and catalogue of plate bending finite elements. *Computers & Structures*, 19(3):479–495, 1984.
101. Shen HS. Thermal postbuckling behavior of shear deformable fgm plates with temperature-dependent properties. *Int J Mech Sci*, 49:466–478, 2007.
102. Shen HS. Thermal postbuckling of shear deformable fgm cylindrical shells with temperature-dependent properties. *Mech Adv Mater Struc*, 14:439–452, 2007.
103. Tung HV and Duc ND. Nonlinear analysis of stability for functionally graded plates under mechanical and thermal loads. *Compos Struct*, 92:1184–1191, 2010.
104. S. Iijima. Helical microtubules of graphitic carbon. *Nature*, 354:56–58, 1991.
105. E. Jaberzadeh, M. Azhari, and B. Boroomand. Thermal buckling of functionally graded skew and trapezoidal plates with different boundary conditions using the element-free galerkin method. *European Journal of Mechanics A/Solids*, 42:18–26, 2013.
106. S. Jafari-Mehrabadi, B. Sobhaniaragh, V. Pourdonya, and A. Taherpour. Mechanical buckling of nanocomposite rectangular plate reinforced by aligned and straight single-walled carbon nanotubes. *Composites Part B: Engineering*, 43:2031–2040, 2012.

107. J.E. Jam and Y. Kiani. Buckling of pressurized functionally graded carbon nanotube reinforced conical shells. *Composite Structures*, 125:586–595, 2015.
108. R. Javaheri and M.R. Eslami. Buckling of functionally graded plates under in-plane compressive loading. *ZAMM*, 82:277–283, 2002.
109. R. Javaheri and M.R. Eslami. Thermal buckling of functionally graded plates. *AIAA Journal*, 40:162–169, 2002.
110. R. Javaheri and M.R. Eslami. Thermal buckling of functionally graded plates based on higher order theory. *Journal of Thermal Stresses*, 25:603–625, 2002.
111. P. Jeyaraj. Buckling and free vibration behavior of an isotropic plate under nonuniform thermal load. *International Journal of Structural Stability and Dynamics*, 13(3):1250071, 2013.
112. Cho JR and Oden JT. Functionally graded material: a parametric study on thermal-stress characteristics using the crank-nicolson-galerkin scheme. *Comput Methods Appl Mech Eng*, 188:17–38, 2000.
113. M. Kaddari, A. Kaci, A.A. Bousahla, A. Tounsi, F. Bourada, A. Tounsi, E.A.A. Bedia, and M.A. Al-Osta. A study on the structural behaviour of functionally graded porous plates on elastic foundation using a new quasi-3d model: Bending and free vibration analysis. *Computers and Concrete*, 25(1):37–57, 2020.
114. R. Kandasamy, R. Dimitri, and F. Tornabene. Numerical study on the free vibration and thermal buckling behavior of moderately thick functionally graded structures in thermal environments. *Composite Structures*, 157:207–221, 2016.
115. T. Kant, J.H. Varaiya, and C.P. Arora. Finite element transient analysis of composite sandwich plates based on refined theory and implicit time integration scheme. *Computers and Strucutres*, 36(3):401–420, 1990.
116. A. Karamanli and M. Aydogdu. Structural dynamics and stability analysis of 2d-fg microbeams with two-directional porosity distribution and variable material length scale parameter. *Mechanics Based Design of Structures and Machines*, 48(2):164–191, 2020.
117. B. Karami, M. Janghorban, and A. Tounsi. Galerkin's approach for buckling analysis of functionally graded anisotropic nanoplates/different boundary conditions. *Engineering With Computers*, 35(4):1297–1316, 2019.
118. B. Karami, D. Shahsavari, M. Janghorban, and A. Tounsi. Resonance behavior of functionally graded polymer composite nanoplates reinforced with graphene nanoplatelets. *International Journal of Mechanical Sciences*, 156:94–105, 2019.
119. E.P. Kasper and R.L. Taylor. A mixed-enhanced strain method, part i : Geometrically linear problems. *Computers and Structures*, 75(3):237–250, 2000.
120. R. Khosravi, S.A. Hosseini, and A. Tounsi. Torsional dynamic response of viscoelastic swcnt subjected to linear and harmonic torques with general boundary conditions via eringen's nonlocal differential model. *The European Physical Journal Plus*, 135:183, 2020.
121. Y. Kiani. Shear buckling of fg-cnt reinforced composite plates using chebyshev-ritz method. *Composites Part B: Engineering*, 105:176–187, 2016.
122. Y. Kiani. Buckling of fg-cnt-reinforced composite plates subjected to parabolic loading. *Acta Mechanica*, 228:1303–1319, 2017.
123. Y. Kiani, E. Bagherizadeh, and M.R. Eslami. Thermal buckling of clamped thin rectangular fgm plates resting on pasternak elastic foundation (three approximate analytical solutions). *ZAMM-J Appl Math Mech/Z Angew Math Mech*, 91:581–593, 2011.
124. Y. Kiani and M.R. Eslami. Nonlinear thermo-inertial stability of thin circular fgm plates. *Journal of Franklin Institution*, 351:1057–1073, 2014.
125. K-D. Kim, G.R. Lomboy, and S-C. Han. Geometrically nonlinear analysis of functionally graded material (fgm) plates and shells using a four-node quasi-conforming shell element. *Journal of Composite Materials*, 42:485–511, 2008.
126. Y.W. Kim. Temperature dependent vibration analysis of functionally graded rectangular plates. *Journal of Sound and Vibration*, 284:531–549, 2005.
127. S. Kitipornchai, D. Chen, and J. Yang. Free vibration and elastic buckling of functionally graded porous beams reinforced by graphene platelets. *Materials and Design*, 116:656–665, 2017.
128. M. Koizumi. Functionally gradient materials the concept of fgm. *Ceramics Transactions*, 34:3–10, 1993.
129. M. Koizumi. Fgm activities in japan. *Composites Part B: Engineering*, 28:1–4, 1997.

130. J. Korelc and P. Wriggers. Improved enhanced strain four-node-element with taylor expansion of the shape functions. *International Journal for Numerical Methods in Engineering*, 40:407–421, 1997.
131. Na KS and Kim JH. Three-dimensional thermomechanical buckling of functionally graded materials. *AIAA Journal*, 43:1605–1612, 2005.
132. D. Hui K.T Lau. The revolutionary creation of new advanced materials-carbon nanotube composites. *Composites: Part B*, 33:263–277, 2002.
133. Shiau LC, Kuo SY, and Chen CY. Thermal buckling behavior of composite laminated plates. *Compos Struct*, 92:508–514, 2010.
134. Y.Y. Lee, X. Zhao, and J.N. Reddy. Postbuckling analysis of functionally graded plates subject to compressive and thermal loads. *Computer Methods in Applied Mechanics and Engineering*, 199:1645–1653, 2010.
135. Z.X. Lei, K. M. Liew, and J.L. Yu. Buckling analysis of functionally graded carbon nanotube-reinforced composite plates using the element-free kp-ritz method. *Composite Structures*, 98:160–168, 2013.
136. Z.X. Lei, K. M. Liew, and J.L. Yu. Large deflection analysis of functionally graded carbon nanotube-reinforced composite plates by the element-free kp-ritz method. *Computer Methods in Applied Mechanics and Engineering*, 256:189–199, 2013.
137. Z.X. Lei, L.W. Zhang, and K.M. Liew. Free vibration analysis of laminated fg-cnt reinforced composite rectangular plates using the kp-ritz method. *Composite structures*, 127:245–259, 2015.
138. Z.X. Lei, L.W. Zhang, and K.M. Liew. Buckling analysis of cnt reinforced functionally graded laminated composite plates. *Composite Structures*, 152:62–73, 2016.
139. U. Leushake, T. Krell, and U. Schulz. Graded thermal barrier coating systems for gas turbine applications. *Materialwissenschaft und Werkstofftechnik*, 28:391–394, 1997.
140. M. Levinson. An accurate simple theory of statics and dynamics of elastic plates. *Mechanics Research Communications*, 7:340–350, 1980.
141. JF. Li, K. Takagi, M. Ono, W. Pan, R. Watanabe, A. Almajid, and M. Taya. Fabrication and evaluation of porous piezoelectric ceramics and porosity graded piezoelectric actuators. *Journal of the American Ceramics Society*, 86:1094–1098, 2003.
142. Q. Li, V.P. Iu, and K.P. Kou. Three-dimensional vibration analysis of functionally graded material plates in thermal environment. *Journal of Sound and Vibration*, 324:733–750, 2009.
143. X.F. Li. A unified approach for analyzing static and dynamic behaviors of functionally graded Timoshenko and Euler–Bernoulli beams. *Journal of Sound and Vibration*, 318:1210–1229, 2008.
144. K.M. Liew, Z.X. Lei, J.L. Yu, and L.W. Zhang. Postbuckling of carbon nanotube-reinforced functionally graded cylindrical panels under axial compression using a meshless approach. *Computer Methods in Applied Mechanics and Engineering*, 268:1–17, 2014.
145. K.M. Liew, Z.X. Lei, and L.W. Zhang. Mechanical analysis of functionally graded carbon nanotube reinforced composites: a review. *Composite Structures*, 120:90–97, 2015.
146. K.M. Liew, X. Zhao, and A.J.M. Ferreira. A review of meshless methods for laminated and functionally graded plates and shells. *Composite Structures*, 93:2031–2041, 2011.
147. K.M. Liew, X. Zhao, and Y.Y. Lee. Postbuckling responses of functionally graded cylindrical shells under axial compression and thermal loads. *Composites: Part B*, 43:1621–1630, 2012.
148. L.S. Ma and T.J. Wang. Nonlinear bending and post-buckling of a functionally graded circular plate under mechanical and thermal loadings. *International Journal of Solids and Structures*, 40:3311–3330, 2003.
149. R.H. MacNeal. Derivation of element stiffness matrices by assumed strain distribution. *Nuclear Engineering Design*, 70:3–12, 1982.
150. K. Magnucki and E.M. Blandzi. Generalization of a sandwich structure model: Analytical studies of bending and buckling problems of rectangular plates. *Composite Structures*, 255:112944, 2021.
151. K. Magnucki, W. Stawecki, and J. Lewinski. Axisymmetric bending of a circular plate with symmetrically varying mechanical properties under a concentrated force. *Steel and Composite Structures*, 34(6):795–802, 2020.
152. K. Magnucki, D. Witkowski, and E.M. Blandzi. Buckling and free vibrations of rectangular plates with symmetrically varying mechanical properties–analytical and fem studies. *Composite Structures*, 220:355–361, 2019.
153. A. Mahmoudi, S. Benyoucef, A. Tounsi, A. Benachour, E.A.A. Bedia, and S.R. Mahmoud. A refined quasi-3d shear deformation theory for thermo-mechanical behavior of functionally graded

sandwich plates on elastic foundations. *Journal of Sandwich Structures & Materials*, 21(6):1906–1926, 2019.
154. P. Malekzadeh and A. Alibeygi-Beni. Free vibration of functionally graded arbitrary straight-sided quadrilateral plates in thermal environment. *Composite Structures*, 92:2758–2767, 2010.
155. P. Malekzadeh and A.A. Beni. Free vibration of functionally graded arbitrary straight-sided quadrilateral plates in thermal environment. *Composite Structures*, 92:2758–2767, 2010.
156. P. Malekzadeh and Y. Heydarpour. Free vibration analysis of rotating functionally graded cylindrical shells in thermal environment. *Composite Structures*, 94:2971–2981, 2012.
157. P. Malekzadeh, S.A. Shahpari, and H.R. Ziaee. Three-dimensional free vibration of thick functionally graded annular plates in thermal environment. *Journal of Sound and Vibration*, 329:425–442, 2010.
158. H. Matsunaga. Thermal buckling of cross-ply laminated composite and sandwich plates according to a global higher-order deformation theory. *Composite Structures*, 68:439–454, 2005.
159. H. Matsunaga. Free vibration and stability of functionally graded plates according to a 2-d higher-order deformation theory. *Composite Structures*, 82:499–512, 2008.
160. M. Medani, A. Benahmed, M. Zidour, H. Heireche, A. Tounsi, A.A. Bousahla, A. Tounsi, and S.R. Mahmoud. Static and dynamic behavior of (fg-cnt) reinforced porous sandwich plate using energy principle. *Steel and Composite Structures*, 24(4):347–367, 2019.
161. K. Mehar and S.K. Panda. Geometrical nonlinear free vibration analysis of fg-cnt reinforced composite flat panel under uniform thermal field. *Composite Structures*, 132:336–346, 2016.
162. K. Mehar and S.K. Panda. Numerical investigation of nonlinear thermomechanical deflection of functionally graded cnt reinforced doubly curved composite shell panel under different mechanical loads. *Composite Structures*, 161:287–298, 2017.
163. K. Mehar, S.K. Panda, and T.R. Mahapatra. Thermoelastic nonlinear frequency analysis of cnt reinforced functionally graded sandwich structure. *European Journal of Mechanics - A/Solids*, 85:384–396, 2017.
164. K. Mehar, S.K. Panda, and T.R. Mahapatra. Large deformation bending responses of nanotube-reinforced polymer composite panel structure: Numerical and experimental analyses. *Proceedings of the Institution of Mechanical Engineers, Part G: Journal of Aerospace Engineering*, 233(5):1695–1704, 2018.
165. M. Mehri, H. Asadi, and Q. Wang. Buckling and vibration analysis of a pressurized cnt reinforced functionally graded truncated conical shell under an axial compression using hdq method. *Computer Methods in Applied Mechanics and Engineering*, 303:75–100, 2016.
166. R. Meksi, S. Benyoucef, A. Mahmoudi, A. Tounsi, E.A.A. Bedia, and S.R. Mahmoud. An analytical solution for bending, buckling and vibration responses of fgm sandwich plates. *Journal of Sandwich Structures & Materials*, 21(2):727–757, 2019.
167. S. Merdaci and H. Belghoul. High-order shear theory for static analysis of functionally graded plates with porosities. *Comptes Rendus Mecanique*, 347:207–217, 2019.
168. E. Milton and P. D. Spanos. Effective elastic properties of nanotube reinforced composites with slightly weakened interfaces. *Journal of Mechanics of Materials and Structures*, 4:887–900, 2009.
169. R.D. Mindlin. Influence of rotary inertia and shear on flexural motions of isotropic elastic plates. *Journal of Applied Mechanics*, 18:31–38, 1951.
170. B. Mirzavand and M.R. Eslami. Thermoelastic stability analysis of imperfect functionally graded cylindrical shells. *Journal of Mechanics and Materials Structures*, 3:1561–1572, 2008.
171. M. Mohammadimehr and R. Rostami. Bending and vibration analyses of a rotating sandwich cylindrical shell considering nanocomposite core and piezoelectric layers subjected to thermal and magnetic fields. *Applied Mathematics and Mechanics*, 39(2):219–240, 2018.
172. A. Mojahedin, M. Jabbari, A.R. Khorshidvand, and M.R. Eslami. Buckling analysis of functionally graded circular plates made of saturated porous materials based on higher order shear deformation theory. *Thin-Walled Structures*, 99:83–90, 2016.
173. M.V.V. Murthy. An improved transverse shear deformation theory for laminated anisotropic plates. *NASA Technical Paper 1903*, pages 1–37, 1981.
174. S.S. Murthy. On the use of the trigonometric ritz method for general vibration analysis of rectangular kirchhoff plates. *Computers Methods in Applied Mechanics and Engineering*, 54:197–222, 1986.
175. K.S. Na and J.H. Kim. Thermal postbuckling investigations of functionally graded plates using 3-d finite element method. *Finite Elements in Analysis and Design*, 42:749–756, 2006.

176. M.M. Najafizadeh and M.R. Eslami. Buckling analysis of circular plates of functionally graded materials under uniform radial compression. *International Journal of Mechanical Sciences*, 44:2479–2493, 2002.
177. M.M. Najafizadeh and B. Hedayati. Refined theory for thermaelastic stability of functionally graded circular plates. *Journal of thermal stresses*, 27:857–880, 2004.
178. M.M. Najafizadeh and H.R. Heydari. An exact solution for buckling of functionally graded circular plates based on higher order shear deformation plate theory under uniform radial compression. *International Journal of Mechanical Sciences*, 50:603–612, 2008.
179. V.H. Nam, N.T. Phuong, and D.H. Bich. Buckling analysis of parallel eccentrically stiffened functionally graded annular spherical segments subjected to mechanic loads. *Mechanics of Advanced Materials and Structures*, 27(7):569–578, 2020.
180. Duc ND and Tung HV. Mechanical and thermal postbuckling of shear-deformable fgm plates with temperature-dependent properties. *Mech Compos Mater*, 46(5):461–476, 2010.
181. Duc ND, Tuan ND, Quan TQ, Quyen NV, and Anh TV. Nonlinear mechanical, thermal and thermo-mechanical postbuckling of imperfect eccentrically stiffened thin fgm cylindrical panels on elastic foundations. *Thin Wall Struct*, 96:155–168, 2015.
182. Duc ND and Thang PT. Non linear buckling of imperfect eccentrically stiffened metal-ceramic-metal s-fgm thin circular cylindrical shells with temperature-dependent properties in thermal environments. *Int J Mech Sci*, 81:17–25, 2014.
183. Duc ND, Kim SE, and Chan DQ. Thermal buckling analysis of fgm sandwich truncated conical shells reinforced by fgm stiffeners resting on elastic foundations using fsdt. *J Therm Stress*, 41:331–365, 2018.
184. Duc ND and Quan TQ. Nonlinear response of imperfect eccentrically stiffened fgm cylindrical panels on elastic foundation subjected to mechanical loads. *Eur J Mech A/Solid*, 46:60–71, 2014.
185. Duc ND and Quan TQ. Transient responses of functionally graded double curved shallow shells with temperature-dependent material properties in thermal environment. *Eur J Mech A/Solid*, 47:101–123, 2014.
186. Khoa ND, Thiem HT, and Duc ND. Nonlinear buckling and postbuckling of imperfect piezoelectric s-fgm circular cylindrical shells with metal-ceramic-metal layers in thermal environment using reddy's third-order shear deformation shell theory. *Mech Adv Mater Struc*, 2017.
187. M. Nejati, A. Eslampanah, and M.M. Najafizadeh. Buckling and vibration analysis of functionally graded carbon nanotube-reinforced beam under axial load. *International Journal of Applied Mechanics*, 08:1–19, 2016.
188. N. M. Newmark. A method of computation for structural dynamics. *Journal of Engineering Mechanics Division, ASCE*, 85:67–94, 1959.
189. Sepahi O, Forouzan MR, and Malekzadeh P. Thermal buckling and postbuckling analysis of functionally graded annular plates with temperature-dependent material properties. *Mater Des*, 32:4030–4041, 2011.
190. Malekzadeh P. Three-dimensional thermal buckling analysis of functionally graded arbitrary straight-sided quadrilateral plates using differential quadrature method. *Compos Struct*, 93:1246–54, 2011.
191. G. Pal and S. Kumar. Fabrication of functionally graded carbon nanotube reinforced aluminum matrix composite. *Advanced Engineering Materials*, 13:323–329, 2011.
192. G. Pal and S. Kumar. Modeling of carbon nanotubes and carbon nanotube–polymer composites. *Progress in Aerospace Sciences*, 80:33–58, 2016.
193. A.N. Palazotto and S.T. Dennis. *Nonlinear Analysis of Shell Structures*. American Institute of Aeronautics and Astronautics,Washington, DC, 1st edition, 1992.
194. J.S. Park and J.H. Kim. Thermal postbuckling and vibration analyses of functionally graded plates. *Journal of Sound and Vibration*, 289:77–93, 2006.
195. H.C. Pham, M.C. Trinh, D.K. Nguyen, and D.D. Nguyen. Nonlinear thermomechanical buckling and post-buckling response of porous fgm plates using reddy's hsdt. *Aerospace Science and Technology*, 128:70–86, 2018.
196. P.Malekzadeh and M.Shojaee. Mechanical buckling of nanocomposite rectangular plate reinforced by aligned and straight single-walled carbon nanotubes. *Thin-Walled Structures*, 71:108–118, 2013.
197. W. Pompe, H. Worch, M. Epple, W. Friess, M. Gelinsky, and P. Greil. Functionally graded materials for biomedical applications. *Materials Science and Engineering A*, 362:40–60, 2003.

198. M.S.S. Prabhu and S. Durvasula. Thermal post-buckling characteristics of clamped skew plates. *Computers and Structures*, 6:177–185, 1976.
199. T. Prakash, M.K. Singha, and M. Ganapathi. Thermal postbuckling analysis of fgm skew plates. *Engineering Structures*, 30:22–32, 2008.
200. X. Qian and D. Dutta. Feature-based design for heterogeneous objects. *Computers-Aided Design*, 36:1263–1278, 2004.
201. P.M. Ramteke, B.Patel, and S.K. Panda. Time-dependent deflection responses of porous fgm structure including pattern and porosity. *International Journal of Applied Mechanics*, 12(9):2050102, 2020.
202. S.N. Rao and N. Ganesan. Vibration of plates immersed in hot fluids. *Computers and structures*, 21:777–787, 1985.
203. B.S. Reddy, J.S. Kumar, C.E. Reddy, and K. Reddy. Buckling analysis of functionally graded material plates using higher order shear deformation theory. *Journal of Composites*, doi:10.1155/2013/808764, 2013.
204. J.N. Reddy. A refined nonlinear theory of plates with transverse shear deformation. *International Journal of Solids and Structures*, 20:881–896, 1984.
205. J.N. Reddy. *Mechanics of laminated composite plates and shells: theory and analysis*. New York: CRC Press, 2nd edition, 2003.
206. J.N. Reddy and C.D. Chin. Thermomechanical analysis of functionally graded cylinders and plates. *Journal of Thermal Stresses*, 21(6):593–626, 1998.
207. J.N. Reddy and C.F. Liu. A higher-order shear deformation theory of laminated elastic shells. *International Journal of Engineering Science*, 23:319–330, 1985.
208. J.N. Reddy and D.H. Robbins. Theories and computational models for composite laminates. *Applied Mechanics Reviews*, 47:147–169, 1994.
209. J. Reinoso and A. Blazquez. Geometrically nonlinear analysis of functionally graded power-based and carbon nanotubes reinforced composites using a fully integrated solid shell element. *Composite Structures*, 152:277–294, 2016.
210. E. Reissner. The effect of transverse shear deformation on the bending of elastic plates. *Journal of Applied Mechanics*, 12(2):69–77, 1945.
211. A.S. Rezaei, A.R. Saidi, M. Abrishamdari, and M.H. Pour Mohammadi. Natural frequencies of functionally graded plates with porosities via a simple four variable plate theory: An analytical approach. *Thin-Walled Structures*, 120:366–377, 2017.
212. Trabelsi S, Frikha A, Zghal S, and Dammak F. Thermal post-buckling analysis of functionally graded material structures using a modified fsdt. *Int J Mech Sci*, 144:74–89, 2018.
213. M. Sahla, H. Saidi, K. Draiche, A.A. Bousahla, F. Bourada, and A. Tounsi. Free vibration analysis of angle-ply laminated composite and soft core sandwich plates. *Steel and Composite Structures*, 33(5):663–679, 2019.
214. A.F. Saleeb, T.Y. Chang, W. Graf, and S. Yingyeunyong. A hybrid/mixed model for non-linear shell analysis and its applications to large-rotation problem. *International Journal for numerical methods in Engineering*, 29:407–446, 1990.
215. Ghannadpour SAM, Ovesy HR, and Nassirnia M. Buckling analysis of functionally graded plates under thermal loadings using the finite strip method. *Comput Struct*, 108-109:93–99, 2012.
216. B.A. Samsam-Shariat and M.R. Eslami. Buckling of thick functionally graded plates under mechanical and thermal loads. *Composite Structures*, 78:433–439, 2007.
217. Ghiasian SE, Kiani Y, Sadighi M, and Eslami MR. Thermal buckling of shear deformable temperature dependent circular/annular fgm plates. *Int J Mech Sci*, 81:137–148, 2014.
218. M. Shariyat. Dynamic thermal buckling of suddenly heated temperature-dependent fgm cylindrical shells, under combined axial compression and external pressure. *International Journal of Solids and Structures*, 45:2598–2612, 2008.
219. H.S Shen. Nonlinear bending of shear deformable laminated plates under transverse and in-plane loads and resting on elastic foundations. *Composites Structures*, 50:131–142, 2000.
220. H.S. Shen. Postbuckling analysis of axially-loaded functionally graded cylindrical shells in thermal environments. *Composite Science and Technology*, 62:977–987, 2002.
221. H.S. Shen. Postbuckling analysis of pressure-loaded functionally graded cylindrical shells in thermal environments. *Engineering Structures*, 25:487–497, 2003.

222. H.S. Shen. Thermal postbuckling behavior of functionally graded cylindrical shells with temperature-dependent properties. *International Journal of Solids and Structures*, 41:1961–1974, 2004.
223. H.S. Shen. Nonlinear bending of functionally graded carbon nanotube-reinforced composite plates in thermal environments. *Composite Structures*, 91:9–19, 2009.
224. H.S. Shen. Postbuckling of nanotube-reinforced composite cylindrical shells in thermal environments, part i: Axially-loaded shells. *Composite Structures*, 93:2096–2108, 2011.
225. H.S. Shen. Thermal buckling and postbuckling behavior of functionally graded carbon nanotube-reinforced composite cylindrical shells. *Composites Part B*, 43:1030–1038, 2012.
226. H.S. Shen. Nonlinear analysis of nanotube-reinforced composite beams resting on elastic foundations in thermal environments. *Engineering Structures*, 56:698–708, 2013.
227. H.S. Shen. Nonlinear bending of nanotube-reinforced composite cylindrical panels resting on elastic foundations in thermal environments. *Engineering Structures*, 80:163–172, 2014.
228. H.S. Shen. Nonlinear vibration of nanotube-reinforced composite cylindrical panels resting on elastic foundations in thermal environments. *Composite Structures*, 111:291–300, 2014.
229. H.S. Shen, F. Lin, and Y. Xiang. Nonlinear bending and thermal postbuckling of functionally graded graphene-reinforced composite laminated beams resting on elastic foundations. *Engineering Structures*, 140:89–97, 2017.
230. H.S. Shen and N. Noda. Postbuckling of fgm cylindrical shells under combined axial and radial mechanical loads in thermal environments. *International Journal of Solids and Structures*, 42:4641–4662, 2005.
231. H.S. Shen and N. Noda. Postbuckling of pressure-loaded fgm hybrid cylindrical shells in thermal environments. *Composite Structures*, 77:546–560, 2007.
232. H.S. Shen and H. Wang. Thermal postbuckling of fgm cylindrical panels resting on elastic foundations. *Aerospace Science and Technology*, 38:9–19, 2014.
233. H.S. Shen and H. Wang. Nonlinear bending of fgm cylindrical panels resting on elastic foundations in thermal environments. *European Journal of Mechanics: A/Solids*, 49:49–59, 2015.
234. H.S. Shen and Y. Xiang. Postbuckling of axially compressed nanotube-reinforced composite cylindrical panels resting on elastic foundations in thermal environments. *Composites Part B: Engineering*, 67:50–61, 2014.
235. H.S. Shen and C.L. Zhang. Thermal buckling and postbuckling behavior of functionally graded carbon nanotube-reinforced composite plates. *Materials and Design*, 31:3403–3411, 2010.
236. H.S. Shen and C.L. Zhang. Non-linear analysis of functionally graded fiber reinforced composite laminated plates, part i: Theory and solutions. *International Journal of Nonlinear Mechanics*, 47:1045–1054, 2012.
237. G. Shi. A new simple third-order shear deformation theory of plates. *International Journal of Solids and Structures*, 44:4399–4417, 2007.
238. J.C. Simo and F. Armero. Geometrically non-linear enhanced strain mixed methods and the method of incompatible modes. *International Journal for Numerical Methods in Engineering*, 33:1413–1449, 1992.
239. J.C. Simo, F. Armero, and R. Taylore. Improved version of assumed enhanced strain tri-linear elements for 3d finite deformation problems. *Computer Methods in Applied Mechanics and Engineering*, 110:359–386, 1992.
240. J.C. Simo and D.D. Fox. On a stress resultants geometrically exact shell model. part:i: Formulation and optimal parametrization. *Computer Methods in Applied Mechanics and Engineering*, 72:267–304, 1989.
241. J.C. Simo and D.D. Fox. On a stress resultants geometrically exact shell model. part:ii: The linear theory: Computational aspects. *Computer Methods in Applied Mechanics and Engineering*, 73:53–92, 1989.
242. J.C. Simo and M. Rifai. A class of mixed assumed strain methods and the method of incompatible modes. *International Journal of Numerical Methods and Engineering*, 29(8):1595–1638, 1990.
243. J.C. Simo, M.C. Rifai, and D.D. Fox. On a stress resultants geometrically exact shell model. part:iii: Computational aspects of the nonlinear theory. *Computer Methods in Applied Mechanics and Engineering*, 79:21–70, 1990.

244. J.C. Simo, M.S. Rifai, and D.D. Fox. On a stress resultants geometrically exact shell model. part:iv: Variable thickness shells with through the thickness stretching. *Computer Methods in Applied Mechanics and Engineering*, 81:91–126, 1990.
245. M. Simsek. Fundamental frequency analysis of functionally graded beams by using different higher-order beam theories. *Nuclear Engineering and Design*, 240:697–705, 2010.
246. A.V. Singh. On vibrations of shells of revolution using bezier polynomials. *ASME: Journal of Pressure Vessel Technology*, 113:579–584, 1991.
247. A.V. Singh and S. Mirza. Asymmetric modes and associated eigenvalues for spherical shells. *ASME: Journal of Pressure Vessel Technology*, 107(1):77–82, 1985.
248. S.J. Singh and S.P. Harsha. Analysis of porosity effect on free vibration and buckling responses for sandwich sigmoid function based functionally graded material plate resting on pasternak foundation using Galerkin Vlasov's method. *Journal of Sandwich Structures and Materials*, 0(0):1–44, 2020.
249. Panda SK and Singh BN. Thermal post-buckling behaviour of laminated composite cylindrical/hyperboloid shallow shell panel using nonlinear finite element method. *Compos Struct*, 91:366–374, 2009.
250. Panda SK and Singh BN. Nonlinear free vibration analysis of thermally post-buckled composite spherical shell panel. *Int J Mech Mater Des*, 6:175–188, 2010.
251. Panda SK and Singh BN. Thermal postbuckling behavior of laminated composite spherical shell panel using nfem. *Mech Based Des Struc Mach*, 41:468–488, 2013.
252. B. Sobhaniaragh, M.Nejati, and W.J.Mansur. Buckling modelling of ring and stringer stiffened cylindrical shells aggregated by graded cnts. *Composites Part B: Engineering*, 124:120–133, 2017.
253. A.H. Sofiyev. Thermoelastic stability of freely supported functionally graded conical shells within the shear deformation theory. *Composite Structures*, 152:74–84, 2016.
254. K. Subbaraj and M. A. Dokainish. A survey of direct time-integration methods in computational structural dynamics—ii. implicit methods. *Computers & Structures*, 32(6):1387–1401, 1989.
255. K. Swaminathan and D.M. Sangeetha. Thermal analysis of fgm plates – a critical review of various modelling techniques and solution methods. *Composite Structures*, 160:43–60, 2017.
256. K.Y. Sze, X.H. Liu, and S.H. Lo. Popular benchmark problems for geometric nonlinear analysis of shells. *Finite Elements in Analysis and Design*, 40:1551–1569, 2004.
257. R. Tanov and A. Tabiei. A simple correction to the first-order shear deformation shell finite element formulations. *Finite Elements in Analysis and Design*, 35:189–197, 2000.
258. H.T. Thai and D.H. Choi. An efficient and simple refined theory for buckling analysis of functionally graded plates. *Applied Mathematical Modelling*, 36:1008–1022, 2012.
259. K.R. Thangaratnam, A. Palaninathan, and J. Ramachandran. Thermal buckling of composite laminated plates. *Computers and Structures*, 32:1117–1124, 1989.
260. B. Thomas and T. Roy. Vibration analysis of functionally graded carbon nanotube-reinforced composite shell structures. *Acta Mechanica*, 227:581–599, 2016.
261. R.P. Thornburgh and M.W. Hilburger. Identifying and characterizing discrepancies between test and analysis results of compression-loaded panels. *NASA/TM-2005-213932*, ARL-TR-3664, 2005.
262. S. Timoshenko and J. Gere. *Theory of elastic stability*. New York : McGraw-Hill, 2nd edition, 1961.
263. T.Kant and C.S. Babu. Thermal buckling analysis of skew fibre-reinforced composite and sandwich plates using shear deformable finite element models. *Compos Struct*, 49:77–85, 2000.
264. Wu TL, Shukla KK, and Huang JH. Post-buckling analysis of functionally graded rectangular plates. *Compos Struct*, 81:1–10, 2007.
265. F. Tornabene, N. Fantuzzi, M. Bacciocchi, and E. Viola. Effect of agglomeration on the natural frequencies of functionally graded carbon nanotube-reinforced laminated composite doubly-curved shells. *Composites Part B: Engineering*, 89:187–218, 2016.
266. Mahapatra TR and Panda SK. Effects of hygrothermal conditions on free vibration behaviour of laminated composite structures. *IOP Conf Ser Mater Sci Eng*, 75:012016, 2015.
267. Mahapatra TR and Panda SK. Nonlinear free vibration analysis of laminated composite spherical shell panel under elevated hygrothermal environment: Amicromechanical approach. *Aerosp Sci Technol*, 49:276–288, 2016.
268. Mahapatra TR, Panda SK, and Dash S. Effect of hygrothermal environment on the nonlinear free vibration responses of laminated composite plates: A nonlinear finite element micromechanical approach. *IOP Conf Ser Mater Sci Eng*, 149:012151, 2016.

269. Mahapatra TR, Kar VR, and Panda SK. Nonlinear free vibration analysis of laminated composite doubly curved shell panel in hygrothermal environment. *J Sandwich Struct Mater*, 17 (5):511–545, 2015.
270. Mahapatra TR, Kar VR, and Panda SK. Large amplitude vibration analysis of laminated composite spherical panels under hygrothermal environment. *Int J Struct Stab Dyn*, 16:1450105, 2016.
271. S. Trabelsi, A. Frikha, S. Zghal, and F. Dammak. A modified fsdt-based four nodes finite shell element for thermal buckling analysis of functionally graded plates and cylindrical shells. *Engineering Structures*, 178:444–459, 2019.
272. L.V. Tran, P. Phung-Van, J. Lee, M. Abdel Wahab, and H. Nguen-Xuan. Isogeometric analysis for nonlinear thermomechanical stability of functionally graded plates. *Composite Structures*, 140:655–667, 2016.
273. L.V. Tran, C.H. Thai, and H. Nguyen-Xuan. An isogeometric finite element formulation for thermal buckling analysis of functionally graded plates. *Finite Elements in Analysis and Design*, 73:65–76, 2013.
274. M.C. Trinh, T. Mukhopadhyay, and S.E. Kim. A semi-analytical stochastic buckling quantification of porous functionally graded plates. *Aerospace Science and Technology*, 105:105928, 2020.
275. E. Tufekci, U. Eroglu, and S.A. Aya. Exact solution for in-plane static problems of circular beams made of functionally graded materials. *Mechanics Based Design of Structures and Machines*, 44(4):476–494, 2016.
276. H.V. Tung. Thermal and thermomechanical postbuckling of fgm sandwich plates resting on elastic foundations with tangential edge constraints and temperature dependent properties. *Composite Structures*, 131:1028–1039, 2015.
277. V.Anumandla and R. F. Gibson. A comprehensive closed form micromechanics model for estimating the elastic modulus of nanotube-reinforced composites. *Composites: Part A*, 37:2178–2185, 2006.
278. E. Viola, F. Tornabene, and N. Fantuzzi. General higher-order shear deformation theories for the free vibration analysis of completely doubly-curved laminated shells and panels. *Composites Structures*, 95:639–666, 2013.
279. E. Viola, F. Tornabene, and N. Fantuzzi. Static analysis of completely doubly-curved laminated shells and panels using general higher-order shear deformation theories. *Composites Structures*, 101:59–93, 2013.
280. Kar VR and Panda SK. Post-buckling behaviour of shear deformable functionally graded curved shell panel under edge compression. *Int J Mech Sci*, 115:318–324, 2016.
281. Kar VR, Panda SK, and Mahapatra TR. Thermal buckling behaviour of shear deformable functionally graded single/doubly curved shell panel with td and tid properties. *Adv Mater Res*, 5:205–221, 2016.
282. M. Wali, A. Hajlaoui, and F. Dammak. Discrete double directors shell element for the functionally graded material shell structures analysis. *Computer Methods in Applied Mechanics and Engineering*, 278:388–403, 2014.
283. M. Wali, T. Hentati, A. Jarraya, and F. Dammak. Free vibration analysis of fgm shell structures with a discrete double directors shell element. *Composite Structures*, 125:295–303, 2015.
284. Q. Wang, S.T. Quek, and X. Liu. Analysis of piezoelectric coupled circular plate. *Smart Materials and Structures*, 10:229–239, 2001.
285. F. Watari, A. Yokoyama, F. Saso, M. Uo, and T. Kawasaki. Fabrication and properties of functionally graded dental implant. *Composites Part B*, 28:5–11, 1997.
286. N. Wattanasakulpong. Thermal buckling and elastic vibration analysis of functionally graded beams and plates using improved third-order shear deformation theory. *Ph.D. thesis, School of Mechanical and Manufacturing Engineering, the University of New South Wales*, 2012.
287. N. Wattanasakulpong and A. Chaikittiratana. Flexural vibration of imperfect functionally graded beams based on timoshenko beam theory: Chebyshev collocation method. *Meccanica*, 50(5):1331–1342, 2015.
288. N. Wattanasakulpong, B.G. Prusty, D.W. Kelly, and M. Hoffman. Free vibration analysis of layered functionally graded beams with experimental validation. *Materials and Design (1980-2015)*, 36:182–190, 2012.
289. N. Wattanasakulpong and V. Ungbhakorn. Linear and nonlinear vibration analysis of elastically restrained ends fgm beams with porosities. *Aerospace Science and Technology*, 36:111–120, 2014.
290. J. Woo, S.A. Meguid, J.C. Stranart, and K.M. Liew. Thermomechanical postbuckling analysis of moderately thick functionally graded plates and shallow shells. *International Journal of Mechanical Sciences*, 47:1147–1171, 2005.

291. Zhao X, Lee YY, and Liew KM. Mechanical and thermal buckling analysis of functionally graded plates. *Compos Struct*, 90:161–71, 2010.
292. X.K. Xia and H.S. Shen. Nonlinear vibration and dynamic response of fgm plates with piezoelectric fiber reinforced composite actuators. *Composite Structures*, 90:254–262, 2009.
293. Ge Y, Yuan W, and Dawe DJ. Thermomechanical postbuckling of composite laminated plates by the spline finite strip method. *Compos Struct*, 71:115–129, 2005.
294. Shi Y, Lee RYY, and Mei C. Thermal postbuckling of composite plates using the finite element modal coordinate method. *J Therm Stress*, 22(22):595–614, 1999.
295. N. Yamaki. Postbuckling behavior of rectangular plates with small initial curvature loaded in edge compression. *Journal of Applied Mechanics*, 26:407–414, 1959.
296. N. Yamaki. Postbuckling behavior of rectangular plates with small initial curvature loaded in edge compression-(continued). *Journal of Applied Mechanics*, 27:335–342, 1960.
297. H.T.Y. Yang, S. Saigal, and D.G. Liaw. Advances of thin shell finite elements and some applications—version i. *Computers & Structures*, 35(4):481–504, 1990.
298. J. Yang and H.S. Shen. Free vibration and parametric resonance of shear deformable functionally graded cylindrical panels. *Journal of Sound and Vibration*, 261:871–893, 2002.
299. J. Yang and H.S. Shen. Vibration characteristics and transient response of shear deformable functionally graded plates in thermal environment. *Journal of Sound and Vibration*, 255(3):579–602, 2002.
300. J. YANG and H.S. Shen. Non-linear analysis of functionally graded plates under transverse and in-plane loads. *International Journal of Non-Linear Mechanics*, 38(4):467–482, 2003.
301. J. YANG and H.S. Shen. Nonlinear bending analysis of shear deformable functionally graded plates. *Composites Part B: Engineering*, 38(2):103–115, 2003.
302. J. Ying, W.Q. Chen, and C.F. Lu. Two-dimensional elasticity solutions for functionally graded beams resting on elastic foundations. *Composites Structures*, 84:209–219, 2008.
303. Y.S.Touloukian. *Thermophysical properties of high temperature solid materials.* Macmillan, New York, 1st edition, 1967.
304. F.Z. Zaoui, J. Ouinas, and A. Tounsi. New 2d and quasi-3d shear deformation theories for free vibration of functionally graded plates on elastic foundations. *Composites Part B: Engineering*, 159:231–247, 2019.
305. A.M. Zenkour. A quasi-3d refined theory for functionally graded single-layered and sandwich plates with porosities. *Composite Structures*, 201:38–48, 2018.
306. S. Zghal, A. Frikha, and F. Dammak. Static analysis of functionally graded carbon nanotube-reinforced plate and shell structures. *Composite Structures*, 176:1107–1123, 2017.
307. S. Zghal, A. Frikha, and F. Dammak. Static analysis of functionally graded carbon nanotube-reinforced plate and shell structures. *Composite Structures*, 176:1107–1123, 2017.
308. S. Zghal, A. Frikha, and F. Dammak. Free vibration analysis of carbon nanotube-reinforced functionally graded composite shell structures. *Applied Mathematical Modelling*, 53:132–155, 2018.
309. S. Zghal, A. Frikha, and F. Dammak. Mechanical buckling analysis of functionally graded power-based and carbon nanotubes-reinforced composite plates and curved panels. *Composites Part B: Engineering*, 150:165–183, 2018.
310. S. Zghal, A. Frikha, and F. Dammak. Non-linear bending analysis of nanocomposites reinforced by graphene-nanotubes with finite shell element and membrane enhancement. *Engineering Structures*, 158:95–109, 2018.
311. D.G. Zhang. Thermal post-buckling analysis of fgm elliptical plates based on high order shear deformation theory. *Mechanics of Advanced Materials and Structures*, 49:283–293, 2014.
312. D.G. Zhang. Thermal post-buckling analysis of functionally graded material elliptical plates based on high-order shear deformation theory. *Mechanics of Advanced Materials and Structures*, 24:142–148, 2017.
313. D.G. Zhang and H.M. Zhou. Mechanical and thermal post-buckling analysis of fgm rectangular plates with various supported boundaries resting on nonlinear elastic foundations. *Thin Walled Structures*, 89:142–151, 2015.
314. L.W. Zhang, Z.X. Lei, and K.M. Liew. Vibration characteristic of moderately thick functionally graded carbon nanotube reinforced composite skew plates. *Composites structures*, 122:172–183, 2015.

315. L.W. Zhang, Z.X. Lei, K.M. Liew, and J.L. Yu. Large deflection geometrically nonlinear analysis of carbon nanotube-reinforced functionally graded cylindrical panels. *International Journal of Non-Linear Mechanics*, 273:1–18, 2014.
316. L.W. Zhang, Z.X. Lei, K.M. Liew, and J.L. Yu. Static and dynamic of carbon nanotube reinforced functionally graded cylindrical panels. *Composite Structures*, 111:205–212, 2014.
317. L.W. Zhang, Z.X. Lei, K.M. Liew, and J.L. Yu. Geometrically nonlinear large deformation analysis of functionally graded carbon nanotube reinforced composite straight-sided quadrilateral plates. *Computer Methods in Applied Mechanics and Engineering*, 295:219–239, 2015.
318. L.W. Zhang and K.M. Liew. Large deflection analysis of fg-cnt reinforced composite skew plates resting on pasternak foundations using an element-free approach. *Composite Structures*, 132:974–983, 2015.
319. L.W. Zhang and K.M. Liew. Postbuckling analysis of axially compressed cnt reinforced functionally graded composite plates resting on pasternak foundations using an element-free approach. *Composite Structures*, 138:40–51, 2016.
320. L.W. Zhang, K.M. Liew, and J.N. Reddy. Postbuckling analysis of bi-axially compressed laminated nanocomposite plates using the first-order shear deformation theory. *Composite Structures*, 152:418–431, 2016.
321. L.W. Zhang, K.M. Liew, and J.N. Reddy. Postbuckling behavior of bi-axially compressed arbitrarily straight-sided quadrilateral functionally graded material plates. *Computers Methods in Applied Mechanics and Engineering*, 300:593–600, 2016.
322. L.W. Zhang, K.M. Liew, and J.N. Reddy. Postbuckling of carbon nanotube reinforced functionally graded plates with edges elastically restrained against translation and rotation under axial compression. *Computers Methods in Applied Mechanics and Engineering*, 298:1–28, 2016.
323. L.W. Zhang, W.H. Liu, K.M. Liew, and J.L. Yu. Geometrically nonlinear large deformation analysis of triangular cnt-reinforced composite plates. *International Journal of Non-Linear Mechanics*, 86:122–132, 2016.
324. L.W. Zhang, Z.X.Lei, and K.M. Liew. Buckling analysis of fg-cnt reinforced composite thick skew plates using an element-free approach. *Composites Part B: Engineering*, 75:36–46, 2015.
325. L.W. Zhang, Z.X.Lei, and K.M. Liew. An element-free imls-ritz framework for buckling analysis of fg–cnt reinforced composite thick plates resting on winkler foundations. *Engineering Analysis with Boundary Elements*, 58:7–17, 2015.
326. X. Zhao and K.M. Liew. A mesh-free method for analysis of the thermal and mechanical buckling of functionally graded cylindrical shell panels. *Computational Mechanics*, 45:297–310, 2010.
327. H. Zhong and C. Gu. Buckling of simply supported rectangular reissner–mindlin plates subjected to linearly varying in-plane loading. *Journal of engineering mechanics*, 132(5):578–581, 2006.
328. P. Zhu, Z.X. Lei, and K.M. Liew. Static and free vibration analyses of carbon nanotube-reinforced composite plates using finite element method with first order shear deformation plate theory. *Composite Structures*, 94:1450–1460, 2012.
329. K.K. Zur. Quasi-green's function approach to free vibration analysis of elastically supported functionally graded circular plates. *Composite Structures*, 183:600–610, 2018.
330. K.K. Zur and P. Jankowski. Multiparametric analytical solution for the eigenvalue problem of fgm porous circular plates. *Symmetry*, 11(3):429, 2019.

Index